DRIVEN BY THE DANE

NINE CENTURIES OF WATERPOWER IN SOUTH CHESHIRE AND NORTH STAFFORDSHIRE

Tony Bonson

THE MIDLAND WIND AND WATER MILLS GROUP

FRONT COVER
Washford Mill on the River Dane at Buglawton near Congleton. Built in the 1760s as a flint mill, it later became a corn mill and silk mill before reverting to a flint mill in the middle of the 19th century. (see pages 65 - 72)

REAR COVER
A view of Congleton Silk Mill (known as the Old Mill) from the air with the town corn mill behind. When the silk mill was built in 1752 it was the fourth textile mill to be erected and set the pattern for textile mill building for the next hundred years. (see pages 123 - 133)

Published by The Midland Wind and Water Mills Group
14 Falmouth Road, Congleton, Cheshire, CW12 3BH

© Anthony Bonson, 2003.

ISBN 0 9517794 4 3

All rights reserved
No part of this publication may be reproduced, stored in a retrieval system or transmitted in any form or by any means mechanical, electronic, recording or otherwise, without prior written permission of the copyright holder.

ACKNOWLEDGEMENTS

Research for this publication started in 1981 and in the intervening twenty two years many people, who are too numerous to mention individually, have provided help and assistance of one form or another which I would like to acknowledge.

I would also like to acknowledge the co-operation of the many mill and site owners involved but at the same time stress that all the mills recorded in this publication are private property and request that their owner's privacy be respected.

During the documentary research I have been ably assisted by all the staff at both the Cheshire and Staffordshire Record Offices. Documents held by the Cheshire and Chester Archives and Local Studies at the county record office are reproduced by the permission of the Cheshire County Council and the owner/depositor to whom copyright is reserved. Documents held by the Staffordshire Record Office are reproduced by permission of Staffordshire County Council. Also the staff at Congleton Library have coped very well over the years with my many requests for obscure and antiquated publications.

I would like to thank all of the various individuals, publishers, museums, etc. who kindly gave their permission for the reproducion of many of the illustrations in this publication. John K. Harrison and his publishers the North Yorkshire Moors National Park Authority deserve a special mention for inspiring the style of this book by their publication "Eight Centuries of Milling in North-East Yorkshire".

I am most indebted to the various members of the Midland Wind and Water Mills Group for their comments and assistance. This is especially true of Tim Booth who has provided a sounding board for my ideas, has undertaken the proof reading of the book, and is responsible for my education with respect to corn mills. However, I hasten to add that any errors are entirely my own work. Chris Bradley has kindly allowed me to make use of his research and survey information concerning the two Washford Mills in Buglawton. Tim Booth and John Boucher generously prepared drawings specially for this publication and Alan Crocker kindly sent me copies of some of his turbine catalogues. Also Robin Clarke has made contributions on some very obscure points and also was unstinting in his pursuit of patent information.

Finally, I wish to acknowledge the tremendous support given by my wife Kate who has accompanied me on all the fieldwork and has checked the script diligently many times, but mainly for her constructive criticism and overall enthusiasm for the project without which I could not have contemplated bringing this publication to fruition.

Tony Bonson, April, 2003

CONVENTIONS

All maps are orientated so that north is towards the top of the page unless otherwise indicated.

The mills are numbered starting at the source of the River Dane and continue downstream as far as the river's confluence with the River Weaver. On encountering a tributary stream the numbering switches from the main river to the source of the tributary, following the tributary downstream until its confluence with the main river is reached. At which point the numbering continues downstream on the main river.

The area covered by this book is divided into seven sections, each of which starts with a map of the section under consideration. On these section maps if a watercourse commences or concludes off the scope of the map then this fact is indicated by arrows on its course near the edge of the map. Watercourses that are shown completely on a map do not have any arrows associated with them.

All sums of money quoted are in £-s-d as that was the currency in use at the time. For the benefit of younger readers 12d = 1s and 20s = £1. All measurements are given in the old Imperial units because, again, they were the units in use at the time, 12 inches = 1 foot, 3 feet = 1 yard. For conversion to the metric system 1 inch = 25.4 mm or alternatively 1m = 39 inch.

During the course of this book a number of abbreviations are used, as follows:-

CBC	Congleton Borough Council
CCALS	Cheshire and Chester Archives and Local Studies (see CRO)
CCC	Cheshire County Council
CRO	Cheshire Record Office
DRO	Derbyshire Record Office
ECTMS	East Cheshire Textile Mills Survey
P.O.	Post Office
PRO	Public Record Office
SRO	Staffordshire Record Office

DRIVEN BY THE DANE

CONTENTS

Introduction 1

Gazeteer of Mills
Section 1. The Source of the River Dane to Hug Bridge 3
Section 2. Hug Bridge to Buglawton 37
Section 3. The Daneinshaw Brook 77
Section 4. The Town of Congleton 109
Section 5. Congleton to the Confluence with the River Wheelock 150
Section 6. The Upper River Wheelock, Kidsgrove to Wheelock 190
Section 7. The River Wheelock and the River Dane to Northwich 222

Technology 239

Chronology 277

Bibliography 291

Index 294

INTRODUCTION

The River Dane rises high in the Peak District on Axe Edge Moor west of Buxton and flows in a westerly direction until it leaves the foothills of the Pennines at Hug Bridge, about halfway between Macclesfield and Leek. This stretch of the river marks the boundary between Staffordshire to the south and Cheshire to the north. Once the river reaches Hug Bridge its westerly progress is halted by a high outcrop called Bosley Cloud which forces the river to flow in a semi-circle round its northern extremity before regaining its westerly direction at the town of Congleton. This westerly flow takes the river past the northern side of the town of Middlewich where it is joined from the south by its only major tributary, the River Wheelock. The combined flow of the two rivers then continues in a northerly direction until it reaches Northwich where the Dane joins the River Weaver (see Figure 1).

In this investigation of the River Dane and its tributaries a total of 87 sites have been discovered that once used waterpower for some purpose or other. The usage of these sites covers a time span of over 900 years, from before the Domesday Book in 1086 until the last water powered mill in the region stopped work in the mid 1980s. During this period this river system was used to provide power for over twenty different applications. Waterpower was used not only by the ubiquitous corn mills that provided flour and other grain products to sustain the population and their animals but also by the great many and various industries that established themselves in this part of Cheshire and North Staffordshire. In the metal trades water was used to power iron furnaces and forges, as well as brass rolling and slitting mills. The textile industry was a major user of waterpower for fulling, silk throwing, cotton spinning, flax spinning and fustian cutting. Other industries came to the area to use waterpower such as flint grinding for the pottery industry, coal grinding for dyeing, bone grinding for agriculture, paper making, saw milling, tanning, brine pumping for salt manufacture, and electricity generation. As well as these major industries, the water of the River Dane and its tributaries was also used to provide power for some specialist applications such as carpet printing, wood flour manufacture, veneer cutting, ice making and even bell ringing!

This array of the industrial application of waterpower in South Cheshire may well be a surprise. However, not only were the streams and rivers put to this wide variety of uses, but this area of Cheshire was once at the "cutting edge" of the industrial revolution, being one of the leading areas for the innovations and developments that changed the country from having an agrarian economy to an industrial one. Today, other locations claim these honours and trade on the glory of their industrial heritage, even in some cases achieving the status of World Heritage Sites. However, while these other areas may be justifiably proud of their contribution to the industrial development of the nation, this forgotten area of South Cheshire had "been there and done that" at an earlier stage or was at least equally abreast of developments taking place elsewhere. As well as these pioneering enterprises, there were many others quick to follow, and these entrepreneurs also deserve to be remembered. If it was the pioneers who showed the way forward, it was the sheer mass of those who came later, applying and improving these innovations, who changed the world.

In doing so they employed the skills of some of the finest engineers of the time. During the 18th century James Brindley, who served his apprenticeship as a millwright nearby and was later to achieve lasting fame for building some of the first canals in the country, played a leading role in the area by designing and constructing waterpowered factories. Later, in the 19th century, William Fairbairn, who brought the design of waterwheels and mill gearing to its zenith and was possibly the leading mechanical engineer of his day, contributing to the building of structures such as the Britannia Tubular Bridge across to Anglesey, was also active in the area.

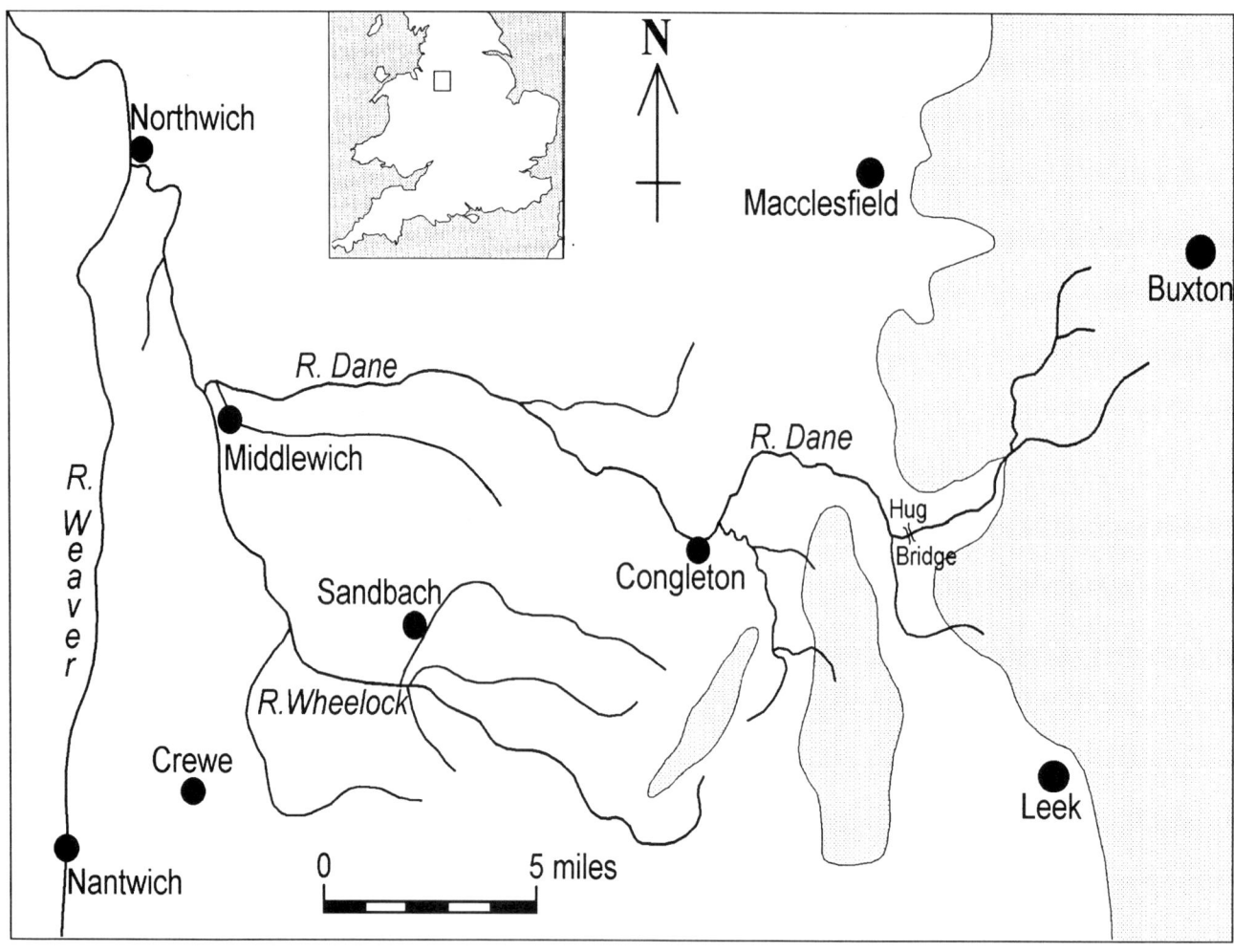

FIGURE 1. MAP OF THE RIVER DANE AND ITS TRIBUTARIES IN SOUTH CHESHIRE AND NORTH STAFFORDSHIRE.
(*The shaded area represents high ground.*)

Although many people associate the steam engine with the major growth of industry, waterpower continued to play a vital role on the River Dane and its tributaries throughout the 19th century and, indeed, the development of waterpower continued well into the 20th century.

Today this area gives hardly any clue to the ground breaking role that waterpower played in earlier times as the greater part of the River Dane basin is now given over to agriculture. The only town of any significance on the River Dane, prior to its confluence with the Weaver at Northwich, is that of Congleton, although both Middlewich and Sandbach lie within the river's catchment area. In its upper reaches, the River Dane flows through typical Pennine upland scenery, partly within the Peak District National Park. This is now characterised by isolated hillside farms and small hamlets with no sign of the factories that were once present. Once the river leaves these uplands, it passes through highly fertile farmlands that specialise in dairy produce. Again, all sign of industry in this area has retreated into the three towns of Congleton, Sandbach and Middlewich. This modern industrial activity is mainly confined to small firms operating in the local area together with retail and service organisations, although there is still a vestige of the salt industry around Middlewich. Otherwise it is probably fair to say that these towns and their outlying villages are largely becoming dormitory areas for the Manchester and Stoke-on-Trent conurbations with a population that no longer appreciates the role they once played in the industrial life of the country.

It is hoped that this study will elucidate the role that the exploitation of the power available from the River Dane and its tributaries has played over the last millennium and emphasise its historical importance not just regionally but nationally and even internationally. Hopefully, a greater knowledge of the historical contribution made by this area will make it easier in the future to preserve and conserve the important structures and other historical evidence that still survive from its past.

GAZETEER OF MILLS

Section 1. The Source of the River Dane to Hug Bridge

During the first part of its life, the River Dane is a fast flowing stream in a deep valley surrounded by the hills of the Pennines. Habitation consists of a few scattered farms and the occasional hamlet such as at Danebridge. The main river is joined from the north by the Clough Brook just above Danebridge, and by a number of small streams that group together in Staffordshire to join the river just below Danebridge.

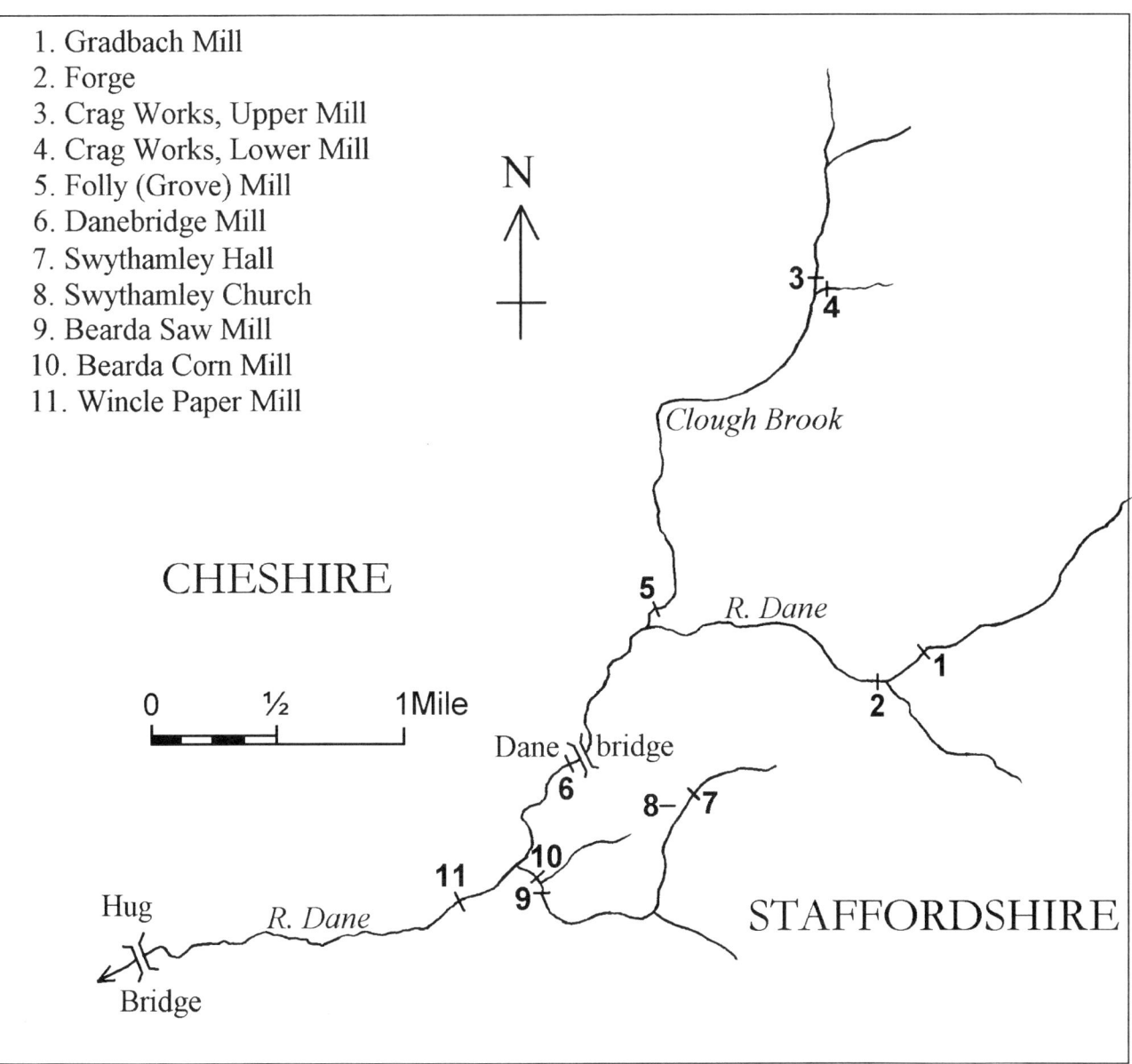

1. Gradbach Mill
2. Forge
3. Crag Works, Upper Mill
4. Crag Works, Lower Mill
5. Folly (Grove) Mill
6. Danebridge Mill
7. Swythamley Hall
8. Swythamley Church
9. Bearda Saw Mill
10. Bearda Corn Mill
11. Wincle Paper Mill

Figure 2. Map of the River Dane and its tributaries from its source to Hug Bridge.

1. GRADBACH MILL   SJ 994661

Gradbach Mill is located on the first area of flat valley bottom on the River Dane about four miles from its source. The mill lies on the south side of the river, in the Parish of Quarnford in Staffordshire. At this point the valley bottom is about 50 yards wide and is about 500 feet below the peaks of the surrounding hills. The mill is built end on to the river with its front facing west, and with its wheelhouse on its southern gable furthest from the river. The access road runs down the valley from the east ending alongside the mill and opposite the mill house which faces east (see Figure 3). There is a footbridge over the River Dane at the mill which leads to a very steep packhorse road up the valley side.

It has been said that the first mill at Gradbach was built in 1640, that Thomas Dakeyne came to the mill in 1780, and that the mill burnt down in 1785 and was then rebuilt.[1] No evidence has been forthcoming to substantiate these claims, nor indeed, to suggest the purpose of such a mill. Certainly there is no indication of a mill at this location on Yates's map of Staffordshire published in the 1770s.

In fact the first solid evidence of the existence of a mill at Gradbach is an indenture for 31 years from Sir Henry Harpur, dated 1792, to Thomas Oliver of Longnor together with James Oliver and Thomas White of Glutton Bridge who were both described as cotton manufacturers. A rent of £20 per year was specified for several closes of land *"at Gradbitch [sic] Farm....to build a mill for working and spinning cotton, wool, or silk into twist or yarn and to make continuous dams, weirs, cuts, sluices, floodgates and wheels upon the premises or across the River Dane."* [2] The wording of this indenture implies that the textile mill was not yet built on what was then a green field site. Also the partners had to borrow £800 at this time in order to finance the building of the mill. It seems quite clear from these documents that no mill existed at the site before this date. A year later, in 1793, another indenture records the mortgaging for £1000 by the Olivers and White to Robert Charles Greaves of *"all the premises and privileges with the mills and buildings then erected and the machinery, engines, wheels, etc"* for 31 years.[3] In the same year they also borrowed £500 from two Manchester cotton merchants called Joseph Tipping and George Walker. Furthermore, in 1794, a lease was taken on another piece of land for 50 years at a rent of 6 guineas per year. This lease is relevant to the mill as the land in question contained the weir and pool for providing the water supply to the mill.[4]

In 1794 Thomas Oliver, James Oliver and Thomas White were bankrupt owing £1502 to Greaves, £500 to Tipping and Walker and £100 to a Thomas Wood. At this time the mill was taken over by the bankruptcy assignees in order to try and recoup some of the outstanding debts.[5] The assignees were unable to sell the mill until 1798 when the Dakeyne family purchased the mill and machinery from the assignees for £402-10s-0d subject to a mortgage for £1202-10s-0d to R. C. Greaves together with the lease for the land including the weir and pool for £180.[6] Under this new lease the Dakeynes were granted the right to scour the river bed for 260 yards downstream of the mill to improve its efficiency. This clause hints that all was not well with the design of the original waterwheel installation at the mill. The Dakeyne family members involved with this lease were Daniel Dakeyne the elder, John Dakeyne, Thomas Dakeyne, Daniel Dakeyne the

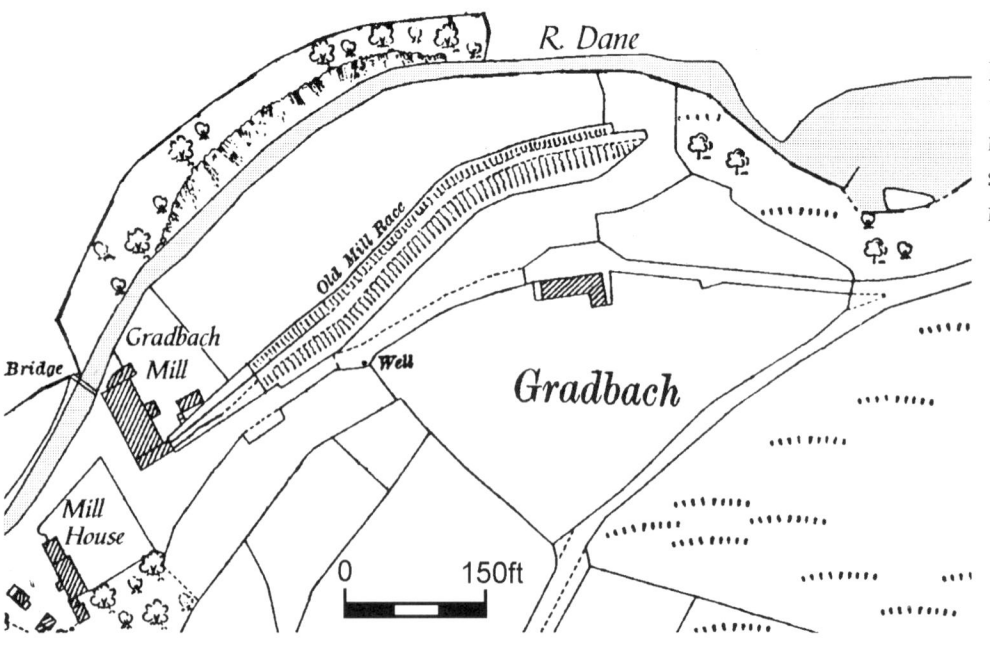

FIGURE 3. PART OF THE 1925 ORDNANCE SURVEY MAP OF GRADBACH SHOWING THE MILL HOUSE, MILL AND MILL RACE (LEAT).

younger, Joseph Dakeyne, Peter Dakeyne, all described as gentlemen, and all of Darley in Derbyshire where they introduced flax spinning using a new patented flax spinning machine called the "Equilinum".[7] This machine had been invented by Edward and James Dakeyne in 1794 and by 1798 the Dakeynes were intent on expanding their flax spinning capability. However by 1798 the mortgage owed to Robert Charles Greaves was still £1200 and the terms of the new lease were particularly onerous on the Dakeynes as they agreed not only to pay 5% interest per annum on the capital but also to a penal sum of £2400 to be owed if the debt of £1200 was not repaid within 18 months.[8] This was an enormous amount for the times and reflects the confidence that the Dakeynes had in their ability to make large profits from their flax spinning business.

Evidence from the Quarnford Land Tax returns indicates Thomas Dakeyne's presence in Quarnford from 1803 onwards as a tenant of Sir Henry Harpur with a levy of 1s-0d per year. This seems a low figure for a commercial concern such as the flax mill and probably only refers to a small parcel of land. The Wincle Land Tax returns show that from 1795 to 1825 the mill pool was owned by a William Shufflebottom (the mill pool was in a detached part of Wincle parish!).[9] In 1821 Thomas Dakeyne's wife Sarah died, and it is only after this, from 1825, that Thomas Dakeyne is listed as the owner of the mill pool and liable for 12s-6d Land Tax on a house and land in Quarnford.[10] It is interesting that the land tax returns make no mention of the mill. The original lease from Sir Henry Harpur expired in 1823 as it only ran for 31 years from 1792. However, the Dakeynes did not renew this lease as they thought they had purchased a 50 year lease from the assignees of the Olivers and White (the deeds show that this was for the pool only and not for the mill). In 1828 they consulted with a solicitor about the confusion. The solicitor then looked into the deeds and wrote the following summary of the situation.

"John & Peter Dakeyne, of Darley, worked the mill for sometime and erected machinery under their patent for spinning flax and built a lodge or warehouse upon the premises at a considerable expense and afterwards let their brother Thomas Dakeyne into possession under certain conditions. The flax concern at Gradbach sometime and is now carried on in the name of John & Peter Dakeyne and as the lease from the late Sir Henry Harpur is expired they think proper to state the nature of the tenancy at Gradbach to prevent mistakes. From the abstract there are 20 years unexpired of a lease from Oliver to them and although the lease from Sir Henry Harpur is expired yet John & Peter Dakeyne presume they still hold the mill at Gradbach under the conditions of such lease, no alterations having taken place, therefore it is thought proper to state to Sir George Crewe for his information who are the real tenants and the persons who might be treated with in case of alterations, conceiving that from lapse of time and other intervening circumstances, they might be strangers to him."[11]

From this it would seem that Sir George Crewe, the successor of Sir Henry Harpur, was unaware of the presence of the Dakeynes at Gradbach and had not realised he was not receiving any rent from them. The very best sort of an absentee landlord! This approach to Sir George Crewe resulted in a new lease being drawn up from 1829 whereby the Dakeynes paid £177-17s-6d for the years from 1823 and a rent of £105 for one year and *"so on from year to year"*.[12]

It must have been around this time that the flax spinning business at Gradbach was trading under the name "John Dakeyne & Co." A price list for linen twist that was supplied to the textile mill at Tean in Staffordshire shows that the thread produced at Gradbach was at the coarser end of the spectrum varying from 3200 to 8000 yards to the pound weight, selling at 1s-5d and 2s-5d per pound weight respectively. The heading on the price list also indicates that the Dakeynes had the use of a warehouse in St. Mary's Gate, Manchester at this time (see Figure 4).[13]

It would appear that Thomas's son Daniel took over the business at the mill around 1830, (in a trade directory the surname being spelt phonetically as Daniel de Coin!).[14] In 1837 Daniel Dakeyne and his partner Thomas Wanklyn were declared bankrupt[15] and the lease of Gradbach Mill was advertised for sale as follows:-

"For Sale by order of the assignees of Daniel Dakeyne and Thomas Wanklyn, bankrupts.

The whole of the machinery for spinning flax and tow belonging to said bankrupts now at work situated at Gradbatch [sic] in Macclesfield. Also the interest in the lease of the premises, seventeen years of which are now unexpired.

The machinery consists of 776 spindles for spinning tow, from 2 to 20 leas; 500 spindles for spinning line, upon the long ratch principle, from 10 to 30 leas; 508 spindles for spinning line or tow, upon the wet principle, from 18 to 60 leas; together with preparing machines for line or tow, drawing and roving frames, cans, straps, bobbins, reels, etc., etc. Slide lathe 12 feet 6 inches long, three turning lathes, grindstone, joiner's bench, taps and dies, turning tools, etc. Smith's bellows, anvil and tools, mill gearing, and steam pipes, steam boiler of four and three quarters horse power, wood and iron patterns, old and new iron, etc.

The lease comprises the mill, a good and substantial building advantageously situated upon and worked by the River Dane, is 25 yards long by 11 yards wide, two storeys

> ## MACHINE LINEN TWIST,
> Spun and Sold by
> ### JOHN DAKEYNE & CO. at Gradbach Works, near Leek, Staffordshire,
> and sold also at their Warehouse,
> ### NO. 4, ST. MARY's GATE, MANCHESTER.
> Fit for Warp, Thread, or any other Purposes.

FIGURE 4. HEADING ON A PRICE LIST SUPPLIED TO TEAN MILL NEAR UTTOXETER BY JOHN DAKEYNE & CO. (SRO)

high; mechanics shop 25 yards by 5 yards and two storeys high; drying store detached, two storeys high; bleach house and yarn room, two storeys high; waterwheel 24 feet diameter and 6 feet wide with about a 30 feet fall, and a good supply of water; a dwelling house, three storeys high, cellared, with dining and sitting rooms to the front, store room and counting house adjoining; and near the works a flax warehouse, two storeys high; heckling shop and warehouse, three storeys high; also about 34 acres of meadow and pasture land with barns, stables, and cottages.

The premises are held under Sir George Crewe, baronet, and are subject during the tenancy to the yearly rent of £110."[16]

This advertisement is the only comprehensive evidence covering the activities at Gradbach and provides considerable information about the mill. The raw material for the mill was prepared flax fibres, as scutching or retting, the first two processes that separate the fibres from the linseed (hence linen) plant are not mentioned. Presumably they took place at the Dakeyne family's mill at Darley in Derbyshire. The processes at the mill started by combing the fibres into parallel strands (heckling), proceeded to spinning the fibres, and concluded with bleaching the yarn. The yarns would then be sold to weavers of linen or calico. Some indication of the maintenance support needed on the site can be inferred from the provision for mechanics, blacksmiths and joiners.

The first two spinning machines mentioned were old and obsolete by 1837, one spun line, the long fibres giving the better quality yarns, and the other spun tow, the short fibres giving an inferior product, with both operating without the benefit of wetting the fibres. This method of spinning flax had been superseded, so consequently the yarns produced by these machines were of a poor quality compared with their competitors, even the yarns produced from line. The third machine was up to date, utilising the wet method of spinning invented about 1820 by Marshall's of Leeds, and being flexible enough to use both types of flax fibres capable of producing thread twice as fine as the two earlier machines. The fact that only a third of the mill's capacity was competitive may have had some bearing on the bankruptcy of Dakeyne and Wanklyn.

In spite of this bankruptcy, evidence shows that Daniel Dakeyne remained at Gradbach until at least 1848.[17] In 1850 his son Bowden Bower Dakeyne, a qualified surgeon (Member of the Royal College of London & Licentiate of Apollo) was managing the flax mill.[18] The 1850s were a boom time for silk spinning so between 1850 and 1860 the mill at Gradbach turned to waste silk spinning as well as flax spinning. This must have been a successful move because the lease from the Crewe estate which expired in 1854 must have been renewed. The process of spinning waste silk was similar to flax spinning and probably the more modern flax spinning machine mentioned in 1837 would have been capable of adjustment for silk spinning as well as flax. This silk spinning was so successful that by 1860 Bowden Bower Dakeyne had leased Danebridge Mill (*q.v.*), lower down the Dane valley, in order to spin silk there.[19]

In 1860, with the passing of the Free Trade Act, the bottom fell out of the British silk industry and many marginal producers, such as Gradbach, went out of business. Gradbach ceased production sometime between 1861 and 1872 when the mill was recorded as "*vacant*".[20] In 1884 it was still "*not in use*" but about that time the mill reverted to the Calke Abbey estate of the Harpur-Crewe family although there was some dispute over the lease for the pool. This dispute was settled in 1887 when Sir Vauncey Harpur-Crewe agreed to pay £62-10s-0d for the pool to James Oliver's executors rather than enter into litigation.[21] The mill was then converted into a sawmill for use of the Harpur Crewe estate.[22] However, even this use had terminated by the turn of the century when the mill was being used as a barn. The Calke Abbey estate finally sold the property in 1951 when the annual rent was £35, a far cry from the £110 p.a. received in 1837![23]

The mill buildings decayed into mild dereliction until in 1978 they were purchased by the Youth Hostel Association who then renovated the mill, turning it into one of their Hostels. Fortunately, the mill was visited in the 1940s and early 1950s by people interested in the mill who recorded some details of what was present at that time prior to the renovation in 1978.

The existing mill building is two stories high and of rectangular shape, 80 feet (nearly 27 yards) long by 37 feet (just over 12 yards) wide (outside measurements) which are close to the dimensions given in the sale notice of 1837. The front of the mill is pierced by ten windows on the top floor and nine windows and the main door on the ground floor. It is constructed of millstone grit with a stone tiled roof (see Figure 5). There is a winding stone staircase in a square tower in the middle of the rear of the mill and there are two small chimney stacks at the top of each gable end.

Inside, the ground floor had a series of drainage channels (which would be consistent with the use of the wet flax spinning process) but they have been covered up in the Youth Hostel conversion. The first floor is constructed of oak beams six feet apart on which there is a planked floor that has been lathed and plastered to seal it. This floor is supported by cast iron columns in the centre of the beams. These columns are circular in form, being three inches diameter at the base tapering to two inches at the top. They have circular head and base plates, stand on raised padstones and have a small torus at their base. Unfortunately these cast iron columns are no longer visible as they have been completely surrounded by plaster during the Youth Hostel conversion. The roof is carried on purlins supported by king post trusses. The building has wooden window frames, the upper windows being 6 feet high by 4 feet 9 inches wide, and the lower ones 7 feet 6 inches high by 4 feet 9 inches wide.[24]

Examination of the large scale maps of the mill shows that originally there was a narrow building attached to the northern gable of the mill building nearest to the river which has, unfortunately, been demolished. It has been suggested that this might have been an earlier wheelhouse.[25] However a photograph taken in the early 1950s shows this building to have been two storeys high with a window on each floor at the front (see Figure 5). There was also a chimney stack and a curved gable end facing the river.[26] There is no trace of any evidence suggesting a leat might have run to this end of the building. In fact, an 1839 enclosure map shows another building situated just upstream of this possible "wheelhouse", situated on the line that any leat would have had to take.[27] From the scale of the maps, this building was the same depth as the main mill building and about 5 yards wide and so it is much more likely that this building corresponds to the *"mechanics shop, 25 yards by 5 yards and two storeys high"* in the 1837 sale notice.

Recorders of the mill in the early 1950s state that the waterwheel then in position was breastshot, being 38 feet in diameter and having 96 buckets with a capacity of 35 gallons per bucket. The wheelshaft position was situated just about at ground level. The wheelhouse on the southern end measures 40 feet by 11 feet wide (see Figure 6) just large enough to fit a 38 feet diameter waterwheel. The wheel was fed through a 12 inch diameter pipe from the leat, and the gearing

Figure 5. Gradbach Mill seen from the west in the 1950s. Note the relatively large size windows typical of mid-19th century practice and the narrow extension on the left hand gable end. (CCALS)

FIGURE 6. GRADBACH MILL VIEWED FROM THE SOUTH SHOWING THE WHEELHOUSE. THE WHEELSHAFT WOULD HAVE BEEN AT GROUND LEVEL SUCH THAT HALF OF THE WATERWHEEL WAS BELOW GROUND LEVEL.

was such that one revolution of the wheel turned the main shaft in the mill 2500 times. This waterwheel was removed for scrap sometime later in the 1950s.[28]

Today the leat still exists, but is dry, and can be followed for about 200 yards up the valley to where the weir and pool were situated. The weir has been removed, although there are some indications in the river as to its position, and some remains of the dam wall and sluices can still be made out in the dense undergrowth alongside the river. A much altered, roofless, two storied building until recently stood on the access road to the mill. which was obviously built earlier than the mill. Judging from its style, this building appeared to have been originally three or four cottages; later the windows had been blocked in and a new set of windows added. This building may be one of the warehouses mentioned in the 1837 advertisement. Unfortunately this ruin was demolished in the late 1990s. Nearer to the mill there is a stone horse trough with stone seats at both ends. The mill house is completely recognisable from the 1837 sale notice. Its window apertures, with stone sides, lintels and ledges, are of the same design as those added to the building on the approach road when it possibly became one of the warehouses, and the same design as those filled in on the mill stair tower. Of the other warehouses, drying store and bleaching house there is now no sign.

The early history of Gradbach Mill as previously published does not comply with the facts now known. Although quoted in a number of publications, the source of the mill's "history" from 1640 to 1785 is traceable to the notebooks of Walter Smith, an early 20th century Macclesfield antiquarian. However, there is no clue as to how he came by this particular information. Certainly the first Harpur lease of 1792 makes it quite clear that a mill did not exist on the site before that date. Although the origin of Gradbach Mill is now clear, there is a major problem concerning its history that still needs to be resolved. That is the difference in waterwheel size between the 38 feet diameter wheel recorded in the 1950s and the 24 feet diameter waterwheel specified in the 1837 sale notice.

The installation of the 38 feet diameter high breastshot waterwheel would have provided a number of benefits over the original 24 feet diameter wheel that was most probably overshot. A high breastshot waterwheel is capable of generating up to 17% more power from a particular fall of water than an overshot waterwheel using the same fall.[29] Also, a breastshot waterwheel will perform better when the tailrace is flooded than an overshot wheel because the waterwheel will be turning in the same direction as the flow in the tailrace. Finally, as a high breastshot wheel will have a greater diameter than an overshot wheel it can accommodate a larger diameter ring gear thereby enabling the main shaft in the mill to achieve its desired speed with fewer steps in the gearing, again reducing potential losses in efficiency.

However, no evidence has been forthcoming so far to suggest when the 38 feet diameter waterwheel was fitted or why. The probability seems to be that it was changed when waste silk spinning was introduced to the processes at the mill in the 1850s. Whenever it was, the wheelhouse was obviously rebuilt to house it, but did this coincide with a rebuilding of the whole mill? The architectural style of the mill, with its large windows and cast iron columns, is more in keeping with the mid-19th century than the date of 1792, when the original mill was built. If the mill was rebuilt after 1837 it would have required considerable capital to be raised, but this is not likely to have been possible for Daniel Dakeyne, who was a declared bankrupt. So was the current mill built in 1792 or has it been rebuilt later, and if so, when, why and by whom? Until this question is resolved then the history of Gradbach Mill cannot be completely told.

References

1. Walter Smith's notebooks, Macclesfield Library; Harris, H. *The Industrial Archaeology of the Peak District*, David & Charles, 1971.
2. DRO, Abstract of leases, 2375M/56/14.
3. *Ibid*.
4. *Ibid*.
5. DRO, Assignment, 1794, D195Z/E4.
6. DRO, Abstract of leases, 2375M/56/14.
7. Wigfull, P., "The Dakeyne Mill and its Romping Lion", *Wind & Water Mills*, 16, 1997.
8. Abstract of leases DRO 2375M/56/14.
9. CRO, Wincle Land Tax Returns.
10. SRO, Quarnford Land Tax Returns.
11. DRO, Abstract of leases, 2375M/56/14.
12. DRO, Lease, 1829, D2375/217/9/6.
13. SRO, Miscellaneous Tean Mill Papers, D 644/1/16.
14. White's Directory of Staffordshire, 1834.
15. *London Gazette*, 15th November, 1839
16. *Macclesfield Courier*, 3rd June 1837.
17. CRO, Wincle Tithe Apportionment.
18. Kelly's Directory of Staffordshire, 1850; SRO Census Returns, Quarnford, 1851.
19. P. O. Directory of Staffordshire, 1860.
20. P. O. Directory of Staffordshire, 1872.
21. DRO, Sale Agreement, 1887, D2375M/88/21.
22. Rathbone, C., *The Dane Valley Story*, 1954.
23. Harpur-Crewe Estate Sale Catalogue, 1951.
24. Massey, J. H., *The Silk Mills of Macclesfield*, Thesis for the Royal Institute of British Architects, 1959.(copy held at Macclesfield Library)
25. Longden, G., *The Industrial Revolution in East Cheshire*, Macclesfield & Vale Royal Groundwork Trust, 1988.
26. Rathbone, C., *The Dane Valley Story*, 1954.
27. SRO, Gradbach Enclosure Map, 1839.
28. Rathbone, C., *The Dane Valley Story*, 1954.
29. Buchanan, R., *Practical Essays on Mill Work and other Machinery*, ed. G. Rennie, 3rd Edition 1841.

---

2. FORGE   SJ 98?65?

Just downstream from Gradbach mill the River Dane is joined by a small tributary flowing in from the Staffordshire side of the river. On the Staffordshire bank there is some flat ground between the river and the escarpment of hills where there is some broken ground which is possible evidence of past human activity. It is here on Burdett's map of Cheshire published in 1777 that a forge is marked. This was probably processing ore produced locally, possible by working a nearby rake. The site is suspiciously close to such a feature in the landscape, known as Lud's Church, which is a deep cleft through the rock. However, there is no further evidence so far to verify the possible source of ore. The forge site has not been confirmed by any evidence other than Burdett's map. No doubt it would be instructive to excavate this site archaeologically to confirm or otherwise its use, and ascertain the likely date of its occupation.

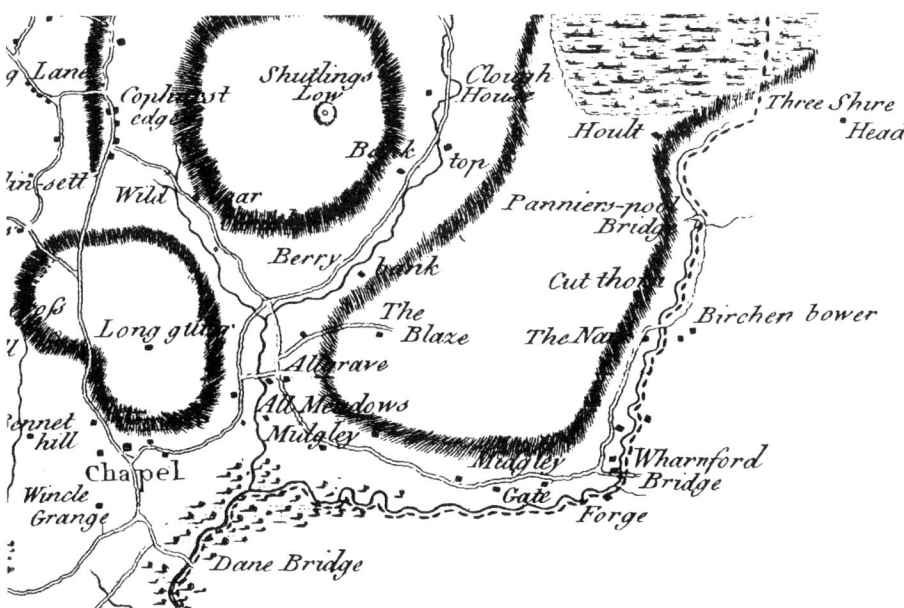

FIGURE 7. PART OF BURDETT'S MAP OF CHESHIRE, 1777, SHOWING A FORGE MARKED JUST TO THE SOUTH OF WHARNFORD BRIDGE. (CCALS, PM 12/16)

3 & 4. CRAG WORKS   SJ 983687

Crag Works were a pair of impressive, stone-built waterpowered factories situated in the parish of Wildboarclough in Cheshire. The site is in a steep sided valley, through which flows the Clough Brook, a tributary of the River Dane, that used to power the works. Although today the valley is quiet, secluded and remote, lying high in the Peak District close to the borders of Staffordshire, Derbyshire and Cheshire, once it was a veritable hive of industry with the works at Crag employing hundreds of people at its zenith. The works consisted of a number of buildings including two driven by waterpower, which were known as the Upper and Lower Mills.

Current local tradition insists that the works were established as a waterpowered paper mill by James Brindley during the period of his apprenticeship with Abraham Bennett, a millwright based at Gurnett, near Macclesfield, around the year 1737. This story should apply to Wincle Paper Mill (q.v.) but, in a classical case of "folk lore" transference, has been attributed both to Folly Mill (q.v.) and, more insistently, to Crag Works. However, the true date for the building of Crag Works can be ascertained from a newspaper advertisement of 1799 which stated:-

"By order of the Assignees of Francis Heywood and George Palfreyman, bankrupts.

For Sale, All those capital premises called Crag Works situated in Wildboarclough in the County of Chester....The premises consist of a very good, new built, stone messuage or dwelling house... Also that large building lately used as a cotton mill and printing shop, 65 yards long by 12 yards wide, four storeys high with two waterwheels, the lower one in the centre of the building, 33 feet diameter by 8 feet wide, the other 24 feet diameter by 6 feet wide, with turning shafts the whole length of the building. Also a dye house, an overlooker's house, twelve good cottages for work people and a house suitable for the accommodation of apprentices...

...The above messuage or dwelling house and building are of free stone, have been erected within the last six years and command a delightful view of the adjacent country of many miles around.

The premises are held by lease under the Earl of Derby for the lives of the said Francis Heywood aged around 34, George Palfreyman aged around 28, and one John Needham aged around 11, subject to a small yearly rent of 50 shillings."[1]

This shows quite clearly that the erection of the mill started in 1793 and it was intended from the outset to be a cotton mill. This date is confirmed by Francis Heywood, one of the leaseholders mentioned in the advertisement as being bankrupt, first appearing in the Wildboarclough Land Tax returns for that year.[2] The specification of the building and its two waterwheels show that the building was purpose built for cotton spinning and was far larger and more powerful than would be required for Brindley's early 18th century paper mill. Obviously the partnership had been made bankrupt by the cost of building the mill before they were able to buy any of the spinning machinery which is noticeably absent from the sale notice.

The Customs and Excise records for the payment of paper duty during the 18th century show that there are no entries for Crag Works. However records do exist for two other mills in the area, one on the Clough Brook near its confluence with the River Dane, known as Folly Mill, and one in Wincle on the River Dane itself.[3] This lack of paper duty records for Crag rule out the possibility that the 1793 cotton mill was built on the site of a paper mill from 1737.

FIGURE 8. AN OLD POSTCARD OF THE UPPER WORKS AT CRAG SHOWING THE ORIGINAL FOUR STOREY MILL BUILT FOR COTTON SPINNING BETWEEN 1793-9. (CCALS)

Although it is a disappointment that Crag Works had no connection with such a famous engineer as James Brindley, it does have other claims to fame. When the Upper Mill was demolished in 1958 it was seen to have been constructed as a fire-proof mill using cast iron pillars and beams. However, the acknowledged oldest iron framed building is considered to be Bage's flax mill in Shrewsbury which is known to have been built in 1796/7. As the building of Crag Works is contemporary with Bage's flax mill it can be said that Crag was one of the first iron framed buildings to be constructed in the world, if not the very first. Certainly the builders of Crag Works came from Derbyshire and the land owner was the Earl of Derby, so it is possible that there was some connection with William Strutt of Belper in Derbyshire who was Bage's collaborator in the design of the first iron framed buildings.[4]

This sale advertisement in 1799 was unsuccessful and the mill was advertised for sale again in the following year.[5] Again interest must not have been very strong and George Palfreyman, one of the original builders of the mill, became its sole owner in spite of his collective bankruptcy with Heywood.[6] He operated the mill as a bleaching and calico printing works. No doubt this section of the textile industry required less capital than equipping the mill with cotton spinning machinery. In spite of using less capital George Palfreyman was bankrupt himself by 1813 and once again the mill was offered for sale. The only difference from the sale notice of 1799 was that there was now a bleaching house, three dye houses, and a blacksmith's shop and "*the premises are capitally adapted for the bleaching & printing business*".[7]

This bankruptcy does not seem to have affected George Palfreyman as he continues to be recorded as the owner and occupier of the works until 1828.[8] In this year Charles Palfreyman took over and is recorded as the owner and occupier until 1831 when his bankruptcy notice was published and there was a grand sale of household furniture and effects.[9] Any moneys raised by this sale only postponed the inevitable for a short time as in 1834 the mill itself was for sale.[10] Once again it proved impossible to sell the mill and it was offered for sale again by Charles Palfreyman in 1839.[11] It is not clear if Charles Palfreyman continued to operate at Crag during the late 1830s but in 1840 the mill was listed as "*not at work*"[12]

Early in the 1840s Joseph Burch came to Crag, no doubt enticed by the bargain price he could negotiate for the lease of the disused mill which had never had any interest shown in its use since it had been built, except by the Palfreyman family. Undoubtedly, Joseph Burch was seen as a godsend to the assignees of Charles Palfreyman. The Burch family lived and worked at the Froghall colour mill in North Staffordshire, six miles south of Leek. Joseph Burch was an inventor, and, like most Victorian inventors, he patented his ideas. In 1839 he took out his first patent, a machine for printing cotton, woollens, and paper.[13] He improved this machine over the years, concentrating on printing heavy materials, such as terry cloth and pile carpets.[14]

His first idea was to develop and manufacture his machines at Crag and market them to the textile industry which was heavily established by this time, especially in Lancashire. Up to 1847 he had a mixed response from his potential customers, selling machines to the firms of Cobden, Brooks and Thompson but was forced into the actual printing of handkerchiefs and delaines to provide some sort of cash flow. In 1847 a report stated that Burch had "*about 35 mechanics with eight machines on the ground floor of the large building and not one working*". Also he employed about six printers and cutters who "*did not have much work and the plant lies quite in the mountains as if at the end of the world*"[15] The remoteness of Wildboarclough obviously made quite an impression on the author of the report.

FIGURE 9. PORTRAIT OF JOSEPH BURCH IN 1851, AGED 26, PAINTED BY CHARLES ALAN DUVAL
(*Science Museum*)

DRIVEN BY THE DANE

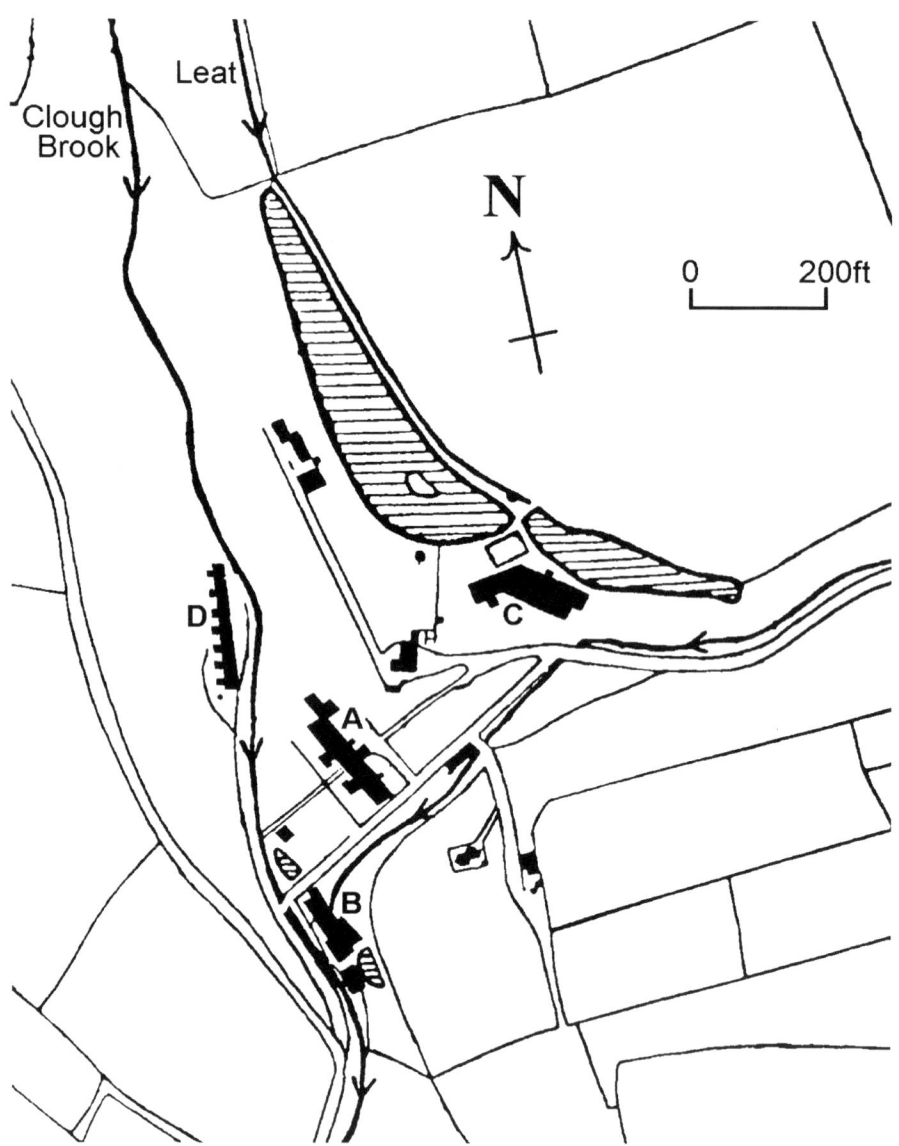

FIGURE 10. PART OF THE WILDBOARCLOUGH PARISH TITHE MAP, 1849, SHOWING THE AREA OF CRAG WORKS AND THE LOCATIONS OF: A. THE UPPER MILL.
B. THE LOWER MILL.
C. THE DYE HOUSE.
D. EDINBURGH ROW.

FIGURE 11. VIEW OF CRAG WORKS LOOKING NORTH-EAST WITH THE LOWER MILL IN THE FOREGROUND. THE BUILDING ON THE LEFT WITH THE TWO TALL, ARCHED, WINDOWS HOUSED THE TWO LARGE WATERWHEELS. (CCC)

Joseph Burch's struggling machine building enterprise took a turn for the better when he made contact with J. Bright & Co. of Rochdale. Initially he had hopes of selling machines to this carpet manufacturer but they proposed a joint venture with Burch whereby they manufactured plain carpets at Rochdale and Joseph Burch used his machines at Wildboarclough to print the pattern onto the carpets. The exact date of this partnership is not recorded but there was an influx of skilled workers into Wildboarclough to operate the mill. The examination of the date and place at which children were born to these migrant workers shows that the partnership must have begun in 1848.[16]

This carpet business was very successful. The machinery at Crag was capable of printing up to six colours in one operation. The carpets produced were imitation Brussels carpets, but they cost 20% less than the real item. During the carpet boom years of the 1840s and 1850s, the carpets made at Crag were in considerable demand, both at home and overseas.[17] New buildings were erected at Crag to cater for the booming business of which the three storey office block testifies by its edifice to the size of the profits, said to have been around £13,000 per year.

The height of Crag's fame and prestige came in 1851 with the Great Exhibition. Local tradition claims that the red carpet used for Queen Victoria to walk on when opening the exhibition was made at Crag. There is no evidence to support this story. However, carpets from Crag were present as exhibits, and were awarded a prize medal in spite of the official exhibition catalogue stating that *"this design is exceedingly bold, though we cannot assign to it much originality of invention"*.[18] During the decade of the 1850s, Joseph Burch continued to patent various improvements to his machinery, indicating that carpet manufacture continued during this period. However, when the 1861 census was taken, it is apparent that there were no longer any skilled workers residing in the neighbourhood, and that the locals, who provided the labourers at the works, had reverted to agricultural activities.[19]

So, what had happened to close such a successful enterprise? One theory was that political enmity between the Prime Minister, the Earl of Derby (landowner of Crag), and John Bright, parliamentarian, whose family firm was in partnership with Joseph Burch [20], spilled over into their commercial activities. Alternatively, John Bright's support of the 1860 Free Trade Act, which had disastrous consequences for the silk trade, could have caused a rift with Joseph Burch, who had relatives in the silk business. The first clue to the demise of carpet manufacture at Crag turns up at the 1862 International Exhibition. Although there was no exhibit from Crag, the Report of the Juries stated:-

'The terry fabrics made some years ago in Rochdale (woven plain by powered loom and then printed) ... and exhibited in 1851, have disappeared, their manifest inferiority leaving no cause for regret at their absence.'[21]

This is an obvious reference to the Bright/Burch carpets, and is a particularly strong statement for the Chairman of the Carpet Jury, J. Crossley, to make about something which was not even exhibited.

At this time the firm of J. Crossley was pre-eminent in the carpet world, a position attained by marketing a good product, and by using patent legislation against possible rivals. In 1858 the Crossley firm had bought a patent from Moxon, Clayton & Fearnley, concerning a "wire motion" loom for carpet manufacture, and in 1859 they started a patent action against J. Bright & Co., due to their (Bright's) use of a similar wire motion machine. It would seem that J. Crossley was not above using his position at the 1862 exhibition to denigrate his commercial rivals whilst the patent case was still in progress! This case lasted until 1864, and it is probable that the production of carpets at Rochdale (and hence at Crag) ceased until the case was settled. This undoubtedly caused a considerable financial

FIGURE 12. ILLUSTRATION FROM THE OFFICIAL CATALOGUE OF THE GREAT EXHIBITION OF 1851 SHOWING A CARPET PRINTED AT CRAG WORKS BY JOSEPH BURCH.

embarrassment to Joseph Burch as this was his only line of business, unlike J. Bright & Co who could continue to manufacture other types of carpets. These financial difficulties came to a head in 1862 when Joseph Burch was forced to sell all his possessions at Crag Hall. Some idea of the wealth created by the production of this type of printed carpet over a period of just over ten years can be gained from the vast range of contents offered for sale including 61 oil paintings, Sèvres and Dresden china, Bohemian glassware, various bronzes together with all the normal household accoutrements.[22] In 1864, after many tribulations, the court found in favour of J. Bright & Co. The main reason for this conclusion was that Crossleys were only holding the "wire motion" patent to oppose their rivals, and were not themselves commercially exploiting the patent. This became obvious during the trial when the loom, supplied to the court by Crossleys as an example, broke down, and Crossleys were unable to repair or replace it. It turned out that this was their only example of this type of loom, and it was only used for fighting patent cases![23] Justice had prevailed in the end.

After the case, J. Bright & Co. decided to consolidate their manufacture at Rochdale, by concentrating on woven carpets only.[24] Consequently, Joseph Burch, with his carpet printing business, was left out in the cold, and Crag Works never reopened for carpet manufacture.

During the late 1860s, Joseph Burch's patents show that he was taking an interest in a variety of areas besides printing machinery, notably in steam engines, and ship hull design.[25] This interest is confirmed by two items in the sale of Joseph Burch's effects in 1862, namely, a model of a screw steamer with a clockwork screw, made by Parley of London and another model of a fully rigged screw steamer that was about 11 feet in length.[26] The local tradition is that Isambard Kingdom Brunel used to visit Crag Hall, and used Crag Pool as a "test tank" for models of the Great Eastern. Whatever the truth of this legend, Brunel was certainly aware of Burch's work, as the lifeboats of the Great Eastern were reputed to be based on one of Burch's patents.[27]

After the finish of carpet manufacturing, the works were never to be operated as a complete unit again. They were still unoccupied in 1874.[28] Various uses were made of some of the buildings from time to time; the Lower Mill became the office and workshop of the Crag estates of Lord Derby; part of the Upper Mill became the telegraph office, and then the post office. One of the floors was used as a Tenants' Hall for local activities, and other areas as storage. In the second world war, munitions were stored at Crag, and a detachment of Dutch soldiers from the Princess Irene Brigade based at Congleton was billeted there as a guard.[29]

In the end, the vandals won, the local council considered that the buildings were unsafe, and they were demolished in 1957, ninety-nine years after they ceased production.

The general arrangement of the buildings at Crag is shown on the 1849 Tithe map (see Figure 10) which clearly shows the Upper and Lower Mills and the dyehouse. The water supply originated from a weir on the Clough Brook about half a mile north of Crag. The water flowed from this weir via a covered leat to Stanley Pool just below Crag Hall. At Crag, the water in the pool is some 55 feet above the brook. The pool was used to supply the Upper Mill and the dye house. Any overflow from the pool went into the small side

FIGURE 13. THE OFFICES AT CRAG WORKS, LATER THE VILLAGE POST OFFICE, BUILT IN THE 1850S. THE STYLE AND SIZE OF THE BUILDING GIVE SOME INDICATION OF THE PROFITS MADE FROM THE CARPET PRINTING BUSINESS.

stream running down into Clough Brook. The waters of this side stream were stored in a small pool alongside the Lower Mill, and used to power the water wheel within that building.

The main building of the Upper Mill was four storeys high, built of squared masonry with prominent quoin work, raised gable ends with a roof of local stone. Internally three rows of cast iron columns were used to support cast iron beams, each 9 feet 6 inch long, that supported the floors. The window frames were also made of cast iron, they all had 30 lights, each one being 9 inches by 6 inches. All the iron work is reputed to have been cast on site during construction.

The Lower Mill was a two story structure built of squared rubble masonry with large quoins, flat stone lintels, and a flagged roof. The guttering was supported on plain stone corbels resembling the protruding ends of timber joists. This mill was entered at first floor level through a doorway situated in the gable end close to the road bridge which spans Clough Brook. The Lower Mill also used cast iron extensively, on the ground floor there was a row of seven columns running centrally along the building supporting the first floor. These columns stood on slightly raised padstones, and their heads contained sleeved bearings through which a 5½ inch diameter line shaft passed. The roof of the Lower Mill was supported by queen posts rising from seven tie beams. The windows in this mill were also made of cast iron but no provision was made for opening them although some ventilation was effected by means of a slot let into the top of each frame.[30]

Although powered by water, by 1839 there were also two steam boilers installed, of 18 and 30 h.p. respectively, to provide heating for the works.[31] Also, coal gas was manufactured and stored in a small gas holder, and was used for lighting the works. The Upper Mill had two waterwheels that worked side by side in a large wheelhouse which had tall arched windows, the whole wheelhouse being similar in style to the steam engine houses built in the 19th century for large cotton mills. Of these two wheels one was 33 feet diameter by 8 feet wide, and the other 24 feet diameter by 6 feet wide.[32] In 1839 they were described as giving about 60 hp, and having a pentway, governor, and gearing.[33] The wheel that powered the Lower works was situated at the southern end of the building in a chamber 5 feet wide by 10 feet deep with its shaft level with the ground floor. It was a breastshot wheel of over 18 feet diameter, and about 4 feet wide, made of iron and built on the suspension principle. It had one inch diameter cross bracing iron spokes and a gear ring mounted on the wheel rim. Although the wheel rotated in the longitudinal axis of the building, the penstock was at right-angles to the wheel.[34] Even more curious was the type of governor used to control the speed of the wheel, (see Figure 14). As the wheel turned, it operated a belt drive to a force pump 'A'. As the speed of the wheel increased, the extra pressure of the water produced by the force pump raised a weighted ram 'C' with teeth. These teeth engaged a geared crank, which in turn closed the sluice on the penstock 'E', causing the speed of the wheel to decrease. If the speed decreased too much the weight of the ram would overcome the pressure of the water from the force pump, and as it fell, it would operate the crank on the sluice, so as to allow more water to the wheel, hence increasing the speed again.[35] The speed of the wheel was controlled by the outlet valve 'B'. The greater the opening then the greater the amount of work required to achieve the water pressure in cylinder 'D' necessary to shut the sluice. Hence the wheel would be governed at a higher speed.

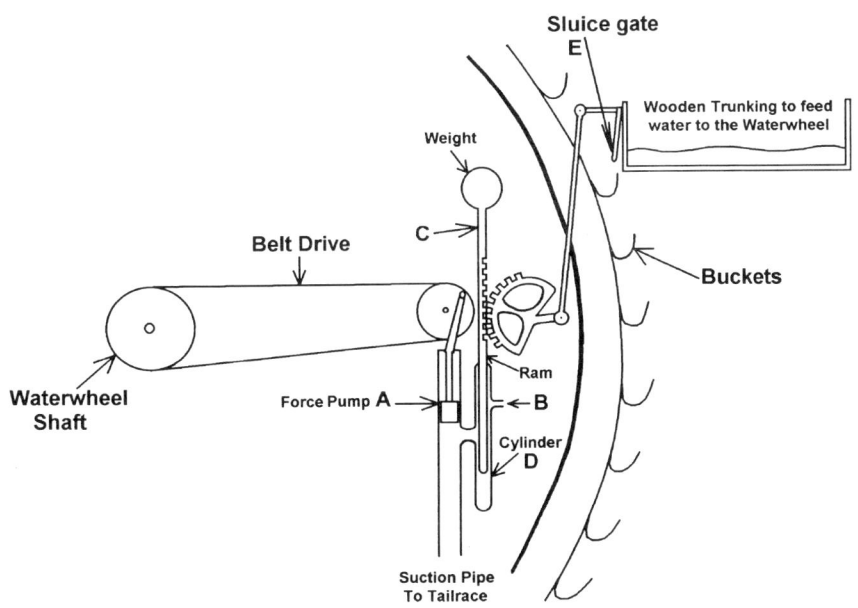

FIGURE 14. DIAGRAM OF THE GOVERNOR CONTROLLING THE SPEED OF THE WATERWHEEL IN THE LOWER MILL AT CRAG WORKS. (*Hanley Reference Library*)

FIGURE 15. WORKERS' COTTAGES AT CRAG KNOWN AS EDINBURGH ROW.

Today, most of the buildings have been demolished, however, the water supply remains virtually intact. The low weir on Clough Brook is still in position close to Clough House Farm, directing a supply of water half a mile through the covered leat to the reservoir, known as Stanley Pool. This is a delightful spot in the summer months, a shimmering sheet of water, favoured by fishermen, perched on the side of an escarpment, surrounded by trees, and overlooked by Shutlings Low and other moorland hills of the Peak District. The two sluices can still be seen, one for the Upper Mill, and one for the dye house, as can the overflow. The dyehouse itself disappeared many years ago, being replaced by the little church of St. Saviour. Although the main buildings of the Upper Mill have been removed, the works offices remain, an imposing three storey building that really emphasises the magnitude of the undertaking at Crag, and the wealth it generated. These offices were built in the carpet heydays, some time after the tithe map survey. They stand between the sites of the Upper Mill and the dye house. One small, two storey block of the Upper Mill remains, today converted into a residence known as the Mill House. Crag Hall still stands in isolated splendour overlooking the pool. Lower down the hill many of the workers cottages and the children's schoolroom can still be seen.

The stone foundations of the Lower Mill, with the tailrace outlet, are still to be seen. The position of the waterwheel is easy to determine due to the presence of the lower half of the mainshaft bearing. This is the last lingering evidence of the highly unusual waterwheel installation that was in use until 1859, and then remained in situ unused for almost another hundred years. The firm of John Bright Brothers was in business in Rochdale until the 1990s and the firm of J. Crossley & Sons is still one of the leading carpet manufacturers in the country. A one eighth scale model of Burch's patented carpet printing machinery is part of the Bennet Woodcroft bequest in the Science Museum, London.[36]

References.

1. *Manchester Mercury*, 13th August 1799.
2. CRO, Wildboarclough Land Tax Returns.
3. PRO, Excise General Letter.
4. Bonson, T., "The First Iron Framed Building?", *Industrial Archaeology Review*, XXII, 2000.
5. *London Gazette*, 24th June, 1800.
6. CRO, Wildboarclough Land Tax Returns.
7. *Macclesfield Courier*, 27th March 1813.
8. CRO, Wildboarclough Land Tax Returns.
9. *Macclesfield Courier.*, 16th June, 1831.
10. *Macclesfield Courier*, 15th February, 1834.
11. *Macclesfield Courier*, 25th May, 1839.
12. List published in *The Textile Colourist*, April 1876, quoted in Turnbull, G., *A History of Calico Printing in Great Britain*, 1951.
13. Patent AD1839 No. 7,937.
14. Patent AD1845 No. 10,657.
15. Graham, J., *The History of Printworks in the Manchester District from 1760 to 1846*, 1847.
16. CRO, Wildboarclough Census Return, 1851.
17. Neville Bartlett, J., *Carpeting the Millions*, John MacDonald, 1977.
18. *The Great Exhibition*, Gramercy Books, 1995.
19. CRO, Wildboarclough Census Return, 1861.

20. Trevelyan, G. M., *Life of John Bright*, Constable, 1913.
21. Juries Report International Exhibition, Class XXII, 1862.
22. *Macclesfield Courier*, 20th September, 1862.
23. Neville Bartlett, J., *Carpeting the Millions*, John MacDonald, 1977.
24. *Ibid.*
25. Patent AD1852 No. 1473; Patent AD1852 No. 1160.
26. *Macclesfield Courier*, 20th September 1862.
27. Patent AD1852 No. 393.
28. Morris's Directory of Cheshire, 1874.
29. Beeken, B., *Wildboarclough*, No date.
30. Massey, J. H., *The Silk Mills of Macclesfield*, B.A. Thesis for RIBA, 1959.
31. *Macclesfield Courier*, 25th May, 1839.
32. *Manchester Mercury*, 13th August 1799.
33. *Macclesfield Courier*. 25th May, 1839
34. Massey, J. H., *The Silk Mills of Macclesfield*, B.A. Thesis for RIBA, 1959.
35. Stoke on Trent Reference Library, S192. (H. Bode).
36. Negative No 20895, Woodcroft Bequest, Science Museum, London.

---

5. FOLLY (GROVE) MILL   SJ 971664

The Clough Brook runs in a north - south direction for about five miles until it joins the River Dane on its north bank about a mile above Danebridge. In the last third of a mile the brook flows through a very steep-sided and wooded gorge. About half way along this gorge, on the western bank of Clough Brook at a place called Gideon (or Gibbon's) Cliff, is the site of Folly Mill which was formerly used to manufacture paper. From Allmeadow Farm, on the heights just above Folly Mill, a defile leads down the valley side to the confluence of the Clough Brook and the River Dane. From this confluence a track used to run along the bottom of the valley of the Clough Brook to provide access to the mill. It was also possible to approach the mill along the top of the cliff to where there were two cottages for mill workers, then a steep flight of steps led down to the mill.

When Folly Mill was built is not known, certainly it did not appear on an estate map of the area in 1774. All it is possible to do is to quote a local historian, Mr. James B. Thornley, who stated in 1923 that he believed the mill was built by Abraham Day, of Allmeadow Farm, possibly about 1780 or 1790. Mr. Thornley also claimed that it was called Folly Mill because two previous attempts at building the mill had ended with the mill being washed away by floods, consequently it was thought folly to build a third. The fact that Abraham Day persisted is said to have caused his wife to take to her bed which she did not leave until she died! This tale has been recounted a number of times but as yet no evidence has been found to support the story.[1] Mrs. Day actually died in 1826 so she must have stayed in bed a good few years. If the mill was built at the end of the 18th century it would have been used to make paper by hand with the waterpower used to drive a "hollander", a vessel containing a roller encircled with sharp edges, which cut and pulverised rags into pulp. Rags were the only raw material capable of being used to make paper prior to the middle of the nineteenth century.

As Folly Mill was a paper mill there has been an attempt to link it to the paper mill built by James Brindley in 1737 when he was apprenticed to Abraham Bennett.[2] This is the same story that is now attached to Crag Works (*q.v.*) without any justification. It is a wonder that the story did not become permanently attached to Folly Mill in view of the lack of hard evidence about the mill's origin. However, throughout the 19th century Brindley's paper mill was always described as being on the River Dane whereas Folly Mill is on the Clough Brook.

In the 18th and 19th centuries the manufacture of paper was subject to a complicated taxing structure which involved the close attention of the Excise collectors, so much so that excisemen lived permanently on site at Folly Mill. In the late 1820s the Bosley parish register records that Jenkin and David, the sons of Jenkin Jones, an exciseman of Wincle, and his wife Maria, were baptised there. The excise records show that Thomas Hope was the master paper maker at both Folly and Whitelee Mill at Wincle (*q.v.*) in 1816, though whether as a tenant of Abraham Day or as the owner of the property himself is not known. Thomas Hope operated the mill up to 1830 when Abraham Day was listed as the occupant by the Excise. As he was 90 years old at this time it is possible that he was not actively engaged in working the mill. Abraham Day lived another five years, dying in 1835 in his 95th year. In his will he left instructions concerning the disposition of five farms but there was no mention of the paper mill. After Abraham Day's death the mill was operated by Thomas Hope for about one year until it was taken by Richard Hope who operated the mill until he died in 1838.[3]

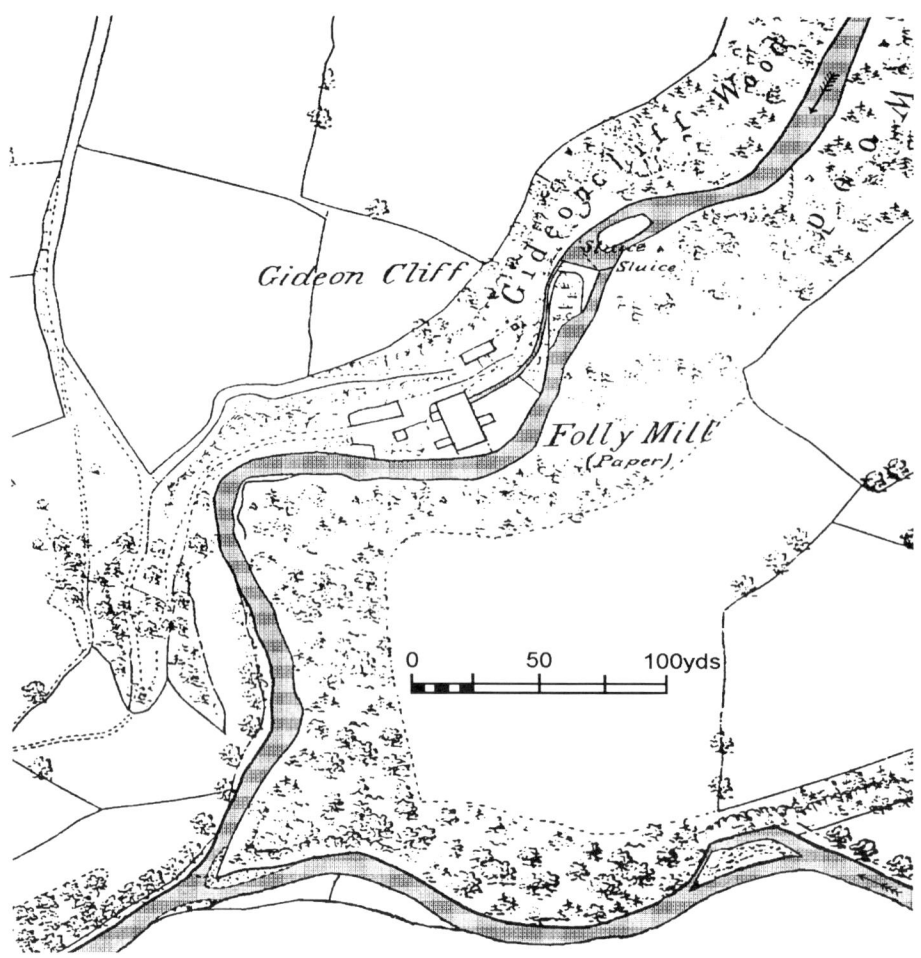

FIGURE 16. ORDNANCE SURVEY MAP OF 1876 SHOWING THE LOCATION OF FOLLY MILL NEAR THE CONFLUENCE OF THE CLOUGH BROOK WITH THE RIVER DANE.

By 1839, Peter Hope was in residence at Folly Grove with his sons Thomas and Samuel having taken over the running of the mill with a partner called John Edge.[4] There were at least another eight paper makers residing in the area in 1841 together with one excise officer living at Folly Mill itself.[5] In December 1848, Folly Grove Paper Mill was to be sold at auction but unfortunately no details of the mill were given.[6] It has been claimed that it was renamed Folly Grove Mill so as not to put off prospective purchasers, but the name "Folly Grove" had been used as early as 1840 by the Ordnance Survey. In January and February 1849 the mill was advertised to let, the contact names given in the advertisement were Mr. Matthew Longden on the premises, or Mr. Thomas Slack, a farmer in Sutton.[7] These two gentlemen were two of the executors of Abraham Day's will so it is possible they were acting in this role in disposing of Folly Mill.

Some years later, in 1850, John Hope and Audrey Ann Hope were described as a paper manufacturers residing at Folly Grove.[8] Samuel Hope, Peter Hope's son, was also living there, described as a servant and paper maker. All told there were twelve paper makers living in the area in 1851, the eldest being 70 years old and the youngest 12 years of age.[9]

In 1857, Folly Grove Mill (Number 250 in the Excise General Letter) was recorded as being operated solely by Audrey Ann Hope, but by 1860 its ownership had changed to John and Matthias Slack who were manufacturing coarse brown paper. It is possible they were related to Thomas Slack, one of Abraham Day's executors. They were also operating another paper mill, the White Hall Mill at Chinley not far from Whaley Bridge in Derbyshire.[10] By this time paper making had become a fully mechanised process with the invention of continuous papermaking using machines invented by the Fourdrinier brothers in Staffordshire and consequently the hand made process as practised at Folly Mill was no longer competitive. By 1861 there were only four people in the area engaged in paper making, three of whom were only youngsters, and the cottages at Folly Grove were uninhabited.[11] According to the Excise General Letter Folly Mill kept operating until 1867 under the Slack family but in 1868 it was listed as "*Not in work*" and in 1869 as "*Unoccupied*".[12] Mr. James Thornley claimed in 1923 that he had visited

FIGURE 17. A PHOTOGRAPH OF THE WESTERN ELEVATION OF THE RUINS OF FOLLY MILL TAKEN C.1930. THE SEMI-CIRCULAR ARCHED CART WAY CAN BE SEEN IN THE CENTRE OF THE MILL WITH WHAT APPEARS TO BE A LOADING BAY ABOVE IT.

the mill in his youth when it was still working and it was making coarse brown and blue glazed paper as used by grocers and ironmongers for wrapping.[13]

The weir was about 50 yards upstream from the mill and was about 7 or 8 feet high. The leat that led to the mill was 6 feet wide and entered the building through a low semi-circular arch. The building was three storeys high but is now in a very ruinous state with some walls still precariously standing and without any floors. The wheel-pit is still discernible being originally about 16 feet long by 8 feet wide, but it had been widened at some time to 11 feet. There is a curve of masonry under the penstock area of the wheel-pit which shows that the wheel was either overshot or pitchback with a diameter of approximately 16 feet. In recent years the site has provided waterpower of a somewhat different kind as the tailrace has been used to power a ram pump which supplies water up the hill to Allmeadow Farm. Where the access track along the valley bottom arrived at the mill there are the remains of some material storage areas, however, this track is now impassable having been washed away by the stream. The only access now is by the track on top of the cliffs to the remains of the two cottages. It must be stressed that the whole site is now in a most dangerous state.

References

1. Smith, W., Notebooks in Macclesfield Local Studies Library.
2. Score, S. D., "Folly Mill", *Cheshire Life*, 1954.
3. PRO, General Excise Letter.
4. *Ibid.*
5. SRO, Census Returns, Wincle, 1841.
6. *Macclesfield Courier*, 30th December 1848.
7. *Macclesfield Courier*, 27th January 1849.
8. PRO, General Excise Letter.
9. CRO, Census Returns, Wincle 1851.
10. Shorter, A. H., *Paper Mills and Paper Makers in England, 1495-1800*, Hilversum, 1957
11. CRO, Census Returns, Wincle 1861.
12. PRO, General Excise Letter.

---

6. DANEBRIDGE MILL   SJ 963651

Where the River Dane passes betwen the parishes of Wincle, on the heshire side of the river, and Heaton on the Staffordshire side, it runs in a fairly steep-sided, wooded valley. The river is crossed by a road at Danebridge where, about a hundred yards below the bridge, the river makes two meanders leaving semi-circles of flat land first on the Staffordshire bank and then on the Cheshire side. It is on this flat land on the Staffordshire side of the river that Danebridge Mill was built.

When the mill was built is not known. Although there was a miller recorded in Heaton in 1327, it is not clear whether he worked at Danebridge Mill or at Bearda Mill (*q.v.*). The earliest record so far discovered confirming a mill's existence at Danebridge is a deed, dated 1652, between William Tonicliffe and his brother Joshua that mentions a paper mill and a corn mill.[1] Nearly twenty years later, in 1671, in a lease between Joshua Tonicliffe and another brother called Joseph, the paper mill at Danebridge is said to adjoin a corn mill and a fulling mill.[2] Obviously these three mills could easily have existed well before being mentioned in these 17th century leases. The location of Danebridge Mill suggests that it might have belonged to Dieu la Cresse Abbey, near Leek, which had a grange at Wincle. The abbey was dissolved in the 1530s. In 1689 Joseph Tonycliffe [sic] raised a thousand year mortgage for £120 on various lands and the mills from Joseph Parker of Derby.[3] This was followed later the same year by another thousand year mortgage to Joshua Tonnycliffe [sic] the Younger, heir of William Tonnycliffe, for £250 on a *"house in occupation of Thomas Leese at Heaton, fulling mill and corn mill in tenure of Joshua Tonnycliffe, William Tonnycliffe his father, and Thomas Leese."*[4]

These mortgages were probably repaid, but in 1716 Joshua Tonycliffe raised £35-9s-8d on mortgage from Thomas Bainbridge, another gentleman of Derby,[5] followed in 1716 with a further mortgage for £40 involving Peter Hope Junior, a clothworker in Warslow, on a dwelling house with paper mill, corn mill, and fulling mill at Danebridge, subject only to the mortgage held by Joseph Parker.[6] These various mortgages seem to have been consolidated around this time for the benefit of Joseph Parker's heirs. Certainly the impression given is of the businesses at the mills being in decline with money having to be raised from a variety of sources. After 1730 the Tonicliffe family were no longer mentioned in connection with the mills when Benjamin Parker, Joseph's son, leased a messuage and three watermills to Peter Hope Junior for 21 years at £16 per year.[7] Around 1740 the estate at Danebridge, including the three mills, had been split up by the marriage of Mary Parker, possibly the daughter of Joseph Parker, to John Fletcher of Stainsby, Derbyshire. It would seem that John Fletcher acquired two thirds of the estate with the other third being in the control of the executors of another Mary Parker, Joseph's unmarried sister.[8]

This was definitely the state of affairs in 1742 when James Brindley, described as a yeoman of Low near Leek, leased a messuage, paper mill, fulling mill, and water corn mill at Danebridge.[9] This particular James Brindley was the person whose eldest son, also called James, was to become a famous engineer and one of the original driving forces involved in the construction of the canal network in Great Britain.[10] However, in 1742 James Brindley junior had just completed his apprenticeship with Abraham Bennett at Gurnett which is less than five miles from Danebridge. He had also been involved in the building of a paper mill on the River Dane while serving his time with Abraham Bennett so would have known the area well and have been in a good position to advise his father about the Danebridge mills.[11] By this time the fulling mill would not have been a paying proposition and the old paper mill, dating from at least the 17th century, would have still been using stamps to pulverise the rags that were the raw material used in the papermaking process. This technology had been superseded recently by the introduction of the hollander for this task, a fact well known by James Brindley junior who had studied and installed one, just downstream in Wincle (*q.v.*), while working for Bennett.[12] Whether James Brindley junior ever worked at Danebridge with his father is not known. At about this time James Brindley is reported to have moved to Leek, taking possession of a mill there.[13] This has been considered to refer to Leek Mill situated in the town (now called the Brindley Mill). However, it is tempting to think this reference to Leek may have been intended to included an area much greater than just the town of Leek. If so, it is possible that Danebridge would fall within this area and that the two James Brindleys worked alongside each other at Danebridge Mill. This would appear to be a logical arrangement because as a millwright James junior would have needed to travel far and wide, especially as his fame spread, which is not compatible with running a mill that needs constant attention. If he was at Danebridge the mill would have been looked after by his father and other brothers. Certainly a number of his siblings became millers but, unlike James junior, it is not known where they received their training unless it was at the most logical place, namely in their father's mill.

Three years after James Brindley senior took the lease on the Danebridge Mill was a time of severe civil unrest, known as the 1745 rebellion, when Charles Stuart, the pretender to the English throne, landed in Scotland from France, raised an army and marched south into England, determined to depose George I, the incumbent monarch. This Scottish army passed through Macclesfield and one column of this army took the road through Danebridge on its way to Derby. Most armies at this time survived by foraging from the land through which they travelled and the Scots were no exception. No doubt the mill at Danebridge would have been a tempting target for them as they passed. Unfortunately for Danebridge, once the Scots reached Derby they decided it was futile to continue and so retraced their path northwards, some of them passing through Danebridge again within a matter of days. Undoubtedly the local population would have been glad to see the back of them, especially the local miller, James Brindley senior.

In 1754, when James Brindley senior let the corn mill at Danebridge to one of his younger sons, Henry, a clause in the lease mentioned *"the site of a paper mill"* and *"the site of a fulling mill"*. Obviously in the fourteen years that James Brindley senior had been at Danebridge these two activities had ceased and the paper mill and fulling mill had been demolished. Another interesting clause in the lease raised a covenant preventing James Brindley senior's wife, Susannah, from having any dower rights to the property.[14] Also it is possible that the corn mill had been modified and improved, a job that would have been well suited to the Brindley brothers.

Later in the 18th century, although Henry Brindley continued to own the mill, it was leased to other millers, notably the Perkins family, with Richard Perkins in occupation until 1783.[15] Although Joseph Perkins took over as miller from his father, the time was opportune for Henry Brindley to increase his income from the mill by being involved with the rapidly expanding cotton spinning business. In 1784 Henry Brindley leased to John Routh, a cotton manufacturer of Stretton, Northwinfield, Derbyshire, the water corn mill at Danebridge for 100 years at £30 per year for first 20 years, £40 per year for second 20 years, £50 per year for third 20 years, £60 per year for fourth 20 years, and £70 for last 20 years, with the option for John Routh to give 6 months notice at end of each 20 years.[16] No doubt Henry Brindley thought he was on to a very good deal for the rest of his life.

Although John Routh converted the mill for cotton spinning he only occupied it for three years. By 1787 it was occupied by John Booth, but in 1788 Edward Maddock, a Macclesfield attorney and property owner, was recorded as the occupier until 1796.[17] In 1791 he advertised the mill at Danebridge to be let, as follows:-

"All that newly erected water mill or factory, four storeys high together with a warehouse, store, smith's shop, joiner's shop, two dwelling houses, stream of water and about two statute acres of land hereto belonging, situated at Danebridge, within Heaton, in the County of Stafford, seven miles from Macclesfield, six miles from Leek, one and a half miles from the turnpike road leading from Congleton to Buxton.

There is about a thirteen feet fall with a stream capable of turning a very considerable number of carding and other machines - The premises are held by lease for the term of 100 years (92 of which are unexpired) and are let for the unexpired term of 12 years at a clear and very low rent of £70."[18]

Obviously the mill described is a small cotton mill and is being sub-let under the terms of the original 100 year lease of 1784. Mr. Maddocks must have been successful eventually as the occupiers of the mill are

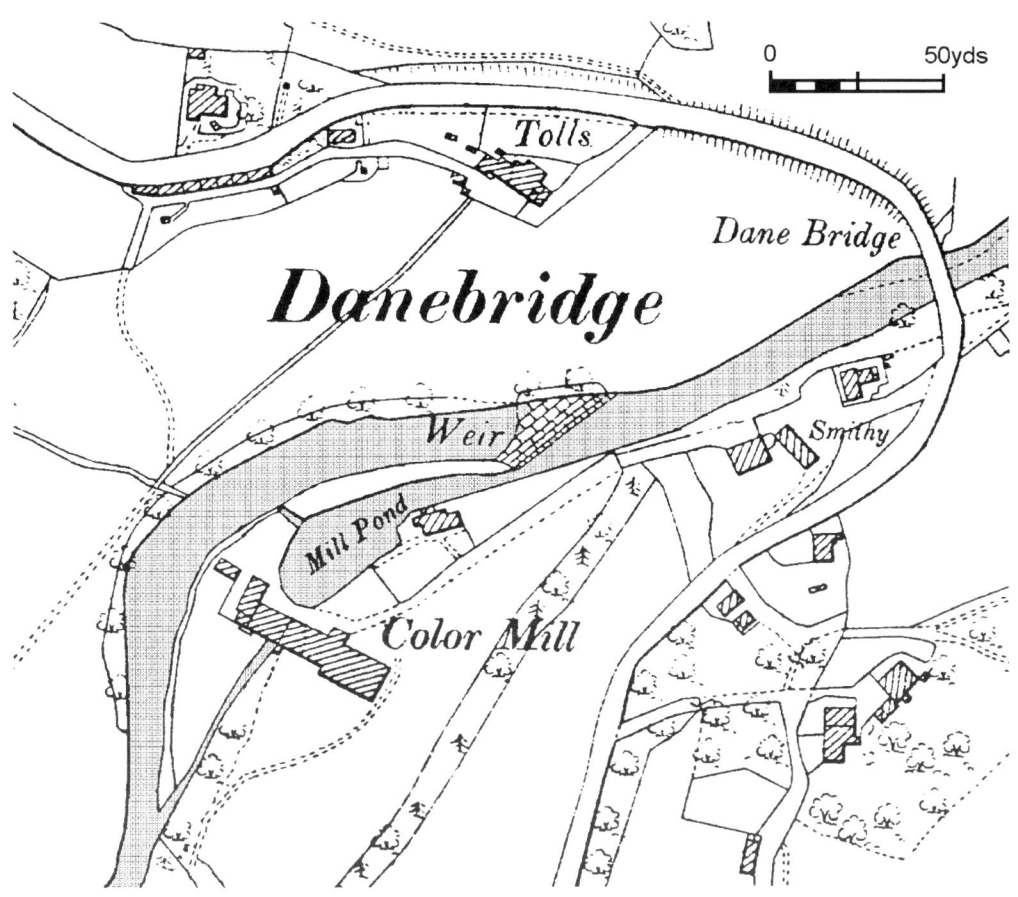

FIGURE 18. ORDNANCE SURVEY MAP, 1897, SHOWING THE LOCATION OF DANEBRIDGE MILL JUST DOWNSTREAM OF THE BRIDGE OVER THE RIVER.

listed as David Whitmore in 1797, Thomas Cockson in 1798-9, and Mr. Rogerson in 1800.[19] However, it is noticeable that the value of the land tax payable remained constant at the 5s-10d of 1781 until it was raised to 11s-8d in 1801. It is possible this increase represents an enlargement of the property at this time.

After 1800, the occupiers were Mr. Cantrell in 1801; Mr. Alsop in 1803; Boardmans & Co. between 1804 and 1806, then George Alsop for 1807-8. In 1809 James Simister became the tenant and occupier the mill for a period lasting up to 1821. It was during this period, in 1811, that a survey of cotton mills was made to determine the nature and number of machines being used for the spinning of cotton in order to establish the level of a government grant payable to Samuel Crompton, the inventor of the cotton spinning mule. The enumerators even reached as far as Danebridge where they listed a Symister [*sic*] of Danebridge as having 4296 mule spindles.[20] This information definitely indicates that the mill was being used to spin cotton, a fact confirmed by the land tax returns after 1825 referring to the mill as a cotton mill. Also in 1831, it was marked as a cotton mill on Bryant's map of Cheshire.

In 1821 James Simister sub-let the mill and John & James Berresford appear in the land tax returns as the occupiers from this date until the returns cease in 1831. In 1834 the mill was still operated by John & James Berresford and is listed as "*a small cotton mill on the River Dane at Heaton*".[21] The mill lease was advertised to let in 1841 when the mill was described as being four storeys high, built of stone, and having rooms 56 feet long by 40 feet wide and 50 feet long by 30 feet wide on each floor. This description of the building at this time giving two distinct floor dimensions, indicates that the mill might have been extended, probably in 1801.

The original lease from Henry Brindley, dating from 1784, was still in force and held by the executors of James Simister. However, the mill was subject to a further sub-let to Joseph Berresford, who was operating Bosley Works (*q.v.*) further downstream on the River Dane. He appears to have been renting Danebridge Mill for £225 per year to his relatives John & James Berresford, who were living with their wives and families together with about 25 other cotton workers at Danebridge in 1841.[22] It would seem that John and James Berresford were giving up spinning cotton at the Danebridge Mill at this time. In the 1841 advertisement Joseph Berresford offered to sub-let the mill for £90 per annum for 1841/2, then £100 for each of the next 20 years, then £110 per year for the remainder of the lease until 1884.[23] After 1841, James Berresford with his wife Hannah and some of his children then went to live at Bosley working in the cotton mill there for Joseph Berresford.[24]

It would appear that Joseph Berresford was unsuccessful in finding another tenant as the mill was unoccupied and empty in 1843. This was at the time that the original owners, the sons of Henry Brindley, were trying to reclaim the property because of non-payment of the rent due under the lease of 1784. This lease was still assigned to James Simister when he died in 1839 and therefore passed to his executors, who stopped paying the rent due to Henry Brindley in 1843 (presumably when the Berresfords stopped paying their rent under the sub-lease). Claiming possession of the mill was complicated by the fact that there was a mortgage outstanding on the property. In an attempt to forestall any legal obligations the executors of James Simister transferred the lease of 1783 to a "man of straw" i.e. a pauper, thereby making it impracticable for the lessor to take legal action for

FIGURE 19. AN EARLY 20TH CENTURY PHOTOGRAPH OF DANEBRIDGE MILL INDICATES ITS ORIGIN AS AN 18TH CENTURY TEXTILE MILL.

the rent due. Eventually, after taking legal advice, the Brindley brothers reclaimed possession of the mill without incurring the liability for the outstanding mortgage.[25]

They must have felt somewhat aggrieved by this situation as they claimed that the last leaseholder, James Simister's executors, were liable for any damage to the mill, especially *"the upright shaft together with the gearing and apparatus which have been removed and taken away from the said premises"*.[26] To substantiate this claim they had an estimate prepared for the repairs, and also for the conversion of the mill back to a corn mill as it had been when originally leased, back in 1784! The estimate shows that the spur wheel was broken and the waste paddle and penstock needed renewing. However they loaded the estimate with a large cost for the renewing of all the floors in the mill, which brought the estimate up to about £390. To reconvert the mill back to a corn mill would have cost about another £300.[27]

In January 1846 the mill was advertised for sale, unsuccessfully, as it was re-advertised in April of that year.[28] This was probably unsuccessful also, because in 1850 the mill was described as being closed.[29] Later in the 1850s the mill was used by Bowden Bower Dakeyne of Gradbach Mill (*q.v.*) for spinning waste silk until after 1860.[30]

After 1860 the mill ceased to be used for yarn production and was acquired by Joseph Burch (or Birch) sometime after his carpet printing business at Crag Works (*q.v.*) had failed. An indication of when Joseph Burch came to Danebridge Mill is given by an exchange of correspondence with Sir Philip Brocklehurst, the owner of nearby Swythamley Hall. In 1870 Sir Philip objected to a carter employed by

FIGURE 20. PHOTOGRAPH OF JOSEPH BURCH IN LATER LIFE WHEN OPERATING DANEBRIDGE MILL

FIGURE 21. JOHN BIRCH & SONS BUSINESS CARD DEPICTING THEIR MILLS AT FROGHALL AND DANEBRIDGE.

FIGURE 22. DANEBRIDGE MILL IN A STATE OF PARTIAL COLLAPSE SHOWING THE TRUSSES IN THE ROOF STRUCTURE.

Joseph Burch using his private coach road to get to Danebridge. In his letter of complaint Sir Philip states *"and if the old mill is worth working, which no one in my time has found it to be, my coach roads are not to be cut up for its convenience."* In Burch's reply he ends with the statement *"hoping we may be good friends and ultimately have good roads and do a good trade at the unfortunate Old Mill."* Which all indicates that Joseph Burch was on the point of starting business at the mill in 1870. A few months later the relationship with Sir Philip must have improved because Burch sent a note to him, as follows;-

"Will you kindly allow Mr. Brassington to go over your private road with four stones from the Roach Quarries to Danebridge paying you acknowledgement, would you also favour him with the loan of your Timber Carriage to bring them on."[31]

The Roaches are a nearby outcrop of millstone grit so it is likely that the four stones mentioned were to be used for the grinding planned at the mill. According to a business card it was operated as a "color" mill in conjunction with Froghall paint and colour mill, run by the family firm of John Birch & Sons (see Figure 21). Danebridge Mill was used to produce a black dye. Coal slack was brought by train to Rushton station each day, then was carried on two carts to Danebridge Mill where it was ground into a fine black powder which was sold as dye.[33] The main market for this dye was the Leek silk makers who were famous for producing "Leek Raven" a well known type of silk made popular in the latter part of the nineteenth century by Queen Victoria's long period of mourning for Prince Albert.[33] In addition the blacking was sold to the firm of Berry's, also in the Leek area, for the manufacture of shoe polish, and it was also used in the production of black

FIGURE 23. A RECENT VIEW THROUGH THE TAILRACE ARCH, DANEBRIDGE MILL, WITH PART OF THE REMAING WATERWHEEL BEARING ON THE RIGHT. THE DEBRIS AND TREE GROWTH RESULT FROM AROUND 100 YEARS OF DERELICTION.

lead pencils, a process patented by a Thomas Birch and a Mr. McGiffert in 1873. Another product that was made using the blacking from Danebridge Mill was stove polish, a popular product considering all the cast iron stove ranges used in late Victorian times.

Towards the end of the 19th century the market for black silk declined and Joseph Burch was also declining in health, finally dying in 1898. Although the mill appears to have been run by Annie Birch, described in 1900 as a paint manufacturer,[34] it was in fact sold in 1898 by the heirs of Joseph Burch and became a blacksmith's and a wheelwright's work shop for a few years before being left empty and unused. Over the following years the mill gradually fell down, the remains being demolished in 1979.

Danebridge Mill was situated about 100 yards downstream from the bridge over the River Dane between Wincle and Heaton. Just downstream of the bridge there used to be a mill weir which ran at an angle across the river. This weir has now been removed by the river authority to ease flooding. Leading immediately off the weir was a small mill pool now dry and overgrown. The mill was four storeys high, built of stone, and 12 bays long, with a king post supported roof. From the evidence of photographs it is obvious that the building consisted of two distinct halves, one half wider than the other such that the roof ridges were not aligned. The original mill probably consisted of the narrower 6 bays and the wheelhouse, with the other six bays added, possibly in 1801. This would explain the two room sizes quoted in the advertisements of 1841 and 1846. There was a stair tower and chimney midway along the front of the mill building and another stair tower at the wheelhouse end at the rear of the building. The waterwheel was situated in the two storey wheelhouse at the end of the building nearest the river, which also probably housed the boiler for the steam heating. The breastshot waterwheel was originally about 18 feet diameter by 14 feet wide and produced about 25 h.p. The position of the outside wheel bearing has been removed and its aperture filled in with stonework. Probably a narrower waterwheel had been fitted at some stage with its outside bearing on a plinth inside the original wheelpit. When this occurred is not known but could well have been in 1870 when the coal grinding business started after a period of about ten years when the mill was disused.

Today there is only a mound of rubble where the mill stood, but the wheelpit and part of a 9 inch diameter shaft bearing are discernible. Running along the centre of the mill, in line with the wheelshaft, there is the remains of a channel 7 feet wide and about 43 feet long which housed the main power system layshaft. Much of the mill stonework has been taken

FIGURE 24. REMAINS OF LINESHAFTING AND A FLAT BELT PULLEY LAY NEAR TO THE MILL SITE UNTIL RECENTLY.

for housebuilding but in the remaining rubble of the mill there are some 42 inch diameter grindstones with square holes in their centres. One of the grindstones is not circular and has four bolts embedded in one of its faces. There is also a 5 inch thick, 48 inch diameter, French burr stone almost completely buried in the rubble and also some relics of lineshafting and belt pulleys. Until recently there were also some remains of the power transmission system in neighbouring fields including two large drive shafts 17 feet 10 inches and 24 feet 7 inches in length and both 5¼ inches diameter.

It is quite probable that the mill site at Danebridge was in use for well over 500 years during which time the waterpower provided by the River Dane has been harnessed to a variety of applications. Initially it was utilised for corn grinding, paper making and the fulling of cloth up to the mid-18th century, followed by cotton spinning, waste silk spinning and then the grinding of coal for the manufacture of black dye, until finally ending its useful life at the dawn of the 20th century. In fact the site is a testament to the flexibility and adaptability of water as a power for industrial processes before the advent of electricity.

References

(Many of the documents of the Swythamley Estate used to be in the Staffordshire Record Office under the designation of D5017, but these documents were returned to the new owner of the estate in the mid-1990s. Fortunately, an abstract of this archive's contents is available at the Record Office and the individual documents are referred to by their Swythamley Archive Bundle number.)

1. Swythamley Archive, Bundle 40/41.
2. Swythamley Archive, Bundle 40/640.
3. Swythamley Archive, Bundle 40/638.
4. Swythamley Archive, Bundle 40/47.
5. Swythamley Archive, Bundle 40/639.
6. Swythamley Archive, Bundle 40/642.
7. Swythamley Archive, Bundle 40/637.
8. Swythamley Archive, Bundle 40/643.
9. Swythamley Archive, Bundle 40/62,63 & 115.
10. Evans, K. M., *James Brindley, Canal Engineer, A New Perspective*, Churnet Valley Books, 1997.
11. Smiles, S., *Lives of the Engineers*, Vol. I, 1862.
12. Phillips, J., *Inland Navigation*, 1805, (reprinted by David & Charles, 1970).
13. Boucher, C. T. G., *James Brindley - Engineer*, Goose, 1968.
14. Swythamley Archive, Bundle 40/64, 122, 123.
15. SRO, Heaton Land Tax Returns.
16. Swythamley Archive, Bundle 40/641.
17. SRO, Heaton Land Tax Return.
18. *Manchester Mercury*, 29th November 1791.
19. SRO, Heaton Land Tax Return.
20. Bolton Central Library, Crompton Spindle List, 1811.
21. White's Directory of Staffordshire, 1834.
22. CRO, Wincle Census Returns, 1841; SRO, Heaton Census Returns, 1841.
23. *Macclesfield Courier*, 4th December 1841.
224. CRO, Bosley Census Return, 1851.
25. SRO, Solicitor's Brief & Associated Letters, 1843, D4974/B/5/69; /5/71; /5/72; /5/73; /5/74.
26. SRO, Letter, 1843, D4974/B/5/69.
27. SRO, Estimate for Repairs, 1843, D4974/B/5/74.
28. *Macclesfield Courier*, 17th January & 18th April 1846.
29. Bagshaw's Directory of Cheshire, 1850.
30. White's Directory of Staffordshire, 1860.
31. SRO, Letter, 1870, D4874/B/5/39.
32. Greenslade, M. W., ed., *Victoria County History of Staffordshire*, Vol. 7, Oxford University Press, 1995.
33. Warner, F., *The Silk Industry*, Dranes, 1921.
34. Kelly's Directory of Staffordshire, 1900.

---

7. SWYTHAMLEY HALL   SJ 973645

A small stream rises on the high ground south of the River Dane between Gradbach and Danebridge in Staffordshire. It flows westward and south-westward to join the River Dane below Danebridge. After about half a mile from its source it flows into a wooded valley, passing below Swythamley Hall, where it is dammed to form a pool. This pool was used to provide the motive force for a water pump which lifted drinking water up to a water tank on the roof of the hall. This may have been driven by a small waterwheel, a turbine, or even a ram pump. Also, according to Philip Brocklehurst in his book *"Swythamley and its Neighbourhood"*, the Swythamley Hall workshops, consisting of a forge, joiner's room and a lathe room, were turned by a turbine which was installed sometime before 1874.

---

8. SWYTHAMLEY CHURCH   SJ 971644

The estate church at Swythamley was built in 1905 as a memorial to Sir Philip Lancaster Brocklehurst of Swythamley Hall. Sir Philip, who was born in October 1827, made his fortune in the silk industry of Macclesfield. In 1884, at the age of 57, he married Annie Lee who was 25 years old at the time. In May 1904 Sir Philip Brocklehurst died at the age of 77 and his wife decided to build a church on the estate dedicated to his memory.

The foundation stone was laid by Lady Brocklehurst in September 1904 and the church was completed in September 1905. Over the years the church was used as the burial place for the Brocklehurst family. One of Sir Philip's sons who was killed on active service in World War II is remembered with a stained glass window and inscription. His other son, who lived until 1975, also has a memorial plaque in the church. The church itself is a small stone built structure with a square tower.

At first sight there does not appear to be any

Figure 25. Ordnance Survey map, 1897, of Swythamley Hall and park with the church at the far left.

accessway up to the tower from inside the church but a section of the nave's wooden roof, 18 feet above the floor, is in fact a counterweighted trap door which leads up into the first room in the tower. In this room is situated a possibly unique water powered installation, namely a set of eight bells driven by an automatic mechanism that had a small water turbine as its prime mover.

The water supply came from the tank on the roof at Swythamley Hall via a pipe across the fields and up a three inch diameter pipe to the turbine. The turbine is probably a small Pelton wheel, having a total diameter of 11½ inches. The outlet is taken away via another three inch pipe, and there is a header tank available for any overflow. Unfortunately there is no maker's plate on the turbine. In later years the water supply from the Hall was cut so an electric motor has been fitted to drive the turbine instead. From the details of the electric motor it is possible to work out that the turbine used to operate at a speed of about 460 r.p.m.

The turbine has a 2½ inch diameter pulley attached to it which drives, via a leather belt, an 8 inch diameter brass pulley wheel on the striking mechanism. The striking mechanism consists of a 13 inch diameter drum, 9 inches wide that is made to rotate by a series of toothed gears driven from the turbine. On the drum there are 8 rows of cams, each row corresponding to one of the bells. When a cam is present, vertical rods are pulled down and released. This vertical motion is transferred via a set of cranks that change the point of action of the vertical motion across the ceiling of the room so as to be underneath the corresponding bell. From the other end of these cranks another vertical linkage transfers the motion up through the ceiling to the bell hammers. The hammers are pulled away from the bells and then released, so striking the bell when the cams on the mechanism dictate. The hammers are reset by a leaf spring. As can be understood, the peel played is determined by the position of the cams on the drum. The striking mechanism has a cast maker's plate fixed to its body giving the maker's name, "J. Smith of Derby".

It is possible to strike seven of the bells simultaneously by a linkage that passes down into the nave of the church near to the main door. The five small bells are contained in the room over the striking mechanism and the three largest bells are situated above the five small ones. The bells are all embossed with the maker's name of John Taylor & Co, Loughborough, and their crest. In addition, all the bells are embossed "P.L.B. & A.L.B. 1905". The largest bell which weighs 13 cwt. is also embossed as follows:-

P.L.B.& A.L.B
So may these bells ring on through coming years
in fond remembrance of our happy married life
A loving memory of my husband dear
They are erected by his loving grateful wife
Oct 23. 1884   May 10. 1904

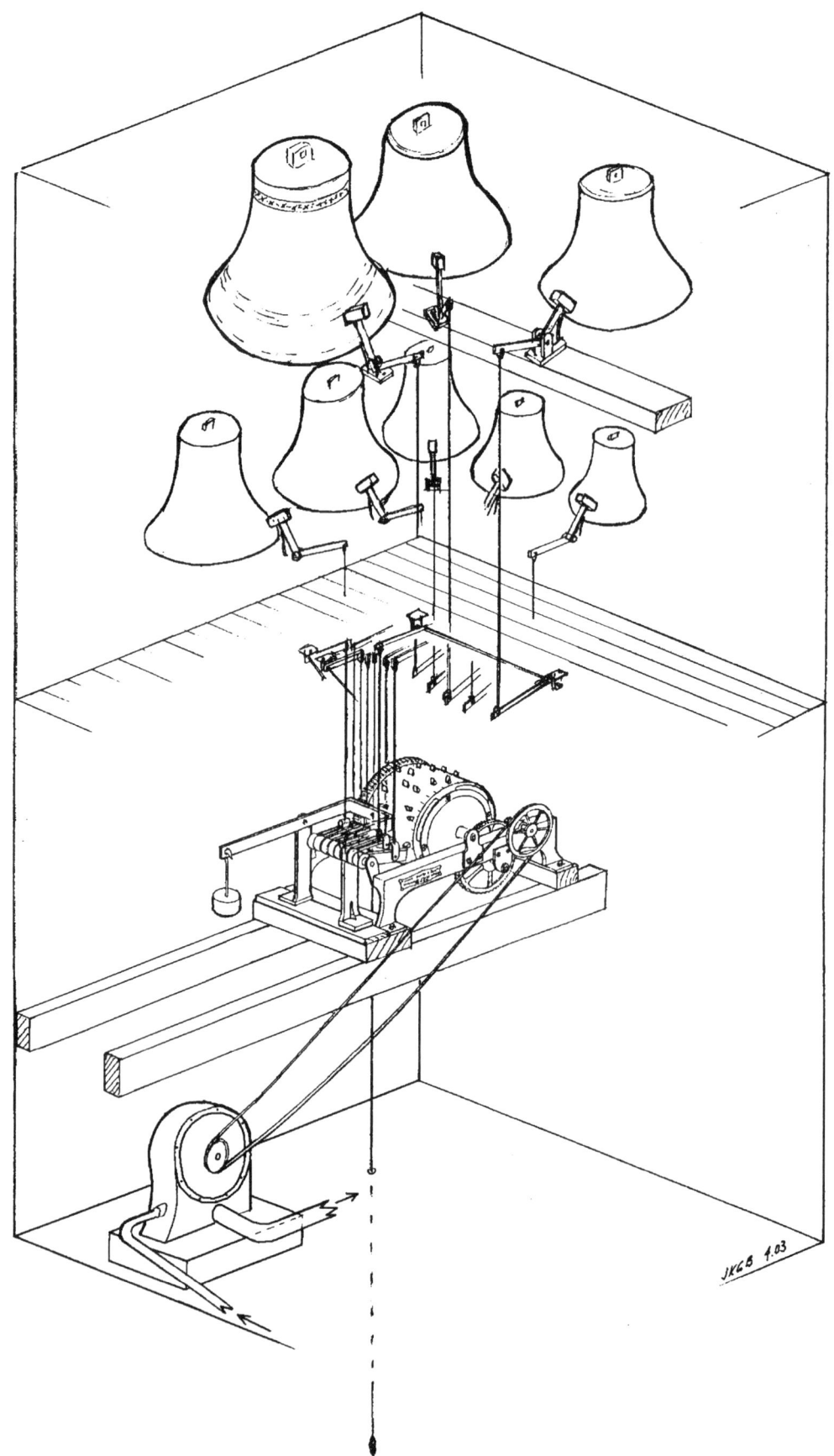

Figure 26. A representation of the automatic waterpowered bell ringing system at Swythamley Church. Only one of the crank arms is shown so that the operation is not obscured. (*John Boucher*).

Figure 27. The turbine with its water inlet pipe at the left. Connected to the turbine by a short belt is an electric motor that was added to the system when the water supply failed.

Figure 28. The main drum with its rows of cams that select the order and sequence of the bells to be rung.

Figure 29. One of the bells with its striking hammer. The vertical rod is pulled down to raise the hammer which then strikes the bell when released.

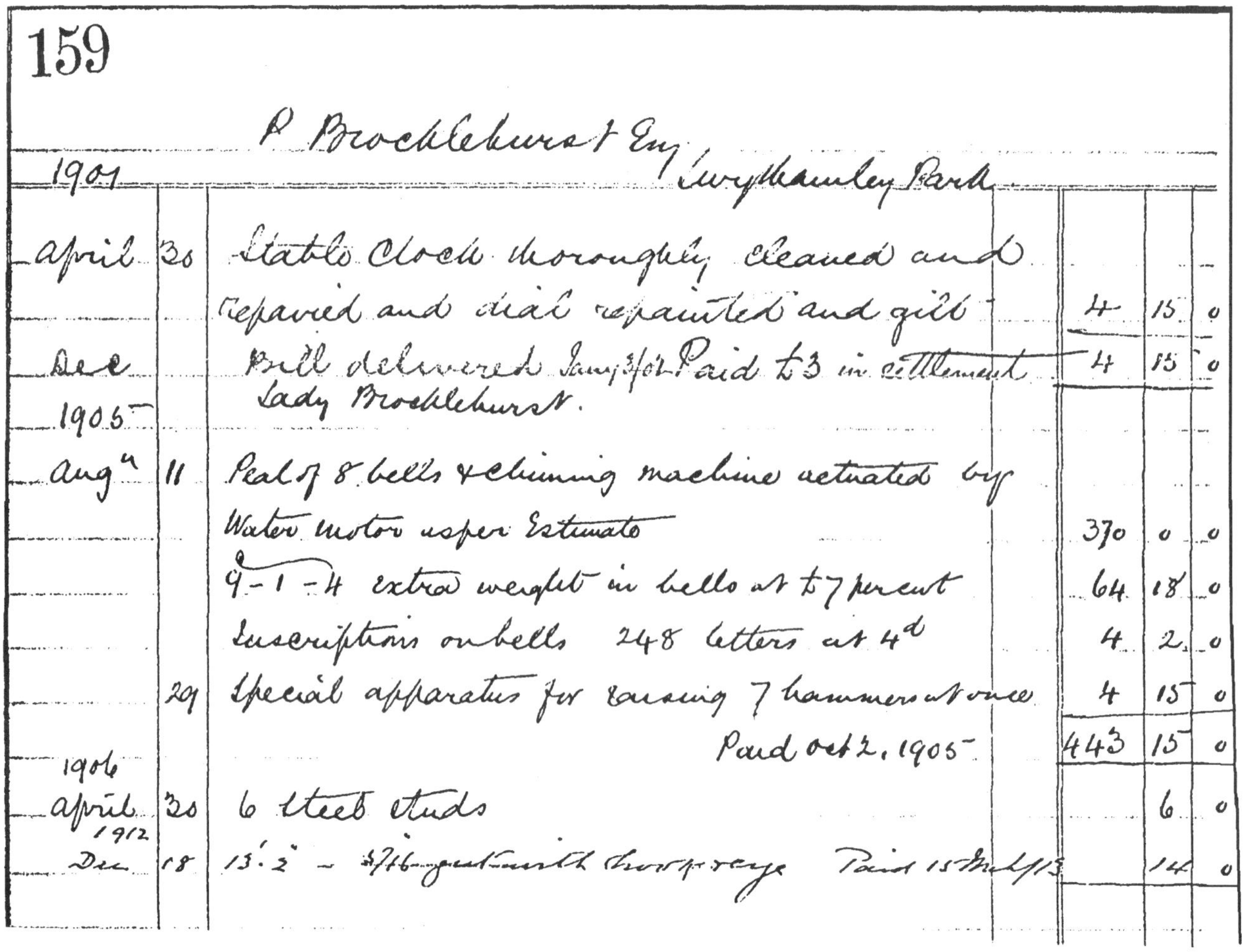

FIGURE 30. THE INVOICE FROM J. SMITH & SONS OF DERBY FOR THE SUPPLY OF THE AUTOMATED BELL RINGING SYSTEM WITHOUT THE TURBINE. (*Smith of Derby Ltd.*)

The firm of John Smith of Derby is still in existence and their records show that the whole installation (without the turbine) cost £443 in 1905 (see Figure 30). Annie Lee, Lady Brocklehurst, died in 1951 and was interred at the church alongside her husband. The Brocklehurst family sold the estate in 1975 to the World Government of the Age of Enlightenment, a sect of transcendental meditators, and from then the church slowly decayed into semi-dereliction. Fortunately, the church was sold in 1990, and was converted into a residence, keeping the turbine, striking mechanism and all eight bells *in situ*. It is thought that there is a covenant in force on the deeds such that if the bells are removed then the property reverts to the Brocklehurst heirs.

---

9. BEARDA SAW MILL SJ 963641

After passing through Swythamley Park, the stream that provided the power for pumping the water supply at Swythamley Hall turns to run westwards until it meets the River Dane. About mile from its confluence with the River Dane the stream is dammed to form quite a large pool. This supplied the water power for Bearda Saw Mill which was owned by the Swythamley Estate.

The only indication of when this saw mill was built is contained in a letter from Thomas Salt, the chairman of the North Staffordshire Railway to Sir Philip Brocklehurst, the owner of the Swythamley Estate in 1886. This was at a time when negotiations were ongoing concerning the use of the weir at Wincle to supply water to Rudyard Lake which was the main reservoir for the Trent & Mersey Canal. Thomas Salt was involved due to the canal being owned at that time by the railway company. One detail of these negotiations concerned the possibility of building a light railway along the Rudyard Lake feeder channel and then over land belonging to the estate as far as Danebridge. In his letter Thomas Salt wrote "*I doubt*

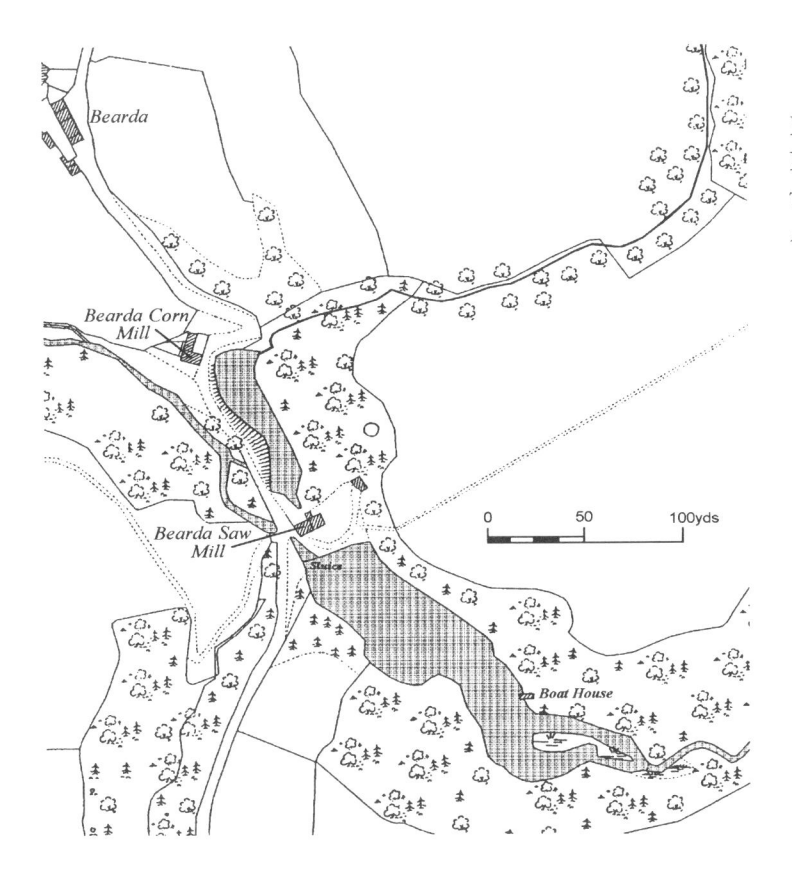

FIGURE 31 (*left*). ORDNANCE SURVEY MAP OF BEARDA, 1897, SHOWING THE LOCATION OF BOTH THE SAW MILL AND THE CORN MILL WITH THEIR ASSOCIATED MILL POOLS.

whether the erection of a saw mill would interfere with us."[1] This suggests that the saw mill was erected sometime after the date of this letter.

Today there is nothing left of the saw mill except for the dam, pool and wheel pit. This wheelpit still contains the cast iron pipe of the headrace together with three iron gears that were used to take the power from the waterwheel into the saw mill. The high breastshot waterwheel had a diameter of about 16½ feet, was 6 feet wide, and operated on a fall of approximately 12 feet.[2] It had a square iron axle with fish belly plates on the four corners.[3] The eight arms appear to have been wooden but the buckets were probably made of iron. Power was taken from the waterwheel by a ring gear attached to the rim of the wheel which meshed with a 22 inches diameter, four armed, iron pinion which had 32 teeth. On the same shaft as this pinion was a 4 feet 6 inches diameter, iron gear wheel, with six arms and 72 teeth, which drove another iron pinion on the mill shaft with 32 teeth.[4] The ratio of gear teeth would have given a shaft speed into the mill of about 200 r.p.m. The pitch of these gears is consistent with a build date of around 1880.[5] There is no indication of how the mill building was constructed but the large scale map of the area shows that the saw mill was around 15 yards long by 7 yards wide with the waterwheel positioned centrally in the mill.[6] The large mill pool has been recently excavated and once again holds water which still flows through the headrace pipe (see Figure 32).

FIGURE 32. THE REMAINS OF THE WHEELPIT AND HEADRACE AT BEARDA SAW MILL.

References

1. SRO, Letter P. L. Brocklehurst to W. Salt, 1886, D4974/B/4/39.
2. Survey measurements, May 2002.
3. Photograph in the possession of Mr. D. Buxton.
4. Survey measurements, May 2002.
5. Stoyel, A., *Perfect Pitch: The Millwright's Goal*, SPAB, 1997.
6. Ordnance Survey, 1893.

10. BEARDA CORN MILL SJ 964642

Immediately after powering the saw mill at Bearda the stream is again dammed to form another pool. This pool is augmented by another small stream which rises near Swythamley Church and flows into the north-eastern end of the pool. Bearda Corn Mill is situated at the north-western corner of this pool.

The date when the first Bearda Mill was built has not yet been discovered but there is a reference to a miller in Heaton in 1327, however it is not known if he worked at Bearda Mill or Danebridge Mill (*q.v.*).[1] The earliest definite record of a mill at Bearda was in the early 1500s when it is known that Dieu la Cresse Abbey, near Leek, owned a corn mill at Bearda.[2] In 1606, the moiety of a mill at Beardhulme [*sic*] was sold by Edward Holinshead to Edward Downes.[3] This was followed in 1614 by the grant to William Tounycliff [*sic*] and William Plant of the manor of Heaton including Beardholme water mill.[4] Later in the century a sale is recorded from William Tonicliffe to his brother Joseph Tonicliffe, in 1652, of a mansion near Danebridge called Beardall or Beardholme which could well have included the mill.[5]

Bearda Mill was marked as a corn mill symbol on Yates's map of Staffordshire of 1775. The earliest definite indication found to date of the occupiers of the mill shows that Joshua Heapy was the miller in 1806.[6] However, he had been superseded by Edward Heapy as the miller when the Swythamley Estate was sold in 1831.[7] It is possible that Edward Heapy was the brother of Joshua because by 1834 Edward Heapy had moved to Rushton Mill (*q.v.*) and Joshua Heapy Junior, presumably the son of Joshua, became the miller at Bearda where he stayed until after 1846.[8] Soon after this date the younger Joshua Heapy left Bearda to become the miller at Bosley Corn Mill (*q.v.*) and by 1851 the miller at Bearda Mill was William Heath who remained there until sometime in the 1860s.[9] Thereafter there was a succession of millers in residence, such as John Wood in 1872, Joseph Brassington in the 1880s, and Charles William Torr, who was described as a farmer and miller, in the 1890s.[10] In 1900, Charles William Torr was describing himself as just a farmer, residing at Bearda Hall Mill, so presumably corn milling had ceased at the mill by this time.

The mill, built of local stone, is still to be seen although it has been converted into a private residence. Inside the building there are the remains of some of the wooden hursting showing the mortises for a bridgetree and brayer arm. Marks and notches in the internal woodwork indicate that the mill possibly had two pairs of millstones at the most, with the second pair being an addition to the original medieval layout. There are the remains of two millstones at the mill. The first is made of local pinkish stone, probably coming from the Roaches or other local rocky outcrop, and measures 4 feet 10 inches but could have been larger as the surviving piece is not a complete millstone. There is also evidence that attempts had been made to balance this stone. The second millstone is a standard French burr stone of 4 feet diameter. During the house conversion work two kiln tiles were discovered. These are about 12 inches square, machine-made of a grey material, with clusters of five small holes arranged as on a dice. The mill pool, headrace and tailrace have all been recently excavated and the owner has installed a new high breastshot waterwheel in its original position. The new waterwheel, which is 12 feet in diameter by 4 feet wide with metal shaft, arms and rims together with wooden buckets and sole, was made in the late 1990s by Harry Baxter.

FIGURE 33. BEARDA CORN MILL, NOW CONVERTED INTO A HOUSE.

FIGURE 34. THE NEW WATERWHEEL AT BEARDA CORN MILL.

References

1. Greenslade, M. W., ed., *Victoria County History of Staffordshire*, Vol. 7, Oxford University Press, 1995.
2. William Salt Library, Stafford, M540.
3. Swythamley Archive, Bundle 40/13 & 40/88.
4. Swythamley Archive, Bundle 1/6.
5. Swythamley Archive, Bundle 40/42.
6. SRO, Swythamley estate map & schedule, 1806, D4974/B/7/1 & B/8/3.
7. Swythamley estate sale catalogue in the possession of Mr. D. Buxton.
8. Merger of Leek Frith and Heaton tithes, 1846, Swythamley Archive, Bundle 29/536.
9. SRO, Heaton Census Returns, 1851/61.
10. Kelly's Directory of Staffordshire, 1872, 1884, 1888, 1892.

---

11. WINCLE (or WHITELEE) MILL   SJ 957642

Wincle Grange was founded around 1400 on the Cheshire or northern bank of the River Dane by the monks of Combermere near Nantwich. It is likely that the Grange had a watermill but the only evidence is from much later when an agreement was made between the monks of Dieu la Cresse and the monks of Crokenden that the mill near Gighall, turned by the waters of the river Dane, should be exempt from tithes.[1] Gighall is the name used for the property on the Staffordshire side of the river adjacent to Whitelee which is in the parish of Wincle on the Cheshire bank. A gig mill is another name for a walk or fulling mill so the name Gighall might imply that the nearby mill was a fulling mill.

Nothing further is known about the fulling mill until the early part of the 18th century when a millwright called Abraham Bennett who lived in Gurnett near Macclesfield was commissioned to build a paper mill. At the time, James Brindley, who was to become a well known canal engineer, was an apprentice of Bennett and his involvement in the building of the paper mill is a well known story worth repeating.

"His master [Abraham Bennett] having been employed to build an engine paper mill, which was the first of its kind ever attempted in those parts, went to see one of them at work to copy after. But notwithstanding this, when he had begun to build the mill, and prepare the wheels, the people of the neighbourhood were informed by a millwright, who travelled the road, that Mr. Bennett was throwing his employers' money away, and could not complete what he had undertaken. Young Brindley hearing this report, was resolved to see the mill intended to be copied; accordingly without mentioning his intentions, he set out on a Saturday evening, after working all day, travelled fifty miles on foot, took a view of the mill, and returned back in time for his work on Monday morning, informed Mr. Bennett wherein he was deficient, and completed the engine to the satisfaction of the proprietors. Beside this, he made considerable improvements in the press-paper."[2]

From this description it would seem that they were building a paper mill that was probably based on the use of a machine called a hollander to prepare the rags instead of the previously used stamps. The use of the hollander was being introduced into the paper making industry in Great Britain at about this time. The hollander was somewhat like a bathtub, inside of which there was a drum with ridges or spikes which was rotated by the power from a waterwheel. Rags and water were placed in the hollander and the motion of the spiked drum macerated the mixture into a wet mulch. The paper was then made by hand from the resulting mulch as in earlier methods. In Samuel Smiles's biography of Brindley published in the middle of the 19th century further details are given. The mill in question was being converted from a fulling mill and was located on the River Dane. James Brindley was recorded as being in his fourth year as an apprentice which would make the date to be around 1737.[3] Certainly Wincle Mill is the only location that satisfies all the elements of the story and was recognised up until the early 20th century as the fulling mill that Brindley converted.[4] Later 20th century writers have located the mill *"on the River Dane in Wildboarclough"*.[5] Unfortunately, this led to the story being attributed to Folly Mill (*q.v.*), which is not on the River Dane.[6] Eventually, following a number of newspaper articles, credence was given to Brindley's paper mill having been built at Crag.[7] Hopefully this curious case of folk lore transference can be discounted and Wincle Mill can resume its role, never questioned in the 19th century, as the mill Brindley and Bennett were involved with in the late 1730s.

Wincle Mill certainly operated throughout the rest of the 18th century as various paper makers appear in the local parish registers from time to time. In 1790 the mill was leased by Charles Clowes of St. John, Horsley Down, Surrey, to Joseph Wagstaff, paper maker of Bollington, Cheshire, for the consideration of £950.[8] Six months later, however, the mill was surrendered to Joseph Wagstaff's trustees as set out in Wagstaff's will.[9] About eighteen months later, in 1792, Wagstaff's executors were able to sell the mill to John Ryle, a silk merchant of Macclesfield, and John Cruso, a gentleman of Leek, to the use of Michael Daintry for the consideration of £1200. Daintry was the vicar of North Rode parish church and Ryle and Cruso were his trustees. Obviously this was an investment to provide income for the vicar. This sale included the paper mill, drying house, farm, etc. adjoining other lands owned by Daintry in Wincle.[10] Towards the end of the 18th century the Hope family were papermakers in Wincle and were in partnership with Jonathon Hughes of Winkhill paper mill which is located in Staffordshire just to the south of Leek. This partnership between Thomas Hope and Jonathon Hughes was dissolved in 1793.[11] It is not clear why the partnership was dissolved or whether Thomas Hope was operating at Wincle Mill or Folly Mill at this time. Certainly by 1816 Thomas Hope was living at Whitelee and was the master paper maker at both Wincle and Folly mills.[12]

In the early 19th century the Trent & Mersey Canal Company were proposing to build a water supply feeder, or leat, from the River Dane to their reservoir known as Rudyard Lake. Initially they envisaged diverting the water from upstream of Wincle Mill on the northern bank of the river. It was intended that the leat would run past Wincle Mill on the Cheshire

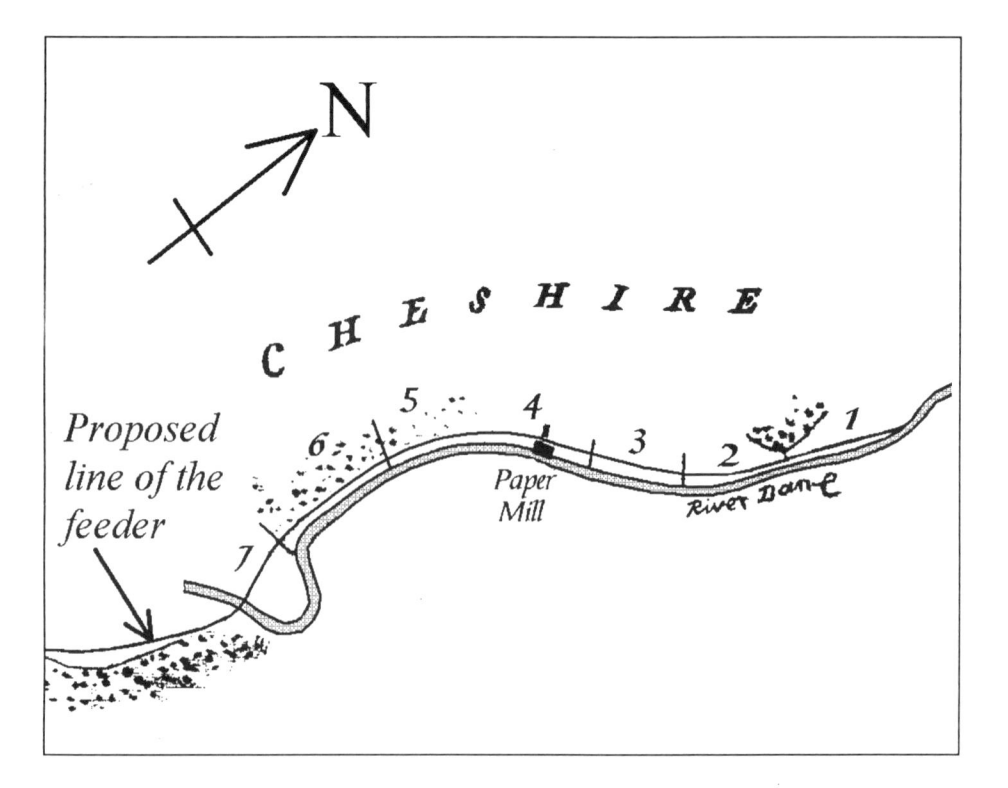

FIGURE 35. THE 1806 PLAN OF THE PROPOSED FEEDER FOR RUDYARD LAKE STARTING UPSTREAM OF THE PAPER MILL, RUNNING ALONG THE CHESHIRE BANK OF THE RIVER DANE PAST THE MILL AND FINALLY CROSSING THE RIVER DOWNSTREAM OF THE MILL. (SRO)

bank and then cross the river on an aqueduct before heading along the southern side of the valley to Rudyard (see Figure 35).[13] However, there must have been a conflict with the owners of Wincle Mill about water rights because when the leat was built it started from a new weir that was built just downstream of the mill (see Figure 36). This leat was not very successful due to it being virtually level. John Rennie, when he reported on this reservoir feeder in 1820, was quite scathing. He recommended that the intake should be from the paper mill's weir so as to collect flood waters at an agreed height above the weir level, thus giving a reasonable fall to the reservoir as well as preserving the mill's normal water supply.[14] This advice was eventually implemented after an Act of Parliament had been passed in 1821 granting the necessary authority for the remedial work and in 1824 the leat was extended upstream to the paper mill weir where it was allowed to collect any flood water over six inches above normal weir height.[15]

Thomas Hope's son Richard eventually succeeded his father in running both paper mills in Wincle,[16] but by this time the technology in use at Wincle was becoming obsolescent due to the invention, in Staffordshire, of continuous papermaking using machines patented by the Fourdrinier Brothers. Because the paper mill was no longer competitive and so could not command a high rent (if at all), the Rev. Daintry's income from the premises was in severe jeopardy. To remedy this situation the Rev. Daintry's trustees entered an agreement with the Trent & Mersey Canal Company in 1834. This agreement transferred the water rights of the mill to the canal company in order to provide a water supply from the weir to their Rudyard reservoir for 50 years until 1884 at £100 per year.[17] The canal company understood that the paper mill would be pulled down after the implementation of the agreement. However, when the first edition of the Ordnance Survey was issued in 1842 it was still described as "*old mill*". Also, in 1860, it was stated that "*The ruins of Whitelee paper mills are situated about one mile south west from the Ship Inn.*"[18] So although the mill ceased to operate after 1834 it was not demolished as planned but simply allowed to fall down.

With the implementation of the agreement with the Trent & Mersey Canal Company the headrace sluice of the mill was used to control the water flow in the reservoir leat which is known as the Dane Feeder. With the sluice closed the water was impounded behind the weir and overflowed into the feeder, once opened the water flowed through the sluice down the river thereby lowering the impounded water and stopping the flow into the feeder. In 1851 the Wincle Grange estate was sold to Philip Brocklehurst who was the owner of Swythamley Hall.[19] After 1884 when the agreement with the canal company terminated there was a dispute between the landowner, now Sir Philip Brocklehurst, and the canal company about the water rights. The canal company expected that at the end of the 1834 agreement the rights would revert to the company but Brocklehurst pointed out that the agreement had no element of a purchase in it. The canal company were most reluctant to renew the lease, claiming that they had no need of the water supply, and, even if they

FIGURE 36. PLAN OF THE FEEDER TO RUDYARD LAKE AS ORIGINALLY BUILT, STARTING AT A WEIR NEAR THE BRIDGE OVER THE RIVER JUST DOWNSTREAM OF THE PAPER MILL AND ITS WEIR. (SRO)

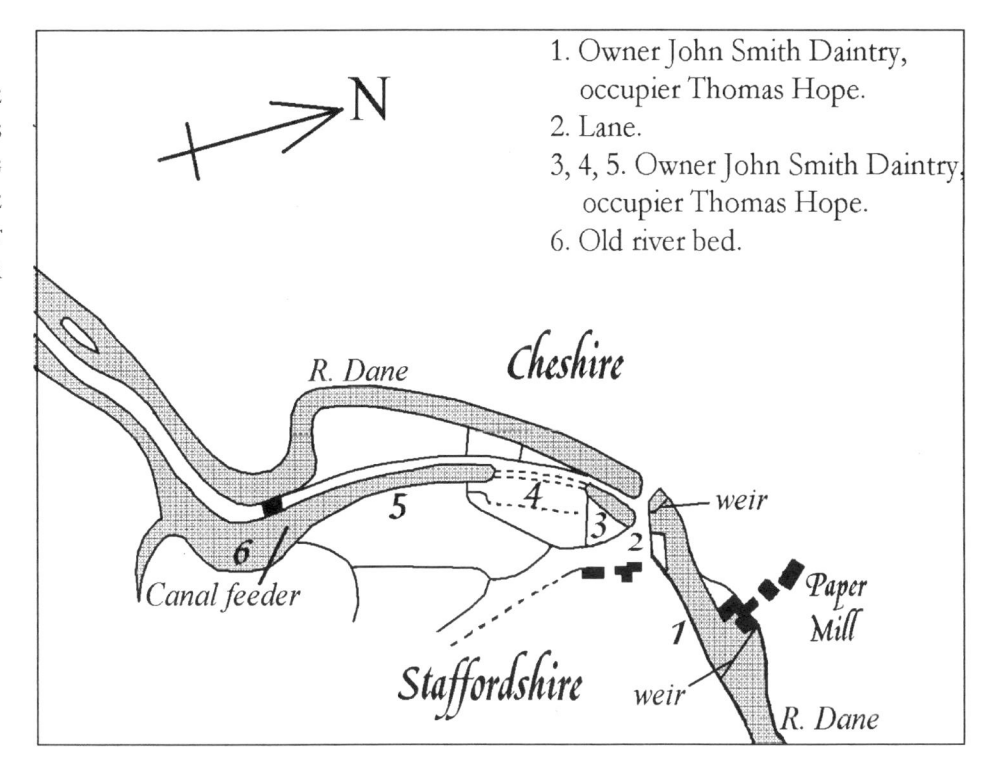

1. Owner John Smith Daintry, occupier Thomas Hope.
2. Lane.
3, 4, 5. Owner John Smith Daintry, occupier Thomas Hope.
6. Old river bed.

did, they were unable to set a value on it. After much argument and a number of long and vinous lunches at Swythamley Hall, the canal company agreed to a trial period of one year at the old rent to determine the supply's value to the company. After this period they renewed their lease of the water rights in 1888 for another 30 years at £50 p.a.[20] This agreement was renewed again in 1909 for 21 years[21] and again in 1930 for 25 years, both leases having a rental of £100 per year.[22] This lease should have lasted until 1955 but when the canal company was nationalised in 1948 it appears that the weir and its water rights were presumed to belong to the canal company and were appropriated by the state. The weir is now assumed to be the property of British Waterways, a position of doubtful legality as the lease held by the company at nationalisation was for the water rights not for the weir itself. Although British Waterways consider that they own the weir they are still making payments of £100 p.a. to the riparian landowner!

Today there is no evidence of Wincle Paper Mill to be seen except the weir across the River Dane and a channel where the headrace and wheelpit must have been. The original Dane Feeder weir is still in existence just downstream of the paper mill weir with the extension leat running in the river bed between the two weirs. Unfortunately the central part of the paper mill weir was washed away during heavy floods in 1998 after more than 260 years of service. This weir has now been rebuilt by British Waterways for about £½M, indicating the value still attached to the water supply from this site which diverts the water originally destined for the Irish Sea (via the Rivers Dane, Weaver and Mersey) into the North Sea (via the Trent & Mersey Canal, the River Trent and the Humber).

References

1. Brocklehurst, P., *Swythamley and its Neighbourhood*, 1874.
2. Phillips, J., *Inland Navigation*, 1792 (Reprint David & Charles, 1970).
3. Smiles, S., *Lives of the Engineers*, Vol. 1, 1862.
4. Visit Report, *Transactions of the Lancashire & Cheshire Antiquarian Society*, Vol. 30, 1912.
5. Boucher, C. T. G., *James Brindley - Engineer*, pub. Goose, 1968.
6. Score, S. D., "Folly Mill", *Cheshire Life*, 1954.
7. Harris, H., *The Industrial Archaeology of the Peak District*, David & Charles, 1971.
8. Swythamley Archive, Bundles 29/529-530.
9. Swythamley Archive, Bundle 29/451.
10. Swythamley Archive, Bundles 29/528-528A & 29/452.
11. Shorter, A. H., *Paper Mills and Paper Makers of England 1495-1800*, 1957; *London Gazette*, 16th November, 1793.
12. PRO, Excise General Letter, 1816.
13. SRO, Plan, Q/RUM/40.
14. Rennie, J., Report on the Rudyard Reservoir, 1820, Institute of Civil Engineers.
15. SRO, Letters, D4974/B/4/39-42.
16. SRO, Letters, D4974/B/2/39-42.
17. SRO, Agreement, 1834, D4974/B/2/9.
18. White's Directory of Cheshire, 1860.
19. Swythamley Archive, Bundle 29/648.
20. SRO, Agreement 1887, D4974/B/2/9.
21. SRO, Agreement 1908, D4974/B/2/10.
22. The Boat Museum Archive, Ellesmere Port, Trent & Mersey Agreement No 345, 1930.

FIGURE 37. WINCLE PAPER MILL WEIR BEFORE BEING DESTROYED BY FLOODS IN 1998. THE EXTENSION TO THE RESERVOIR FEEDER CAN BE SEEN ON THE RIGHT.

Section 2. Hug Bridge to Buglawton

By the time the River Dane reaches Hug Bridge on the main Macclesfield to Leek road its valley has widened out and it has reached the level of the Cheshire plain. However, there is one more obstacle directly in its path preventing its westward flow, namely the massive outcrop known as Bosley Cloud. In order to circumnavigate this obstacle the river turns to the north-west just after passing under Hug Bridge. It is at this point that a small stream enters the Dane from the south. About two miles north of Hug Bridge the main river is joined by the Bosley Brook which rises on the high ground to the east. Having rounded the northern extremity of Bosley Cloud, the River Dane then flows south westerly, entering Buglawton, a suburb of Congleton, where there is the confluence with the Daneinshaw Brook.

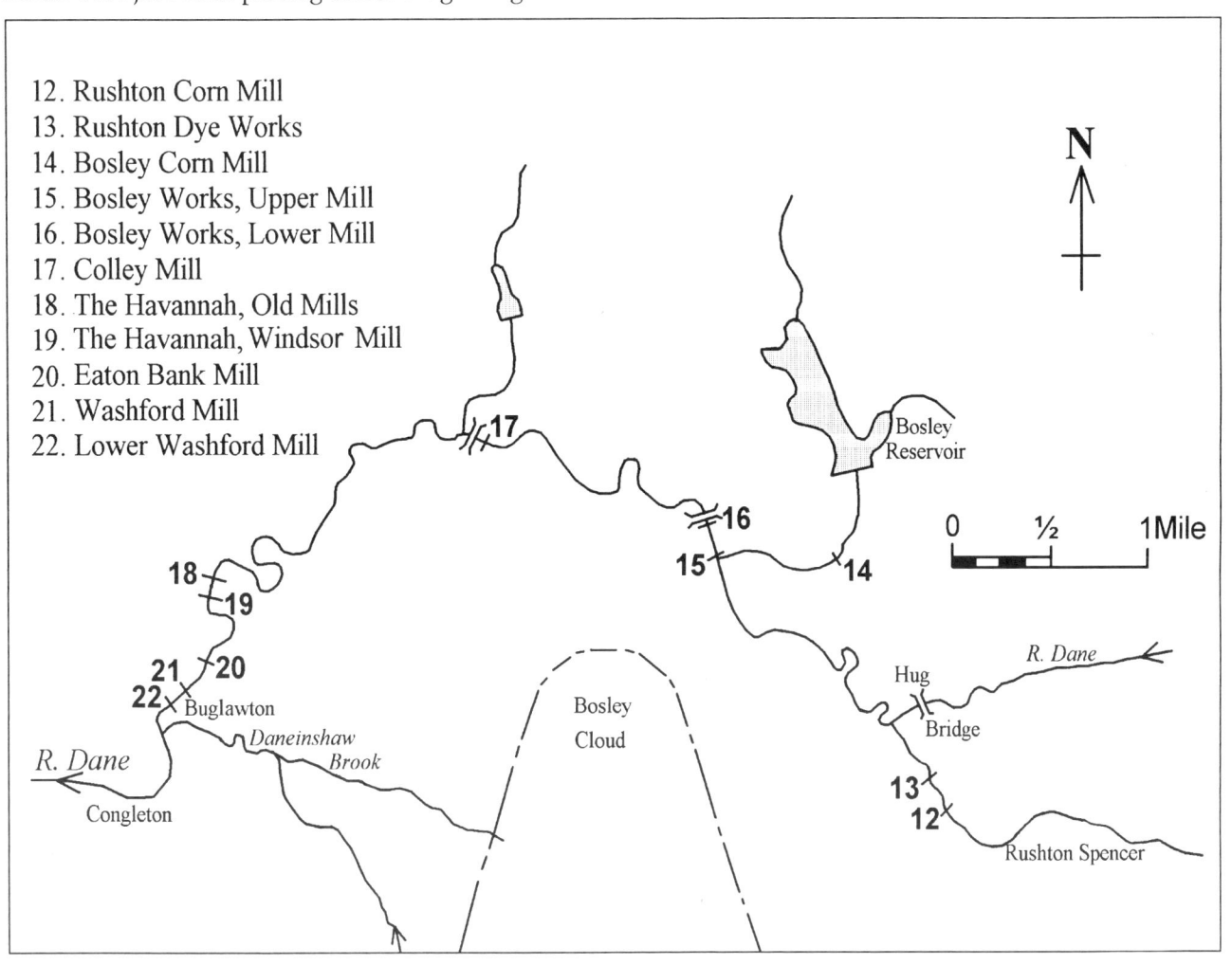

12. Rushton Corn Mill
13. Rushton Dye Works
14. Bosley Corn Mill
15. Bosley Works, Upper Mill
16. Bosley Works, Lower Mill
17. Colley Mill
18. The Havannah, Old Mills
19. The Havannah, Windsor Mill
20. Eaton Bank Mill
21. Washford Mill
22. Lower Washford Mill

FIGURE 38. MAP OF THE RIVER DANE AND ITS TRIBUTARIES FROM HUG BRIDGE TO BUGLAWTON.

12. RUSHTON CORN MILL SJ 932627

Close to Hug Bridge the River Dane is joined by a small stream which rises about three miles to the south-east, on the slopes of Gun Hill, which marks one of the major watersheds of the country. Despite its short length, this stream was sufficient to accommodate two mills, the first of which, situated just to the north-west of Rushton Spencer village, is Rushton Mill which was sometimes known as Nether Lee Mill.

The origins of Rushton Mill are obscure but there is a reference to a mill in the parish as early as 1329.[1] Its existence during the 17th century is confirmed by deeds mentioning the mill in 1600. In 1645 it was occupied by Alice Eardley, a widow, and her son Jonathon, and in 1679 it was again mentioned in another deed.[2] William Nixon was the tenant at the mill in 1703 and by 1729 the tenancy was held by Thomas Nixon.[3] Its appearance on Yates's map of Staffordshire published in 1775, and on the Rushton Enclosure map of 1777, confirm that this 18th century mill occupied the same location as the current mill. From 1781 to 1784, according to the Land Tax Returns, the owner of the mill was a Mr. Armitt and the miller was Samuel Hargreaves. After 1784 the mill was operated by John Hargreaves, possibly Samuel's son, as a tenant of Mr. Armitt until 1800 when it appears that John Hargreaves bought the mill. A few years after the start of the new century John Hargreaves became involved with the cotton factory erected just downstream of Rushton Mill. However, he continued to own and operate the corn mill until 1820. Around 1820 Richard Gaunt, a silk manufacturer from Leek, probably purchased the cotton mill and also leased the corn mill. This state of affairs lasted until 1825 when Richard Gaunt purchased Rushton Mill from John Hargreaves. By owning both mills he could better control the water supplies to what was now the silk mill. Strangely Gaunt's first tenant, from 1825 to 1827, was a John Hargreaves but whether this was the previous owner or some relation is not clear. From 1828 to 1830 John Wild was operating the silk mill as well as leasing the corn mill from Richard Gaunt. When John Wild gave up silk spinning in 1831, Richard Gaunt leased the corn mill to William Heapy, a corn miller from Bearda Mill.[4]

It is interesting to note that in 1841 William Heapy lived at the mill with the family of William and Sarah Hargreaves. William Hargreaves was also a miller, as was his father who also lived with them, although at the age of 80 he probably was not very active in the mill. At this time the corn mill would appear to have been quite prosperous as it also employed another miller and a miller's servant.[5] William Heapy remained in occupation of the mill until some time in the 1840s but he had been superseded by Joseph Vernon by 1850.[6] It was about this time that James Cook became the tenant at the silk mill and, presumably, the corn mill. His interest lay in converting the silk mill to a dyeing concern so the corn mill was leased to a succession of millers. From at least 1860 to 1872 the miller at Rushton Mill was Thomas Whittaker, followed by John Brassington who occupied the mill in the first half of the 1880s. He was replaced by John Lockitt junior in the 1890s. By 1940 the mill was being worked by Jonathon Cook, a descendant of the James Cook who became the tenant at the silk mill in 1850. The mill remains in the hands of the Cook family to this day, the current incumbent being John Cook, corn merchant.[7]

As the water supply for this mill is derived from quite a small stream, water storage has been necessary to provide for the continuous use of the mill. About half a mile away at Heaton, in the hills above Rushton, the stream was dammed to provide the first level of storage. It is possible that this dam was built as a speculative venture rather than for the corn mill as, in September 1822, an advert appeared in the Macclesfield Courier offering to let *"a stream and dam (silk mill to be erected)* at Heaton"[8] which may refer to this dam. However, the dam is now breached and no longer serves its purpose. In Rushton itself there was another storage pool which was filled in sometime in the late 1950s. It used to be alongside the Macclesfield to Leek road, on the western side, at Lane Ends. This pool was fed from a small weir just west of the main road by an underground conduit. The pool's discharge was taken back into the stream just upstream of the weir which is situated across the road from Rushton Station. From this weir a leat runs under the railway embankment and widens out as it reaches the mill to form the mill pool.

From the lie of the land it is possible that the waterwheel was a high breastshot or even overshot type. The current mill buildings are in three sections. The centre section is the oldest having a datestone of 1874 with the initials EC, whereas the north and south extensions are relatively modern, having datestones of 1941 and 1946 respectively. At the rear of the northern section is another datestone "JC 1943". The mill is one of only two sites on the River Dane and its tributaries still commercially involved in the milling trade.

References

1. SRO, D(W)1761/A/4/149.
2. SRO, Deeds, D3359/Condlyffe.
3. *Ibid.*
4. SRO, Rushton Land Tax Returns.
5. SRO, Rushton Census Return, 1841.
6. Kelly's Directory of Staffordshire, 1850.
7. Kelly's Directories of Staffordshire, 1860, 1872, 1884, 1892, 1900, 1940.
8. *Macclesfield Courier*, 30th November 1822.

---

13. RUSHTON DYEWORKS   SJ 931630

About two hundred yards downstream of Rushton Corn Mill (*q.v.*) is situated the group of buildings, once worked by water, known as Rushton Dyeworks. The water powered mill was built in 1791 when a pool south of Lane Ends Farm, on the western side of the Leek to Macclesfield turnpike road (now the A523), was let to John Burgess of Wilmslow and Thomas Percival of Bosley with licence to use the water to power a cotton mill.[1] In 1797 the mill was run by William Maskrey who was described as a cotton manufacturer, dealer

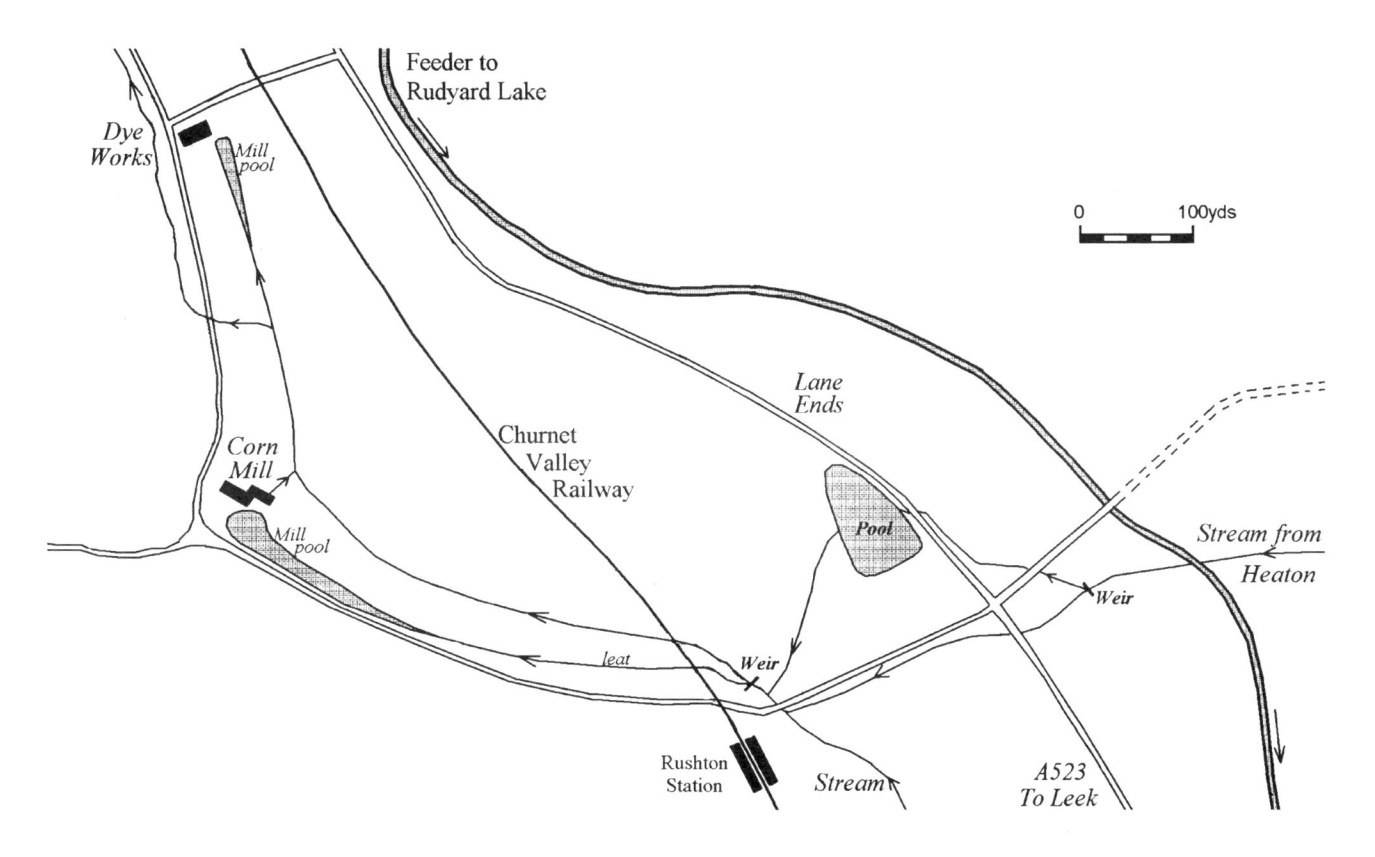

FIGURE 39. MAP OF THE WATER SUPPLY ARRANGEMENTS IN RUSHTON SPENCER FOR THE CORN MILL AND THE DYEWORKS.

and chapman.[2] However, by 1803, it was Peter Goostry who was supplying cotton weft to the Manchester market from the mill.[3]

In 1811, when the mill was offered for sale at auction, it was described as *"All that newly erected cotton factory... well supplied with a stream of water capable of constantly supplying a wheel"*. At the time it was in the occupation of Thomas Percival and John Hargreaves who also occupied Rushton Corn Mill.[4] Four years later, when the mill was again offered for sale, it had *"a newly opened reservoir that is of considerable advantage to the premises which are in good repair, and provided with preparations for turning machinery and a water wheel nearly new"*.[5] Pitt recorded that the mill was concerned in the manufacture of cotton twist,[6] and in 1818 the mill was still being operated by Peter Goostry.[7] Shortly after this, in 1821, the pool was purchased by Richard Gaunt, a Leek silk manufacturer, who also leased Rushton Mill at the same time so as to have full control of the water supply to the cotton mill. This cotton mill shared a water storage system with Rushton Corn Mill. Water from both the storage pools eventually arrived at a small weir, which is now near Rushton railway station, where it could be diverted into the mill pool of Rushton Mill. The stream continued from this small weir, passed the site of Rushton Mill where it was joined by that mill's tailrace, which was just upstream of where it was diverted into the race for the cotton mill.

Under Richard Gaunt cotton spinning at the mill came to an end as it was superseded by silk throwing.[8] In 1825 Richard Gaunt became the owner of Rushton Mill as well as the silk mill. When the silk mill was offered for sale in 1831 it was said to be in the occupation of John Wild, but the owner appeared to be Nathan Percival, the heir of Thomas Percival, the original owner of the mill.[9] However, two years later, when the premises were offered for letting, they were described as *"formerly used as a silk mill but latterly used as a silk dye house and now in the occupation of Messrs Bayley and Adshead, silk dyers."* Again, the owner appeared to be Nathan Percival of Bosley.[10]

In 1834 it was offered again *"as a silk mill which has just undergone a thorough repair with new waterwheel and everything complete."* Presumably the silk dyeing business had left the mill in a somewhat unattractive state![11] Ralph Tatton was listed as a silk spinner living with his wife and family at the mill in 1841.[12] In 1851 the mill was once again being used for silk dyeing, the owner being James Cook, aged 29, listed as a silk dyer, who presumably purchased Rushton Corn Mill at the same time. He lived at what was now the dye works with his two unmarried nieces, Harriet and Caroline Booth, aged 20 and 21 respectively, together with a lodger, John Plem, a 19 year old dyer's labourer.[13] The identification of the premises on the Tithe Map of 1852 as a silk mill was, presumably, based on outdated information.

In 1860 the mill was to be sold at auction, at which time it was *"a silk dyehouse with reservoir, waterwheel and water privilege thereto belonging"* with application invited to be submitted to Mr. Bullock of Gawsworth Hall.[14] It is probable that the mill was bought by the tenant at the dyehouse, James Cook, because after this date the premises were always occupied by a member of the Cook family. The only changes were to describe the site as a chemical works, and the occupier as a manufacturing chemist. In the 1880s and 1890s the occupier was William Cook, by 1900 it was John Cook, and by 1940 it was Jas. Cook,[15] the current occupier being John Cook who also operates Rushton Corn Mill.

The living quarters are very much how they were when built, at least on the outside. However, the silk mill itself, described in 1831 as *"two rooms 17 yards long by 11 yards wide and three rooms 6½ yards wide by 11 yards long with an attic over the whole"*, has been demolished and replaced by a single story brick building erected in 1943.[16] The leat is now dry, but extremely obvious, and its position would suggest that the waterwheel had a low breastshot configuration.

References

1. Greenslade, M. W., ed., *Victoria County History of Staffordshire*, Vol. 7, Oxford University Press, 1995.
2. SRO, Deed of 1797, D3359/Cruso.
3. SRO, Abstract of title, D3359/Phillips.
4. *Macclesfield Courier*, 23rd February; 22nd June, 1811.
5. *Macclesfield Courier*, 4th February, 1815.
6. Pitt , W. *History of Staffordshire*, 1817.
7. Nightingale, J., *Beauties of England and Wales*, xiii, 1851; Parson & Bradshaw Directory of Staffordshire, 1818.
8. William Salt Library, Deed of 26th March 1821, WSL 25/6/46 .
9. *Macclesfield Courier*, 26th February, 1831.
10. *Macclesfield Courier*, 28th October, 1833.
11. *Macclesfield Courier*, 13th September, 1834.
12. SRO, Rushton Census Return, 1841.
13. SRO, Rushton Census Return, 1851.
14. *Macclesfield Courier*, 11th February, 1860.
15. Kelly's Directory of Staffordshire, 1884, 1892, 1900, 1940.
16. Datestone on building.

---

14. BOSLEY CORN MILL   SJ 922648

The site of Bosley Corn Mill is to be found about two miles north of Hug Bridge, at a location known as Smithy Green, where the Bosley Brook passes under the Macclesfield to Leek turnpike, now the A 523. The Bosley Brook and its tributaries are only about eight miles long in total but provide a plentiful supply of water. These streams rise on the high ground in the east of the Bosley parish, known as Sutton Common and Bosley Minn.

The first indication of a mill in Bosley occurs when a *"molendinum aquaticum"* is mentioned in 1400.[1] The mill must have been well established on the death of John Bradbury, the miller at Bosley in 1699.[2] On Burdett's Map of Cheshire in 1777 and on Aitken's map of 1795, it was given its full name of "Bosley Mill" in addition to the usual mill symbol. At the end of the 18th century the mill was owned by the Earl of Harrington who owned most of the parish of Bosley. In 1784 the miller was Samuel Malkin but he gave way in 1785 to William Malkin who was probably Samuel's son. William Malkin remained the miller at Bosley Mill until 1788, when Samuel Malkin died and William gave up the tenancy of the mill. From 1789 to 1796 the tenant was John Leah (or Lea) who was a member of a milling family resident in Swettenham at this time. Presumably, moving to Bosley enabled him to acquire his own mill. However, when the opportunity presented itself in 1796, he moved back to take over Swettenham Mill (*q.v.*). In 1796-7 the miller was John Thompstone, from the Gawsworth milling family, who was followed by Robert Goodwin from 1798 to 1804. A period of stability then followed from 1805 to 1820 during which the tenancy was held by another memnber of the Malkin family, namely Thomas Malkin.[3]

When the mill was offered for sale in 1818 it was, no doubt, a very attractive proposition with its proximity to the turnpike and the combination of the two professions of miller and innkeeper. The sale advertisement, besides providing details of the mill, its miller and the terms of the lease, indicates that the mill was already quite old, as follows:-

"To be sold at auction by Mr. Wayte.

All that ancient and well accustomed corn mill situated in Bosley in the said county of Chester and on the road leading from Macclesfield through Leek to London, containing two pair of grey stones, one pair of French stones and a wheat machine together with the drying kiln, a stable, shippon, pig-stye, and a garden to the said corn mill belonging in the occupation of Mr. Joseph Hargreaves as tenant at will. Also a public house in Bosley near to the corn mill known as The Traveller's Inn occupied by Mr. William Davenport as tenant at will. The above premises are held for the unexpired term of 52 years on 29th September 1817 subject to an annual chief rent of £10 which will be apportioned to lots at the sale.

Further particulars from Thomas Malkin, Bosley."[4]

It is noticeable that at this time Thomas Malkin, who was the leaseholder from the Earl of Harrington, had sub-let the property and the miller was, in fact, Joseph Hargreaves. In spite (or because) of this arrangement Thomas Malkin was declared bankrupt in 1823.[5] From 1821 to 1830 the occupier of the mill was Stephen Massey.[6]

In 1825, parliament passed the necessary Act to allow the building of the Macclesfield Canal. This canal was built between 1826 and 1831 and to provide the water needed by the canal the Macclesfield Canal Company decided to dam the valley of the Bosley Brook upstream of the mill to form Bosley Reservoir. Because the mill had prior claim to the water of the Bosley Brook and its tributaries the Canal Company had to divert the stream around the eastern side of the reservoir. At the confluence with the tributary flowing down Swallowdale, mid-way along the reservoir, a small pool with an overflow dam was built such that the normal flow of Bosley Brook and its tributary continued to the mill but any excess was deposited in the reservoir. Obviously this source of water was not considered to be sufficient by itself so the Canal Company built two reservoir feeders. One from the Radcliffe Brook at Oakgrove, a mile or so to the north, followed the contour line around the hills until it connected with the now disused bed of Bosley Brook and then flowed down the valley into the reservoir. The second feeder came from the Shell Brook, a tributary of the River Dane away to the south-east. This feeder had to enter the reservoir at the eastern end of the actual dam. Consequently it was necessary for the diverted Bosley Brook to cross this Shell Brook feeder to get to the mill. This was achieved by using an inverted siphon which allowed the water of the Bosley Brook to pass under the feeder. After rising up the inverted siphon the water then flowed down the edge of the reservoir dam wall and into the existing channel of Bosley Brook downstream of the reservoir. Because this water supply was so dependable, the mill pool further down the valley was not much more than a widened leat.

In order to carry out this major civil engineering project, the Macclesfield Canal Company had to buy out the lease of the mill which subsequently led to them offering the remaining 35 years of the lease for sale towards the end of 1833.[7] By 1841 the miller was John Booth[8] but by the time the Tithe Map and Apportionments were drawn up in 1850 the occupier was Joshua Heapy who had previously been the miller at Bearda Mill (*q.v.*). This was the start of a long association between the Heapy family and Bosley Mill. It is interesting to note that the Traveller's Inn of 1818 had become the Millstone Inn by 1850.[9]

Joshua Heapy remained at Bosley Mill until 1851 when he appears to have purchased the farm at Colley Mill (*q.v.*) in nearby North Rode to which he retired, leaving his son John, aged 25, and his wife Sarah Ann,

FIGURE 40. A SKETCH MAP OF THE WATER SUPPLY ARRANGEMENT FOR BOSLEY MILL DUE TO THE CONSTRUCTION OF THE MACCLESFIELD CANAL RESERVOIR IN 1831.

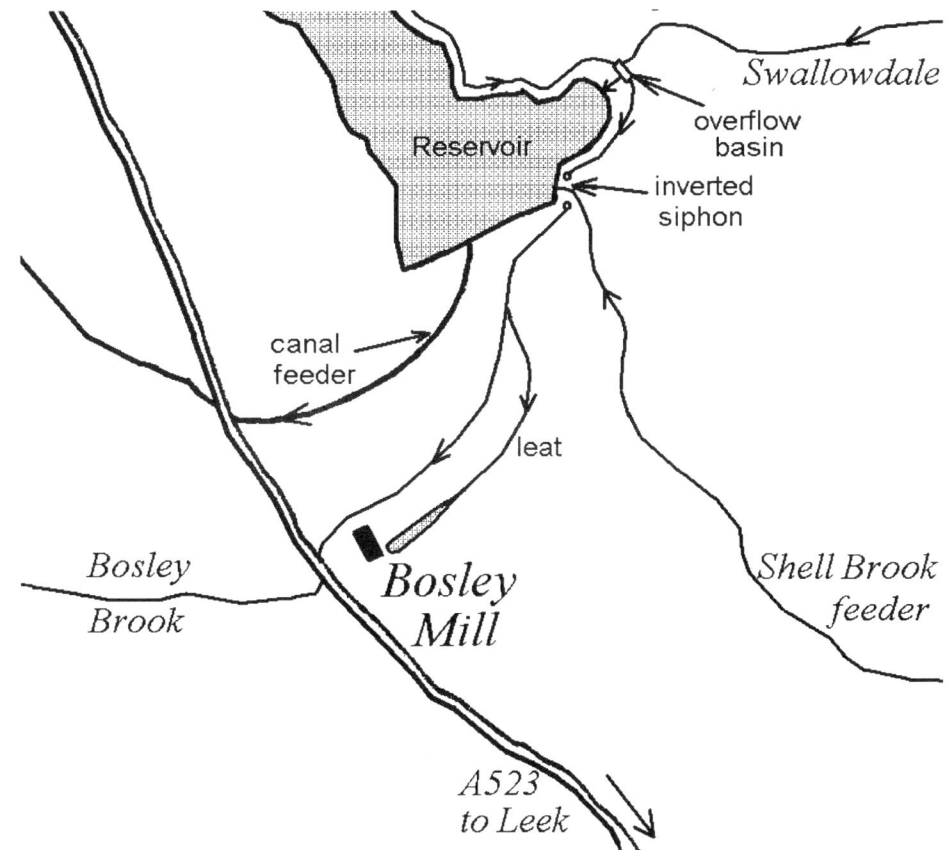

aged 21, to run the mill with the help of a paid employee, George Read.[10] This young family was augmented by the birth of a son in 1853 who was christened Arthur Horatio Heapy but known thereafter as Arthur H. Heapy (and who can blame him). Twelve years later John Heapy died prematurely at the age of 39.[11] After John Heapy's death, his widow continued to operate Bosley Mill with the help of George Read. No doubt the young Arthur helped his mother in the mill as soon as he was able. Certainly Arthur had taken over the business by 1896 although his mother did not die until 1913 when she was 83 years of age.

Arthur H. Heapy appears to have run a successful business as a miller, corn merchant and baker, not to mention his lease of the nearby public house. The profits from these businesses enabled him to purchase a nearby residence called Boggins Hill in the early years of the 20th century.[12] There are stories of him and another local eccentric, John Erlam, being noted for playing marbles on the stretch of road near the mill, a pastime not to be recommended these days due to the ever-present high speed traffic.[13]

The mill itself was three storeys high on the south side where the waterwheel was situated, and two storeys to the north side. The motive power was provided by a large iron breastshot waterwheel. In the 1920s, after the mill had ceased producing flour, it was converted to making ice by F. R. Thompstone & Sons of Bosley Works.[14] The waterwheel was removed and replaced by an 13.5 inch diameter Series IV turbine, producing 14.4 h.p. at 430 r.p.m., which was supplied by Gilbert Gilkes & Co. to drive the compressor used to make ice. The turbine was designed to operate with a head of 25 feet and a flow of 392 cubic feet per minute.[15] This was at a time when mains electricity supplies were not available and there was a trade in ice to supply the many hotels and restaurants in the Cheshire and South Manchester area. After the coming of mains electric in the 1930s the customers of the ice business were able to make their own ice and so the turbine and compressor were adapted for use in the garage trade.

When the garage closed in the 1960s, the mill was used for storage and the pool became silted up and overgrown.

The water supply arrangements at Bosley Reservoir remained in operation until the 1980s when the reservoir's current owner, British Waterways, abandoned the compensation leat around the reservoir, restoring the Bosley Brook to its original course, flowing into the reservoir. Below the reservoir dam, the Bosley Brook was kept flowing by a direct feed from the outlet of the dam. The original bed of the Bosley Brook below the reservoir has been abandoned with the water now flowing along the mill leat as far as the overflow at the mill site where it is diverted back into the original stream bed. Although most of the corn grinding equipment was removed along with the waterwheel early in the 20th century, the buildings remained intact until the 1990s when the roof and floors were destroyed by fire causing the remains of the mill to be demolished.

References

1. Dodgson, J. McN., *The Place Names of Cheshire*, Part 2, Cambridge University Press, 1970.
2. CRO, Will of John Bradbury, 1699.
3. CRO, Bosley Land Tax Returns.
4. *Macclesfield Courier*, 17th January 1818.
5. *London Gazette*, 18th March 1823.
6. CRO, Bosley Land Tax Returns.
7. *Macclesfield Courier*, 14th December 1833.
8. CRO, Bosley Census Return, 1841.
9. CRO, Bosley Tithe Map & Apportionments.
10. CRO, Bosley Census Return, 1851.
11. Gravestone, Bosley Parish Church, St. Mary the Virgin.
12. Kelly's Directories of Cheshire, 1896-1914.
13. Women's Institute, *Cheshire Village Memories*, 1961
14. Bosley Wood Treatment Ltd., Turbine Records.
15. Gilbert Gilkes Ltd., Turbine Records.

---

15 & 16. BOSLEY WORKS   SJ 914648

Bosley Works are situated on the River Dane as it flows in a northerly direction to the east of Bosley Cloud. The Higher Works are located at the confluence of the Bosley Brook and the River Dane and used the water from both streams, whereas the Lower Works, which are about 100 yards to the north and downstream from the Higher Works, only derived their power from the main river. The history of these water powered sites starts in the middle of the 18th century and they have been in continuous use, for a variety of purposes, up to the present day.

In the 1740s & 50s Charles Roe had prospered by being involved in the early blossoming of the silk industry in Macclesfield. However, in 1758, he acquired the lease to land in Macclesfield in order to build a smelting works to exploit the copper ores mined at nearby Alderley Edge, using the local coal available in Macclesfield.[1] By 1762, to provide the necessary finishing processes for the copper and brass production Charles Roe acquired a "greenfield" site in Buglawton which he called the Havannah (*q.v.*).[2] This expenditure caused him to realise his assets in the silk industry which he finally sold in 1764. The money from this sale was then invested in another site suitable for water powered processing of brass and copper at Bosley.[3]

The land and water rights at Bosley were leased from the Earl of Harrington, the local landowner, for 99 years from the 25th September 1766 at an annual rent of £25. The lease granted Roe the powers to *"To erect a mill or mills, dwelling houses, barns, stables etc; get stones from River Daven* [Dane], *get clay and make bricks for buildings only, make weir or weirs across River Daven or any other brook or stream, make cuts to carry water out of River Daven."*[4] About 100 years later, it was suggested that James Brindley was responsible for the arrangements for harnessing the great water power of this original establishment.[5] Certainly the scope and size of the works would suggest a millwright of Brindley's stature being involved, and his millwrighting business was locally based, but unfortunately there is no direct evidence to support this claim. Indeed there are factors that mitigate against James Brindley's involvement. Firstly, by 1766 Brindley was absolutely overwhelmed by canal engineering work. Secondly, Charles Roe was a leading promoter of a scheme for a canal from Macclesfield to the River Weaver, in opposition to the Duke of Bridgewater. On the 18th April 1766 both James Brindley and Charles Roe gave evidence to the House of Lords, Brindley in support of the Duke of Bridgewater against the scheme supported by Charles Roe. Such was Brindley's reputation that the Duke of Bridgewater won the day.[6] Thirdly, earlier in 1765 James Brindley had married and moved to Turnhurst in the Potteries so weakening his local links with Leek. The identity of the engineer at Bosley therefore remains unknown.

The Macclesfield Copper Company, as Charles Roe's enterprise became known, went from strength to strength, having mining interests in Anglesey, Cornwall, the Lake District and Ireland, and other large smelting works in Liverpool and Neath Abbey.[7] The largest customer of the works was the four eastern dockyards of the Royal Navy which the company supplied with copper sheets for sheathing ship's hulls and also copper nails. Later in the 1780s they were supplying two of the Navy's western yards with brass bolts. The company was also supplying its products to merchants in Birmingham and to the African trade. Another large investor in copper was the East India Company.[8] Although production continued at Bosley after Roe's death in 1781,[9] the fortunes of the Company declined steadily, with money becoming scarce during the 1790s,

until in 1801 the three sites at Macclesfield, Buglawton, and Bosley were advertised for sale.[10]

It is from this sale notice that it is possible to determine the scale of the operation at Bosley. There were two distinct areas of operation; the Higher Works having five waterwheels, with a 25 feet 3 inch head, developing a total of 40 h.p. These water wheels powered one rolling and four battery mills for brass working. Also there was a blacksmith's, wheelwrights, carpenter's, and cooper's shops, together with a handsome brick dwelling house with stables, a good garden and sixteen acres, plus eight cottages. About 100 yards downstream, the Lower Works, having one waterwheel with a 14 feet 9 inch head and developing 20 h.p., were used for rolling and hammering copper. Also at the Lower Works there was a warehouse and two other dwelling houses with gardens.

So as the 19th century started, the copper and brass working came to an end at Bosley, no doubt due to the cost of shipping the raw materials of ore and coal such long distances to this remote part of Cheshire after the local source had earlier become exhausted. However there was substantial water power available awaiting harnessing to a new use.

In 1806 the Macclesfield Copper Company again advertised the property for sale in much the same terms as in 1801 but with a significant addition.[11] The Higher Works had been occupied by Messrs. Berresford for cotton spinning sometime since 1801, although the Lower Works were unoccupied. As the Berresfords' lease had seven years to run in 1806 it is probable that the lease started in 1803 (10 year leases were common at this time). This was the start of a fifty year association between the Berresford family and Bosley Works.

In the late 1820s the Lower Works were advertised to let as a silk factory by Warrington & Hambleton, who probably acquired the lease to the complete site from the Macclesfield Copper Company as an investment for sub-letting.[12] This silk factory was being worked earlier in the 1820s by Francis Johnson but he had ceased to operate by 1827.[13] At that time the Lower Works had about 300 dozen spindles plus swifts and doubling machines, with considerable room for expansion. The mill itself was a three storey building, 24 yards long by 8 yards wide, and was heated by steam. Also there was a warehouse, wash house, and seven cottages offered with the mill.[14] Warrington & Hambleton seem to have been unsuccessful in letting the Lower Works as it was still advertised to let in 1830.[15]

Meanwhile, at the Higher Works, the Berresford family had taken on a sleeping partner and continued their cotton spinning business as tenants of Warrington

& Hambleton. In 1811 the firm of Berrisford & Warring[*sic*] were recorded as having 4404 mule spindles and 720 throstle spindles[16] and by 1828 they were known as Warren & Birrisforth [*sic*].[17] Shortly after 1830 part of the Lower Works became a corn mill, although it is not clear whether this was a new building or an adaptation of the existing buildings.[18] About this time the other part of the Lower Works was occupied by the Berresfords who continued to operate it for silk throwing.[19]

In 1833 and 1834 both the Higher and Lower Works were advertised to let, even though the Berresfords still had another ten years outstanding on their current lease. At this time one of the waterwheels was described as a *"powerful iron waterwheel of a superior construction which has a supply of water from the River Dane and is equal to 22 h.p."* It is interesting to note that one of the contacts for further particulars concerning the premises was Messrs. Fairbairn & Lillee, Engineers, Great Ancoats Street, Manchester. This suggests that the wheel may have been recently installed by them and may have been a suspension wheel with ring gearing, a type for which they were particularly recognised.[20]

The lease was sold in 1834 to Thomas & Archibald Templeton for £1070 but the Templetons never settled the contract. In 1835 the Templetons offered to pay £300 on account but this did not materialise. As late as 1838 they had still not paid any money for the mill. By 1838 the shareholders in Roe & Co. were getting impatient at the delays in winding up their company and the amount outstanding on Bosley Works had risen to £1123-10-0. The company wrote to its shareholders in August 1838 explaining that The Templetons had paid £450 on account and were trying to raise the rest of the purchase money on mortgage and that if the money was not forthcoming in a fortnight they would institute bankruptcy proceedings.[21] In 1840 the Templetons were declared bankrupt and the Lower Works was advertised for sale including the corn mill.[22] The outcome of this sale is not known but the Berresfords continued with their cotton and silk enterprises. In 1841 there was a total of 40 workers employed in cotton spinning and 12 in silk throwing. Out of these 52 people, only 13 were female and the youngest worker was aged 13.[23]

By 1850 the Berresfords had given up the struggle with the silk business, concentrating entirely on cotton spinning. The Lower Works were now occupied by Thomas Pimlett, a wood turner, as well as continuing to be used for corn milling.[24]

At the mid-point of the century the whole cotton spinning enterprise had all the signs of being extremely successful. In 1851 there were 62 workers employed at the cotton factory, of these, 36 were female (a higher proportion than in 1841) and the youngest employee was twelve years of age (lower than in 1841). There were seven separate Berresford families at Bosley, all involved in the cotton works. Many of the offspring of these families worked in the factory and even the wives who had to look after large families of young children are described as *"cotton reeler at home"*. The factory also provided support and employment for the members of eleven other families as well as the Berresford clan. The patriarch of this community was Joseph Berresford, born in 1782, and described as a

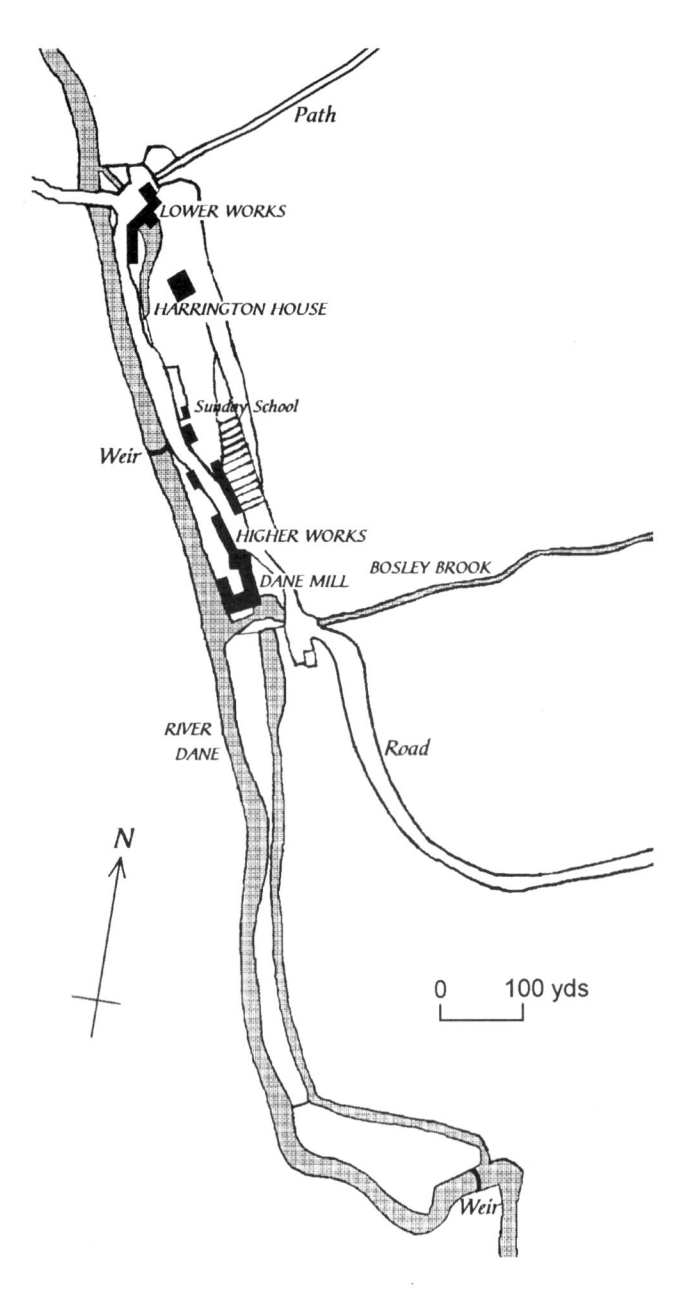

FIGURE 41. PART OF THE BOSLEY TITHE MAP, 1850, SHOWING THE UPPER AND LOWER WORKS AND THEIR ASSOCIATED WEIRS AND WATER SUPPLY ARRANGEMENTS. (CCALS, EDT 57/2)

"*master cotton spinner*"; his eldest surviving son, John, was the shopkeeper; and his three other sons and a grandson, William, Thomas, Isaac and James, were all warehousemen in the factory. (The seventh Berresford family was that of Hannah, the widow of Joseph's eldest son James.)[25]

However, in 1853, at the age of 70, and after about 50 years of spinning cotton at Bosley, Joseph Berresford decided to retire and liquidate his assets. Consequently, all the equipment used in the business was offered for sale, thus giving a complete inventory of a small 19th century cotton factory, as follows:-

"All that valuable machinery for the preparation and spinning of cotton. A double beater blowing machine (nearly new) for 40 inch cards with fan and dust pipes, by Kaye of Bury; fifteen single carding engines 40 inch on the wire, with rollers, cleaners, lickers-in, and plungers by Hibbert & Platt and others; grinding machine with extra rollers for do.; two drawing frames, four heads each, and two deliveries to each head with plungers; presser slubbing frame 48 spindles and 8 inch lift by Higgins & Sons nearly new; do. do. 60 spindles and 6 inch lift by Higgins; four do. pressure bobbins do., 72 spindles and 6 inch lift by do.; new pressure roving frame 7 inch lift and 120 spindles by Kaye; soft bobbin roving frame 6 inch lift and 72 spindles by Higgins; soft bobbin do. 6 inch lift and 120 spindles by do.. The following hand mules all by Gore of Manchester: pair of hand mules 648 spindles; three pair do., 720 spindles in each pair; three pair most excellent do., 864 spindles in each pair; and three pair hand mules nearly new 1004 spindles in each pair. The following throstles by Walker & Co. and others are chiefly in first rate working order and are 2 inch lift and 17 spindles, viz. four throstles 144 spindles ea.; one throstle 168 spindles; two throstles 120 spindles ea.; eight throstles 136 spindles ea.; one throstle 128 spindles; six throstle 108 spindles ea.; one throstle 132 spindles; and two do. 120 spindles ea.; thirty cop reels and seven bobbin reels; two wrap reels, all the roving, stubbing, throstle and other bobbins, spools, skewers, etc.; all the drawing and card cans, straps, skips, etc.; mill clocks, lamps, etc.; capital strong scale beams, scales and weights; sets of new cards; tallow, oil, colours and all the stores.

The contents of the carpenters, mechanics, and smith's shops embracing chest of joiners tools, grindstones and frames, new woodwork benches and vices, stocks, taps and dies; 6 inch single speed lathe with driving apparatus, 8 inch do. with do.; wheel cutting engine; 26 inch bellows and frame, anvil, swage block, smithy and mechanics tools; capital narrow wheel pack cart, pair of extra wheels and axle for do.; gig; all the warehouse and counting house fixtures counters, desks, shelving, etc.; together with the valuable mill gearing, wrought and cast shafting, wheels, pulleys, pedestals, couplings, hangers, and brass steps; also all the steam piping from 2 inch to 3 inch diameter etc.; and a vast assemblage of other valuable items and effects." [26]

This list of eqiupment shows that the Berresfords were using both mules and throstles, producing weft and warp threads respectively, but noticeably the mules were hand driven. These mules were definitely obsolete by this date as a fully automatic, self-acting mule had been perfected as early as 1830.

After 1853 most of the Berresford families moved away from Bosley and the Higher Works were converted into corn mills. Only John Berresford, the eldest son of Joseph the master cotton spinner, remained operating the Lower Works as a silk throwster until the middle of the 1860s.[27] This enterprise was not very successful in view of the slump in the silk trade caused by the 1860 Free Trade Act. Even though there were fourteen people employed in silk working

FIGURE 42. A RECENT VIEW OF THE COTTAGES WHERE THE VARIOUS MEMBERS OF THE BERRESFORD FAMILY LIVED AT BOSLEY WORKS.

in 1861, they were made up of John's immediate family or the young offspring of families involved in the corn milling business.[28] In fact business was so bad that in 1860 John Berresford had to supplement his income by working in the corn mills some of the time.[29] At this time when the whole of the English silk trade was decimated, silk throwing came to an end at Bosley in 1864.

About 1830 part of the Lower Works was converted to corn milling and was being offered for rental in 1832 for £125 per annum.[30] At that time the mill had a drying kiln, stable, and a dwelling house with garden. In the mill there was *"one pair of french stones, one pair meal, one pair shulling, one pair batch with room for two pair others, turned by two powerful wheels."* In 1841 the miller here was Samuel Malkin who ran it as a family business, eventually employing his son William in 1851.[31] Samuel's father Thomas Malkin had earlier been the miller at Bosley Mill when he went bankrupt in 1823.

As stated previously, after 1853 the Higher Works was converted to commercial corn milling, being taken over by Robert Brindley, trading as Francis Brindley & Co.,[32] who had been corn millers at the steam mills in Marple, Cheshire, and also at the steam mills in Macclesfield until 1852/3.[33] Certainly by 1861, Francis Brindley & Co. were employing six millers in their corn merchant's business at Bosley, rising to eight millers and eight miller's labourers by 1871.[34]

About the same time that Robert Brindley took over the Higher Works, Francis Rathbone Thompstone arrived at the corn mill at the Lower Works employing two men and two boys.[35] The Thompstones were a well known family of Cheshire corn millers, members of the family operating both Gawsworth and Prestbury mills around this time.[36]

These two corn merchants existed side by side at Bosley for about twenty years until about 1873 when Francis Rathbone Thompstone took over the lease for the whole of the works at Bosley, operating the whole complex for commercial corn milling.[37] In 1881 he employed seven millers, ten miller's labourers, three flour carters, and one flour sack mender.[38]

Both Francis Rathbone Thompstone and his son Frank, who inherited the business in 1902,[39] made many improvements to the mills during their tenure. They were very independent minded and generally undertook any engineering improvements themselves. At one stage, towards the end of the 19th century, they had a dispute with the North Staffordshire Railway, whose tracks ran past the Bosley Works, over the terms of their freight charges. To circumvent the North Staffordshire Railway's monopoly position, the Thompstones built a 2 feet 8 inch narrow gauge railway running from the mills to a wharf on the Macclesfield Canal about half a mile away.[40] They then used the canal for the transportation of the corn and flour, with a small steam engine hauling the materials over the narrow gauge railway. Originally they owned their own fleet of narrowboats on the canal but these were superseded eventually by a number of Foden Sentinel steam wagons.

In order to remain competitive, the Higher Works was converted to roller milling, probably around the turn of the century, and to provide power more efficiently, water turbines were used to replace the water wheels. In 1888, W. Günther & Sons of Oldham were quoting to install a 130 h.p. Girard type turbine for the Higher Works.[41]

In 1894 the Thompstones purchased a "Titan" sprinkler system, from George Mills & Co. at the Globe Iron Works, Radcliffe near Manchester, to guard against fire. This was installed using a 20 h.p. Jonval type water turbine, from W. Günther & Sons, that automatically provided the water pressure for the sprinkler system.

FIGURE 43. THE HIGHER WORKS DURING THE TIME IT WAS BEING USED AS A CORN MILL. THE INSCRIPTION ON THE TOWER WAS MADE OF WHITE BRICKS AND READ "DANE CORN MILLS". (*George Bowyer Collection, courtesy of Basil Jeuda*)

FIGURE 44. THE NARROW GUAGE ENGINE AND ROLLING STOCK USED BY THE THOMPSTONES TO CONNECT BOSLEY WORKS TO THE NEARBY WHARF ON THE MACCLESFIELD CANAL. (*Max Birchenough Collection, courtesy of Basil Jeuda*)

The system provided two automatic fire alarms. There was a brass hammer attached to the turbine shaft so that when the turbine operated to maintain water pressure (i.e. when there was a fire and the sprinklers operated) the hammer would repeatedly hit a steel gong. Also, when the turbine operated it opened a valve to a steam whistle.[42] They also installed a 50 h.p. Girard turbine from W. Günther & Sons at this time.

Later, in 1901, the Thompstones were in correspondence with Gilbert Gilkes & Co. Ltd. of Kendal about replacing the water wheel at the Lower Works with a double-vortex turbine of 25 h.p. In 1924 the water feed to this turbine was improved to give 35 h.p.[43] The size and details of the waterwheels replaced is, unfortunately, not recorded, but part of a wheel scrape on the side of the old mill at the Lower Works shows that a wheel of about 10 feet diameter, probably overshot, once worked there.

The whole of the Harrington Estate (the original landowner) was advertised for sale in 1920 and the Thompstone family purchased the Bosley Works site, including the cottages and the superior dwelling house known as Harrington House. The sale brochure gives a good description of the works at this time as:-

"*The highly important and well-known Dane Corn Mills, Bosley, with an attractive residence and land, a house and nineteen cottages with gardens and allotments .... The works adjoin Bosley Station on the North Staffordshire Railway. The Macclesfield Canal bounds the north side of the property, and there is a short Steam tramway connecting the wharf with the Mills .... The River Dane provides an ample head of water from the Derbyshire Hills. Gas Engines, steam Drying Apparatus and Electric Light are installed, and the Mills are equipped with modern Corn-milling Machinery.*"[44]

The Higher Works were described as a substantial five-storeyed structure of brick with a slated roof, having a west wing, water tower, and north wing. In the west wing, the ground and first floor housed the roller mills, the second floor was the purifying room, the third was the centrifugal floor, and the fourth was the scalper floor. In the north wing, the ground and first floors were the screen and sack room, the second floor was the wheat warehouse, the third was the packing floor, and the fifth floor was the flour warehouse. The first three floors of the water tower was the turbine house, with the third and fourth floors being part of the packing room. At the top of the water tower was a 10,000 gallon water tank, supplying the sprinkler system.

The Lower Works consisted of the Old Mill, a four-storeyed, brick built structure with slated roof comprising a grinding floor, delivery floor, warehouse and small office. There was also another three-storeyed mill built of sandstone with slated roof adjacent to the Old Mill.

The whole of this property was purchased by Frank Thompstone from the Harrington Estate at this time. Thus the Thompstones became the owners of Bosley Works after about seventy years as tenants, a situation that continues to this day.[45]

The next generation of Thompstones, four brothers, decided to convert the mills from corn milling to wood flour milling, trading as Bosley Wood Treatment Ltd. This conversion was complete by 1930[46] the process being very similar to corn milling but using woodshavings and sawdust as its raw material, producing a variety of grades of inert wood flour. This is used as a filler by many industries, initially for linoleum and bakerlite production, and now for cosmetics, paints, adhesives, polishes, synthetic rubbers and all manner of plastics.[47]

This change of business was very successful and as the business expanded there was a need for more and more power. In 1933, a Gilkes 268 h.p. Series R horizontal turbine was installed in the Higher Works to provide power for grinding. Later, in 1936, a 110 h.p. Series R vertical turbine from the same supplier was

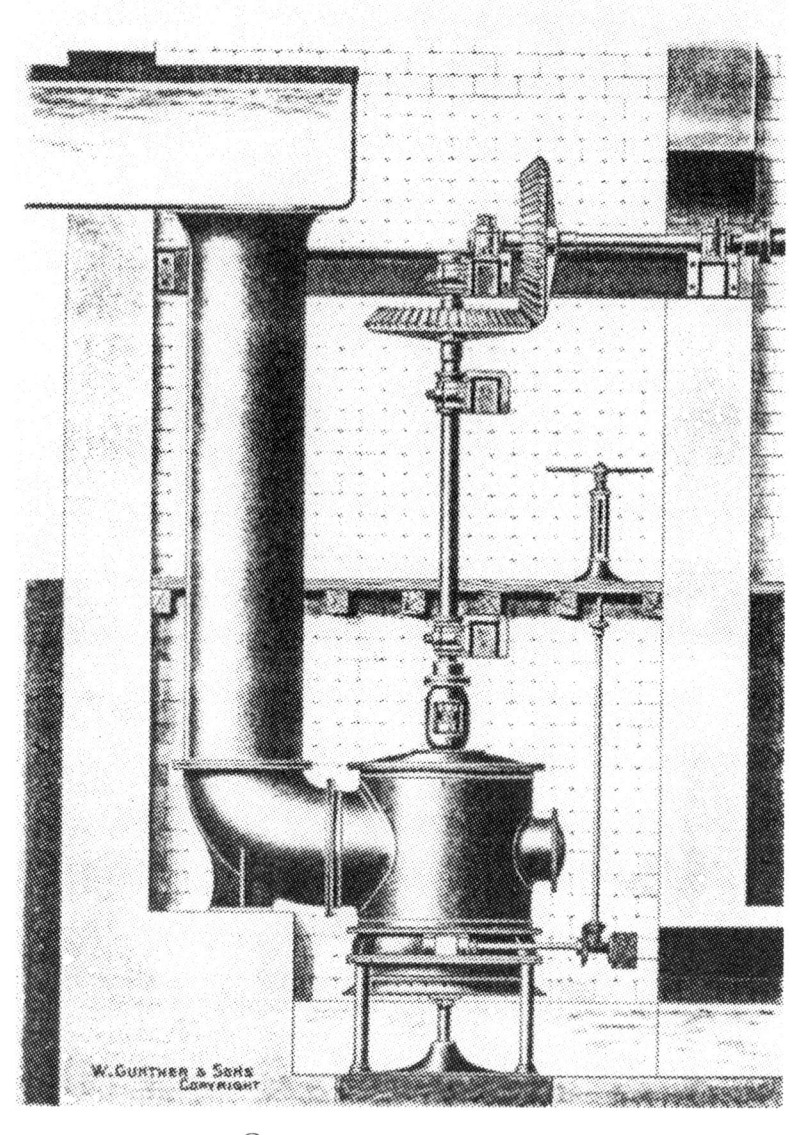

FIGURE 45. A TYPICAL INSTALLATION OF A GIRARD TURBINE SIMILAR TO THE ONES SUPPLIED TO BOSLEY WORKS IN 1888 AND 1894 BY W. GÜNTHER & SONS. (*W. Günther & Sons catalogue*)

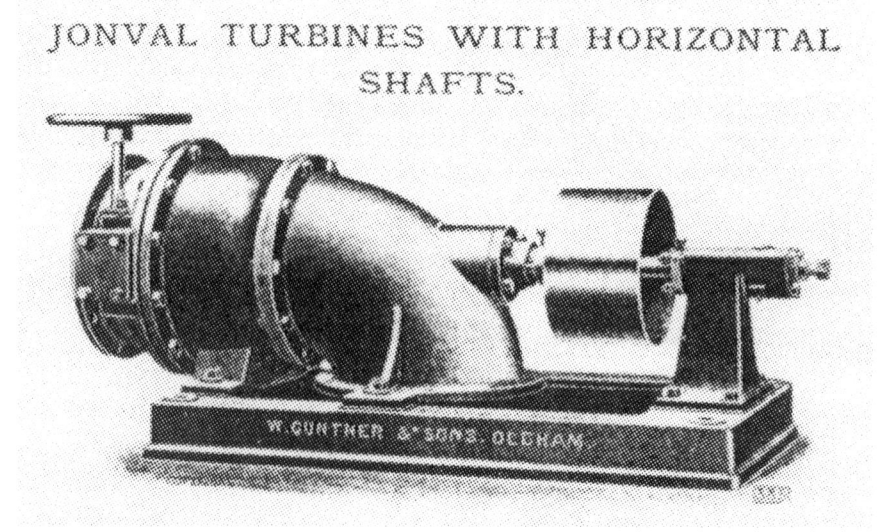

FIGURE 46. A HORIZONTAL JONVAL TURBINE SIMILAR TO THE ONE INSTALLED AT BOSLEY WORKS IN 1894. (*W. Günther & Sons catalogue*)

installed in the Lower Works to generate electricity. In order to maintain the frequency within suitable tolerances, an oil pressure governor was used on this installation.[48]

In 1939 the grinding of mica was added to the range of operations. Mica is a mineral material that is inert in the presence of most acids and alkalis. The ground mica is used as a dielectric in the electrical industry, and for the manufacture of fireproof plasterboard, insulators, welding rod coating, etc. As the business expanded, the grinding of other inert vegetable materials was added to the repertoire of products; coconut shells in 1945, olive stones in 1970, guar and vegetable gums in 1979, and almond shells in 1986.

These products have a wide application. Olive stone flour is used as an extender in glues, as a filler in plastics, and is suitable for mixing with resins. Coconut shell flour has similar properties and can also be used as a carrier of pesticides. Guar gum is used in the well-drilling industry to thicken the drilling mud and in the food industry as a thickener, binder or stabiliser in pet food meat, sauces, salad dressings, pie fillings and ice cream. Almond shell flour provides a filling medium where the final product has to take on dyes and stains, such as plastic mouldings where a wood grain effect is required.[49]

The last turbine to be installed was in 1967 when the company purchased a second-hand 32.5 h.p. Gilkes turbine from the Miningaff Saw Mills, Newton Stewart, Wigtown in Scotland, installing it in the Higher Works.[50] However, the business was growing at such a pace that the water supply was incapable of delivering the level of power now being required. Also the River Authority started demanding payment for the use of the water as a power source.[51] These factors caused the demise of the use of water power and the switch over entirely to electric power from the National Grid System. Consequently the use of water power at Bosley came to an end after just over two hundred years of continuous operation. Its longevity and flexibility in adapting to the needs of copper production, cotton spinning, silk throwing, corn grinding and wood grinding is a tribute to the entrepreneurial talents of its many owners and the durability of water as power source in the face of competition from steam and electricity.

About six hundred yards upstream of the Higher Works there is a weir across the Dane which is about three feet high. From this weir water is diverted into the leat which widened out as it approaches the Higher Works to form the mill pool. The northern part of the mill pool, nearest the mill, has been filled in and built over but the water supply is culverted and still reaches the mill. Of the older mill buildings the turbine house and brick water tower still survive but the inscription "DANE CORN MILLS", which used to be picked out with white bricks, has been altered to read "DANE WOODMILLS". Part of the old corn mill building attached to the turbine house has also survived but is surrounded by modern appendages. Inside the turbine house the massive Gilkes turbine from 1933 and the spiral cased turbine, purchased second hand in 1967, can still be seen. It is possible that a thorough survey would reveal the remains of some of the earlier turbines that were installed here. In the mill building there are still some of the vertical millstones and plan sifters from the corn milling days now used for processing other products. The row of stone built cottages alongside the river have been converted into offices, etc, for the modern business but the row of ten brick built cottages across the road are still occupied as dwellings.

Just downstream of the Higher Works, the slightly concave weir for the Lower Works lies across the main river. This weir is about 14 feet high. The leat fed from this weir is very short ending in a curve shaped mill pool that is nothing more than a slightly widened leat. The oldest mill building at the Lower Works is built of brick, being about 9 yards wide and slightly longer.

FIGURE 47. THE SECOND-HAND SERIES R SPIRAL CASED TURBINE PURCHASED FROM MININGAFF SAW MILLS IN 1967.

FIGURE 48. THE TOWER ABOVE THE TURBINE HOUSE AT THE HIGHER WORKS TODAY.

Just behind this mill there is a small pool fed from the weir. The waterwheel used to be external on the eastern side of the building. Nearer to the river there is a two storey sandstone building approximately 27 yards long by 7 yards wide with a datestone which reads "FRT 1867". In the northern end of this building, the vertical turbine and associated generator installed in 1936 are still *in situ*. These buildings are also surrounded by the machinery and buildings of the present business.

The track for the narrow gauge steam tramway through the village has disappeared under tarmac but north of the Lower Works the line of the tramway can still be seen. It ran parallel to the North Staffordshire branch line as far as the canal where there is a small wharf. In a cutting the sleepers for the tramway are buried under considerable vegetation but still show the holes where the rails were mounted. There is now no sign of the corrugated iron covering that once sheltered the wharf at the canal.

References.

1. Chaloner, W. H., "Charles Roe of Macclesfield, *Transactions of the Lancashire & Cheshire Antiquarian Society*, Vol. 62, 1950/1.
2. SRO, Water rights sale deed, D(W) 1909/J/5.
3. Chaloner, W. H., 1950/1, *op. cit.* 1.
4. DRO, Deed of 1767, D664M/E1.
5. Sutcliffe, Rev. W., *Records of Bosley*, 1865.
6. *Journal of the House of Lords*, Vol. 31.
7. Chaloner, W. H., 1950/1, *op. cit.* 1.
8. CRO, Busfield Papers, D 4792.
9. Aikin, J., *A Description of the Country for Thirty to Forty Miles around Manchester*, 1795.
10. *Manchester Mercury*, 20th October 1801.
11. *Manchester Mercury*, 1st April 1806.
12. *Macclesfield Courier*, 10th March 1827.
13. Pigot's Directory of Cheshire, 1828.
14. *Macclesfield Courier*, 10th March 1827.
15. *Macclesfield Courier*, 17th July 1830.
16. Bolton Library, Crompton Spindle List, 1811.
17. Pigot, Directory of Cheshire, 1828.
18. *Macclesfield Courier*, 12th May 1832; CRO, Bryant's Map of Cheshire, 1831.
19. CRO, Bosley Tithe Apportionments.
20. *Macclesfield Courier*, 2nd February 1833, & 6th September 1834.
21. CRO, Busfield Papers, D 4792.
22. *Macclesfield Courier*, 19th December 1840.
23. CRO, Bosley Census Returns, 1841.
24. Bagshaw's History & Directory of Cheshire, 1850.
25. CRO, Bosley Census Returns, 1851.
26. *Macclesfield Courier*, 29th October 1853.
27. Morris's Directory of Cheshire, 1864.
28. CRO, Bosley Census Returns, 1861.
29. White's Directory of Cheshire, 1860.
30. *Macclesfield Courier*, 12th May 1832.
31. CRO, Bosley Census Returns, 1841/51.
32. White, Directory of Cheshire, 1860.
33. *Macclesfield Courier*, 9th October 1852, 9th April 1853.
34. CRO, Bosley Census Returns, 1861/71.
35. CRO, Bosley Census Returns, 1861.
36. Richards, R., *The Manor of Gawsworth*, Scolar Press, 1957; Slater's Directory of Cheshire, 1848.
37. Morris's Directory of Cheshire, 1874.
38. CRO, Bosley Census Returns, 1881.
39. Date on gravestone in Bosley churchyard.
40. On-site field investigation.
41. Bonson, T., "The Turbines Used at Bosley Works, Cheshire", *Wind and Water Mills*, 17, 1998.
42. *Ibid.*
43. *Ibid.*
44. SRO, Harrington Estate Sale Catalogue, 1920.
45. Bonson, T., "The History of Bosley Works, Cheshire", *Wind and Water Mills*, 14, 1995.
46. Bosley Wood Treatment Ltd, Publicity Brochure.
47. *Ibid.*
48. Gilbert, Gilkes & Gordon Ltd, Turbine Records.
49. Bosley Wood Treatment Ltd, Publicity Brochure.
50. Bosley Wood Treatment Ltd, Turbine file.
51. *Ibid.*

17. COLLEY MILL   SJ 892658

At about the most northern point of its course around Bosley Cloud, just prior to swinging southwards towards the town of Congleton, the River Dane is crossed by a narrow bridge constructed for the Congleton to Buxton turnpike, now the A54. On the south-eastern side of the bridge can be found the buildings of the corn mill known as Colley Mill.

Colley Mill was certainly in existence in 1750, when the miller at that time, Randle Plant died.[1] Later in the 18th century, according to the land tax returns, it was still owned by the Plant family, but the miller was Ralph Pointon. Ralph Pointon remained the tenant at Colley mill, with his wife Mary, until his death in 1812,[2] although ownership of the mill passed from Jonathon Plant to Lord Crewe in 1796 and then to the Rev. John Daintry in 1811.[3] Thereafter the tenancy passed to Samuel Pointon. He married his wife Ellen in 1812 and had three children priot to 1817.[4] In 1817 Samuel was joined at the mill by Thomas Pointon who was possibly his brother. After 1817 Thomas became the sole tenant until 1822 when it passed to John Pointon, probably another brother.[5] John Pointon died in 1836 and is the last occupant of Colley Mill to be described as a miller.[6] Although the next tenant was also a Pointon, namely William, he was described as a farmer. Certainly on his death in 1845 there was an advert for the sale of his farm animals and implements but no mention of any milling equipment.[7]

The Tithe Apportionments of 1845 for the parish of North Rode show that the owner was still the Rev. Daintry, minister at North Rode, and Andrew Wright & George Goddard were the occupiers of Colley Mill homestead. It is not known when the milling machinery was removed from the mill. In the late 1840s it was bought by Joshua Heapy, the miller at the nearby Bosley Mill (*q.v.*), for his retirement and possibly to ensure that the mill never was able to recommence milling in competition with Bosley Mill.[8] After the late 1850s the mill and mill house became two dwellings, one of which was occupied by a wheelwright, John Holland, until about the end of the 19th century, but it is not known if he used waterpower in his business.[9]

The tithe map shows that there was just a short leat from the river to the mill so, as there is not much fall in the river at this point, the waterwheel was most likely to have been undershot.[10] Today the mill and mill house still stand close to Colley Bridge. The mill is a two storey, half timbered building with a slate roof and small stone buttresses either side of the main door. The mill house which is attached to the mill is also two storeyed, built of brick and also with a slate roof.

References

1. CRO, Will of Randle Plant, 1750.
2. CRO, Will of Ralph Pointon, 1812.
3. CRO, Land Tax Returns, North Rode.
4. CRO, Gawsworth Parish Register.
5. CRO, North Rode Land Tax Returns.
6. CRO, Will of John Pointon, 1836.
7. *Macclesfield Courier*, 15th November 1845.
8. CRO, Census Return, North Rode, 1851.
9. Kelly's Directory of Cheshire, 1857, 1896.
10. CRO, North Rode tithe map.

FIGURE 49. THE HALF TIMBERED MILL BUILDING AND ATTACHED MILL HOUSE AT COLLEY MILL, BOTH NOW USED AS HOUSES.

18. & 19. HAVANNAH MILLS SJ 869646

Having swept around the northern end of Bosley Cloud, the River Dane then heads south-west towards the town of Congleton, forming the boundary between the parish of Eaton, to the west, and the township of Buglawton to the east. At a point where the river comes nearest to Eaton Hall, about one mile north west of Congleton, there is a bridge over the Dane, where on the Buglawton side, the water powered factory and associated community of the Havannah was established.

The origin of this water powered site is traceable to 24th July 1762 when Charles Roe, a Macclesfield industrialist, acquired the water rights by leasing land from George Lee and Elizabeth Ford for a term of 99 years *"with liberty to erect a mill thereon and to impound the water in the River Dane or Davon running through the said tenement for driving and working the wheels of said mill...and shall and may have liberty to stop and impound the water in the said river for the purposes aforesaid"*.[1] This acquisition was followed on the 27th April 1763 by Charles Roe leasing from George Lee for a term of 99 years at an annual rent of £30-10s-0d *"the liberty and privilege of erecting one or more water mill or mills and to make such wears [sic], mill dam or reservoirs for water and likewise such subterranion [sic] passages for water as Charles Roe should think fit"*. Among the parcels of land leased by Charles Roe were two acres intended to be covered by water. George Lee retained the right to the hunting, fishing, forestry, mining and quarrying but agreed not to mine or quarry in any part of the field where the mills were to be erected or *"within ten yards of the ware [sic] to be erected"*. Charles Roe indemnified George Lee against *"all actions or damage to any ancient mill or mills or overflowing of land on account of erecting any water mill or mills or making any wear, mill dam, or reservoir or raising or impounding the water"*.[2]

As has been stated, Charles Roe's career started in the silk industry, diversifying into copper and brass production in 1758. To extended this copper and brass business he acquired this land at Buglawton which he then called The Havannah in commemoration of the siege and capture of Havannah in 1762 by the Royal Navy.[3] Having acquired the necessary leases at The Havannah, Charles Roe built a factory on the site for making brass wire and for rolling copper sheet and bolts. This factory had a large and impressive weir across the River Dane to provide the necessary water power. Charles Roe's continued investment in the copper industry caused him to sell his interest in his silk throwing factory at Macclesfield in 1764. He then invested in another site at Bosley (*q.v.*) where he built rolling and hammer mills for processing brass and copper.

By the 1770s the Alderley mines had been superseded by the development of copper ore deposits elsewhere and the Macclesfield coal deposits were also exhausted. This meant that both Havannah and Bosley had no commercial advantage due to their location, in fact the very opposite was probably true. One of Charles Roe's instincts as a businessman was to sell at the top of the market, so in 1777 he had a lease drawn up to sell the Havannah factory to Francis Berrisford of Ashborn [*sic*] in Derbyshire but for some reason this lease was not executed.[4]

Charles Roe, the driving force in the Macclesfield Copper Company, died in 1781 and his son, William Roe, and Edward Hawkins, one of the partners originally from Neath, took over the running of the Company. In 1783 the partners must have been considering expansion at Havannah because they leased more land from a Mr. Armett for 99 years at a rent of £25 per annum. This land was known as Armett's Meadow, or Dean's Meadow, together with the Long Field and the lane leading to the factory.[5] Again in 1785 the partners had an eye to increasing production for they contacted Boulton & Watt in Birmingham with a view to purchasing one of their steam engines. They enquired about the terms *"for engines of several degrees of power with calculations of the daily expenditure of coals"* and were particularly interested in *"returning all the water now running, with moderate velocity, to waste through a space containing nine square feet"*. James Watt visited the Havannah and received the hospitality of Edward Hawkins and Brian Hodgson, another of the partners, in Congleton and on 16th August he wrote to Matthew Boulton as follows:-

"I was at Messrs. Roe & Co.'s brass works; their consumption of water is amazing. They have five wheels; the one belonging to the wire mill requires constant water and may be replaced by a 20 horse rotative engine."[6]

However, the Macclesfield Copper Company did not proceed with this proposal. In 1787 in a letter from Edward Hawkins to one of the investors in Roe & Co. he stated that *"they do not wish to do much in the wire trade which is the least profitable of all branches of our manufacture"* which probably explains why they did not purchase the engine from Boulton & Watt.[7]

From this high spot of the early 1780s the fortunes of the Macclesfield Copper Company started to falter, especially with respect to its operations in Cheshire. In 1801 the factory at Havannah was put up for sale together with the other factories at Macclesfield and Bosley. The Havannah premises were described as follows:-

"All that pile of buildings situated at Havannah within two miles of Congleton in the said County and adjoining the turnpike road leading from thence to Macclesfield and

now used for making brass wire and rolling copper sheet and bolts having five waterwheels from 18 feet to 25 feet diameter with the head and fall of 13 feet 3 inch and the whole of the River Dane for supply. The premises consist of a brass rolling mill; a wire mill and long chamber over; annealing houses and oven; copper mill, blacksmith's, millwright's and carpenter's shops; two warehouses; seven cottages with gardens, yards for wood, charcoal, etc., together with eight acres of land the whole under lease from Richard Ayton Lee Esq. for 99 years from 14th May 1763 at a very low rent of £30-10s-0d per annum."[8]

Unfortunately this sale was not successful and the whole works were again advertised in 1806.[9] The only additional information given was that the waterwheels of the Havannah were equivalent to 30 horses and that the property was now occupied by Messrs. Keys & Co., presumably tenants that had arrived after the attempted sell-off in 1801. This sale was no more successful than the one in 1801 with the Havannah remaining the property of Messrs. Roe & Co. for another seven years.

In 1813 the eleven partners in the company decided to dissolve the partnership, liquidating their assets and sharing out the proceeds. The Havannah was valued at £3850 and became the property of Edward Hawkins as his share of the company. He recouped £450 by letting the parcel of land acquired in 1783 to Peter Whitely for one year and immediately mortgaged the rest of the Havannah to John Jeffries (or Jefferis), a Gloucester merchant who was one of the other partners in the Macclesfield Copper Company, for £4300 at 5% interest per annum. This debt was to be repaid by August 1818.[10]

This repayment did not take place because Edward Hawkins died in 1816. At the time of his death, besides owing the original sum of £4300, he owed a further £430 in interest payments and John Jeffries was out of pocket for further sums for ground rent and repairs that he had paid for. As Edward Hawkins' heirs were unable to pay these amounts, John Jeffries took over the ownership of the Havannah in November 1816.[11]

His first move was to advertise for a tenant for one of the buildings, but in January of the following year he made over the ownership of all the lands and buildings of the Havannah to Charles Jeffries, one of his sons who was a lieutenant in the Royal Navy, and Rowland Cockson, identified as gentleman of Havannah. Although the transfer price was £3998, this was not paid but was secured to John Jeffries by a bond of Jeffries & Cockson and guaranteed by a Charles Weaver, also of Gloucester, another of the partners in the defunct Macclesfield Copper Company. The condition of this loan was that Jeffries and Cockson, although they had to pay 5% interest per annum, did not have to repay the capital until five years after John Jeffries death.[12]

The layout of the manufactory can be seen in a survey made at this time, it is likely that this layout of buildings is as used by the Macclesfield Copper Company but it is unlikely to have been used for brass and copper manufacturing since at least 1801. After

FIGURE 50. PART OF AN ESTATE MAP OF 1817 SHOWING THE VARIOUS BUILDINGS AT THE HAVANNAH DURING THE PERIOD OF OWNERSHIP OF MESSRS. ROE & CO. (SRO)

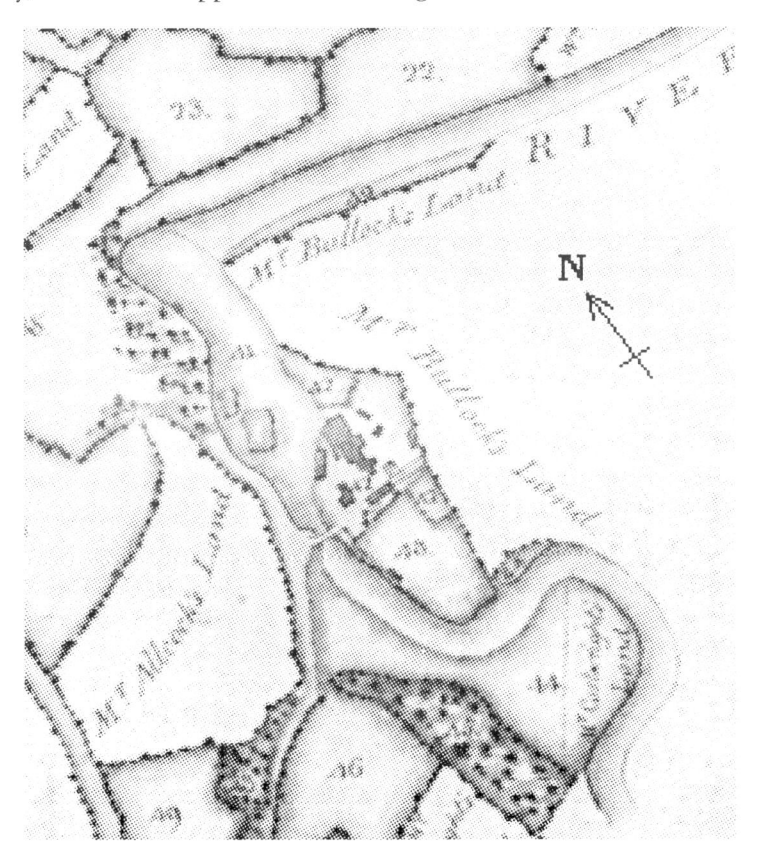

this time it is possible that the buildings were leased for other purposes or remained idle. Certainly by 1817 part of the buildings had been converted to a corn mill with George Goodall, a miller, in occupation. However, the buildings were over 50 years old and were recorded as *"going much out of repair"*.[13] Once Jeffries & Cockson took over the whole site they continued with the corn milling and started a silk throwing business in the other buildings. In 1818 they advertised to let *"A capital mill situated at the Havannah near Congleton lately occupied as a tobacco manufactory capable of being made serviceable for carrying on any business where powerful water turning is required."* Again it is not clear how long this tobacco business had been established at the Havannah, possibly only from 1816.[14]

With the future looking very promising Jeffries & Cockson decided to invest heavily in developing the site. They refurbished the buildings and installed the silk throwing machinery and they also decided to expand by building another water powered mill and more cottages. Unfortunately, John Jeffries died in 1818 which meant that the loan of £3998 had to be repaid in 1823, five years later. When this time had elapsed they were not in a position to repay the loan as their plans were only part implemented, any profits having been ploughed into the development. The heirs of John Jeffries, his other seven children, were not prepared to wait for their inheritance and so a William Fowler of Bristol, another of the partners in the Macclesfield Copper Company, paid them their outstanding moneys and gave Jeffries & Cockson another three years to pay off the £3998 due.[15]

By 1825 it was becoming necessary to complete the development and recoup the profits. Jeffries & Cockson decided to relinquish the silk throwing business and let the Havannah silk mills, with 21 cottages, to Hobson & Ford for 21 years at a rent £505 per annum. Quite a sum for those days! Jeffries & Cockson were *"obliged to maintain the weir or mill dam, waterwheel, first shaft, main shaft, light gearing and machinery so that the mill could be worked to any extent, both night and day"* and Hobson & Ford were *"obliged to repair the*

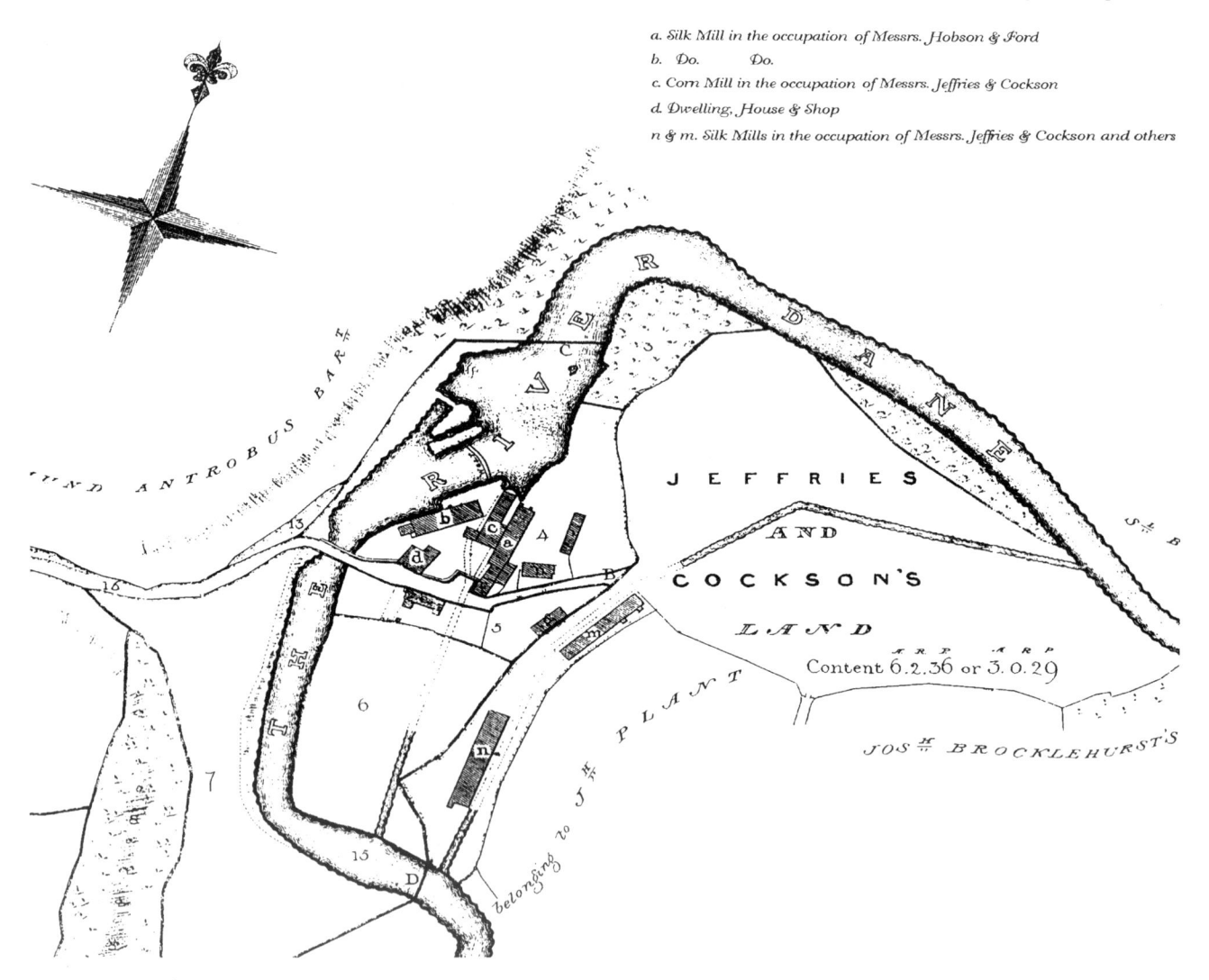

a. Silk Mill in the occupation of Messrs. Hobson & Ford
b. Do. Do.
c. Corn Mill in the occupation of Messrs. Jeffries & Cockson
d. Dwelling, House & Shop
n & m. Silk Mills in the occupation of Messrs. Jeffries & Cockson and others

FIGURE 51. AN ESTATE MAP OF THE HAVANNAH, 1825, SHOWING THE NEW LEAT FROM THE RIVER RUNNING TO THE NEW MILL BEING BUILT TO THE SOUTH-EAST OF THE ORIGINAL MILLS AT THE WEIR. (SRO)

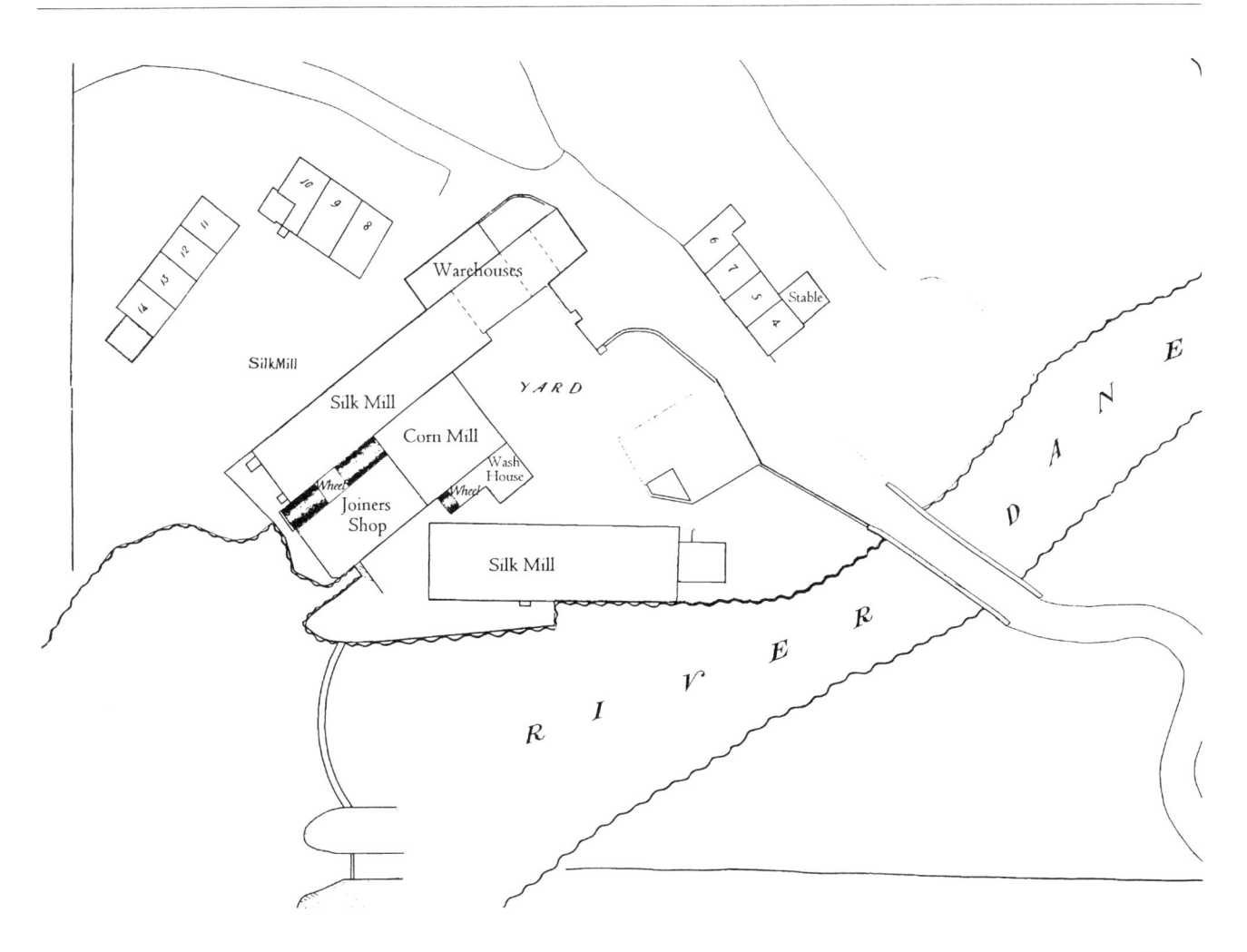

FIGURE 52. A SITE PLAN OF THE HAVANNAH MILLS SHOWING THE CORN MILL IN BETWEEN THE TWO SILK MILLS. NOTE THAT ONLY TWO WATERWHEELS ARE INDICATED, POSSIBLY THE SILK MILL NEAREST THE RIVER WAS UNPOWERED. (SRO)

premises, paint and whitewash the mill every three years".[16] As part of the development of the estate Jeffries & Cockson had excavated a new leat on the land to the east of the original factory which provided water power to the new mill that they had then completed, together with more cottages for the expected work force, bringing the total number of cottages to thirty nine. On 1st January 1826 they leased this new mill and cottages to Henry & Edward Tootal for 10 years.[17] Even so they were not in a position to repay their original debt later in the year. William Fowler recouped his money by selling the mortgage to a group of businessmen with connections in the silk trade in London and Cheshire. The leading member of this group was James Hogg, described as a silkman of Cheapside, London but who was, in fact, part of the family that operated Davenshaw Mill (*q.v.*) in Buglawton.[18] This consortium would know that the value of the Havannah was now much greater than the debt outstanding, and they immediately foreclosed on Jeffries & Cockson, taking over possession of the whole of the Havannah complex.

A number of plans were drawn up around this time clearly showing the layout of the old mills and the leat, or New Cut as it was called, leading to the two new mills south-east of the old mills. Although shown on the map (see figure 51) as two completely separate mills, when built they were actually adjacent to each other. The old mills consisted of three main buildings with the corn mill operated by Jeffries & Cockson in the middle (see Figure 52), one silk mill is situated alongside the eastern side of the corn mill while the other silk mill is alongside and parallel to the river on western side of the corn mill. Two waterwheels are clearly shown, one on the corn mill and the other on the adjoining silk mill. This is a considerable reduction from the five waterwheels used by Roe & Co. for brass and copper working, and it is difficult to see from the layout where these may have been situated. Certainly the westernmost silk mill would appear to have no waterwheel at all.

The change in ownership in 1826 gave rise to many changes at the Havannah. Hobson & Ford, the tenants of the old silk mills split up, with William Ford taking over the corn mill from Jeffries & Cockson and James Hobson continuing the silk throwing business. The

Tootal brothers were persuaded to give up their 10 year lease on the new mill which was then advertised for new tenants as follows:-

"To be let - Two completely and recently erected silk mills situated at Havannah in the township of Eaton in the County of Chester known by the names of the Windsor Mills, the larger being 30 yards long by 9 yards broad, the smaller being 20 yards long by 9 yards wide, turned by water derived from the River Dane and by a wheel of great value and improved principle from which immense power may be derived. The above mills are new and constructed upon the best and most modern plan - both are five storeys high and are well lighted."[19]

This advert gives a good picture of the considerable size of the Windsor Mills which were the new mills erected by Jeffries & Cockson. The new mills and the many new cottages built give some idea of the extent of the large investment that they made at the Havannah. One cannot help speculating whether this represented the profits made by their businesses in such a short time of nine or ten years or was possibly the use of prize money from the Napoleonic Wars by Charles Jeffries, the one-time lieutenant in the Royal Navy! Whatever the source of their funds they were not able to benefit from their investment, that was realised by the group which purchased the mortgage and then sold the property around this time to the owner of the freehold, Gibbs Craufurd Antrobus, who resided at nearby Eaton Hall.[20]

At the end of 1828 James Hobson, who lived at Eaton Cottage nearby, renegotiated his lease on the silk mills, replacing it with a lease to *"all those two silk mills called or known by the name of the Old Havannah Silk Mills and situated within Eaton and all that water corn mill near to the said silk mill with the drying kiln thereto"*. Apart from the usual references to the waterwheel, shafts and gearing the lease also mentioned *"a large steam boiler and large steam pipes"* which would have provided heating for the mills.[21]

A few years later, In 1831, G. C. Antrobus leased the Windsor Mills to Messrs. Wood & Westhead, who were tape manufacturers from Manchester, for 21 years at a rent of £145 per annum. The lease included *"that sluice or watercourse which has lately been cut from the River Dane to the said mills for the supply of water to the waterwheel whereby the machinery in the said mill is intended to work"*. Antrobus also had to erect *"a mechanic's shop at the said mills, cover in the waterwheel, and erect pillars in the mill for the support and strengthening of the floors"*. The phraseology of the lease gives the impression that this was the first time that the Windsor Mill had actually been used, especially as the floors had to be strengthened to take the weight of the tape weaving machines.[22]

After this date life at the Havannah became less hectic with James Hobson running the Old Havannah silk mills, sub-letting the corn mill to William Ford, and Wood & Westhead weaving tapes at the Windsor or New Havannah mills, employing a manager, William Clarke, plus 22 male and 26 female employees.[23] Then in 1840 calamity struck when the corn mill caught fire and was destroyed. A good corn mill was an essential part of the commerce of a district and was a useful hedge against the vagaries of the silk industry in the Antrobus business portfolio so it was rebuilt immediately. Fortunately some of the plans of the rebuilding have survived to show something of the layout of the new, three storey, mill. It had a high breastshot waterwheel about 15 feet in diameter with six arms and about 40 buckets. The penstock was designed with three separate water feed hatches in order to cope with different water levels in the headrace (see Figure 53).

This wheel, situated on the western side of the mill, had a ring gear driving a pinion and associated shaft into the mill. Once in the mill the drive was transferred to a lineshaft running parallel to the waterwheel that had five bevel gears spread along its length driving five bevelled stone nuts. These stone nuts drove five pairs of millstones on the stone floor (i.e. the first floor), three pairs were French burr stones, 4 feet 4 inch diameter, for producing flour; one pair of 5 feet

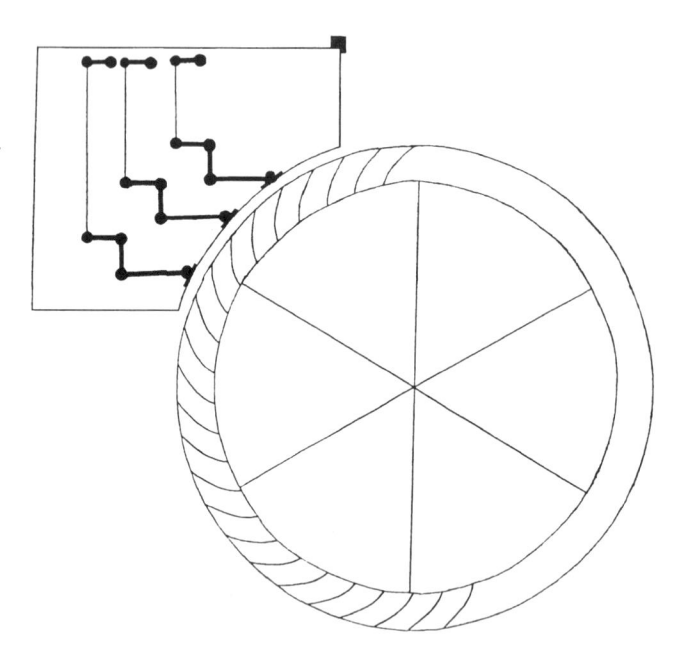

FIGURE 53. SKETCH OF THE NEW WATERWHEEL IN THE REBUILT CORN MILL, 1840, AND THE THREE WATER FEED HATCHES AND ASSOCIATED CRANK ARM CONTROL RODS. (SRO)

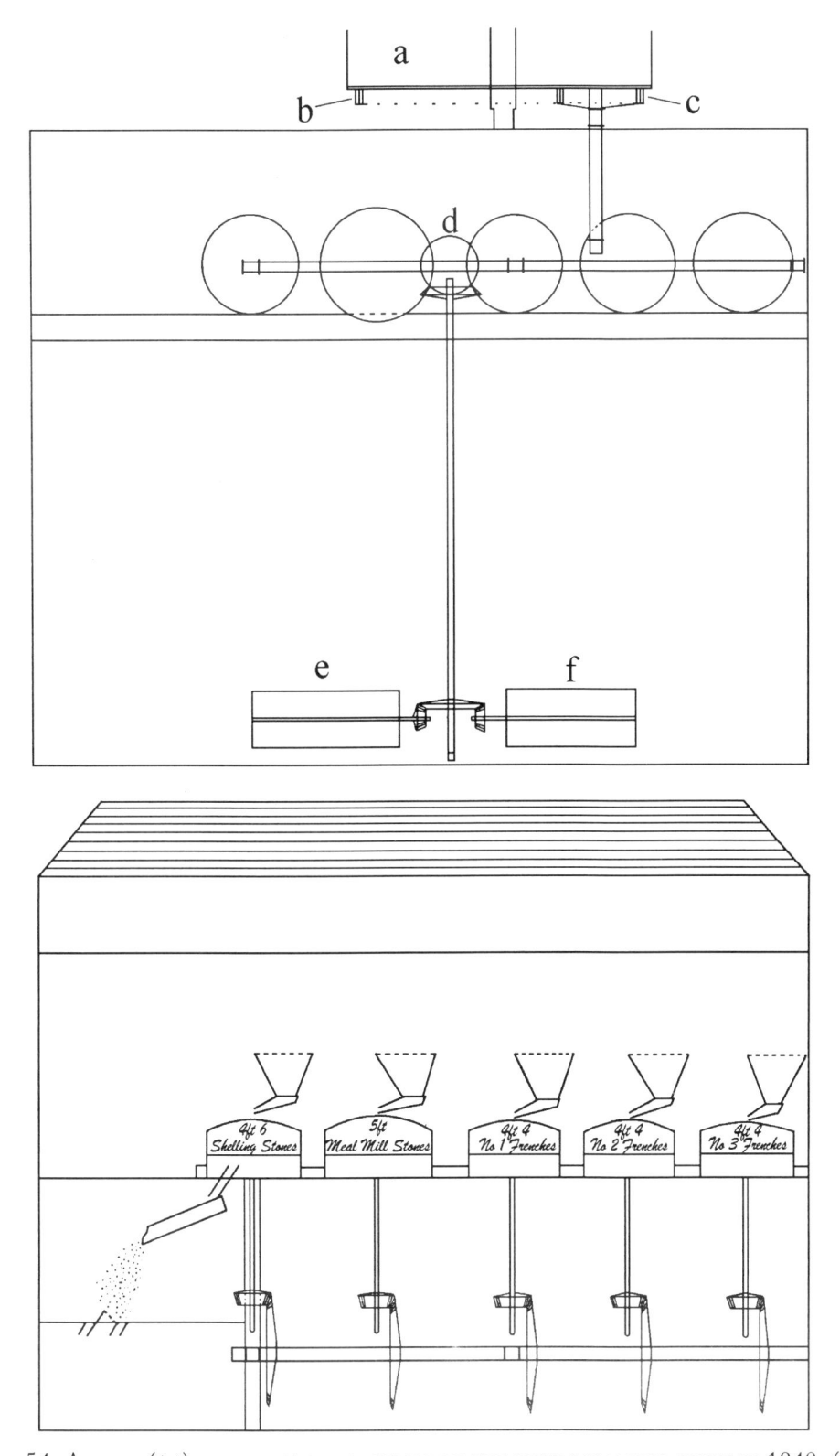

FIGURE 54. A PLAN (*top*) AND VERTICAL SECTION OF THE NEW CORN MILL BUILT IN 1840. (SRO)
a, WATERWHEEL. b, RING GEAR. c, PINION. d, BEVEL GEAR. e & f, GRAIN CLEANER & FLOUR DRESSER.

diameter Peak stones for producing meal, and one pair of Peak stones, 4 feet 6 inch diameter, for shelling oatmeal. The drawings also show the drive being taken across the mill to drive via bevel gearing two reels on the stone floor on either side of a drive shaft. There was undoubtedly a vertical shaft driven from the main layshaft to take power up to a sack hoist at the top of the mill which would have gearing at ceiling height on the stone floor to drive the lineshaft across the mill to the two reels. These were probably a grain cleaner and a flour dresser, as it was usual to fit these items on the stone floor (see Figure 54).[24] The local millwright, John Edwards, was contracted to provide the new waterwheel for the corn mill at a price of £153-3s-7d

57

provided that he could keep the old material. There was a discount of £17-8s-0d to G. C. Antrobus if he provided the oak for the arms for the waterwheel and its segments. The removal of the old wheel and the installation of the new one was planned to take two weeks and three days altogether. The work must have gone more or less to plan as John Edwards was paid £178-10s-0d for the work two months later.[25] Once rebuilt the new corn mill was advertised to let as "*all that powerful water corn mill situated at the Havannah...together with a drying kiln*". The fire of 1840 in the corn mill must have caused James Hobson to think about his situation in the adjacent silk mills for in 1842 he insured both Havannah silk mills for £900, the stock in the mills for £1700 and the goods in his nearby house for £500, with the Guildhall Insurance Company.[26]

In 1841 the partnership between James Wood and Joshua & Edward Westhead expired and the two Westhead brothers continued the tape business at Havannah on their own.[27] Unfortunately the first millers in the new corn mill, John and Edward Herdman, went bankrupt in 1842.[28] By 1850 John Spragg had taken over as miller from the Herdmans.[29] Then in 1855 James Hobson died and the Westhead brothers also vacated their premises leaving G. C. Antrobus to find new occupants for the three silk mills. In the same year, the Antrobus's agent wrote to Charles Thornton, a Nottingham silk lace manufacturer, about the mills and the terms of their possible lease. This letter gives an excellent description of the mills at that time, so a draft is reproduced in full as follows:-

"*I duly received your letter of the 7th inst. the New Mills are about 50 yards long by 10 yards wide, five stories high with small attic & well lighted & up to March last was used as a smallware manufactory by Messrs. J. P. & E. Westhead & Co of Manchester, there is a smithy & stable attached, turned by a water wheel of about 18 horse power, the Old Mills are in two buildings one of them is 50 yards long by 8½ yards wide the other is 38 yards long by 9 yards wide all outside measure one part three stories high and well lighted the other two stories.(but not so well lighted)[crossed out]*

It is now in the hands of the late Mr. Hobson's representatives (who we give up at any time) with warehouse, staff room, etc. used as a silk mill, this is driven by a wheel of about the same power as the other, with an engine for occasional use of about 11 horse, in the village there are 65 cottages & the house held with the above is a very commodious residence for a respectable family with outbuildings, etc., gardens & about 24 acres of lands. The rent asked for the foregoing is as follows:-
Eaton cottage with lands 80- 0-0
65 cottages etc. @ £3-15-0 ea 243-15-0
Old factory with engine & waterwheel 140- 0-0
New factory... 280- 0-0
743-15-0
Say £740 per annum
This is rental asked for the concern if taken as whole & off which take £40 for repair etc. making the net rent £700 per annum for the whole concern to be due payable on the 29th Sept. and 25th March. The landlord to put the buildings in tenantable repair, the tenants to keep them so having materials found in the rough. The landlord to put & keep the waterwheels in repair. Should the steam engine be required the tenants to find all fuel & other necessaries & to keep it in repair. I may add here that the River Dane is a very constant stream and coals plentiful. A lease of the above can be given if required. The rights of the water corn mill will have to be secured to it in any lease or agreement that may be made hereafter. There may be a few minor stipulations which I have no doubt would be easily arranged."[30]

FIGURE 55. A POSTCARD VIEW OF THE OLD MILL AT THE HAVANNAH. THE THREE STOREY BUILDING IS THE CORN MILL. NOTE THE CHIMNEY INDICATING THE USE OF THE STEAM ENGINE.

It is clear that the two new mills of 1826 were now considered to be one mill driven by one waterwheel, the old mills were also still driven by water power but had been augmented by a steam engine. It is noticeable that the total number of cottages in the village now numbered 65. A month later Thornton replied with a counter offer of £600 per annum for the whole site less the corn mill and £500 for the silk throwing machinery of James Hobson. However he was not willing to occupy the mills immediately as new machinery would have to be purchased for the Windsor Mill and requested that he should only pay half the offered rate for the first six months. The delays in occupying the mills got longer and by mid 1856 Antrobus advertised the mills for let in the local paper. Maybe this pressured Charles Thornton into action for in February 1857 he and his brother John eventually signed a 14 year lease on the terms outlined in his reply of 1855.[31]

After 1857 the tenancies at Havannah entered another settled period with John Spragg as the corn miller and John & Charles Thornton occupying Havannah Old Mills and Windsor New Mill as lace silk throwsters. As their main centre of business was still in Nottingham their premises at the Havannah were overseen by their agent John Gaskin for many years. The silk lace business had a market niche that appears to have been unaffected by the great depression in the general silk industry caused by the Free Trade Act of 1860. Right from the beginning of their tenure at Havannah they also used part of Eaton Mill (*q.v.*) when more production was required and this practice continued through to the beginning of the 1890s. At the end of the 1870s the firm became known as Thornton & Whitely.[32]

According to the trade directories Thornton's silk lace business ceased at the Havannah sometime between 1878 and 1881. Also John Spragg who had occupied the corn mill since before 1857 ceased to appear in the directories after 1883 but in all probability had finished trading some years earlier. It is possible that both concerns finished trading in 1872 when the mills and the village were inundated when Bosley Reservoir overflowed. In 1881 the corn mill at Havannah was refitted by Mr John Thornton, a millwright of Worksop in Nottinghamshire for Messrs Fitton & Sons. Fittons were the miller at the steam mill in Macclesfield which was severely damaged in an explosion in September 1881.[33] Obviously they wanted to continue their business whilst their mill in Macclesfield was being rebuilt. The new fittings consisted of two large size patent flour dressers, or silk machines with improved brushes, a silk cleaner, bran dusters, centrifugal machines, purifiers, rolls, a complete new grain scouring arrangement, exhaust to millstones, and complete new fittings to the stones, with necessary elevators, conveyors, etc for the grain, flour, and meal. The whole of the above works were erected under the direction of Mr. C. Redwood, who was the manager at Havannah corn mill for Messrs Fitton & Son.[34] During the two years in which the Havannah Corn Mill was being operated on behalf of Messrs Fitton the manager was a Mr Samuel Hickson, a Cheshire man who had learnt his trade of roller milling at the Usk Shore Mills at Newport in South Wales.[35] Two years later, in 1883, when Fittons had rebuilt their steam mill in Macclesfield they offered to let the corn mill at Havannah and Samuel Hickson went to work for Messrs. J. Thompstone & Co. at Prestbury Mill. At this time the Havannah Corn Mill still contained five pairs of millstones capable of producing 300 sacks of flour per week and the waterwheel was described as providing 20 h.p.[36]

FIGURE 56. MR. SAMUEL HICKSON, MANAGER OF HAVANNAH CORN MILL, 1881-3. (*The Miller*)

After 1883, with no work at the mills the occupants of the cottages drifted away from the hamlet leaving the premises empty so giving rise to the phrase "the deserted village" used to describe Havannah towards the end of the 19th century. However the corn mill came back to life fitfully when it was being run by John Orme by 1896, but he had left the Havannah by 1900 to run Congleton Mill (*q.v.*). Due to the dramatic changes in the flour industry, with the massive importation of foreign wheat which led to the provision of steam powered roller mills located at the ports at the end of the 19th century, it seems that the corn mill at Havannah was not used for milling again after the departure of John Orme. The silk mills were not used again until 1899 when the Havannah was taken over by the Andiamo Tobacco Company for the manufacture of cigarettes and "genuine Havannah" cigars which were marketed under the brand name of Marsuma. This proved to be a very successful and profitable trade. In the early 1900s around 400 people were employed at the Havannah and the firm had to expand, first into Albion Mill in Macclesfield and then in 1906 into New Mill (now known as Riverside) in Congleton . Although the company attempted to manufacture cigars by machinery this was soon abandoned in favour of hand rolling. Because electricity was not available at the Havannah (electricity only reached Congleton in 1931) the waterwheels were arranged to generate electricity for the mills. Although the tobacco company itself thrived up to the 1950s, it ceased production at the Havannah shortly after the First World War.[37]

FIGURE 57. ADVERTISEMENT FOR MARSUMA CIGARETTES IN THE EARLY 20TH CENTURY. (*Congleton Chronicle*)

When the tobacco company moved out of the Havannah it left a large hole in the finances of the landowner, C. J. Antrobus. Although the mills were leased to John Gibson & Co in 1920 at a rent of £75 for the first year, £225 for the second year and £525 per year thereafter, this arrangement did not survive its first year. The state of the property at this time can be inferred from a condition of the lease that John Gibson & Co were required to complete the repair of both the old silk mills and the Windsor mill "*in such a manner that the same may be in all respects fit for occupation as factories*". They were also expected to repair at least 24 cottages to be fit for habitation by the end of four years. They had to make good any damage to the bridge over the River Dane and to strengthen it, as well as maintaining a supply of water from a spring to Eaton cottage. It is interesting to note that they were also expected to supply C. J. Antrobus with at least 1500 watts of electricity for lighting, although Antrobus would pay for the wiring needed. It is no wonder that these conditions proved to be too onerous for the times. The schedule attached to the lease indicates that the waterwheel in the mill at the western side of the corn mill was being used to drive a pump to fill a tank on the top floor with water, probably as a supply for the cottages. Also the waterwheel in the Windsor mill was still present together with a steam boiler for heating.[38]

After the arrangement with John Gibson & Co ceased no other tenants were forthcoming, so in 1922 Antrobus decided that he would have to go into business himself. After the massive decline in the silk throwing trade in the area after 1860 a new textile industry was introduced into Congleton to utilise the empty mills and unemployed textile workers, this was the fustian trade. Fustian was a type of velvet made by cutting the loops of thread in the woven cloth, an operation carried out traditionally by hand. In 1918 a machine had been invented by a local man to mechanise this process, a development that C. J. Antrobus hoped to exploit. He set up a partnership, under the name of the Havannah Mills Company Limited, with George Rogers, who was already in the fustian trade.[39] Considerable capital was needed by the new company as it had to manufacture all its own machinery before it could start producing any velvet. This required the usual range of machine tools necessary to equip a mechanical workshop. These tools required a larger electricity supply than provided previously by the waterwheels, which had only to power lighting in the buildings, so a turbine was purchased from Gilbert Gilkes & Co. of Kendal which was installed in the Windsor Mill to replace the waterwheel. This turbine was an Vortex turbine, Serial No 2994, which was designed for a 13 feet head with a water flow of 69½ cu. ft. per minute to give 80 h.p. at 290 r.p.m.[40] This turbine took over the responsibility for electricity generation which enabled the supply to be connected to the cottages in the village as well as to Eaton Hall.

However the supply was only available from 6 a.m. to 11 p.m. The first floor of the two storey mill was demolished to become the fitting shop with an oil engine installed to drive overhead lineshafting to three lathes, three drilling machines, a metal planer, power saw and grinding wheels. This building also included a blacksmith's forge. The first floor of the three storey mill became the joiner's shop with a circular saw, a bandsaw, wood planer, and lathe which were used to produce the patterns for the castings used in manufacturing the fustian cutting machines. A total of 66 machines were built and installed in Windsor Mill. At the Windsor Mill a twin cylinder steam engine was eventually installed to produce further electric power and an asbestos sheeted stiffening shed was attached to the mill. This was needed because fustian had to be limed and stiffened before it could be machine cut. The process consisted of passing the cloth through a trough filled with a lime mixture. The cloth was then carried round six 5 feet diameter, steam heated, copper drums to be dried. These drums were rotated by a belt driven from the steam engine. Once dried the cloth was passed through a stiffening mixture and then round the drums again. The steam was provided by a Lancashire twin fire boiler with a 100 feet high chimney.[41]

During the start up years the Havannah Mills Company was short of money with C. J. Antrobus lending the company £2000 in 1924 and a similar amount in 1927.[42] The Havannah Mills Company was the first in the world to introduce machine cutting of pile fabrics. The patent for the velvet cutting machine (invented in 1918) which was applied for in 1924 was granted the following year, but improvements continued to be necessary with a further five patents granted.[43] Undoubtedly the partners in the company considered that the design of their fustian cutting machine was a valuable asset which they sought to protect by taking out patents in the France, Germany and the U.S.A.[44] The company's final patent was issued in 1939.[45] During the Second World War the velvet cutting ceased. The cutting machines were dismantled and stored while the machine shop worked on government contracts for the war effort. The Ministry of Food used the three floors of the old buildings for storage, paying a rent of £145 per annum; the Ministry of Works requisitioned the Windsor Mill, paying compensation of £180 per annum.[46] During this period the mills manufactured parts for tanks and armoured cars, pumps for submarines and machine tool parts for the manufacture of shell cases, as well as undertaking subcontract work for Rolls Royce. This work involved changes in the mills as two more milling machines and another power saw were installed in the fitting shop. In 1942 a contract was obtained to manufacture bearing housings for S.K.F., a Swedish company. This contract needed the conversion of the joiner's shop to house extra milling machines.[47]

In 1943 Crawfurd John Antrobus died and as part of his estate the works at Havannah were valued at £5846 for tax purposes. After paying death duties on his brother's estate, the next incumbent at Eaton Hall, Ronald Henry Antrobus, a lieutenant colonel in the army, looked for ways of raising money. Early in 1945 he was in correspondence with Alfred Simons & Sons of Whitworth Street in Manchester concerning selling the fustian cutting machines at the Havannah, which were stored in a dismantled state. R. H. Antrobus wanted a price for them as working machines rather than a scrap value. He thought the Ministry should pay for them to be re-erected and that it would be better to wait six months until "*after an armistice*" for better commercial conditions. However, Simon's interest in the machines caused Antrobus to completely over-react. He made them a number of offers on the freehold of the Havannah, namely:-

a) The whole concern for £72,000;

b) The whole concern for £60,000 plus £550 p.a. rent;

c) £20,000 for whole concern if R. H. Antrobus appointed a local director;

d) Individual machines were not for sale;

e) The purchaser to pay cost of transport.

All these offers were on the condition that Mr. Rogers was not to be allowed near the mills as long as Antrobus was the landlord. This last phrase sums up the level that the relationship between the directors of Havannah Mills Company had sunk to by this time. Alfred Simons & Co. replied extremely politely that they had been completely misunderstood and were only interested in a few of the machines, thus ending Antrobus's hopes of one last, glorious, payout from the ownership of the Havannah Mills.[48] Later in 1945 he sold the property to Fred Jackson, a local fustian cutter for £10,000, after which the Havannah Mills Company went into receivership.

At the time of this sale in late 1945 a plan and inventory of the Havannah site was made. On the plan the mills appear to be exactly as shown on the plans drawn up in 1826 indicating that there had not been any change to the buildings since the Windsor Mill had been built and that the old mills were probably the same buildings as erected by Charles Roe, the corn mill having been rebuilt in 1840. The inventory shows that the corn mill still had a waterwheel, albeit only 5 feet 6 inch diameter (this may be a mistake and could refer to the dimension of the radius) and a hydraulic ram pump providing a source of water. The rest of

FIGURE 58. WINDSOR MILL, HAVANNAH, ON FIRE IN 1972. (*Congleton Chronicle*)

the corn mill and the old silk mills consisted of workshops, fitter's shop, precision grinding room, foreman's office, blacksmith's shop, pattern maker's shop, three store rooms, manager's office or "board room", billiard room for use of the staff's billiard and rifle club!, secretary's office and the main velvet shed. These premises contained a variety of machine tools, saws, hand tools etc., including 58 dismantled fustian cutting machines. In the Windsor mill there was a 10 h.p. steam engine, another steam engine by Allan of Bedford, a Lancashire steam boiler; an oil engine by Allan of Bedford, and a water turbine. The water turbine was connected to an electric dynamo for charging a set of batteries. The building also housed a fire engine, a staff canteen, and an oil refining house together with tools and patterns for Admiralty work and 37 more dismantled fustian cutting machines.[49] All in all a sad reflection of the state of British industry in the middle of the 20th century.

In 1946 Fred Jackson restarted fustian cutting by machinery at the Havannah and in 1949 this business became the English Velvet Company. Fustian cutting continued up to 1962 when the mills at Havannah were finally closed. The mill buildings survived until 1972 when the five storey Windsor Mill caught fire and was destroyed. The old mills remained standing until they were demolished in 1988. Today there is little to be seen except for the weir which is still an attractive sight, the land where the mills stood is derelict, but some of the cottages have been modernised and are still inhabited.

References

1. SRO, Deed of 1762, D(W) 1909/J/5.
2. SRO, Deed of 1763, D(W) 1909/A/8/1.
3. Chaloner, W. H., "Charles Roe of Macclesfield", *Transactions of the Lancashire and Cheshire Antiquarian Society*, Vol. 62, 1950/1.
4. SRO, Lease of 1777, D(W) 1909/A/8/1.
5. CRO, Lease of 1783, DCB/1179/7.
6. Chaloner, W. H., 1950/1, *op. cit.* 3.
7. CRO, Busfield Papers, D 4792.
8. *Manchester Mercury*, 20th October 1801.
9. *Manchester Mercury*, 1st April 1806.
10. SRO, Mortgage of 1813, D(W) 1909/A/8/2.
11. SRO, Deed of 1816, D(W) 1909/A/8/6.
12. SRO, Mortgage of 1816, D(W) 1909/A/8/7.
13. SRO, Estate Survey, 1817, D(W) 1909/D/1/1.
14. *Macclesfield Courier*, 25th July 1818.
15. SRO, Agreement of 1823, D(W) 1909/A/8/9.
16. SRO, Lease of 1825, D(W) 1909/J/4/1.
17. SRO, Lease of 1826, D(W) 1909/A/8/10.
18. SRO, Mortgage of 1826, D(W) 1909/A/8/11.
19. *Macclesfield Courier*, 1st December 1827.
20. SRO, Deeds of 1826, D(W) 1909/A/8/15-18.
21. SRO, Lease of 1828, D(W) 1909/J/4/2; CRO, DCB/1595/27/8.
22. SRO, Lease of 1831, D(W) 1909/J/4/3.
23. CRO, Buglawton & Eaton Census Returns, 1841.
24. SRO, Plans for corn mill of 1840, D(W) 1909/E/21/7.
25. Documents held by Congleton Museum Trust.
26. Guildhall Insurance Documents, MS 11937/281 No 1395110.
27. *London Gazette*, 19th November 1841.
28. *London Gazette*, 27th September 1842.
29. Bagshaw's Directory of Cheshire, 1850.
30. SRO, Letter of 1855, D(W) 1909/K/3/1.
31. SRO, Lease of 1857, D(W) 1909/J/4/4/2.
32. Morris's (1874), Slater's (1883), & Kelly's (1892), Directories of Cheshire.

33. Job, B., "A Mill Explosion at Macclesfield, Cheshire", *Wind and Water Mills*, 21, 2002.
34. *The Miller*, 7th Nov. 1881.
35. *The Miller*, 3rd September 1883, pg 435.
36. *The Miller*, 3rd Sept 1883.
37. Day, A. R., "Congleton's Lost Tobacco Industry," *Journal of the Congleton History Society*, Vol. 4, 1980.
38. SRO, Lease of 1920, D(W) 1909/A/8/20.
39. SRO, Agreement, 1922, D(W) 1909/K/3/7.
40. Records held by Gilbert Gilkes Ltd, Kendal.
41. *Congleton Chronicle*, 28th January 2000.
42. SRO, Letters, D(W) 1909/K/3/11 & 13
43. Patent 213,723, 1924.
44. Brevet d'Invention No. 713529, 1931; Patentschrift No. 564051, 1932; U.S. Patent 1850692, 1932.
45. Patent 511,410, 1939.
46. SRO, Letters, D(W) 1909/K/3/32
47. *Congleton Chronicle*, 28th January 2000
48. SRO, Letter, D(W) 1909/K/3/15
49. SRO, Sale Plan & Inventory, 1945, D(W) 1909/K/3/22

---

20. EATON BANK MILL   SJ 867639

As the River Dane runs south west from the Havannah towards Congleton it runs close to the centre of Buglawton on its eastern bank. This used to be an industrial township with a considerable number of manufactories, many of them waterpowered. However, the first factory downstream of the Havannah is on the western bank of the river at the bottom of Eaton Bank.

It is not known when the mill here was first established. It does not appear on Moorhouse's map of Congleton published in 1818 and it is first mentioned as part of the Broadhurst family's estate when they became bankrupt in 1829. At this time the Broadhursts owned a number of silk mills and other property in the Buglawton and Congleton district. The mill was described as a four storey silk mill, 124 feet by 27 feet 6 inch, together with a boiler house and two counting houses. The boiler house must have been used for heating rather than for power as the mill was described as having "*a waterwheel supplied from the rapid River Dane from which an almost unlimited supply of power is available*". The weir across the river was specified to have a height of 3 feet 8½ inch above the next weir downstream at Washford Mill (*q.v.*). As part of the lot on offer there was also a wooden bridge over the river which any prospective purchaser would have to maintain. Obviously the mill must have been built some time in the years between 1818 and 1829.[1]

Most of the Broadhurst estate was purchased by the Wallworth brothers, Matthew and Charles. Charles Wallworth was listed as a silk throwster at Eaton Mill in 1834, whereas his brother Matthew was more directly involved with Washford Mill.[2] However, by 1838 Charles Wallworth was attempting to sell the property which he described as "*newly built*". The sale advertisement gives some indication of the scale of silk production at Eaton mill by stating that the machinery comprised of 2146 swifts and 1020 cleaning bobbins. Charles Wallworth had obviously made some investment in the property since its purchase in 1829 because the sale advertisement also states that the factory was lit by gas and turned by a powerful waterwheel and a steam engine.[3]

It is not clear what happened at this sale in 1838. Matthew and Charles Wallworth were listed as both owners and occupiers in the Eaton tithe apportionments of 1840 but in November of that year they were attempting to sell the property again. This time a certain H. Hogg is also mentioned as being an interested party in the sale and Messrs Joseph Bridgett & Co were operating as silk throwsters at the mill.[4] It is possible that the Wallworths raised a mortgage on the property from Henry Hogg, a well known silk mill owner in Buglawton, and gave up silk throwing themselves at Eaton Mill in favour of renting the mill to another throwster. This attempted sale was probably unsuccessful as it was advertised for sale again in February 1842 when Messrs Bridgett & Co were still the tenants. These sale advertisements provide some additional details about the mill, as by this time the machinery consisted of 824 dozen spindles and the steam engine assisting the waterwheel is stated to be a 8 h.p. beam engine made by Walkers of Burslem.[5]

By 1848 Messrs Joseph Bridgett & Co. had relinquished the tenancy at the mill in favour of David Clarke, also a silk throwster who employed a William Ford as his manager.[6] In 1849 the then only surviving Wallworth brother, Matthew, died (Charles having died in 1846), and the mill was again offered for sale by his son William. This sale was again unsuccessful and David Clarke continued renting the silk mill until 1852 when he gave up the tenancy. This prompted William Wallworth and Henry Hogg to try and recover their capital again through selling the mill, even adding a number of cottages to sweeten the package, all to no avail. Their final attempt came in 1853 without achieving any more success than previously.[7] The failure to sell the mill probably caused William Wallworth to try a different strategy to exploit his ownership of the mill by going into partnership, as Wallworth & Daniel, to operate the mill as silk throwsters. This state of affairs lasted until around

63

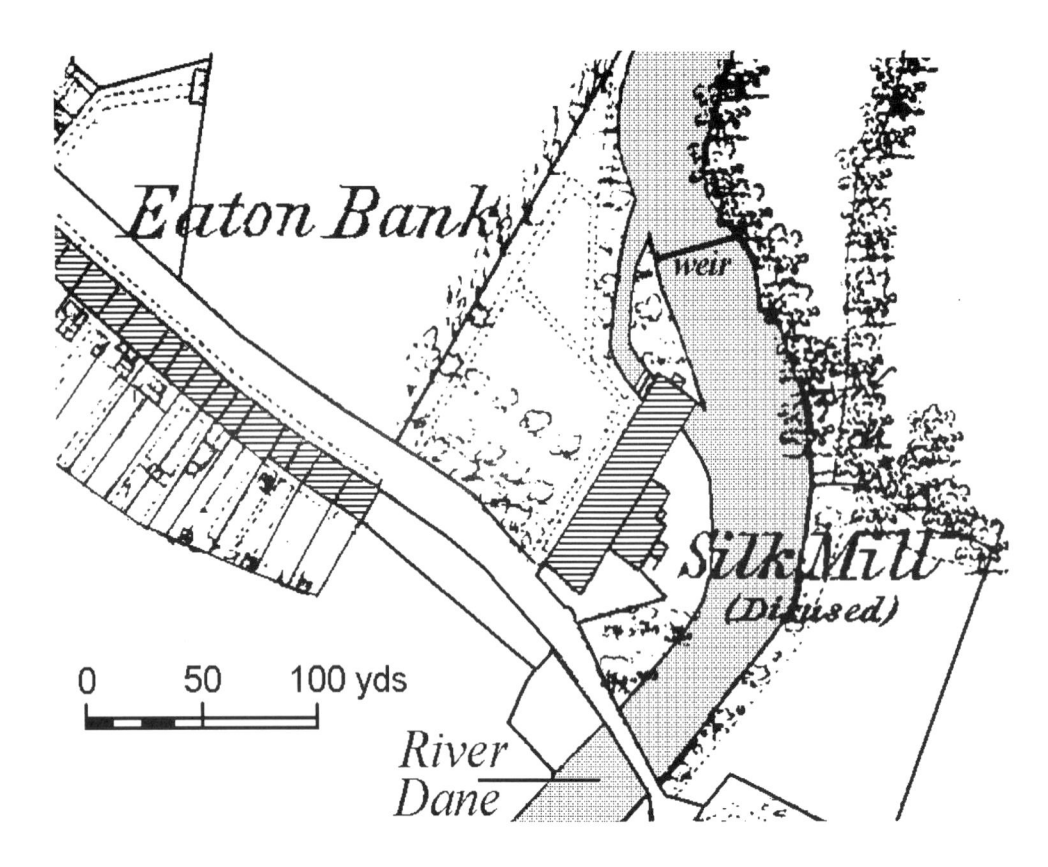

FIGURE 59. ORDNANCE SURVEY MAP, 1875, WITH EATON BANK MILL, WEIR AND LEAT TOGETHER WITH THE ADJACENT FOOTBRIDGE OVER THE RIVER DANE.

1870, when Elmy & Co took over part of the silk throwing at the mill with John & Charles Thornton working the other part. The Thorntons were silk lace manufacturers in Nottingham who were operating both the silk mills at the Havannah (*q.v.*) just upstream at this time.[8]

After 1860, though, the silk industry was in terminal decline due to the passing of the Free Trade Act, with many of the steam powered silk mills closing down altogether during the 1860s. In this kind of financial climate it is only surprising that it took until the 1880s for silk throwing to yield to a new manufacturing process at Eaton Mill with the arrival of Carson & Bradbury as paper manufacturers.[9] Although this may seem to be a complete change of manufacturing process from the silk industry, it was, in fact, a spin-off from the making of textiles in that Carson & Bradbury were initially involved in producing Jacquard cards for the control of weaving looms. From this start in the paper industry they branched out into other applications of paper and especially cardboard, mainly for use as packaging. This proved to be a very successful business with the firm occupying Eaton Mill until well after World War II, expanding into other premises in the neighbourhood before finally moving to a purpose built new factory on the site of Bank House, alongside the main Congleton to Buxton Road, in Buglawton.[10]. One interesting news item from 1935 reported the death of one of the workers at Eaton Mill. He had been caught up in the lineshafting in the mill which, apparently, was still driven by the waterwheel and sometimes by electricity. This late date for the use of such an industrial waterwheel is no doubt due to the late arrival of mains electricity in the Congleton area in the 1930s.[11]

Today there is no sign of Eaton Mill as the site is now part of the modern Eaton Bank Trading Estate but the weir across the River Dane is still to be seen, no doubt still 3 feet 8½ inch in height.

References

1. *Macclesfield Courier*, 21st November 1829.
2. Pigot's Directory of Cheshire, 1834.
3. *Macclesfield Courier*, 24th March 1838.
4. *Macclesfield Courier*, 24th October 1840.
5. *Macclesfield Courier*, 26th February 1842.
6. Slater's Directory of Cheshire, 1848.
7. *Macclesfield Courier*, 7th February 1852; and 29th October 1853.
8. Directories of Cheshire, 1857, 1960, 1864, 1869, 1874, 1878.
9. Slater's Directory of Cheshire, 1883.
10. Kelly's Directories of Cheshire, various years from 1892 to 1939.
11. *Congleton Chronicle*, 1935.

21. WASHFORD MILL SJ 865637

As the river flows in a south-westerly direction from Eaton Bank towards the town of Congleton it passes through the township of Buglawton, which was an independent borough until 1935 when it merged with the neighbouring borough of Congleton. In Buglawton, on the river's left (or south-eastern) bank is the building called Washford Mill or Dane Mill. In the late 20th century the mill consisted of a central section with a long wing on either side.

Fortunately information concerning the early history of Washford Mill has survived. In 1775 and 1803, surveys were made of Buglawlon and Congleton, presumably to assist in the assessments for the Land Tax. Contained in the 1775 survey there is a sketch map of the area around Washford Mill that is entitled "The Flint Mill Estate". It is evident from this map that at this time Washford Mill consisted of just the central section without any extensions. From the information attached to the 1803 survey, entitled "Flint Mill Situation", it would appear that the land was purchased by Philip Antrobus in 1751 for £40 and that sometime between then and 1775 he built Washford Mill for about £400. Certainly in 1775, according to the survey, it was a flint mill, presumably used to grind flint for use by the pottery industry, and was probably built for this use in the first place.[1]

According to the list of transactions appended to the 1803 survey, Antrobus sold Washford Mill to J. Booth in 1786 for about £400.[2] In Tunnicliffe's Directory of 1789 a John Booth of Congleton was described as a cotton manufacturer.[3] The survey map of the Washford Mill area in 1803 shows that the mill then consisted of the original centre section plus a small wing to the south-west which covered about the same area as the centre section. In 1790 John Booth sold the mill for £400 to someone called Burne who owned the mill for 12 years, but unfortunately, nothing is currently known about Burne or his business at Washford Mill. In 1802 Burne sold the mill to Wedgwood for £459, a name that suggests that the mill may still have been grinding flint for pottery making at this time, but there is no indication as to whether this was the famous pottery firm or some other less prominent Wedgwood. Wedgwood re-sold the mill in the same year but appears to have made a significant capital investment in the mill as the list of transactions shows:-

"*Wedgwood sold it to Broadhurst for................ £1000 having erected two new wares [sic] so he lost by it as they cost each abt. £300, sold in 1802.*" [4]

The "*wares*" mentioned in this statement probably refer to weirs, but which two weirs and why, is unknown.

The Broadhurst family, who bought the mill in 1802, did not have a background in flint grinding or the pottery industry, in fact John Broadhurst was listed in 1789 as a breadmaker working in Congleton. It is probable that the Broadhursts converted the mill to corn grinding to produce flour for their established bakery business. By 1813 Jonathan Broadburst claimed to own "*a corn mill in the said township of Buglawton*",[5] and in the 1828 directory he is listed as a miller at Buglawton. Having acquired the whole of Washford Mill they also set up a silk throwing business, presumably in the small south-west wing, where they were listed as the firm of J. & J. Broadhurst (Jonathan & James) in 1828.[6]

Unfortunately, this tale of ownership is not confirmed by the land tax returns themselves. According to the tax records, the mill was owned and occupied by Thomas Johnson from before 1782 to 1787 and from 1787 until the end of the tax records in 1831 the owner was William Johnson. Various occupiers were recorded such as Thomas Amson from before 1782 to1785, William Shufflebottom from 1790 to 1793, John Reece from 1704 to 1797, and Randle Wallworth in 1798 and 1799. The only common point between the tax records and the information in the "Flint Mill Situation" in the surveys occurs when they agree that John Broadhurst was the occupier after 1802. These discrepancies may well be due to complicated sub-letting, but the evidence does confirm that the Johnson family were the original landowners for much of this part of Buglawton and Congleton.[7]

In the first half of the 19th century the Broadhurst brothers profited greatly from their use of Washford Mill, re-investing in the purchase of other silk mills in the area, and in other pieces of property in the neighbourhood. At this time the silk industry was oscillating between periods of boom and recession, and the over-exposure of the Broadhurst fortunes to the silk industry led to their bankruptcy in the depression of the late 1820s. In 1829 the sale of their whole estate listed nineteen separate auction lots including the King's Head public house in Lawton Street in Congleton, eighteen cottages in Congleton and Buglawton, a smithy, an allotment on Congleton Moss, a silk factory and seven cottages occupied by Matthew and Charles Wallworth, Eaton silk mill, Lower Washford silk mill with sixteen cottages and a dye house, seven various lots of land in Buglawton, and also Washford Mill which was described as a corn and silk factory with the corn mill having five pairs of stones and the silk factory was unoccupied but "*recently erected and is most advantageously situated for water turning*".[8]

Most of the Broadhursts' estate was purchased by their tenants at one of the silk mills, namely the

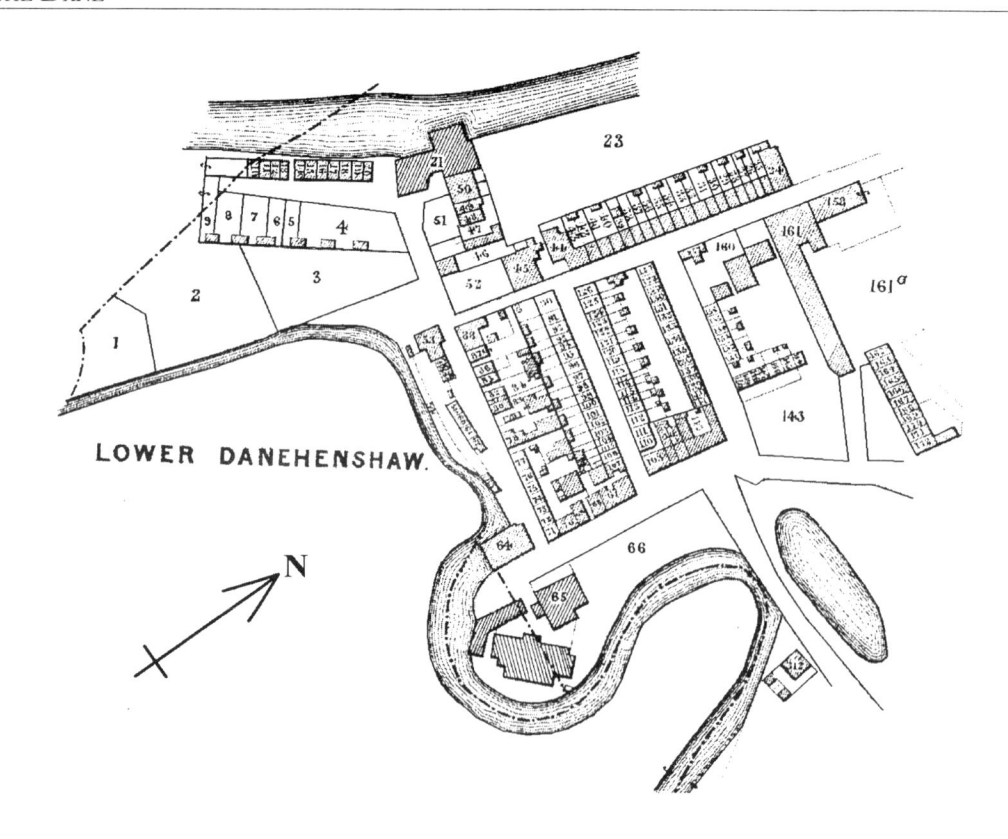

FIGURE 60. PART OF THE BUGLAWTON TITHE MAP, 1840. WASHFORD MILL WITH ITS SMALL SOUTH-WESTERN WING IS SHOWN ON THE RIVER DANE AT THE TOP OF THE MAP. DAVENSHAW MILL (*q.v.*) IS ALSO SHOWN INSIDE ONE OF THE LOOPS OF THE DANEINSHAW BROOK NEAR THE BOTTOM OF THE MAP. (CCLS, EDT 75/2)

Wallworth family. According to the deeds of Washford Mill the mortgagees transferred the mill to Randle and Matthew Wallworth in August 1830 and in December of that year the Wallworths raised a mortgage on the property from Samuel Higginbotham who was acting in trust for Messrs. Brocklehurst.[9] In 1834 Thomas Wallworth was listed as a miller at Buglawton and the 1840 tithe apportionment shows that the mill was then owned by Charles and William Wallworth, who carried on a silk throwing business there, and occupied by Thomas Wallworth.[10] The tithe map shows that at this time Washford Mill still consisted of the centre section plus a small wing of similar size to the south-west, as shown in 1803.

In the same year that the tithe apportionments were recorded, Washford mill was again offered for sale as both a corn and silk mill. By this time the corn mill had six pairs of stones and was "*turned by a powerful waterwheel on the River Dane and is now in full working order*", whereas the silk factory was "*fitted up with proper machinery for carrying on the silk throwing business and is turned by the waterwheel and steam engine jointly*".[11]

This advertisement appears to have been unsuccessful as the Wallworths were again trying to sell the mill in 1842 when the advertisement confirmed that there were two waterwheels.[12] This may have been more successful because in 1844 the deeds show that Washford Mill was conveyed from the late Randle and Matthew Wallworth to John Johnson who owned a steam powered silk mill nearby in Buglawton called Throstle's Nest Mill and who lived alongside this mill at Throstle's Nest House. The plan prepared for this conveyancing shows that the mill still consisted of the centre and small south-west extension and was still operating as a silk and corn mill. The north-eastern part of the corn mill was labelled as the "old flint mill". The owner of the mill had a right of access over the old ford just upstream of the mill for the repair of the mill's weir from the other side of the river.[13] It is likely that John Johnson bought Washford Mill, recovering a previous family holding, and providing an investment to be let to tenants rather than to work the mill himself. In 1850 a Mr. Diggory was a tenant at the mill employing 60 hands at silk throwing.[14]

In 1850 John Johnson, as the owner of Washford Mill, instructed solicitors to act against Richard Latham at Lower Washford Mill (*q.v.*) for raising his weir so causing back water to affect the waterwheel at Washford Mill. The amount of backwater in question was 7½ inches. A plan prepared for this case shows that at this time Washford Mill consisted of the central corn mill powered by a single internal low breastshot waterwheel approximately 14 feet in diameter by 9 feet 6 inches wide together with a new building replacing the earlier small south-eastern wing.[15] When John Johnson advertised that the corn mill at Washford was

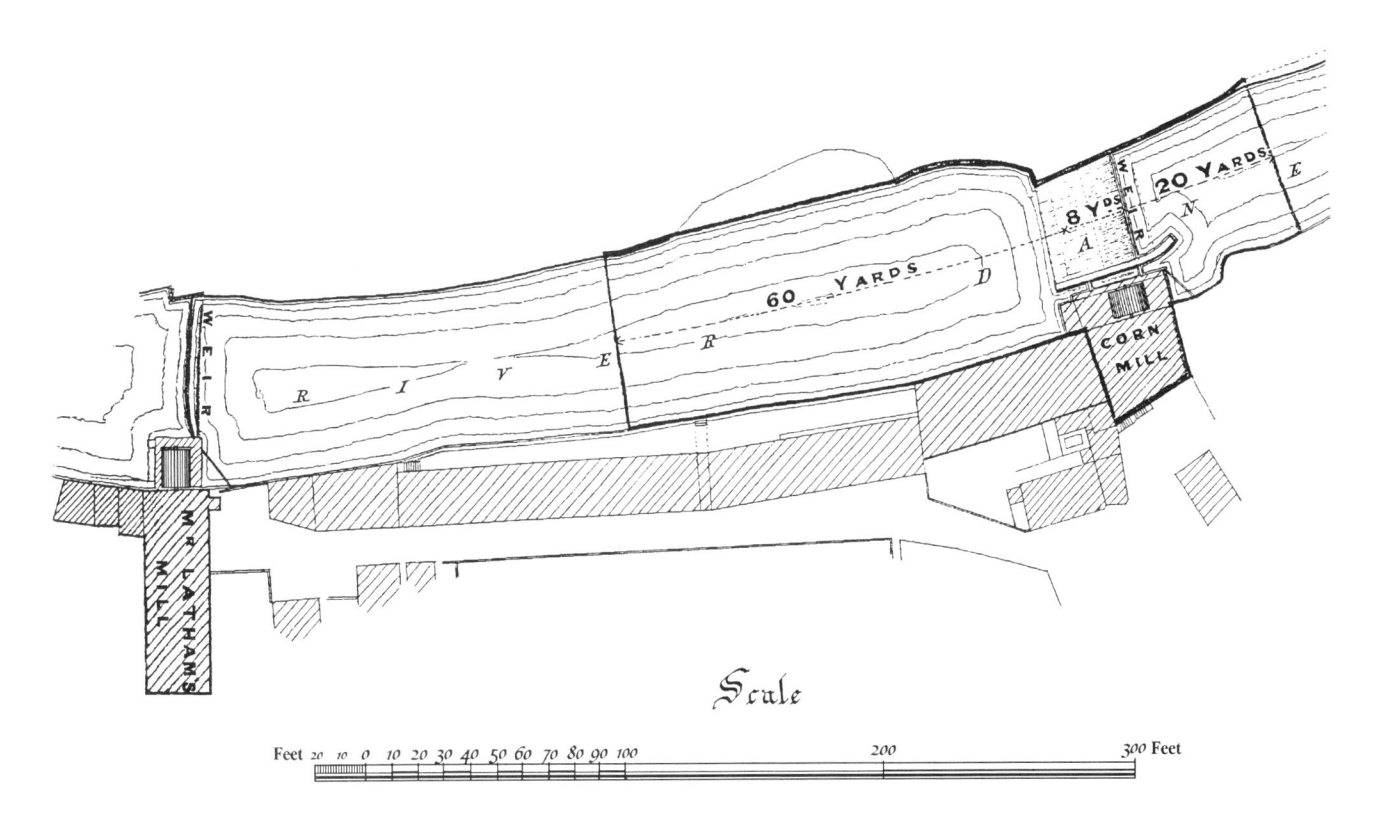

FIGURE 61. THE MAP DRAWN IN 1850 FOR THE LEGAL DISPUTE OVER THE WEIR HEIGHT AT LOWER WASHFORD MILL (*left*) CAUSING TAILWATER TO AFFECT THE WATERWHEEL AT WASHFORD MILL (*right*). NOTE THAT THE SOUTH-WESTERN WING IS A DIFFERENT SIZE FROM THAT SHOWN ON THE TITHE MAP. (*Private deposit on loan to Congleton Town Museum*)

to let in 1852, the miller was Mr. Joseph Wood.[16] Two years later, in 1854, John Johnson was again trying to let the corn mill, extolling the fact that the mill "*had just been repaired and a new waterwheel put in*".[17] This advert must have been unsuccessful thereby causing John Johnson to decide that Washford Mill would be a better investment if it reverted to its original occupation of flint grinding. Consequently he advertised the contents of the corn mill for sale eight months later, which provides an interesting inventory of the main machinery in the mill at that time, as follows:-

"*To be sold at public auction. All the excellent machinery, French Buhr [sic] stones, driving gears, straps, etc. comprising:-*
Cast iron vertical shaft 12 feet 6 inches by 8½ inches.
Cast iron cog pit wheel 10 feet diameter, 8 inches wide, 2¾ inches pitch.
Cast iron crown wheel 4 feet 6 inches diameter, do, do.
Cast iron crown spur wheel 12 feet 6 inches diameter, do, do.
One pair French Buhr stones 4 feet 4 inches diameter with vertical shaft, cog wheel, and wooden boxing.
One pair ditto, 4 feet 4 inches diameter, do, do, do.
One pair ditto, 4 feet 6 inches diameter, do, do, do.
One pair grey stones 4 feet 6 inches diameter, do, do, do.
One screen complete with driving gear.
One dressing machine with 16 inches cylinder.
Hoisting tackle,
Vertical and horizontal wrought iron shafting with bevel wheels, seven pulleys, leather driving straps.
The mill is now at work and may be inspected any day prior to the sale. The sale of machinery, stores, etc. is imperative to afford room for the necessary fittings for grinding Potters' Materials."[18]

This advertisement shows that the power transmission in the corn mill used one waterwheel to drive a great spur wheel operating four pairs of millstones. It is not known how the other two pairs of millstones, mentioned in the advertisement of 1840, were driven. It is possible that they were driven from a second waterwheel but it was much more probable that any second waterwheel was exclusively used to provide power for the silk throwing business in the wing of the building.

The conversion to flint grinding in 1854 must have been of interest to entrepreneurs in the Potteries as the central part of the mill was occupied in 1857 by William Webberley, a flint, stone, and bone grinder from Longton. He appears to have other interests that kept him in the Potteries because although still operating the mill up to the late 1870s, a Joseph Pickin or Perkin was named as manager or agent at Washford.[19] After William Webberley, the flint grinding business was operated by William Edward Cartledge in 1878 and by the 1880s it was being operated by Thomas Ford and his son William.[20] It was William Ford who, in 1887, took out a mortgage of £900 from

FIGURE 62. THE CENTRE SECTION OF WASHFORD MILL WITH THE WEIR AND WATERWHEEL

the Congleton Equitable Benefit Building Society to buy Washford Mill. In 1886 Henry Barton was silk throwing at the mill but he only employed 30 hands, considerably fewer than previously employed at the mill when the industry was at its height.[21]

Fortunately the spaces provided by the two wings could be utilised by a variety of trades which led to them being occupied as "industrial units", with leaseholders occupying only one floor or even part of a floor. In 1896 the firm of W. Hoyle & Co were making turkish towels in the mill, a trade that lasted until after 1939. Also the two wings at Washford Mill were ideal for the use of fustian cutters such as Joseph Slater in 1902, Charles Henry Cobb in 1902 and 1906, and William Barton from 1912 to the mid 1930s.[22]

In the mean time the flint grinding mill in the central section of the mill had passed from the Ford family to Thomas Malkin, but his death shortly before 1910 ensured yet another change of occupier in the shape of Jonathan Hopkins who maintained the flint grinding operation at the mill until the start of World War II, as well as operating Davenshaw Mill nearby on the Daneinshaw Brook.[23] After the Second World War the mill was taken over by Mr. J. M. Goodwin who operated the flint grinding equipment until 1968 when all production was transferred to Lower Washford Mill. Even after 1968 the flint grinding machinery in the centre section of the mill remained intact for many years and the two wings were used as industrial units for a variety of tenants. This presented ample opportunity to survey both the machinery and the buildings prior to the destruction of the south-west wing by fire in 1988.[24]

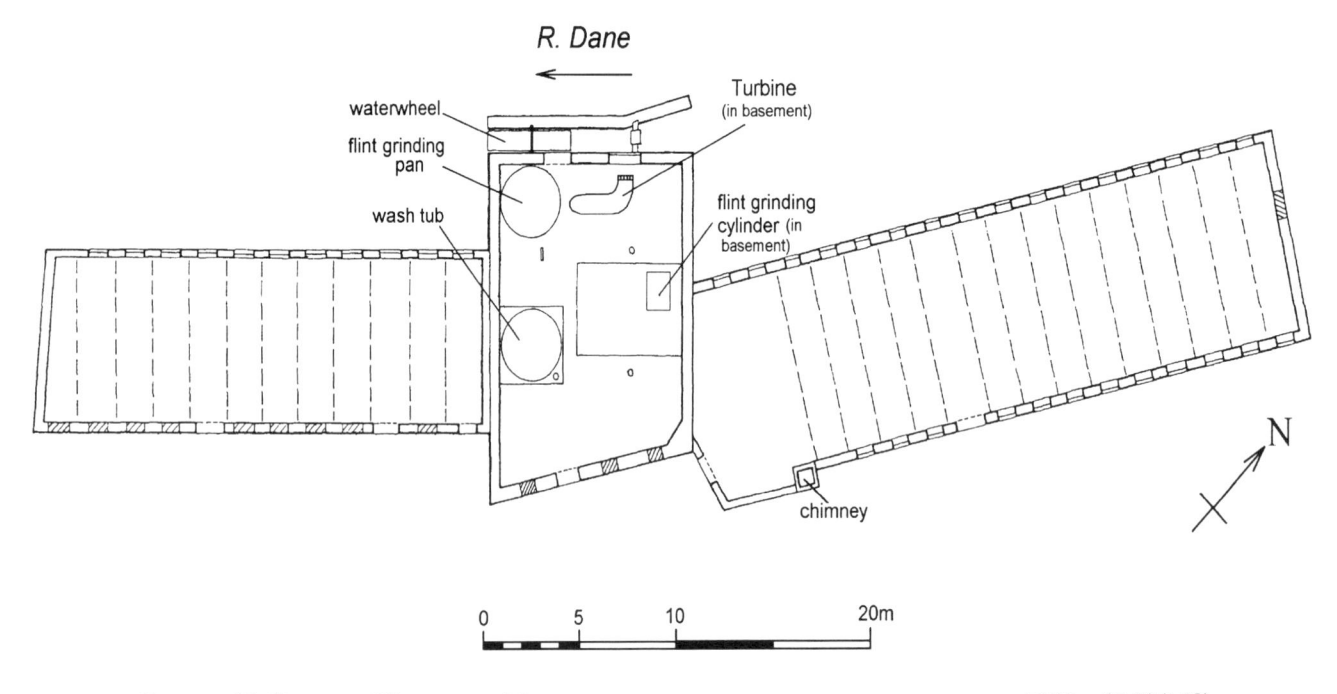

FIGURE 63. PLAN OF WASHFORD MILL WITH BOTH WINGS AS COMPLETED IN THE 1850S. (ECTMS)

FIGURE 64. WASHFORD MILL PRIOR TO THE FIRE IN 1988 WHICH DESTROYED THE NEAREST WING.

The south-west wing was four storeys high with eleven bays, being 75 feet long by 27 feet wide and the north-east wing is three storeys high with 13 bays as it is 105 feet long by 31 feet wide. Both wings and the centre section of the mill had slate roofs supported by queen post trusses. Minor differences in the brickwork, especially in the detailing of the window appertures, suggest that the two wings were not erected at the same time. These two wings were not shown on the 1840 tithe map (see Figure 60) or on the 1844 conveyancing map. However, the new south-western wing is shown on the map of 1850 (see Figure 61) confirming that this wing was built between 1844 and 1850.

At this time there is no indication of the north-east wing nor is it mentioned in the 1852 sale advertisement although there is mention of space available in the south-west wing. Although the north-east wing does appear on the 1875 large scale Ordnance Survey map of the area, identified as a silk mill, it would be amazing if this wing was built after 1860 because the Free Trade Act, passed by parliament that year, discouraged further investment in the British silk industry. Although steam power had been introduced before 1840, this was in the small wing to the south-west. Whenever the north-east wing was built a new steam plant was installed with its chimney at the south corner of the new wing. The level of investment and the need to provide steam power sugessts that this wing was built not long after 1852 and certainly before 1860.

The central section of the mill is not quite square, being 35 feet long and about 40 feet long wide on one side and 50 feet wide on the other side. There are three storeys above street level and a basement below. The lower part is built of stone up to about the first floor level and the rest of the building is made of brick with stone quoins at the second floor level. The north-west end of the building extends about 12 feet over the river, with the former wheelhouse below, and beyond this is a channel with a second, external, waterwheel still in place. Beyond this channel a stone weir extends across the river, giving a head of about 6 feet. Prior to 1999 the basement and ground floor contained the stone grinding equipment with a large opening in the ground floor to give ready access to the basement, but the upper floors were derelict.[25]

The breastshot waterwheel is made entirely of iron with a round shaft and is about 14 feet in diameter and 4 feet wide with 30 buckets. There is no maker's name or date, though it has been suggested that it may have been made by Booth's of Congleton.[26] It is located at the end of the weir with a control sluice and a short headrace. The waterwheel is placed near the downstream end of the mill wall, presumably so that its shaft was clear of the internal waterwheel that was mounted in the wheel chamber beneath the end of the mill. The outer bearing of the waterwheel is on the wall separating the wheelpit from the weir and a wooden staircase leads up from here to a door into the ground floor of the mill. Just outside this door is a handwheel controlling the sluice which admits water to the wheel.

The wheelshaft passes through an opening in the wall of the old wheelhouse and the inner bearing is carried on an iron frame which also carries the footstep bearing of the grinding pan shaft. The two shafts are linked by iron bevel gears which give a slight speed reduction. The position of these bevel gears may be adjusted so that they can be put out of mesh, though not while the machinery is running. A door from the mill basement gives access to these gears and bearings. The grinding pan shaft passes up through the mill floor and the grinding pan, and is located above the pan by

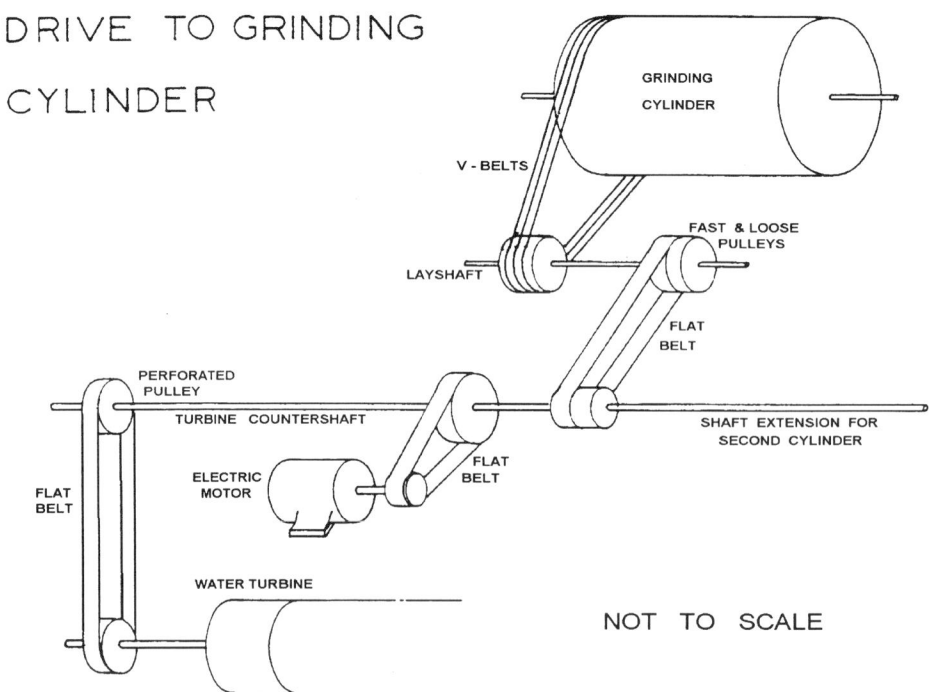

FIGURE 65. A SCHEMATIC DRAWING OF THE TRANSMISSION SYSTEM FROM THE TURBINE TO THE GRINDING CYLINDER IN THE CENTRE SECTION OF WASHFORD MILL UNTIL 1999. (*A. C. Bradley*)

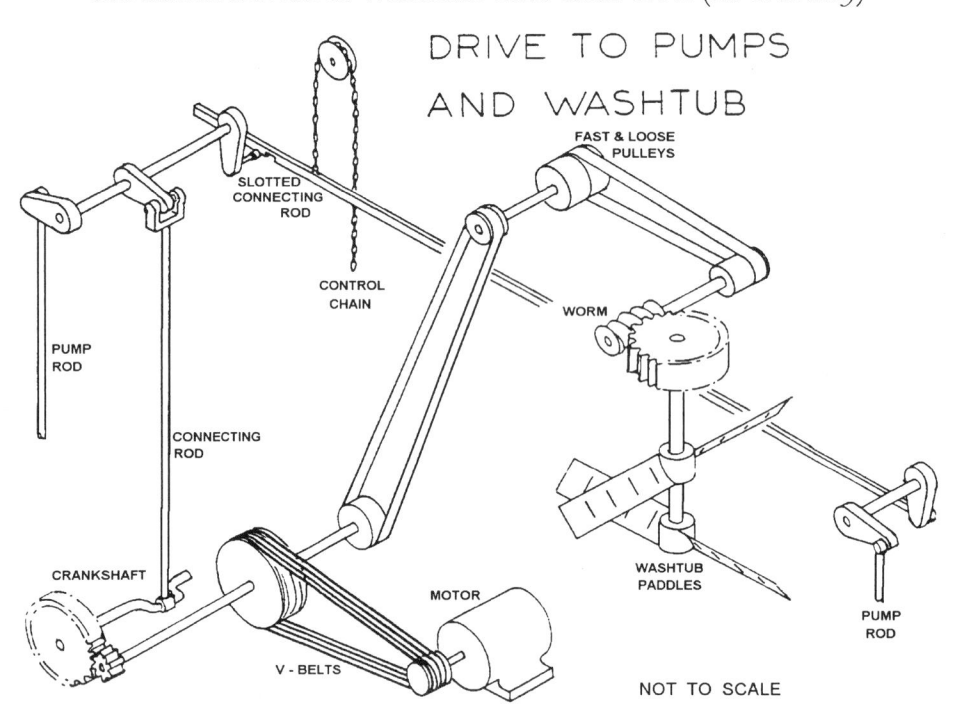

FIGURE 66. A SCHEMATIC DRAWING OF THE TRANSMISSION SYSTEM TO THE PUMPS AND WASH TUB IN THE CENTRE SECTION OF WASHFORD MILL UNTIL REMOVED IN 1999. (*A. C. Bradley*)

a bearing carried on iron brackets attached to the wall. There is an extension to the shaft supported by another bearing. This is probably a remnant of the former drive from this shaft to the pumps and washtub.

On the ground floor of the mill is the grinding pan made of of steel plates. It is about 10 feet in diameter and 3 feet deep with a small central sleeve around the pan shaft. There are four radial sweep arms, slightly curved, attached to the pan shaft through a device which allows the set of arms to slip on the pan shaft if the arms meet sudden resistance, and thus avoid damage. A cock in the side of the grinding pan, at the level of its floor, allows slop (ground material and water) to be run into a wooden channel beneath the floor which leads it into a holding tank in the basement.

The internal waterwheel was replaced by a water turbine in 1936 after a child had been killed by the wheel, though this was not necessarily the reason for

FIGURE 67. THE GRINDING PAN AND ASSOCIATED CRANE ON THE GROUND FLOOR, WASHFORD MILL.

FIGURE 68. THE GRINDING CYLINDER IN THE BASEMENT WITH ITS LOADING HOPPER ON A SHORT RAIL TRACK ABOVE.

the change. A 27 inch diameter Series R Francis turbine installed by Gilbert, Gilkes & Gordon Ltd. in 1936 produced 15.82 h.p. when used on the six feet head available. This turbine consumed water at the rate of 1700 cubic feet per minute when running at its rated speed of 126 r.p.m.[27] It was designed to run submerged, so two transverse concrete walls were built across the old wheelpit to divide it into a headrace, a turbine pit and a tailrace into which the turbine draught tube projects below water level. A sluice gate, controlled by a handwheel and worm gear just above it, admits water into the turbine pit, which is normally flooded to weir level well above the turbine. Water flows straight into the turbine from this pit. The turbine speed was controlled by inlet vanes linked by rods and cranks to a handwheel on a pillar in the ground floor of the mill above the turbine, until one of the links broke and immobilised the control mechanism. The turbine, which is an inward flow type and is mounted with its axis horizontal and at right angles to the river bank (i.e. parallel with the waterwheel shaft), is now almost covered by silt. The draught tube runs horizontally towards the mill wall, then bends through 90 degrees to pass through the concrete wall into the old tailrace where it is now hidden by silt.

The turbine was connected via belts, pulleys and shafts to drive a grinding cylinder, or ball mill, situated in the basement of the mill. The system was designed so that a second grinding cylinder could be added, if required, at a later date. The 9 feet diameter washtub was 4 feet deep and its vertical shaft carried slotted radial arms to stir the contents. The washtub, which was located 3 feet above the ground floor, was driven by a number of belts and pulleys and, although it had no nameplate, was said to have been made by the Manor Engineering Company.[28] In the basement there were two brick built arks used as the holding and settling tanks. The holding tank was about 10 feet by 8 feet, and the settling ark about 20 feet by 15 feet. They were both about 5 feet deep and of irregular shape. On one side of the settling tank there was a "plug plank", an iron plate with a vertical row of plugs stopped by wooden plugs. These were removed one by one to allow water to run to waste from above the slop as the ground material settles through it.

There were also two mechanical pumps, one of which lifted slop from the holding tank in the basement to the washtub and the other lifted thickened slop from the settling tank and discharged it through a trough just outside the mill door. River water was pumped into a tank on the first floor from where it was distributed to the washtub, grinding pan and grinding cylinder. There is a small hand-operated crane, made of wood and steel, near the grinding pan which was used to lift runner stones when necessary and probably also to lift containers of stone into the pan for grinding. There was a short length of rail track, which spanned the opening in the ground floor and ran directly over the grinding cylinder. On this track was a small hopper truck used for loading the grinding pan with material to be processed.[29]

The machinery in Washford Mill had not been used since 1968, all production since then being at Lower Washford Mill. At the time of the miners' strike in the 1970s, when industry was forbidden to use electricity for two days a week, Mr. Goodwin started to prepare the machinery for use; the main task was to clear silt from the waterwheel and turbine. However, the strike ended before this was completed. Washford Mill and its machinery were intact (in spite of the fire in the wing next door) until 1999 when the central block was sold. Unfortunately at this time most of the machinery was stripped out of the mill leaving only the turbine, grinding pan and crane, notwithstanding the "protection" provided by the building's Grade II listing. The structure of the central block and north-east wing were in reasonable condition and the current owner has converted the mill space for use as offices, retaining the waterwheel, turbine and grinding pan which is the only machinery still present in the mill.

References

1. Survey of Congleton, 1775, held by the Congleton Chronicle.
2. Survey of Congleton, 1803, held by the Congleton Chronicle.
3. Tunnicliffe's Directory of the County of Chester, 1789.
4. Survey of Congleton, 1803, *op. cit.* 2.
5. CRO, Buglawton Inclosure Claim, 1813, DCB/73.
6. Pigot's Directory of the County of Chester, 1828.
7. CRO, Land Tax Returns, Buglawton.
8. *Macclesfield Courier*, 3rd October 1829.
9. Deeds of Washford Mill, held by Mr. Goodwin.
10. CRO, Buglawton Tithe map and Apportionments, 1840.
11. *Macclesfield Courier*, 24th October 1840.
12. *Macclesfield Courier*, 26th February 1842.
13. Deeds of Washford Mill, held by Mr. Goodwin; Conveyancing map of Washford Mill to Mr. John Johnson, 1844, held by Congleton Town Museum.
14. Head, R., *Congleton Past and Present*, 1887.
15. Plan of Washford Mill and Lower Washford Mill for legal dispute, 1850, held by Congleton Town Museum.
16. *Macclesfield Courier*, 4th December 1852.
17. *Macclesfield Courier*, 21st January 1854.
18. *Macclesfield Courier*, 16th September 1854.
19. Kelly's Directory of Cheshire, 1857.
20. Slater's Directory of Cheshire, 1883.
21. Head, R., *Congleton Past and Present*, 1887.
22. Kelly's Directories of Cheshire, 1896 - 1939.
23. *Ibid.*
24. Bonson, T., "The History of Washford Mill, Buglawton, Cheshire", *Wind and Water Mills*, 15, 1996.
25. East Cheshire Textile Mills Survey Records, held at The Heritage Centre, Macclesfield.
26. Wailes, R., "Water Driven Mills for Grinding Stone", *Transactions of the Newcomen Society*, Vol. 39, 1966-7.
27. Gilbert, Gilkes & Gordon Ltd., Turbine files; CRO, Gilbert, Gilkes & Gordon Ltd. Dwg Nos 4157 & 9625, D 4931.
28. Wailes. R., 1966-7, *op. cit.* 26.
29. Bradley, C., "Stone Grinding at Washford Mill, Buglawton", *Wind and Water Mills*, 14, 1995.

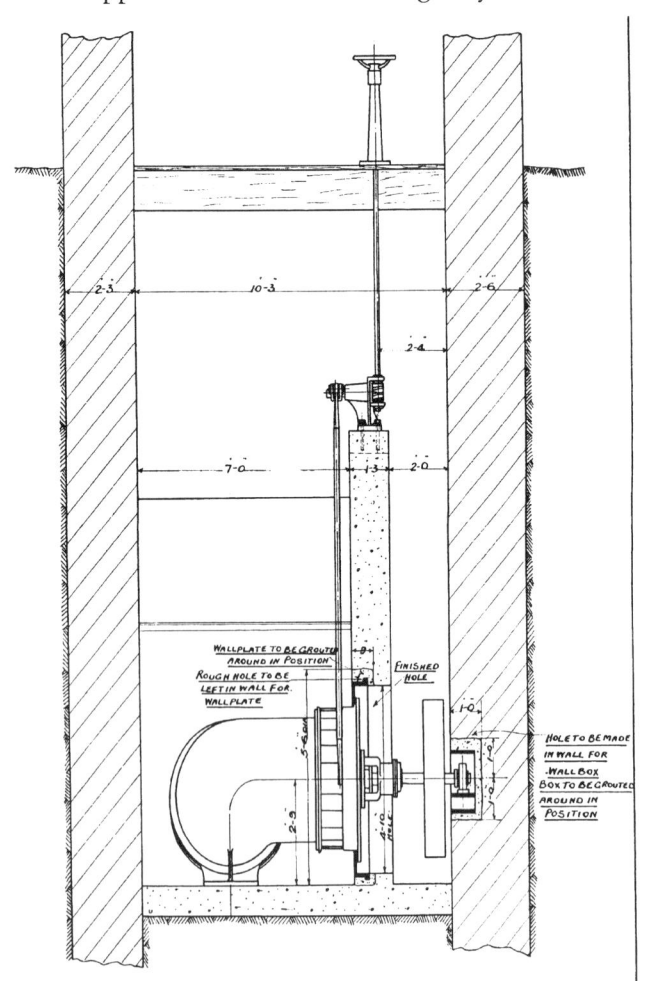

FIGURE 69. GILBERT, GILKES & GORDON LTD. INSTALLATION DRAWING FOR THE SERIES R TURBINE AT WASHFORD MILL, 1936. (CCALS, D4931)

22. LOWER WASHFORD MILL   SJ 864636

About a hundred yards downstream of Washford Mill, on the same side of the River Dane, is Lower Washford Mill. On Moorhouse's map of Congleton, dated 1818, there is no indication of a mill on this site but the area is shown as belonging to William Johnson. The mill was probably built by a John Johnson as there is definite evidence that he built the weir for the mill.[1] The date carved on the datestone in the wheelhouse wall can probably be interpreted as 1818 (the third number is not clear).[2] By 1829 when the Broadhurst estate was for sale due to bankruptcy, one of the lots was described as Lower Washford silk mill at Cutler's Croft. The mill must have been in existence for a few years at least by this time because the remains of a waterwheel were said to be attached to the mill. There were also sixteen cottages and a dye house occupied by Smallwood & Moss.[3] From the position of the mill on the river and the height of its weir, which was only about a foot, it would seem that this waterwheel must have been undershot.

No further information is forthcoming until the 1840s when the tithe apportionments for the Buglawton township show that the owner of the mill was a Richard Latham and it was occupied by Simon Krinks who was said to employ around 70 hands at this time.[4] The mill was offered for sale in 1847 which gives a reasonable description of the mill and its machinery as follows:-

"For Sale - All that silk mill or factory, three storeys high, 68 feet long by 27 feet wide with a staff room and warehouse 32 feet long by 18 feet wide with engine house and excellent condensing engine of 8 h.p. Also a wash house, gig house, stable, and other outbuildings together with 160 feet of steam piping and machinery consisting of six swift engines, 76 bobbins each; two cleaners, 84 bobbins each; two doublers, 73 bobbins each; three spinning mills, 24 dozen spindles each; three throwing mills, 10½ dozen spindles each; one throwing mill 24 dozen spindles; now in the occupation of Mr. Krinks."[5]

Of interest is the fact that silk spinning and silk throwing were both carried on at this time but in very small amounts, the total number of spindles being only around 125 dozen, making its output one of the smallest for the silk mills in the Congleton area. Also of note is that there is no mention of waterpower being used only the condensing steam engine. That waterpower was still being used at Lower Washford Mill is shown by the legal dispute in 1850 between Richard Latham, who was still the owner of Lower Washford Mill at that time and John Johnson owner of Washford Mill. This dispute concerned the allegation that Latham had raised the height of the weir at Lower Washford Mill thereby raising the river level which was adversely affecting the waterwheel at Washford Mill. The rise in question was 7½ inches. This dispute shows that although Lower Washford Mill was using steam power it was thought to be advantageous to attempt to improve the power of its waterwheel, even by such a small amount, as late as 1850.[6]

Although Simon Krinks is recorded up to 1850 as a silk throwster at the mill there is no further information about the mill until 1890 when the mill was producing turkish towels.[7] The first recorded turkish towel manufacturer was W. Hoyle but he had moved to Washford Mill by 1892 and Robert Calvert took over at Lower Washford Mill. Four years later, in 1896, turkish towels were being manufactured at the mill by

FIGURE 70. VIEW OF LOWER WASHFORD MILL FROM ACROSS THE RIVER SHOWING THE LOW WEIR AND WHEELHOUSE.

Pears Moffat & Co. who were also operating at Primrose Vale Mill (q.v.).[8]

At the turn of the century towel manufacturing ceased and Joseph Slater, a fustian cutter, occupied the mill until the start of the First World War.[9] From 1914 to 1918 the mill became an ammunition factory. The history of occupation in the mill during the 20th century is obscure until it came into the ownership of Goodwin & Sons, potters' material suppliers, who moved here from Hanley. The mill then had grinding cylinders installed together with their associated settling tanks, etc. necessary for the grinding and separation of bulk material for the manufacture of pottery. This business continued until Mr. Goodwin's retirement from the business in 2000.

The mill is a simple rectangle in shape lying with its short side parallel to the river bank, with the wheelhouse still attached to the western end of the building at the end of the small weir across the River Dane. The brick building is two storeys high, not the three storeys mentioned in 1847, with a slate roof supported by king post trusses. It is 75 feet long by 27 feet wide. The first floor is supported by cast iron pillars running along the centre of the mill which have lineshaft brackets attached to them. The lower courses of the wheelhouse are constructed of sandstone blocks and there are arched openings on either side where the waterwheel bearings were positioned. The original single storied wheelhouse has been extended upwards at some time to provide a first storey room over the wheel position.[10] This wheelhouse is about 18 feet square in plan and used to house an undershot waterwheel about 10 feet diameter and 14 feet in width.[11] This waterwheel was only removed about 1940, which would suggest that it was of mainly metal construction being removed for scrap as part of the war effort. A gas engine was also removed at this time.

There are seven grinding drums (ball mills) on the ground floor of the mill, four large ones, about six feet diameter, are driven through belts by separate electric motors. The remainder of the drums, being smaller, are all driven from one electric motor. One of the large drums is said to be the first cylinder mill installed in this country, which occurred in 1910 at a factory in Hanley and subsequently moved to Lower Washford Mill. The grinding drums are lined with Silex blocks about 4 inch by 6 inch that come from Belgium, Jugoslavia, or China. On the first floor there are three circular washtubs, two of which are driven by their own electric motor. There are a number of hoist rails

FIGURE 71. THE POWER TRANSMISSION TO GRINDING CYLINDER SEVEN AT LOWER WASHFORD MILL.

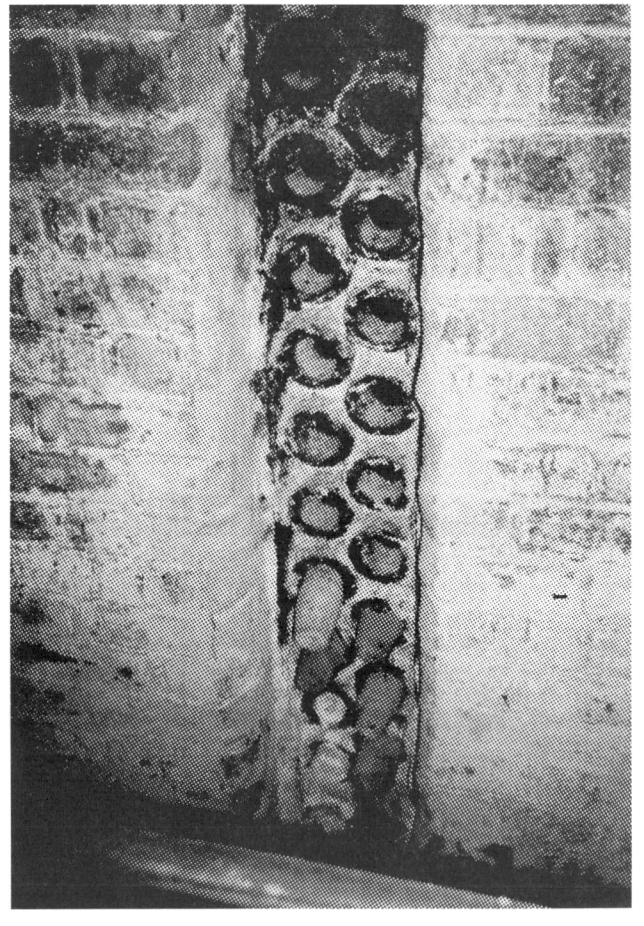

FIGURE 72. A PLUG PLANK ON ONE OF THE SETTLING TANKS OUTSIDE LOWER WASHFORD MILL.

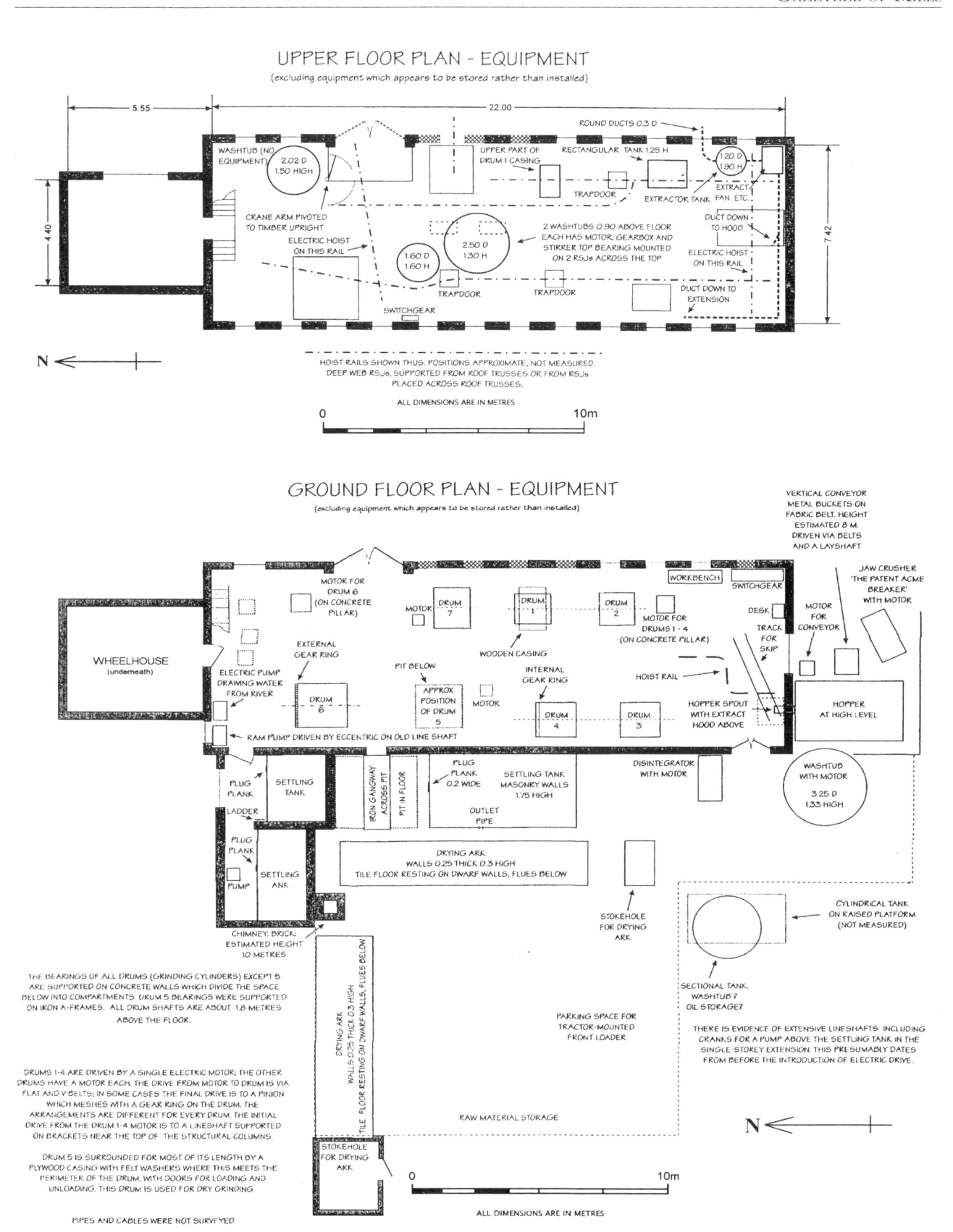

FIGURE 73. Floor plans of the two floors at Lower Washford Mill with the seven grinding cylinders on the ground floor and the washtubs at first floor level. Some of the settling tanks and drying arks were outside to the west under a lean-to roof. (*A. C. Bradley*)

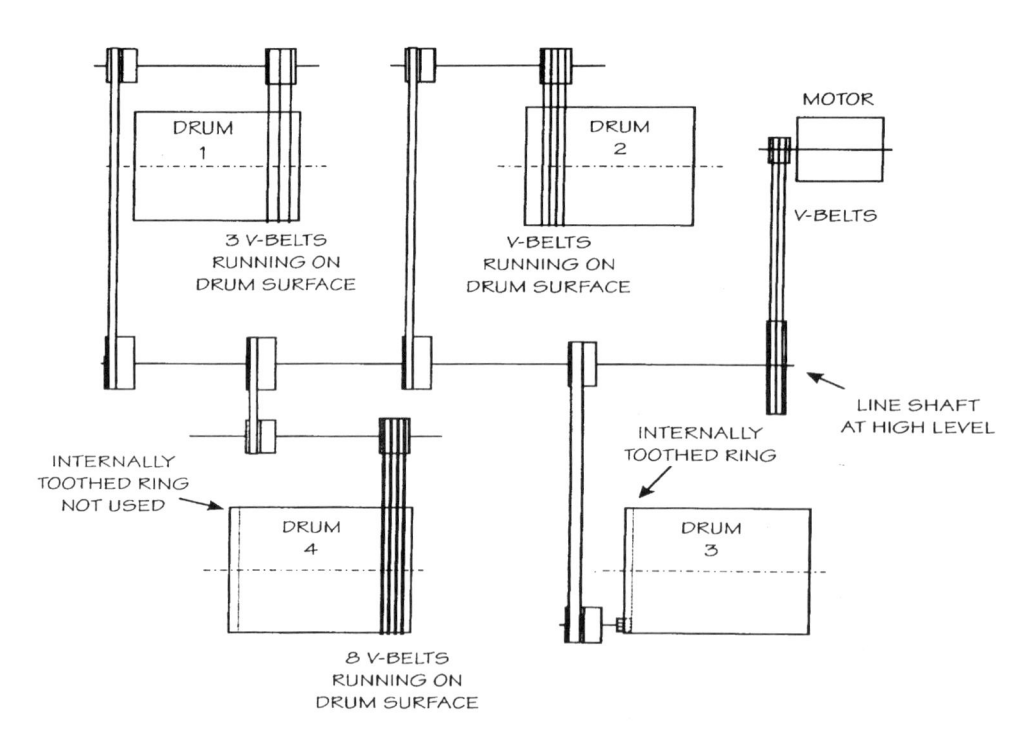

FIGURE 74. A SCHEMATIC LAYOUT OF THE POWER TRANSMISSION ARRANGEMENT OF LINESHAFTS, PULLEYS AND BELTS DRIVING GRINDING CYLINDERS 1 - 4 AT LOWER WASHFORD MILL. (*A. C. Bradley*)

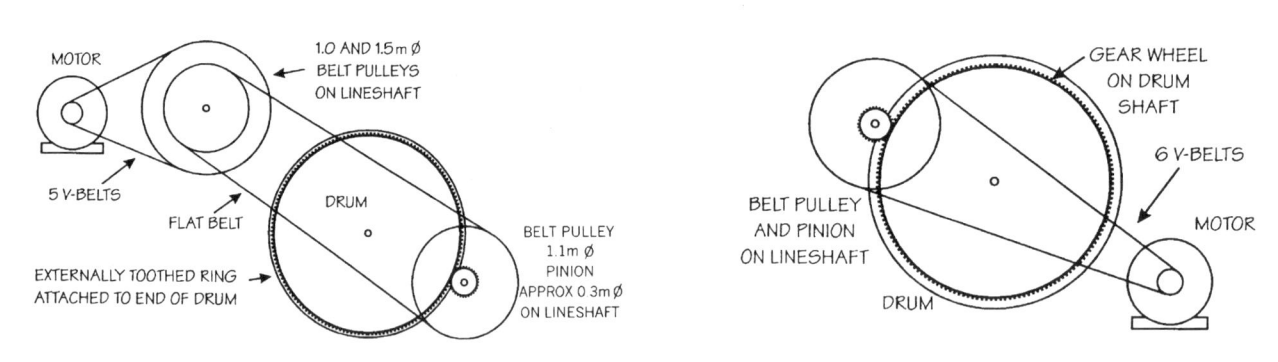

FIGURE 75. THE POWER TRANSMISSION ARRANGEMENTS FOR GRINDING CYLINDERS 6 (*left*) AND 7 (*right*). (*A. C. Bradley*)

made from RSJs to move the material around. There are also two crushers to prepare the material. One of the crushers is at the end of the mill, driven by an electric motor which also drives a bucket chain that lifts the crushed pieces into a large external hopper with a spout protruding into the mill. This fills a skip on a truck which moves the material until the skip can be lifted from the truck by a hoist which then conveys the skip to the entrance of a grinding drum. All the pumps are all electrically driven. Outside the mill on its western side there are three settling tanks with their associated plug banks and two drying arcs with their stoking holes.[12]

When Mr. Goodwin retired in 2000, the oldest flint grinding cylinder was removed to the Stoke Industrial Museum at Jesse Shirley's Etruscan Bone and Flint Mill to become an exhibit.[13] The future of Lower Washford Mill is uncertain but it will probably be demolished so that the site can be "developed" for housing.

References

1. Plan of Washford and Lower Washford Mills, 1850, held at Congleton Town Museum.
2. Bradley, A. C., Notes made on visits to the mill, 1978 - 1980.
3. *Macclesfield Courier*, 21st November 1829.
4. Head, R., *Congleton Past and Present*, 1887.
5. *Macclesfield Courier*, 18th August 1847.
6. *Op. cit.* 1.
7. Slater's Directory of Cheshire, 1890.
8. Kelly's Directory of Cheshire, 1892, 1896.
9. Kelly's Directory of Cheshire, 1914.
10. Survey of the Lower Washford Mill by A.C. Bradley, B.Job, & T. Bonson, 2001.
11. *Op. cit.* 1.
12. Bradley, A. C., *op. cit.* 2.
13. Bradley, A. C., "Potter's Milling - an early ball mill rescued", *Industrial Archaeology News*, 117, Summer 2001.

Section 3. The Daneinshaw Brook

The Daneinshaw Brook, or Biddulph Brook as it is known in Staffordshire, rises on the high ground of the watershed near Knypersley in Staffordshire. This watershed is one of the main dividing features of England, the stream on the other side of this divide is the infant River Trent flowing southward and eventually eastward to the North Sea whereas the water of the Daneinshaw Brook eventually ends up in the Irish Sea. As it flows northwards towards the River Dane the brook is joined on its eastern side by two small streams that rise on Biddulph Moor. Closer to Buglawton it is also joined by the Timbersbrook which rises on the western slope of Bosley Cloud. After considerable twisting in its lower reaches the Daneinshaw Brook joins the River Dane on its eastern side just downstream from Lower Washford Mill.

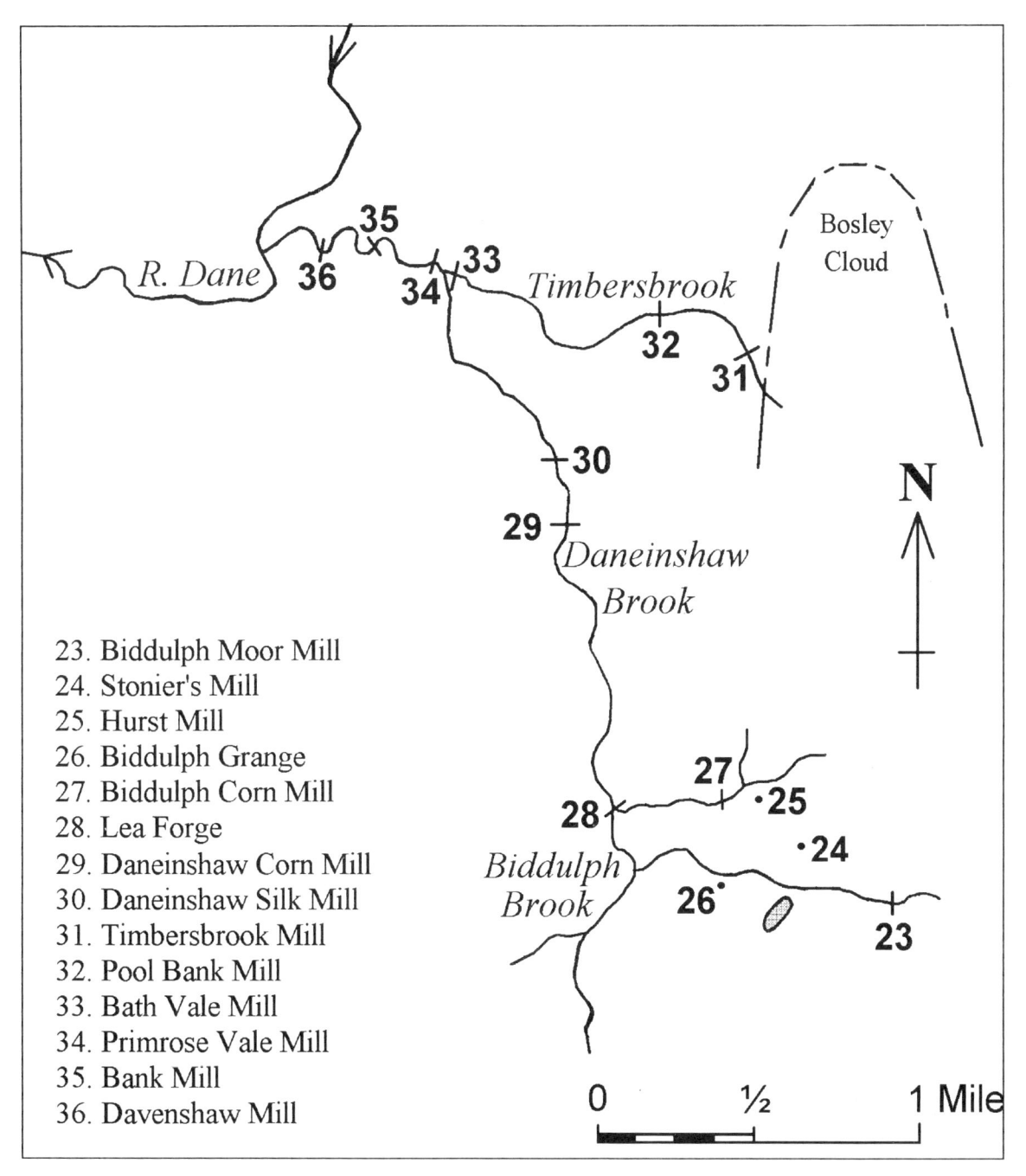

23. Biddulph Moor Mill
24. Stonier's Mill
25. Hurst Mill
26. Biddulph Grange
27. Biddulph Corn Mill
28. Lea Forge
29. Daneinshaw Corn Mill
30. Daneinshaw Silk Mill
31. Timbersbrook Mill
32. Pool Bank Mill
33. Bath Vale Mill
34. Primrose Vale Mill
35. Bank Mill
36. Davenshaw Mill

FIGURE 76. MAP OF THE DANEINSHAW BROOK FROM ITS ORIGIN ON BIDDULPH MOOR TO ITS CONFLUENCE WITH THE RIVER DANE IN BUGLAWTON INCLUDING ITS MAIN TRIBUTARY THE TIMBERSBROOK.

23. BIDDULPH MOOR MILL   SJ 903591

In Staffordshire, close to Biddulph Grange, the Biddulph Brook is joined by a small stream that rises about a mile to the east on Biddulph Moor. Not far from its source, at a point where the stream flows into a natural defile, is the site of Biddulph Moor Mill. Very little information has been found so far concerning this mill. It appears on a map produced in 1838 for a newly proposed tramway between the Macclesfield Canal in Buglawton and various mines in and around Biddulph. This tramway was intended to have many branches, one of which was to run to the stone quarries in the Biddulph Moor area where the mill is clearly shown.[1]

Two years later the mill was also shown on the Biddulph tithe map under the name of Folly Silk Mill, with Jane Daniels listed as both the owner and occupier. However, the small mill pool was recorded as being owned by William Stonier.[2] Although there is an impressive fall available at this mill site the evidence shows that this mill was extremely small. No doubt it was a very marginal proposition, built when times were good in the silk trade, but unable to compete when trade was difficult, due to its small size and remote position. Today there is no sign of the mill building itself but the outline of the pool is still clearly visible.

References

1. CRO, Map of Buglawton - Biddulph Tramway, QDP 155.
2. SRO, Biddulph Tithe Map & Apportionments.

---

24. STONIER'S MILL   SJ 899594

After leaving Biddulph Moor Mill, the stream passes through a steep wooded ravine where, some 50 yards or so downstream, there is a small weir. This weir provided the water supply for a leat on the stream's northern bank that powered two mills before returning its water to the stream. After leaving the weir, the leat branched into two water courses, the higher leat led to Stonier's Mill, the nearest of the mills to the weir. The lower leat ran below Stonier's Mill, receiving that mill's tailwater, to a pool which was used by Hurst Mill (*q.v.*) some 750 yards from the weir. Stonier's Mill was used mainly for cotton spinning whereas Hurst Mill was operated mainly as a silk throwing mill. The details of these two mills are very difficult to separate as they were both owned by William Stonier who lived at The Hurst, close by, and both mills may have been known as "Hurst Mill" at some time. It is possible that both mills were originally built by William Stonier.

Stonier's Mill was in existence in January 1813 when it was advertised to let as a cotton factory containing 1248 throstle spindles producing cotton twist. This is quite a small number of spindles for a cotton mill which reflects the small size of the mill. This machinery was said to be turned by a powerful waterwheel, with a steam engine to assist when short of water. It is perhaps indicative of William Stonier's sensibilities that he ended the advertisement in 1813 with the statement "*None need apply but people of respectability who can, if required, give security for the machinery and rent.*"[1] Later in 1813 it was advertised for sale again, giving more details of the machinery as follows:-

"*To be sold by private contract in a factory at the Hurst near Congleton.*

six throstles 168 spindles ea; and two ditto 120 spindles ea; 16 one inch rollers; one drawing frame six heads and two skeletons four heads ea; two carding engines 40 inch; one ditto 32 inch; and ditto 18 inch; one stretcher 102 spindles; one making up press; one pincher; reels, straps, cans, skips, cards, drums, etc."[2]

The steam engine had been installed in 1811 because the advertisement states "*...also a steam engine, five horse power, has worked about two years.*" Both advertisements recommended contacting William Stonier at the Hurst. Whether the mill was built with both power sources, or the steam engine was added some time after the mill was built, is not known. This date of 1811 is the earliest traceable evidence of the mill although it is possible that a previously existing building was converted to a mill rather than the mill being newly built at that time.

The mill was advertised to let again in March 1816[3] and appears to have been worked after this date by William Watts who was described as a cotton spinner in 1818.[4] It appears that William Watts was related to William Stonier but whether this relationship existed before Watts came to the mill or arose after he had been in occupation for some time is not known. On the death of William Watts eight years later the mill was once again advertised to let.[5] This was followed by a further advertisement in 1827, and in 1828 both this mill and Hurst Mill were offered to let in the same advertisement.[6] These attempts to let the mill were probably unsuccessful considering the state of trade at the time, coupled with the fact that William Stonier

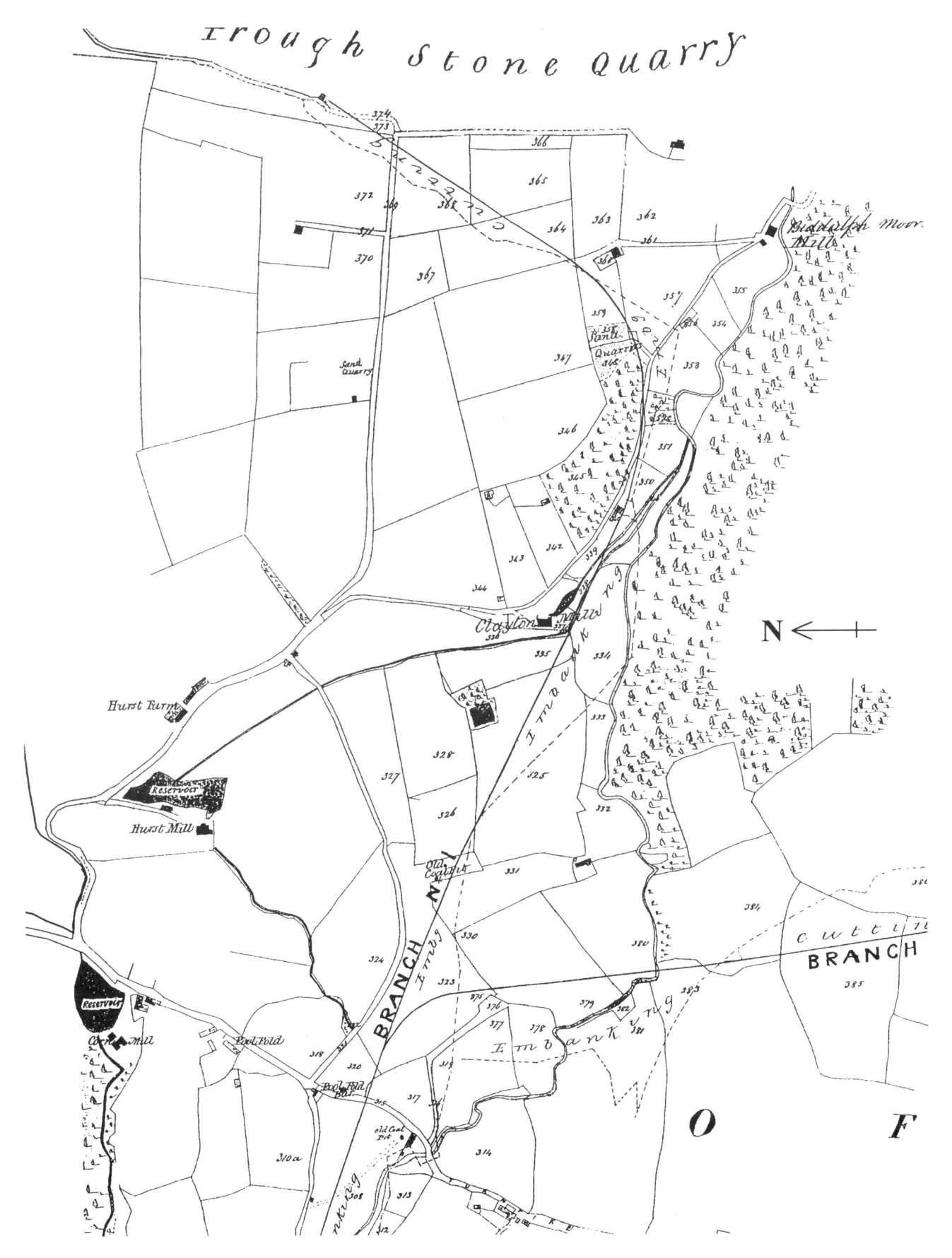

FIGURE 77. PART OF THE PLAN FOR A PROPOSED TRAMWAY IN 1838 FROM THE CANAL WHARF IN BUGLAWTON WITH BRANCHES TO COAL MINES AND QUARRIES IN THE BIDDULPH AREA. FOUR MILLS ARE SHOWN ON THIS PART OF THE PLAN, NAMELY BIDDULPH MOOR MILL, STONIER'S MILL (noted as Clayton's Mill), HURST MILL AND BIDDULPH CORN MILL. (CCALS, QDP 155)

FIGURE 78. THE REAR OF STONIER'S MILL AT THE HURST, BIDDULPH, CONVERTED TO A HOUSE AROUND 1850.

was himself registered as both owner and occupier of the mill from William Watts's death until at least 1831.[7]

A further attempt was made to sell the mill in 1830 when the mill was described as

"...a cotton mill or factory upon the Hurst estate at present unoccupied measuring 48 feet by 30 feet inside with a waterwheel of 30 feet diameter and a pool containing upwards of ¾ acre, plentifully supplied with water."[8]

Unfortunately the width of the waterwheel was not specified but its diameter is quite considerable for a mill of this small size.

In 1834 James Clayton was listed as a silk throwster at Hurst Mill (the name of the mill lower down the water supply system).[9] However, in 1838, when there was a proposal to build a tramway from Buglawton to Biddulph with numerous branches, the cotton mill was marked on the tramway map as "Clayton's Mill".[10] This gives rise to some confusion; the official deposited map for the proposed tramway is unlikely to be in error as to the occupant of the mill, so is the directory of 1834 incorrect in locating James Clayton at Hurst Mill? Alternatively, Clayton may have moved from Hurst Mill to Stonier's Mill sometime between 1834 and 1838. Certainly by 1840 when the tithe map was drawn up, William Stonier was again identified as both owner and occupier of the mill which was described as a silk mill.

It would seem that not long after the date of the tithe map the mill was converted into a dwelling house. In 1851 there was no evidence of industrial activity when William Stonier was living in the property (described as a surveyor of roads) together with his wife Martha and their eight children, plus Mary Watts, aged nine, described as a niece.[11]

Possibly because of its use as a dwelling house the building still stands today, a two storey sandstone building, seven bays long, although some of the windows are now filled in as they would have been redundant when the building ceased to be used for industry. Today only part of the leat system can be traced. The weir and the initial sections of the leats are now in Biddulph Grange Country Park and so can be followed relatively easily. From the weir the flowing water follows the course of the higher leat as used by Stonier's Mill. The first section of the lower leat to Hurst Mill is dry and has disappeared in many places. As the leat approaches the boundary of the Country Park the water is diverted from the higher leat into the lower leat and then runs across the fields to the mill pool at what was Hurst Mill. The continuation of the higher leat towards Stonier's Mill has been obliterated due to a modern bungalow being built on the site of the small pool which was adjacent to Stonier's Mill.

References

1. *Macclesfield Courier*, 9th January 1813.
2. *Macclesfield Courier*, 5th June 1813.
3. *Macclesfield Courier*, 9th March 1816.
4. Parson & Bradshaw, *Directory of Staffordshire*, 1818.
5. *Macclesfield Courier*, 30th July 1825.
6. *Macclesfield Courier*, 15th July 1826, 10th March 1827, & 5th January 1828.
7. SRO, Biddulph Land Tax Returns.
8. *Macclesfield Courier*, 7th August 1830.
9. White, Directory of Staffordshire, 1834.
10. CRO, Buglawton - Biddulph Tramway, QDP 155.
11. SRO, Biddulph Census Returns, 1851.

25. HURST MILL   SJ 897597

The leat from Stonier's Mill (*q.v.*) runs across the fields in a north westerly direction for about ¼ mile until it enters the pool for Hurst Mill. Again it is not obvious when this mill was first built but logically it must have been planned at the same time or before Stonier's Mill as their water supply arrangements are interdependent. However the mill can only be identified in the land tax returns from 1823 when it is listed as owned by William Stonier and occupied by Ash & Pyatt.[1] An advertisement offering the mill to let in 1824 states that at that time it was a silk factory and gives some indication of the size of the mill.

"To be let and entered on immediately, A new and desirable silk factory situated near the Hurst, in Biddulph, in the County of Stafford, three miles from the town of Congleton and within a quarter of a mile of the turnpike road leading into the Potteries. The factory, now in work by Messrs Ash & Pyatt, consists of three rooms 15 yards by 9 yards, is turned by a new waterwheel, with power sufficient to turn 250 doz. spindles."[2]

This offer was probably unsuccessful as Ash and Pyatt continued to be listed as the occupiers until 1826. Thereafter William Stonier is listed as both owner and occupier so presumably the mill was not in work at this time.[3] A further attempt to let the mill in 1828 appears to have been equally unsuccessful. However a new tenant was recorded in 1831 when the occupier of the mill was Paul Bailey.[4]

In 1840 William Stonier offered the mill for sale giving details of the machinery, as follows:-

"All the valuable silk machinery, steam engine of 4 h.p., water wheel, shafting, steam piping, and other effects comprising three throwing mills containing 26½ doz. spindles ea.; one ditto 15 doz. spindles; seven spinning mills 24 doz. spindles ea.; one doubling frame 84 bobbins; one ditto 72 bobbins; one ditto 98 bobbins; one ditto 36 bobbins; four hand wind silk engines 84 swifts ea.; a large quantity of frames, shaft and engine bobbins; silk press, tram and organ boxes, cleaners, silk skips, mill horses, cast & wrought iron shafting, pedestal hangers, bearers and wheels; quantity of four inch steam piping, straps, old iron, etc. Also an excellent steam engine of four horse power and a water wheel 24 feet diameter, 3 feet wide with trunk, penstock, pit wheel, etc.

The above factory is three storeys high, 14 yards long by 8 yards wide, is firmly built of stone, with a plentiful supply of water and in a good neighbourhood for hands, and the whole presents an excellent opportunity to anyone desirous of embarking in the silk business."[5]

By the time that the tithe map and apportionments were made, also in 1840, William Stonier must have successfully sold the mill because it is listed as a silk mill owned by Reverend William Henry Holt, who was the vicar of the Biddulph parish church of St. Lawrence, and occupied by Samuel Hawthorn [*sic*], who had just moved from the silk mill at Timbersbrook (*q.v.*).[6]

In 1841 the mill was still occupied by Samuel Harthan [*sic*], a silk throwster, with his wife Judith together with two sons. Nearby lived William Harthan (presumably Samuel's brother), also listed as a silk throwster, with his wife Sarah and their five children.[7] The Harthan family (sometimes spelt Hawthorn, Harthern, etc depending on the decodability of the local accent) belonged to the border land between Staffordshire and Cheshire centred around Overton, which until 1819 lay along the main turnpike road from Tunstall to Buxton via Biddulph and Bosley.

Some time before 1845 Samuel Harthan died and his widow Judith continued in business at Hurst Mill

FIGURE 79. THE SURVIVING POOL AND MILL HOUSE AT THE SITE OF HURST MILL.

as a silk throwster, together with her brother-in-law William.[8] However, before 1860 Judith decided to retire from the business and moved back to Timbersbrook where her two sons were engaged in silk throwing at the mill there. This left William Harthan and his family in sole occupation at Hurst Mill.[9] The census returns from these days show that a number of local people were employed in the silk industry and most of those listed as being workers in silk in this district must have been employed at Hurst Mill. It is noticeable that these employees were all young girls with ages ranging from ten to seventeen, with only one adult employee (a spinster) being in her twenties.

In 1862 catastrophe struck when Hurst Mill caught fire and was totally destroyed. A newspaper published at the time reported on the fire as follows:-

"Hurst Mill, a massive stone building situated at Biddulph near Congleton occupied by William Harthan, a silk throwster, was totally destroyed by fire at an early hour on Wednesday morning together with valuable contents of silk and machinery. The roof fell in and by morning the whole building was level with the ground. The mill belonged to the Rev. W. H. Holt, vicar of Biddulph, and was formerly used as a cotton mill the flooring having thus become saturated with oil. The value of the silk and machinery was £500 or £600 and was insured with the Manchester Fire Office. The building was uninsured."[10]

It later turned out that the fire was deliberately started by William Harthan's son, Edwin, who was arrested by the police for robbery and arson. The reference to the mill formerly being used as a cotton mill is puzzling as no other evidence for this use of Hurst Mill has been found.

Thus Hurst Mill came to an end in 1862. Today the water supply and the pool can be traced as can the tail water which runs south west to rejoin its original stream near to Biddulph Grange. Unfortunately the pool is now empty although it held water until the 1990s. Recently a new house has been built on the site of the mill and presumably problems with damp has caused the occupier to drain the pool. This is a shame because with the water present it was a very attractive location, as no doubt the house builder perceived.

References

1. SRO, Biddulph Land Tax Returns.
2. *Macclesfield Courier*, 10th July 1824.
3. SRO, Biddulph Land Tax Returns.
4. SRO, Biddulph Land Tax Returns.
5. *Macclesfield Courier*, 20th June 1840.
6. *Macclesfield Courier*, 2nd May 1840.
7. SRO, Biddulph Census Returns, 1841.
8. *Macclesfield Courier*, 1st March 1845; SRO, Biddulph Census Returns, 1851.
9. SRO, Biddulph Census Returns, 1861.
10. *Staffordshire Sentinel*, 22nd February 1862.

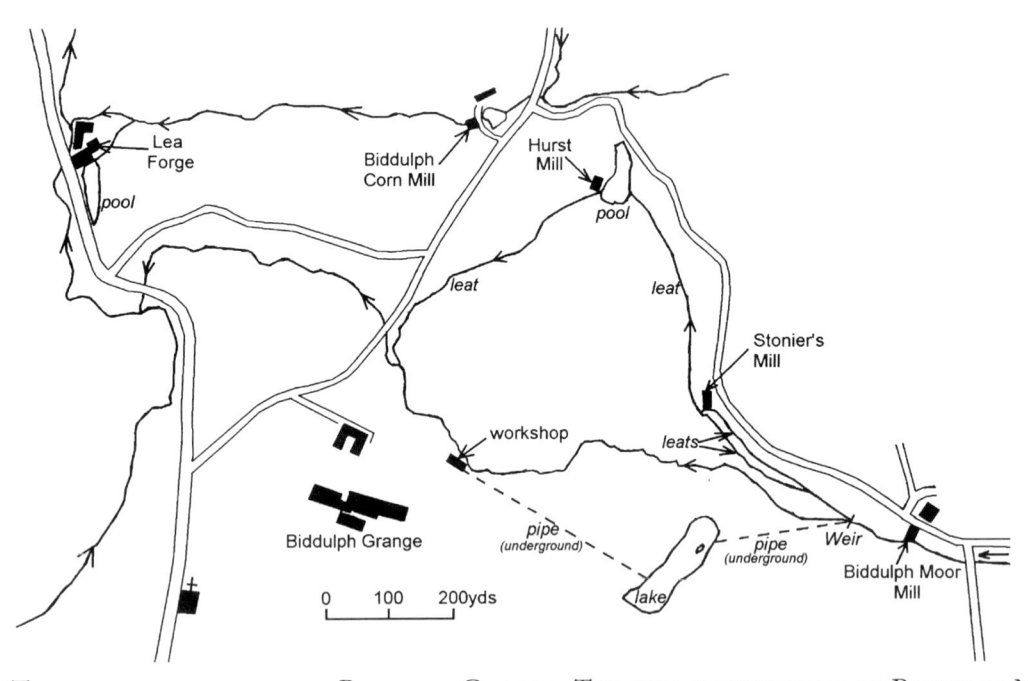

FIGURE 80. THE WATERCOURSES AROUND BIDDULPH GRANGE. THE WEIR DOWNSTREAM OF BIDDULPH MOOR MILL SUPPLIED TWO LEATS, ONE TO STONIER'S MILL AND ONE TO HURST MILL. FROM HURST MILL ANOTHER LEAT CARRIED THE WATER BACK TO THE ORIGINAL STREAM DOWNSTREAM OF THE SITE OF THE ESTATE WORKSHOP. CURRENTLY THE WEIR ALSO PROVIDES THE WATER SUPPLY FOR THE ELECTRICITY GENERATION TURBINE IN THE COUNTRY PARK VIA TWO PIPES, ONE FROM A VALVE IN THE WEIR TO THE LAKE AND ONE FROM THE LAKE TO THE SITE OF THE ESTATE WORKSHOP. A SEPARATE STREAM SUPPLIED BIDDULPH CORN MILL AND LEA FORGE.

26. BIDDULPH GRANGE　　　　SJ 893593

When Biddulph Grange was extensively rebuilt in 1903/4 by Robert Heath, a cast iron pipe, 8 inches in diameter, was inserted into the existing weir used to supply Stonier's and Hurst Mills (*q.v.*). This pipe diverted water to a lake on the high ground to the east of Biddulph Grange. This lake, as well as being a feature in the extensive grounds of Biddulph Grange, also provided water for the fire hydrants in the Grange, power for the estate workshop and also for the generation of electricity.[1]

The workshop was a small two storey building situated to the north east of the main building and 400 yds. from the outlet from the lake. It had a 15 h.p. Vortex turbine (serial number 1671) supplied by Gilbert Gilkes & Co. for £63 that provided power for a circular saw and drilling machine on the ground floor. It also drove a generator which provided electricity to charge batteries for use in the house. The top floor of the workshop was used by the painters and decorators of the estate. The building and equipment was demolished in the 1960s because of extensive vandalism.[2]

In the year 2000 a new hydroelectric scheme was installed using the water from the weir via the lake and the original 8 inches diameter cast iron piping. The turbine chamber of the demolished estate workshop was excavated and a second hand turbine, also manufactured by Gilbert Gilkes & Co., was installed driving a generator to produce an A.C. electric supply for use in the new visitor centre. The turbine used is a No. 15 Pelton wheel (serial number 5481) which was supplied new by the Gilkes company in 1957 to Salford University (then Salford Technical College). Its wheel is only 12 inches in diameter but working on a 89 feet fall it can generate a maximum of 3Kw. of electricity.[3] The system was installed by Derwent Hydropower Ltd. and is currently the only water powered site on the whole of the River Dane and its tributaries that is still in use.

References

1. Kennedy, J., *Biddulph, By the Diggings*, University of Keele, 1980.
2. Information supplied by Mr. E. Bowers, head gardener at Biddulph Hall, 1984.
3. Hydroelectric Walk leaflet, Staffordshire Moorlands District Council, 2001; and communication from Derwent Hydropower Ltd.

FIGURE 81. A VORTEX TURBINE WITH PART OF ITS CASING REMOVED TO SHOW THE MOVABLE GUIDE VANES AND THE RUNNER. (*Gilbert, Gilkes & Co. catalogue*, 1901)

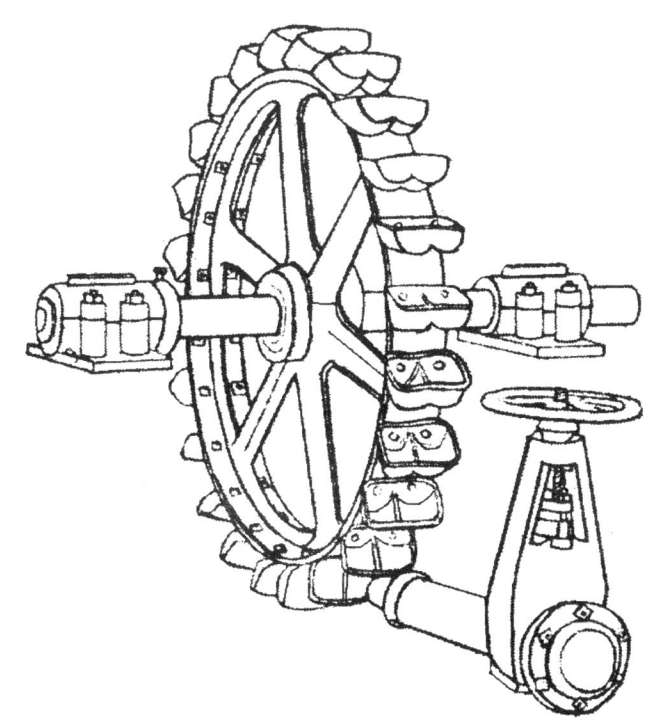

FIGURE 82. THE BASIC CONSTRUCTION OF A PELTON WHEEL (*without its casing*) SHOWING THE WHEEL WITH ITS DIVIDED BUCKETS WHICH RECEIVE THE JET FROM THE NOZZLE CONTROLLED BY THE VALVE IN THE FOREGROUND.

27. BIDDULPH CORN MILL　　　SJ 895599

Another small stream rises on the high ground north west of Biddulph Hall, running southwards initially but then turning westwards near the Talbot Arms, to flow through a wooded valley eventually to join the Biddulph Brook. On the north side of the Talbot Arms where the stream turns westwards is a small pool, at what is known as Pool Fold, that provided the necessary water storage for Biddulph Corn Mill. The mill was situated at the western edge of the pool.

The origins of this mill are lost in antiquity but it does appear on numerous large scale maps from 1770 onwards, and a mill was mentioned in a document confirming the Biddulph family holdings in 1644.[1] The earliest mention of a miller was in 1824 when William Stubbs appeared in a trade directory.[2] William Stubbs was still the miller at the age of 61 in 1851 when he lived with his wife, Judith, together with his daughter, Elizabeth, and her husband, James Plant, in the mill house.[3] The landowner for the mill around this time was Lord Camoys.[4] Sometime during the 1850s James Plant took over as miller from William Stubbs but in 1861 William Stubbs and his wife, who were both in their seventies, were still living at the mill house with their daughter and son-in-law.[5]

Although James Plant was the miller until the 1870s when his son Benjamin took over, the mill and the Biddulph Hall estate had become the property of the Stanier family by 1865.[6] By 1892 the miller at Biddulph Corn Mill was Thomas Walley who usually described himself as a farmer and miller until 1932 when he was listed solely as a farmer.[7] From this information it is likely that the mill went out of use in the late 1920s or early 1930s.

The only evidence of the mill building is the survival of two old postcards showing the mill. It was a small, two storey stone building with a stone tiled roof and an external breastshot waterwheel on its south-eastern wall. This waterwheel appeared to be about 10 feet diameter and about 2 feet 6 inches wide. The small size of the mill and its height of only two storeys indicates that there probably would have been no more than two pairs of millstones at the most. Certainly the photograph shows that the mill had the same size and shape as shown on all the earlier maps and was most likely unimproved since at least the 18th century if not earlier.

Unfortunately the mill has been demolished but the pool remains as does the mill house, although in a much altered state.

References

1. Kennedy, J., *Biddulph, by the Diggings*, University of Keele, 1980.
2. White, Directory of Staffordshire, 1824.
3. SRO, Biddulph Census Returns, 1851.
4. SRO, Biddulph Tithe Map and Apportionments, 1840.
5. SRO, Biddulph Census Returns, 1861.
6. SRO, Stanier Estate Map, 1865, D 997/7/2.
7. Kelly's Directories of Staffordshire, 1892, 1912, 1932.

FIGURE 83. A VIEW OF BIDDULPH CORN MILL FROM THE NORTH-WEST IN THE LATE 19TH CENTURY. (*D. J. Wheelhouse*)

FIGURE 84. AN OLD POSTCARD VIEW OF BIDDULPH CORN MILL FROM THE SOUTH-WEST WITH THE BREASTSHOT WATERWHEEL ON THE NEAREST GABLE-END AND THE MILL HOUSE ON THE HILL AT THE REAR. (*D. J. Wheelhouse*)

28. LEE FORGE   SJ 888598

After leaving Biddulph Corn Mill the stream runs in a westerly direction down a wooded valley until it joins the Biddulph Brook which then runs northwards eventually to join the River Dane. At the confluence of these two streams is the site of Lee Forge. Coal mining and iron working have a long history in the Biddulph area due to the seams of coal outcropping on the hillsides either side of the town. Interspersed with these seams of coal are bands of ironstone. Certainly an estate map of 1597 shows a bloomsmithy not far from the area of Lee Forge with what appears to be an artificial watercourse or leat. This bloomsmithy appears to have been still in use in 1644 when Mrs. Biddulph, wife to Capt. Francis Biddulph, was granted by the Staffordshire County Committee *"to have use ocupie, possesse and enjoy the Biddulph demenses the Mill, Colepits and Bloomsmithy"*.[1]

As shall become apparent, Lee Forge has had a long association with the Gosling family whose name first appears in the district in 1529 when a John Geslynge (or Gosling) applied to Chancery for confirmation of his right to lease 80 acres of land which he had occupied for over thirty years under a verbal agreement, paying a rent of 31s per year.[2] Unfortunately it is impossible to confirm which land was involved in this transaction, or even that he was involved in iron making at this time. The earliest confirmation of a forge on the site in question occurs on Yates's Map of Staffordshire in 1770. The ownership of the forge in the latter part of the 18th century is mentioned in Jonathan Wilson's diary with entries during 1775 and 1776 such as *"... stayed a while at the forge"* and *"... stayed a couple of hours at Frank Goslings at Lee Forge House."* Jonathan Wilson was the local vicar in Biddulph and also headmaster of a school in Congleton so would often be passing the forge on his travels between these two locations.[3] Also, the land tax returns confirm that the owner and occupier of the forge up to 1785, paying a tax of £1-7s-10d, was Francis Gosling.

Francis Gosling was also involved in operating Pethills Forge, near Winkhill, a few miles south-east of the town of Leek. Possibly the Pethill Forge was a better proposition at this time as it was situated alongside the main road between Leek and Ashbourne whereas Lee Forge was about ½ mile from the original road from Biddulph to Congleton which went via Overton. The access to the forge from this road was along a track down a steep sided valley. Whatever the reason, in 1786 Lee Forge was converted into a flint mill and leased to a Joseph Smith and Anthony Keeling. A fatal accident caused Jonathan Wilson to remark in his diary for 31st August 1787 that *"William Eardley drowned in flint mill pond."* From 1788 to the turn of the century Anthony Keeling was the sole occupier of the flint mill but was joined by his brother Enoch in 1803. Some indication of the type of raw materials ground at the mill can be seen in another entry in Jonathan Wilson's diary for 1797 which read *"Looked at flint mill now grinding lithage"*. In 1804 and 1805 the Keeling brothers were succeeded by Jeremiah Ginders, who was himself replaced as the operator of the flint mill in 1806 and 1807 by Benjamin Challinor.[4]

In 1812, Francis Gosling of Pethills, iron master, and George Gosling, only son of Francis, leased the flint mill together with the water and other wheels, machinery and all appurtenances to Joseph Machin and Jacob Baggaley, china manufacturers, of Burslem for eleven years from 25th March 1811. This lease included *"Lee House, Mill buildings, ways, waters, dams, weirs, goits, sluices, watercourses, streams, particularly the water running and flowing to the mill for the turning and working of the machinery"* but not the Miller's Meadow or the Clough (which was fenced out for purpose of growing crate wood for the Goslings). Machin and Baggaley were to pay a rent of £130 per year in two, half yearly payments with all parliamentary and other taxes, church and parish duties, and had to cleanse and scour the watercourse in the Clough together with repairing the weir when necessary, plus keeping the property in good tenantable repair. The Goslings were allowed to inspect the premises twice per year and reserved the timber, mines and quarries of coal, stone, ironstone and other minerals, with access to dig for, sink sough, and erect engines so as to get minerals or timber and make charcoal.[5]

This lease ran its course until 1819 when a plan was put forward to build a new road from Biddulph to Congleton which would pass right by the flint mill.[6] This coincided with developments on the Leek to Ashbourne road where eventually the turnpike was to take an improved route, by-passing Pethills Forge. Consequently, Francis Gosling decided that it would be advantageous to rescind the lease of the Biddulph property and convert the flint mill back into a forge so that he could transfer his business from Pethills. Joseph Machin and Thomas Baggaley drove a hard bargain, but finally on 21st April 1819 an agreement was signed whereby they agreed to give up the mill in three months time but would continue to lease the farm until 25th March 1820, for which they would pay a rent of £50. Francis Gosling was to pay them £400 as compensation, spread over two years, plus interest from when he entered the mill. Also he had to take over the £50 share in the new road that Machin and Baggaley had subscribed.[7] Once this agreement was made, the

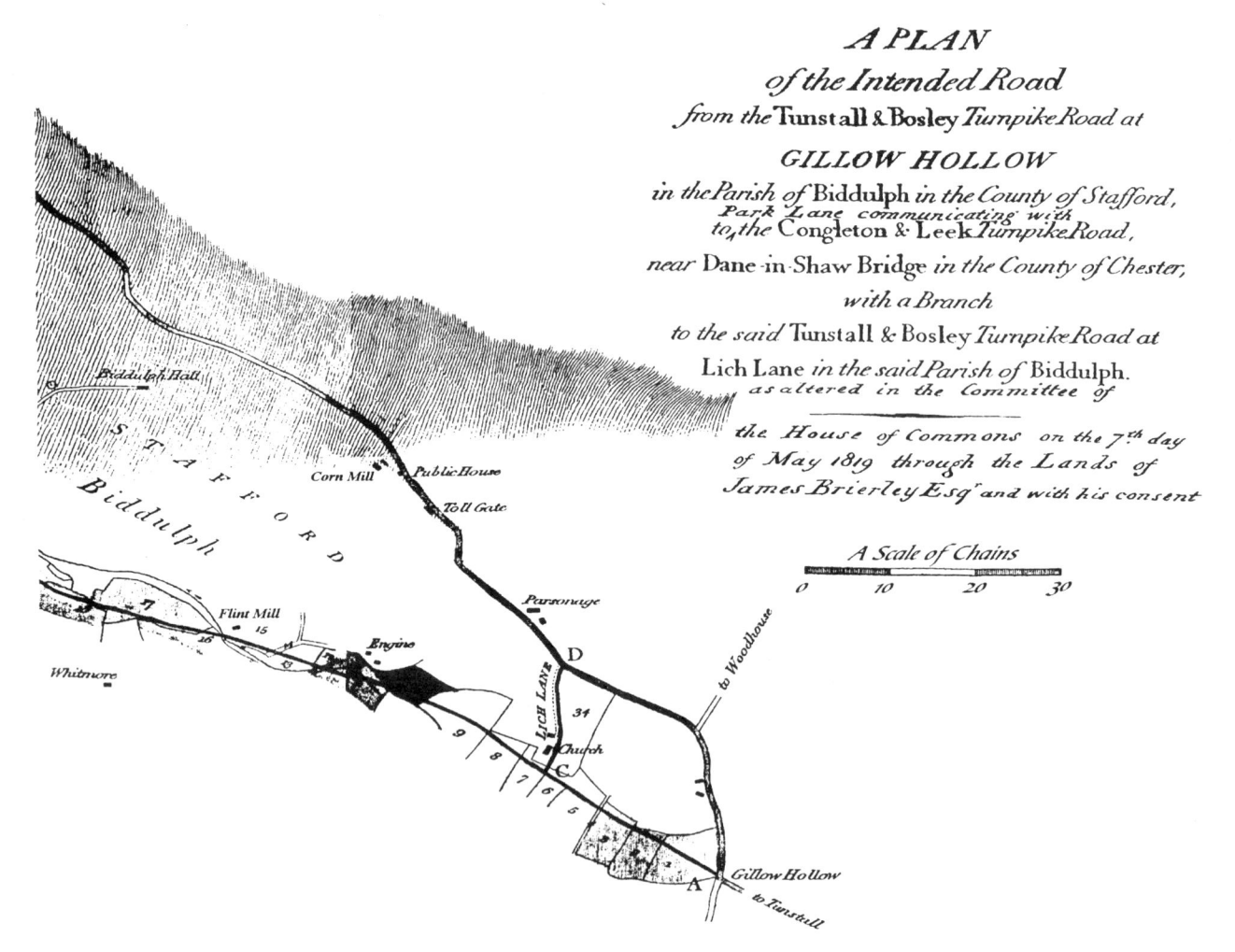

FIGURE 85. PART OF THE PLAN FOR THE NEW TURNPIKE ROAD BETWEEN BIDDULPH AND CONGLETON IN 1819 (*north to the left*). THE OLD ROAD IS SHOWN RUNNING TO THE EAST ON THE HIGH GROUND PAST THE CORN MILL WHEREAS THE PROPOSED ROAD RUNS PAST THE MARKED LOCATION OF THE FLINT MILL. (CCALS, QDP 46)

mill was converted back into a forge and early in 1823 its products were advertised as follows:-

"Francis Gosling & Son beg leave most respectfully to acquaint their friends and the public that they have on sale a constant supply of IRON AXLE-TREES, turned and fitted up in a very superior manner, with the best warranted Lancashire bushes, at very reduced prices. Contact Lee Mills, Biddulph, near Congleton."[8]

Sometime in the next ten years both Francis Gosling and his son George died and their estate was inherited by George's son Samuel Franceys Gosling who was born in 1823. As Samuel Franceys was a minor, the forge was run as a scrap iron, spade and shovel manufactory by the executors of George Gosling's will, namely by George Cambell and Samuel Franceys.[9] Once Samuel Franceys Gosling finished his education as a surgeon he took over the running of the forge, but in a less "hands on" capacity than his forebears, as he employed a forge manager for the day to day management. In the 1870s the forge manager was Robert Forrester, who happened to be Samuel Franceys Gosling's brother-in-law (or perhaps Samuel Franceys Gosling became his forge manager's brother-in-law).[10] Some idea of the type of business that the forge conducted in the latter part of the 19th century can be seen from their entry in the various trade directories, namely:-

"Samuel Franceys Gosling, manufacturer of uses, spades, shovels, draining tools, cable & crane chains, iron pans, ladles, iron arms, axletrees, etc. scrap, tyre & wrought iron, wrought iron shafting, edge tools, etc." [11]

Samuel Franceys Gosling died in 1882 but his brother-in-law, Robert Forrester, continued operating the business up to at least 1907. Samuel Franceys Gosling's widow eventually died at Lee House in 1912.[12] By the end of the 19th century the forge consisted of around seven different buildings and a triangular shaped pool.

In 1917, Mr. H. N. Ashton, who originally ran a silk manufacturing business in Manchester, brought his workers from Manchester and set up at the forge in the colour business, concentrating solely on the development of prussian blue. In 1918, Mr. Ashton and his partner Mr. J. M. Bennett bought the entire property from the Gosling estate. The site was ideal for this new venture, having an abundance of water and an artesian well on the premises. The forging equipment was removed and the buildings altered to suit the new process, however water power continued to be used, driving a turbine that turned a generator in order to make electricity.

The firm prospered and at one time employed forty people. In 1934 there was another fatal accident at the factory. This time the dead man, a long time employee, had fallen into a 12 feet deep vat containing about four feet of dye in which he drowned. Henry Ashton died in 1953 and was succeeded by his son, Reginald, who continued the prussian blue business. In 1960 illness forced him to sell the business to Hardman & Holdman and thereafter there were a number of take-overs until the business ended up with Manox Ltd who continued operating it until its final closure in 1981. Right up to the end the business continued to generate its own electricity rather than rely on the standard electricity supply. Apart from continuity of supply while processing a batch of dye they also needed to use direct current which was much simpler to use for controlling the speed of the mixers. Unfortunately the turbine was replaced at some time by three gas engines for which the firm made its own gas. [13]

After closure the buildings at Lee Forge, which were all tinged with a blue coating, were demolished and the site has become overgrown, derelict, and no doubt heavily contaminated by the dyestuffs.

References

1. Kennedy, J., *Biddulph, By the Diggings*, University of Keele, 1980
2. *Ibid.*
3. *Ibid.*
4. *Ibid.*
5. SRO, Lease, D 1065/1/10.
6. CRO, Biddulph - Congleton Turnpike Diversion, QDP 46, 1819.
7. SRO, Lease, D 1065/1/11.
8. *Macclesfield Courier*, 22nd February 1823.
9. SRO, Biddulph Tithe map & Apportionments.
10. SRO, Biddulph Census Returns, 1871.
11. P.O. Directory of Staffordshire, 1872.
12. Kennedy, J., *Biddulph, By the Diggings*, University of Keele, 1980.
13. *Congleton Chronicle*, 12th June 1981.

---

29. DANEINSHAW CORN MILL   SJ 884615

The Biddulph Brook runs northwards from the Biddulph area and changes its name to the Daneinshaw Brook as it reaches the Staffordshire/Cheshire border at a point where it is joined by a small side stream known as the County Brook. Just downstream of this confluence is the site of a small weir where a leat was

FIGURE 86. THE MILL POOL AT DANEINSHAW CORN MILL LOOKING SOUTH. THE MILL IS ON THE WESTERN SIDE OF THE POOL (*right*). THIS PHOTOGRAPH DATES FROM BEFORE 1860 WHEN THE RAILWAY WAS BUILT ON AN EMBANKMENT ALONG THE EASTERN SIDE OF THE POOL. (*Town Hall Studio, Congleton*)

taken from the brook on its eastern side. This leat led to Daneinshaw Corn Mill which was situated just to the south of Daneinshaw chapel, on the eastern side of the track from the Castle Inn to the Congleton/Leek road.

The origins of this corn mill are unknown but it did appear in the earliest records of the Land Tax from 1783 to 1785, listed as "Daneinshaw Mill", owned and occupied by Samuel Delves at a tax of 8s-0d. It was also recorded in 1786 under the same ownership but occupied by Peter Delves when it was recorded as "New Mill" at the same tax of 8s-0d.[1] It would seem then that the mill could well have been rebuilt in 1786.

It did not remain long in the possession of the Delves family as from 1788 to 1794 the owner was recorded as Thomas Rowley with the miller being James Leavesley.[2] When Thomas and John Whitfield purchased the Daneinshaw estate of Richard Martin on his death in 1795, they also purchased the corn mill from Thomas Rowley. James Leavesley was replaced as the miller in the following year by Samuel Brindley who was one of the sons of Henry Brindley who owned Danebridge Mill (*q.v.*) at this time.[3]

In 1798 the occupier of the mill became Joseph Williamson and then in 1799 and 1800 the owner was recorded as Samuel Whitfield. However, on the death of Thomas Whitfield in 1800, the mill was part of his estate as described in his will. It would seem that the Whitfields were rather elastic in their allocation of the ownership of property within the family as the corn mill reverted to being recorded as being owned by John Whitfield from 1801 to 1808 when there was a succession of occupiers such as Ralph Johnson in 1803, Josiah Bayley in 1804, Matthew Goodwin in 1805, John Plant in 1806 and Thomas Bigley in 1807/8.[4]

In 1808, when the Whitfield estate was advertised for sale the corn mill was described as

"... *a water corn mill in complete repair containing four pairs of stones and which is now let to a respectable tenant at the yearly rent of £75.*"[5]

This sale does not appear to have been successful as the mill remained in the ownership of Whitfield & Co., with a Mr. Glover as occupier until 1812, when the owner was recorded as Jonathan Broadhurst with John Galley as the miller. John Galley remained at the mill until 1818 when he died; in the meantime ownership was transferred from John Broadhurst to Broadhurst, Hall & Co. in 1812/1816 and then to Hall & Co.[6]

On the death of John Galley the milling was taken over by Oakes and Cookson until 1826 when Daniel Oakes continued the business on his own. Changes in the fortunes of Hall & Co. meant that the ownership was transferred to Hall & Johnson in 1822. In 1827 the Daneinshaw estate was again advertised for sale by Hall & Johnson when the mill was described as

"...*a very powerful water corn mill, two houses & outbuildings, 20 to 30 cottages, chapel, schoolrooms, meadow land, etc.*"[7]

The Hall family continued to own the mill, with Daniel Oakes as miller, until 1846 when the death of Philip Hall occurred (see Daneinshaw Silk Mill). When Philip Hall's estate was advertised for sale the following description of Daneinshaw corn mill was published

" ... *all that well accustomed water corn mill situated on the brook called Dane in Shaw Brook near to the said last described premises* [Daneinshaw Silk Mill] *on the opposite side of the said Congleton to Leek turnpike road with the two waterwheels, millstones, machinery and working gear and also the commodious drying kilns, carthouses, stables and other outbuildings attached to the*

FIGURE 87. VIEW OF DANEINSHAW CORN MILL FROM THE CONGLETON TO LEEK ROAD LOOKING SOUTH. (*Town Hall Studio, Congleton*)

FIGURE 88. THE COURSE OF THE LEAT TO DANEINSHAW CORN MILL, NOW DRY AND GATED, WITH THE WEIR IN THE BACKGROUND.

said corn mill. The mill house is occupied by Daniel Oakes."[8]

By merging the information from the various sale advertisements it becomes clear that the new Daneinshaw Corn Mill which was probably built in 1786 had two waterwheels driving four pairs of millstones (probably two millstones per waterwheel) and at least one drying kiln.

In 1846 Daniel Oakes was 65 and by 1850 he had been replaced by Thomas Jepson as the miller, who then worked the mill with his younger brother John. Thomas Jepson had married a widow some years older than himself with three children all under the age of ten. It is probable that his wife was in fact the daughter of Daniel Oakes but as he was her second husband this relationship is not clear.[9] This marriage may have been based on true love but is more likely to have been more to do with expediency with respect to the tenancy and operation of the mill. The Jepsons only remained at the mill for about ten years when they were superseded by Mark Thompstone as the miller.[10] He was part of a milling family that was involved with both Gawsworth Mill and Bosley Works (*q.v.*). Mark Thompstone and his wife remained at the mill until about 1880. As they only had a family of daughters with no sons to help out in the mill, Mark Thompstone employed three journeyman millers to share the workload at the mill.[11]

When the Thompstones left the mill around 1880 the tenancy was taken by Samuel Harrison.[12] This was at a time when small country watermills were feeling the pressure from the introduction of large rolling mills using imported wheat, so to supplement his income Samuel Harrison had to diversify and was listed not just as a miller but also as a cheese factor.[13] He remained at Daneinshaw Corn Mill for just over twenty years until his death around the turn of the century. The mill was then run by his executors until the end of the First World War when the last miller, Sam Burgess, took over.[14] The mill ceased milling altogether in 1939 and was used for storing rations during the Second World War. After the war the mill was not used and deteriorated to such an extent that it was finally demolished in 1963.[15]

Although the mill is no longer standing, the remains of its weir on the Daneinshaw Brook can still be made out together with its leat which is now dry. In 1860, when the Biddulph Valley Railway was built, the tracks crossed the valley at Daneinshaw and ran behind the mill on an embankment. At the time, arrangements had to be made, not just to pass the brook under the railway, but also the mill leat. This separate tunnel is still to be seen. The pool for the mill was also affected by the coming of the railway as the embankment ran over some of its area. The outline of the mill pool is still traceable but getting quite overgrown. The evidence of a small chimney suggests that at some stage the mill had been supplied with steam powered assistance. The mill itself was four stories high, with the pool to its rear with circular overflow. There was an overshot wheel in the centre of the mill, with another wheel externally at the northern end of mill with the drying kiln at the south end.[16] The mill building was quite narrow, when viewed today there hardly seems room for a mill between the boundary of the pool and the track that ran in front of the mill, however the old photographs of the mill confirm that this was in fact its position.

References

1. CRO, Buglawton Land Tax Returns.
2. *Ibid.*
3. Evans, K. M., *James Brindley, Canal Engineer, A New Perspective*, Churnet Valley Books, 1997.
4. CRO, Buglawton Land Tax Returns.
5. *Staffordshire Advertiser*, 10th September 1808.
6. CRO, Buglawton Land Tax Returns.
7. *Macclesfield Courier*, 1st April 1827.
8. *Macclesfield Courier*, 28th February 1846.
9. CRO, Congleton Census Returns, 1851.
10. White's Directory of Cheshire, 1860.
11. CRO, Congleton Census Returns, 1871.
12. CRO, Congleton Census Returns, 1881.
13. Slater's Directory of Cheshire, 1888.
14. Kelly's Directories of Cheshire, 1896-1914.
15. *Congleton Chronicle*, 10th May 1963.
16. Norris, J. H., "The Water-Powered Corn Mills of Cheshire", *Transactions of the Lancashire & Cheshire Antiquarian Society*, Vol. 75, 1965.

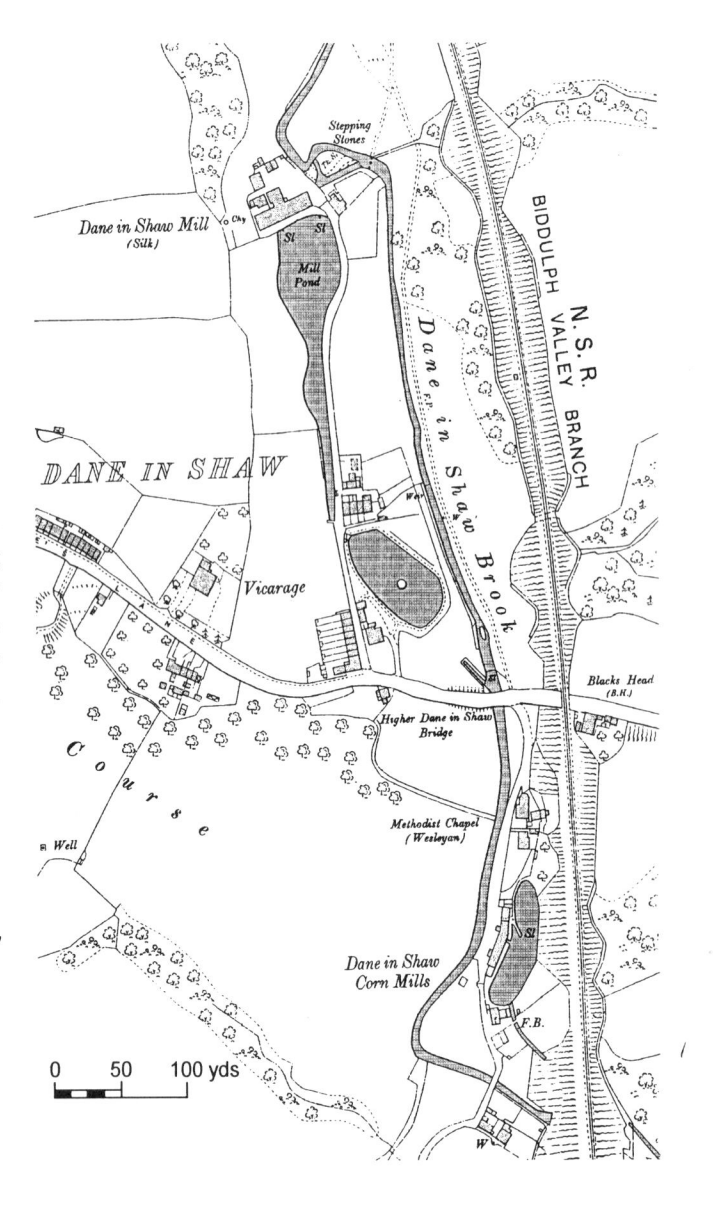

FIGURE 89. ORDNANCE SURVEY MAP OF DANEINSHAW, 1909. THE CORN MILL IS SOUTH OF THE ROAD WITH THE RAILWAY EMBANKMENT RUNNING ALONGSIDE THE POOL. THE LEAT LEADING FROM THE WEIR ON THE DANEINSHAW BROOK AWAY TO THE SOUTH-EAST PASSES UNDER THE EMBANKMENT. NORTH OF THE ROAD CAN BE SEEN THE WATER SUPPLY ARRANGEMENTS AND THE LOCATION OF DANEINSHAW SILK MILL.

---

30. DANEINSHAW SILK MILL   SJ 883620

The Daneinshaw Brook runs northwards from the site of the Daneinshaw Corn Mill under the Congleton to Leek road. About 200 yards beyond the road on the western side of the brook is the site of the Daneinshaw Silk Mill.

A survey of the Congleton area in 1770 shows that this land in Daneinshaw belonged to the Martin family but there is no sign of any mill building on the associated map.[1] In 1788 Richard Martin joined in a partnership with two brothers, Thomas and John Whitfield, to manufacture cotton yarns at two mills known as Nearer Daneinshaw Mill and Further

Daneinshaw Mill. John Whitfield was to operate Nearer Daneinshaw Mill with Richard Martin operating at Further Daneinshaw, each of them receiving £100 per year more than Thomas Whitfield, who was to sell the cotton thereby receiving a 2.5% commission.[2] Nearer Daneinshaw Mill was a considerable way downstream near the bridge on the Congleton to Buxton road (now the A 54) in Buglawton and has been identified as later becoming known as Davenshaw Mill (*q.v.*). There is definite evidence that the Nearer Daneinshaw Mill existed before 1783 (see Davenshaw Mill). Richard Martin was a cotton spinner on his own account prior to the setting up of the partnership with the Whitfields but at which mill is not entirely clear, although there is some indication that he might have been at Nearer Daneinshaw. In January of 1788 Richard Martin wrote to his main customer, James Brown in Glasgow, urging the payment of outstanding bills as he was "*...sinking a considerable sum of money in erecting a twist mill. We mean to work about 1500 spindles*".[3] Later that year there are entries in Richard Martin's notebook for payments on building materials, such as bricks, in some quantity. As there is no evidence of any connection between the partnership of Richard Martin and the Whitfield brothers with any mills other than at Nearer and Further Daneinshaw it seems likely that it was Further Daneinshaw Mill that was being built in 1788 and that Richard Martin then moved so as to manage this new mill.

The partnership continued in business until Richard Martin's death in 1795 when the survivors, Thomas and John Whitfield, wished to purchase Richard Martin's mill from his executors. This caused something of a problem as it was not easy to set a value on a cotton mill in these early days of cotton spinning and the executors, Ralph Deane of Macclesfield and Charles Fielder of Bath, were not familiar with the industry. However, by December 1795 George Reade, a family friend with his own cotton interests, was able to write to Charles Fielder as follows:-

To Mr. Fielder
Buglawton, 24th December 1795

Dear Sir,
 Agreeable to your desire I have made inquiry subjecting a proper person to value the cotton concern at Congleton and am happy to inform you that I have met with a man (in my opinion exactly to your purpose). I enquired of some of my acquaintances in Manchester and they all agreed in recommending this businessman as the proper person by much in Manchester or that they could think of, if he would consider. I have seen him and informed him as much as I could what was to be done, he thought with you that it was a very serious affair and said his own business confined him so much that he could but ill spare the time to go from home, but if you could not engage with any other person to your satisfaction he could come. His name is Barton he is a very respectable man, has a concern in a large cotton mill at Manchester and was formerly by connection with the late Sir Richard Arkwright. He has a thorough knowledge of the business and has been several times engaged in valuing cotton mills - I shall see him again before the time that you will want him and in the mean time if you have any objection I shall be glad to inform you I am with best respects to Mr. Fielder,

George Reade[4]

Arrangements were brought to a close in 1797 when the Whitfields raised a mortgage of £2700 to purchase the mill, land and a number of cottages at Further Daneinshaw.[5] They also purchased Daneinshaw Corn Mill but it is not clear where the money came from for that purchase.

When Thomas Whitfield died in 1800 his will confirmed that he was the owner of both cotton mills and the corn mill in Daneinshaw together with land, houses and about 30 cottages. He had also bought land in nearby Rushton Spencer and in Congleton town. His will does not mention any provision for John Whitfield, only for his wife and daughter, so presumably the partnership concerned sharing profits only, not the tangible assets. In spite of that, John Whitfield is recorded as the owner of Further Daneinshaw Mill in the land tax returns until 1808 and he raised a mortgage with Ellis Needham of £1700 on the property in 1805. In 1808 the Whitfield estate was offered for sale, giving a description of the cotton mill at Daneinshaw as follows:-

 "*Lot 1. All that newly erected cotton factory (with or without) the machinery situated on the River Dane in Shaw near the town of Congleton aforesaid, being three storeys high, each room being 60 feet long, 30 feet wide and 10 feet high exclusive of warehouses, counting house, joiners, smiths, and clockmaker's workshops, with an excellent waterwheel and 21 feet 6 inch of fall capable of working a large quantity of silk or cotton machinery together with 15 acres of land or thereabouts (little more or less) including the woodland and reservoirs together with eight dwelling houses.*"[6]

The map made in 1803 during a survey of Congleton for the land tax, and also Moorhouse's map of Congleton published in 1818, show a much smaller and squarer mill at Daneinshaw than the one that stands there today. The dimensions given in the advertisement above no doubt relate to this earlier mill. In spite of the advertisement the mill continued to be recorded as owned either by Mrs. Whitfield or Whitfield & Co. up to 1825. However, it was recorded as occupied from

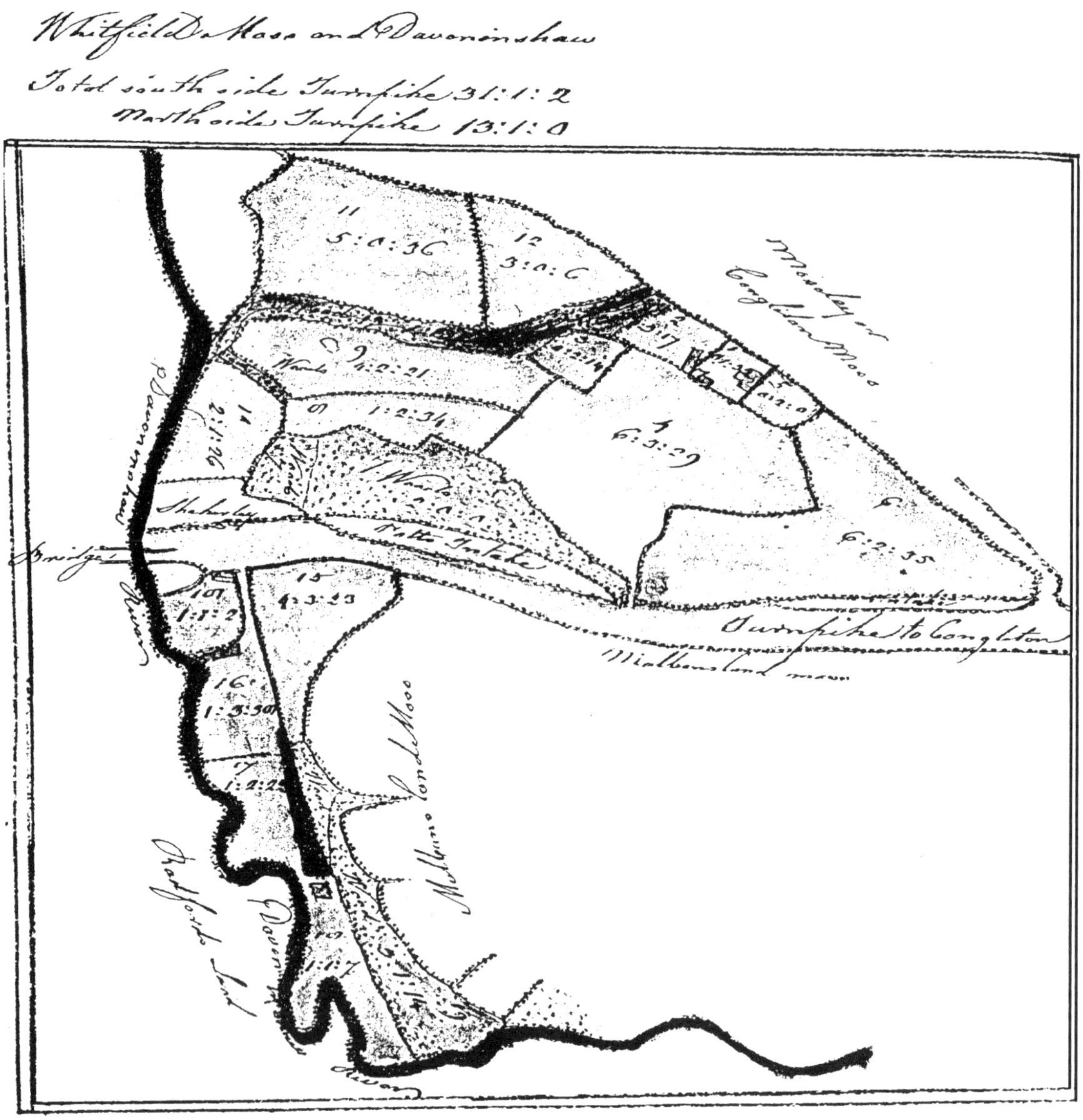

FIGURE 90. A SKETCH MAP OF "WHITFIELD MOSS AND DAVENINSHAW", PART OF A SURVEY OF 1803 (*north is towards the bottom of the page*). THE TURNPIKE ROAD FROM CONGLETON TO LEEK CROSSES THE MIDDLE OF THE MAP WITH THE SILK MILL TO THE NORTH. NOTE THE SHAPE OF DANEINSHAW SILK MILL AND THE SIMPLICITY OF ITS WATER SUPPLY COMPARED WITH FIGURE 89. (*Congleton Chronicle*)

1810 to 1812 by Johnson & others, from 1813 to 1816 by Johnson & Gent and from 1817 to 1831 by Hall & Johnson, who were also recorded as owners from 1825 to 1831.[7] All these firms were involved in the silk trade as throwsters rather than being cotton spinners, so it seems likely that the cotton spinning machinery was removed from the mill about 1808 and the mill converted to silk throwing. In 1827 an attempt was made to sell the mill which was advertised as:

"*The valuable estates belonging to the late firm of Hall & Johnson will shortly be offered for sale by public auction, including a very extensive silk mill with machinery complete possessing peculiar advantages of steam and reservoirs, etc. situated on the river Dane in Shaw near Congleton close to the line of the Macclesfield Canal, on the turnpike road from Congleton to Leek near excellent collieries, stone quarries & lime kilns.*"[8]

This advertisement highlights that steam power had been added to the mill sometime between 1808 and 1827.

It was at this juncture that Philip Hall & Son took over at the mill as silk throwsters, with Thomas Hall taking over from his father sometime before 1845. In that year Thomas Hall's death took place in tragic circumstances, with a report in the local paper which read as follows:-

"*The town was thrown into great consternation at a report that Thomas Hall, Esq. of Daneinshaw, one of the Borough Magistrates, had committed suicide.*

R. L. Ginder, a nephew of the deceased, had been with him for 26 years, had seen him at 11.45 in the counting house, when returning at 1 o'clock accosted by an apprentice who could not open the door latch (it was fastened inside by a nail) so a private rear door was tried,

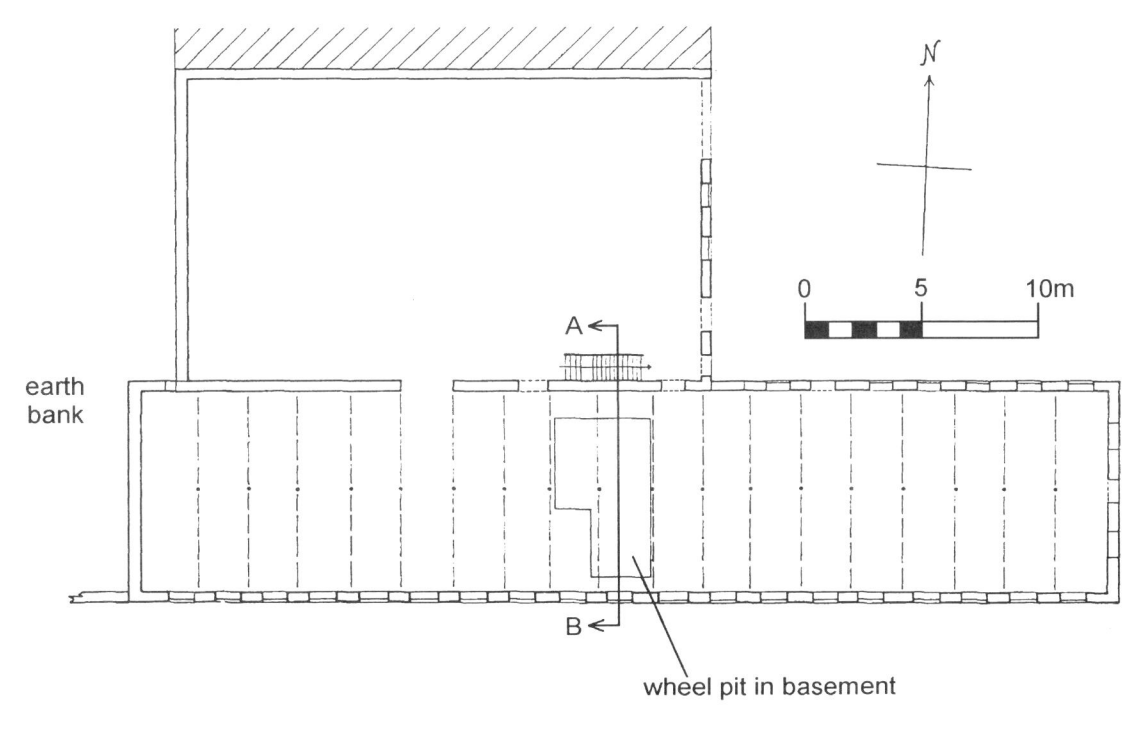

FIGURE 91. PLAN OF DANEINSHAW SILK MILL PRIOR TO CONVERTION INTO APPARTMENTS. (ECTMS)

this was taken off its hinges. The deceased was suspended by his neck from a beam over 6 feet high."
Evidence was given by the cook and also by business colleagues that Thomas Hall was depressed, suffered bilious attacks, acted unnaturally and was probably insane.[9]

This sad occurrence precipitated yet another sale of the mill which was described as:-

"All that extensive silk mill formerly used as a cotton factory situated at Higher Daneinshaw aforesaid. It is four storeys high and 43 yards 6 inches long by 10 yards 2 feet wide with a steam engine of 10 h.p., boiler, powerful waterwheel, main shafts and heavy gearing whereby the machinery and the said mill is worked and also that newly erected smithy behind the mill and a large room over the same used as a joiners shop with the lathe connected with the waterwheel, which said silk mill and premises with the appurtenances were late in the possession or occupation of Thomas Hall, deceased, but now in the occupation of Mr. R. L. Ginder. The whole mill is heated by steam."[10]

It is obvious that the dimensions listed in the advertisement above are very different from those quoted in the advertisement of 1808 and are in fact those of the mill to be found at Daneinshaw today. Sometime between 1808 and 1846 the mill had been demolished and completely rebuilt. This might have been after the Whitfields relinquished the ownership of the mill in 1825.

After the sale in 1846, R. L. Ginder operated the mill as a silk throwster. He remained in business on his own account until 1869 when he went into partnership with Arthur Myott for a short time and then Arthur Myott continued on his own into the 1880s. In spite of the decline in silk throwing in the area in the latter half of the 19th century Daneinshaw Mill was able to continue into the 20th century under the ownership of William Tagel, up to around the turn of the century, and William Ward up to the First World War.[11]

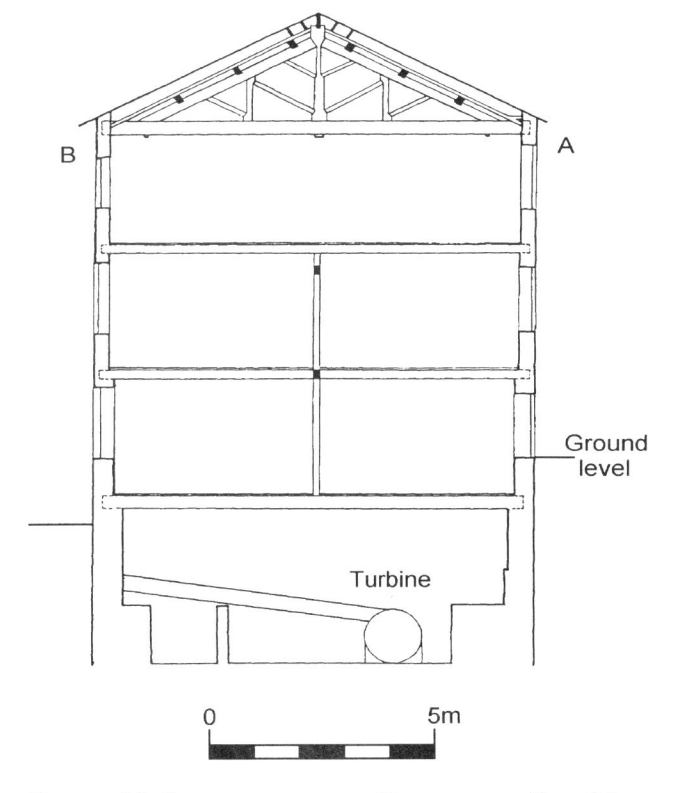

FIGURE 92. SECTION THROUGH DANEINSHAW SILK MILL WITH THE TURBINE INSTALLATION OF 1927. (ECTMS)

FIGURE 93. VIEW OF THE MILL POOL AND THE SOUTH FACE OF DANEINSHAW SILK MILL PRIOR TO CONVERSION INTO NINE APPARTMENTS.

After the First World War the mill was taken over by Shaws Ltd., a fibre and brush manufacturer and then in 1927 by Baxters Ltd. of Littleborough (near Rochdale) who were manufacturers of bedspreads. Even in the 20th century waterpower continued to be used at the mill; in 1927 Baxters replaced the waterwheel with a Series 1 turbine from Gilbert Gilkes & Gordon Ltd. of Kendal. The turbine was quite small relative to the waterwheel it replaced, being only twelve inches in diameter. Working off a head of 20 feet, it used a flow of 712 cu. ft./min. to develop 21 h.p. at 526 r.p.m. This turbine gave excellent service and, in 1948, it was fitted with a governor. Presumably by that time it was more appropriate to use the turbine to generate electricity rather than use its output directly and so a governor would be necessary to maintain the voltage and frequency within the appropriate limits. The turbine was still running in 1961 when it had a major repair, having a new runner fitted.[12] From that date until the early 1980s the mill was used for various manufacturing processes. After that the mill stood empty for a number of years and it has now been converted into nine separate appartments.

Daneinshaw Mill is one of the finest examples of a "Georgian" textile mill that has survived in the whole of South-East Cheshire. It is a four storey building, 19 bays long, with a central pediment housing a clock. The main building faces south and there was a north running wing at the back of the main building which has now been demolished. The rear face of the mill and its northern wing are nowhere near as attractive as its southern face, having been repeatedly altered over the years with no thought for its appearance. To the west, part way up the valley side, is a small octagonal chimney that provided the draft for the steam boilers.

There are two weirs still to be seen on the Daneinshaw Brook, one just north of the road from Congleton to Leek which feeds into two pools on either side of the mill lane via a culvert (see Figure 89). The second weir is about 100 yards downstream of the first weir and was probably used for the original mill when there was only a single, smaller mill pool. When the current mill was built the new weir raised the water level of the mill pools which is probably why the pool embankment in front of the mill obscures most of the ground floor. This gives the mill the appearance of being only three storeys high. Although the changes due to converting the mill into appartments has removed the turbine and resulted in some changes to the mill's appearance it is still a very handsome building and deserves to have survived even if in the guise of living accommodation.

References

1. Survey of Congleton, 1770. Held by the editor of the Congleton Chronicle.
2. CRO, Partnership agreement between Whitfield Bros. and Martin, 1788, DCB/1595/27/14.
3. CRO, Richard Martin's notebook, DCB/1595/7/11.
4. CRO, Martin family papers, DCB/2114/19.
5. CRO, Indenture, 1797, DCB/1595/2/5; CRO, Lease, 1797, DCB/1595/8/6.
6. *Staffordshire Advertiser*, 10th September 1808.
7. CRO, Buglawton Land Tax Returns.
8. *Macclesfield Courier*, 31st March 1827.
9. *Macclesfield Courier*, 5th April 1845.
10. *Macclesfield Courier*, 28th February 1846.
11. Trade Directories of Cheshire, Slater, Morris, Kelly
12. Gilbert Gilkes & Gordon Ltd. of Kendal, Turbine Records.

31. TIMBERSBROOK MILL   SJ 895627

The Timbersbrook rises from a spring on the southern slopes of Bosley Cloud and it runs in a westerly direction to join the Daneinshaw Brook at Bath Vale. Where the stream leaves the slopes of Bosley Cloud, not far from the site of the spring, it is dammed to form a small pool that provided the power for a mill in the village of Timbersbrook.

The mill at Timbersbrook was built some time prior to 1817. In that year an advertisement announced the sale of some cotton spinning machinery at a mill in Timbersbrook. The amount of machinery offered for sale consisted of *"four spinning frames for twist, five carding engines, two roving frames, one stretching frame, one drawing frame, one lathe cutting engine, quantity of cans & various articles."*[1] This sale may have been caused by a change from cotton spinning to silk throwing but the amount of machinery involved was quite small and may not have represented the whole contents of the mill. However, it was about this time that Jonathan Broadhurst became involved in silk throwing at Timbersbrook. When he, in turn, offered the mill for sale in 1825 it was described as *"turned by a powerful wheel with a plentiful supply of water & is heated by steam, containing 1300 swifts, 274 doz. spindles with doubling engines, bobbins, trams, etc."* The mill was accompanied by a house, eight cottages and four acres of land.[2]

In 1829 the mill was still involved in silk throwing when it was again offered for sale. Fortunately this advertisement gave some indication of the size of the mill as it was described as a *"...factory well adapted for any kind of manufacturing business containing in length 15½ yards and in breadth 13 yards being four stories in height with a regular supply of water sufficient to turn any kind of machines."* Among the additional properties to be sold with the mill was a house, five cottages, five acres of land and a stone quarry. An extra attraction was considered to be that the mill was *"only a few hundred of yards from the new canal connecting the Peak Forest and Red Bull canals which will shortly be complete"*, a reference to the Macclesfield Canal that was completed in 1831. Although the owners were also attempting to rent the silk throwing machinery they were hedging their bets by declaring that *"The stream of water running by the mill is a constant & powerful one and the place altogether is suitable for a cotton, paper, silk, or flax factory or a bleaching ground."*[3]

In 1840 when the tithe map and apportionments were drawn up, the silk mill and reservoir was listed as being owned by John Washington. The Washington family were farmers and landowners in the Timbersbrook and Daneinshaw area and they may well have been the original builders of the mill. Once again in 1840 the mill was advertised for sale, this time by John Washington, as follows:-

"To be sold or let by private treaty, a silk mill or factory at Timbersbrook in Buglawton near Congleton with the water wheel, engine house, and other conveniences.

The mill is four stories high, substantially built of stone, and turned by water of which there is a constant supply by means of a spring and reservoir. It is well adapted for silk spinning or manufacturing, as well as silk throwing, for which last purpose it has been worked for some time by Mr. Samuel Harthan whose tenure therein expires on 12th May next when possession can be given.

A considerable quantity of suitable machinery for silk throwing stands in the factory and will be sold or let with the premises.

Together with a respectable dwelling house suitable for the residence of a small manufacturer, six cottages, and about thirteen acres of meadow, pasture or common land. The property is freehold and will be disposed of on modest terms."[4]

FIGURE 94. THE FOUR STOREY, ALMOST SQUARE, MILL AT TIMBERSBROOK AS DESCRIBED IN THE 1829 SALE ADVERTISEMENT.

As can be seen from the terms of this advertisement, an engine house is mentioned indicating that steam power had been installed at sometime after 1829 to supplement the use of water power. This sale was undoubtedly unsuccessful as John Washington advertised it again in 1845. Although Samuel Harthern moved from Timbersbrook to Hurst Mill (*q.v.*) at Biddulph in 1840, the 1845 advertisement for Timbersbrook Mill quotes the last occupant as "*the late Samuel Harthern*" so presumably the mill had not been at work between 1840 and 1845. To increase its potential desirability it was thought that "*at a trifling expense it may be converted to a corn mill.*"[5] However, it continued to be used for throwing silk after 1845 by the Harthern family, in the guise of Samuel's brothers John, Joseph and Ezra Harthern who operated the mill until 1860. Judith Harthern, who was Samuel's widow, had continued to operate Hurst Mill with her sons until the age of about 65 when she moved back to Timbersbrook. She was described as a "*retired silk throwster*" living with her son Isaac, a surgeon, and her other son Samuel who operated Timbersbrook Mill. The depression in the silk industry caused by the Free Trade Act of 1860 does not appear to have immediately affected Timbersbrook and by 1871 Samuel had been joined by brother Isaac in running the silk throwing business at the mill.[6]

The connection between the Harthens and Timbersbrook Mill was broken by 1878 when William Dutton took over the mill to continue throwing silk. Some idea of the employment offered by this small and out of the way silk throwing mill can be gained from the fact that in 1881 William Dutton was listed as employing 24 hands.[7] He continued to operate the mill right up to 1899 when the owner, a Mr. H. H. Baker, sold the mill to a Manchester businessman called Thomas Royle which brought the silk throwing trade at Timbersbrook to an end.[8] Thomas Royle had another trade in mind when he bought the mill and in 1902 the Silver Springs Bleaching & Dyeing Co. Ltd. was set up to operate from Timbersbrook Mill with Thomas Royle as managing director.[9]

Over the years there were a number of managing directors but eventually in the 1930s the company was managed by F. E. Mason who continued the business until 1961 when the factory was closed under the government cotton reorganisation scheme. In 1962 the mill and its estate were conveyed to the Piece Dyeing, Raising & Finishing (only) Realisation Co. Ltd. for £45,000, but in 1963 the mill was sold at auction to Timbersbrook Mill Ltd. for only £9,000. After that it passed through various hands until in 1973 it was purchased by Congleton Borough Council at auction for £32,750.[10] The Borough Council then proceeded to demolish the mill and associated settling tanks etc. and converted the whole area into a car park, so ending a long association between Timbersbrook Mill and the textile trade.

References

1. *Macclesfield Courier*, 19th April 1817.
2. *Macclesfield Courier*, 8th January 1825.
3. *Macclesfield Courier*, 25th April 1829.
4. *Macclesfield Courier*, 2nd May 1840.
5. *Macclesfield Courier*, 1st March 1845.
6. CRO, Buglawton Census Returns, 1851, 1861 & 1871.
7. CRO, Buglawton Census Return 1881.
8. *Congleton Chronicle*, 1st September 1899.
9. Kelly's Directory of Cheshire, 1902.
10. Deeds held by Congleton Borough Council.

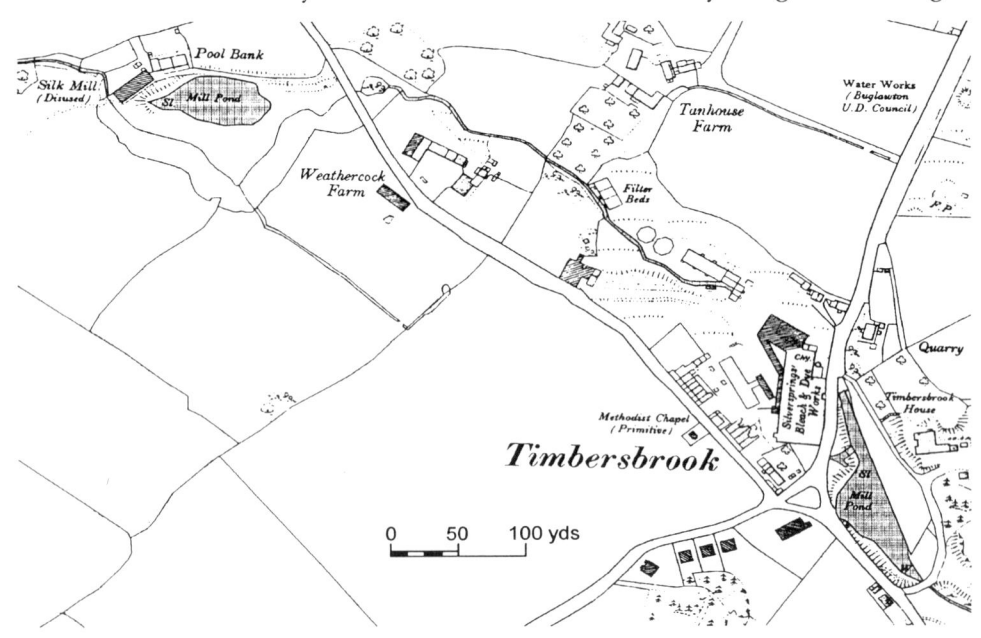

FIGURE 95. PART OF THE ORDNANCE SURVEY MAP OF 1925 SHOWING THE CLOSE PROXIMITY OF TIMBERSBROOK AND POOL BANK MILLS.

32. POOL BANK MILL SJ 891629

About ¼ mile downstream from Timbersbrook Mill, on the same stream, is the site of Pool Bank Mill, or Woodhouse Mill as it was sometimes known. This is a considerably older mill than the one at Timbersbrook being originally a corn mill that appears on Burdett's map of Cheshire in 1775. Just how long before this date the corn mill existed is not known. It appears in the Land Tax returns for 1783 to 1785 as being owned by Joseph Perkin when the mill was called "Brookes Mill" having a value of 10s-0d. The ownership details and the value of 10s-0d can be traced over the years in these taxation records to confirm that Brookes Mill is indeed the same mill as Pool Bank Mill.[1]

In 1786 the ownership changed from Joseph Perkins to the Porter family who continued to own the mill until after the Land Tax records end in 1831. During this period there were a number of occupiers of the mill listed, such as Mr. Stubbs, John Read(e), Edward & George Comberledge, and Thomas Harden followed by his son Samuel. By the time Samuel was in occupation in 1825 the mill was known as Pool Bank Mill and must have been converted to silk throwing as both the Hardens (or Hartherns) are known to have been in that trade. When this change occurred is not known but could easily have been shortly after the mill was bought by John Porter in 1786 as it is then that the name Brookes Mill was dropped.[2]

To further complicate the name situation, when the mill was advertised for sale in 1832 it was called Woodhouse Mill.

"*Lot 2 For Sale. A waterpowered mill called Woodhouse Mill also situated in Buglawton (formerly used as a corn mill but recently used for silk throwing) and two dwelling houses with outbuildings and land surrounding the same, now in the occupation of Mr. Samuel Harthern.*"[3]

Two years later when an attempt was made to let the mill by Mr. Lewis Porter further details were provided with the mill described as "*being two storeys high, 30 yards long and 10 yards wide with counting house, staff room, and waterwheel complete with a good supply of water capable of turning 200 doz. and a large reservoir belonging to the same.*"[4]

In 1839, when the mill was offered for sale at auction, it was occupied by Charles Ginders & Dennis Bradwell who were silk throwsters.[5] They probably arrived at the mill in response to the advertisement in 1834 but by September 1840 they were "*declining the silk throwing business*" and liquidating their assets with a sale of their machinery as follows:-

"*To be sold at Auction. That very excellent silk mill containing 31 doz. spindles, one throwing mill containing 10½ doz. spindles, two hard silk engines containing 76 swifts in each, one cleaning engine containing 50 bobbins, one doubling engine containing 70 bobbins, four ribbon power looms, warping mill, rubbing machine, turn off capital blocking machine, doubling and quill wheels, one excellent soft silk engine nearly as new containing 76 spindles, counter with mahogany top, desk, beams, scales & weights, joiner's bench, lathe complete, engine & cleaner, bobbins, trams, warp and shute bobbins, with numerous other articles of the trade.*"[6]

A month later the mill itself was offered to let being described as

"*The mill is stone built, two storeys high, 33 yards long and 9 yards wide with good lights and turned by waterpower whose stream never fails and the waterwheel is in excellent condition. In the mill is a quantity of machinery of modern make in the highest state of preservation, and the mill is fitted with a boiler, steam piping, gearing, and every convenience for carrying on the trade of silk throwing and manufacture.*"[7]

It seems that the mill was purchased by Thomas Templeton, a speculator who bought a number of mills in the Congleton/Buglawton area but who went bankrupt in 1841, causing the mill, house, cottages and household goods to be auctioned off "*at no reserves*".[8] The next occupant was George Brown who presumably bought the mill at the auction in 1841. In the 1851 census he was described as a silk manufacturer who lived with his family by the mill, his parents occupying one of the cottages. He ran the mill under the firm of Brown & Kennerley until 1858 when "*A waterpowered mill called Woodhouse or Pool Bank Mill*" was sold at auction to Daniel Barrow for £905.[9] An estate map drawn up in the same year shows that the waterwheel was centrally positioned in the mill building.

Daniel Barrow bought the mill as an investment rather than a place of work. He raised £550 on the property and tried to sell off the machinery and let the empty mill. Once again the sale notice gives a good inventory of the silk throwing machinery which consisted of

"*three winding engines and two single sides of winding containing together upward of 330 good lancewood swifts, two cleaning engines containing 176 spindles, one doubling engine containing 56 spindles, three spinning mills 25 doz. each, one mill half spinning half throwing containing 8 doz. spinning 14 doz. throwing, and one throwing mill containing 28 doz., with mill horses, reals, shafting, steam piping, bobbins, shafts, etc.*"

The mill building itself was described as "*14 windows long, two stories high and 9 yards wide*"[10]

When Daniel Barrow died in 1861 the mortgage on the property devolved on his widow Mary Jane Barrow and his sister Martha Barrow who had to have a guarantor in the shape of a Richard Timmis. The role

FIGURE 96. THE REAR OF POOL BANK MILL TODAY.

of guarantor was taken over in 1869 by Richard Low Ginder of Daneinshaw Mill (*q.v.*).[11] A number of firms continued the silk trade at the mill such as William Wood & William Souter who were listed as silk waste dressers in 1864.[12] Even in the depressed state of the silk trade after 1860 William Tagel, another silk throwster, was employing 62 hands in 1881.[13] At some stage after this date the mill was occupied by G. & W. Dutton operating as silk throwers as well as at Timbersbrook Mill (*q.v.*). Silk throwing came to an end with the demise of the Dutton firm in 1899. The mill was then always used in conjunction with Timbersbrook Mill, initially with the Silver Springs Bleaching & Dyeing Company until 1962, and then with the Piece Dyeing, Raising & Finishing (only) Realisation Co. Ltd. until they sold the mill for £1250 in 1964 to the Brookdale Development Company Limited.[14] In 1972 the mill was sold by Walter Wolfson to Irene Rowland for £2500.[15]

Today Pool Bank Mill or Woodhouse Mill is still to be seen, much as described in the 19th century advertisements, being used by a light engineering company. It is a two storey, stone built, building with a slate roof, stretching for 14 bays along its long side.

The roof is supported by simple wooden king posts and beams. Pool Bank House and Pool Bank Cottage, next to the mill, are still lived in but the pool just to the east of the mill has been filled in.

References

1. CRO, Buglawton Land Tax Returns.
2. *Ibid.*
3. *Macclesfield Courier*, 21st July 1832.
4. *Macclesfield Courier*, 13th December 1834.
5. *Macclesfield Courier*, 7th September 1839.
6. *Macclesfield Courier*, 12th September 1840.
7. *Macclesfield Courier*, 10th October 1840.
8. *Macclesfield Courier*, 25th December 1841.
9. CRO, Mortgage 1858, DCB/1595/33/4.
10. *Macclesfield Courier*, 10th July 1858.
11. CRO, Deeds, DCB/2114/143
12. Morris's Directory of Cheshire, 1864.
13. CRO, Congleton Census Return, 1881.
14. Deeds of Pool Bank Mill in the possession of the owner Mr. P. Surtees.
15. *Ibid.*

---

33. BATH VALE MILL   SJ 873633

The small stream known as the Timbersbrook joins the Daneinshaw Brook about half a mile from its meeting with the River Dane. Bath Vale Mill was located on the land between the Timbersbrook and the Daneinshaw Brook at their confluence. In the survey made in 1770 for the Land tax this land was owned by John Vaudrey but there was no suggestion of a mill on the site at that date.[1]

John Vaudrey was involved in cotton spinning on Lawton Street in Congleton in 1788 and was described in a directory of 1789 as a cotton merchant.[2] Richard Martin's notebook of that time also mentions John Vaudrey with respect to the purchase of raw cotton.[3] However, it is not until 1800 that John Vaudrey is listed in the Land Tax returns as owner and occupier of a cotton mill at Bath Vale valued for a tax of £1-0s-0d which gives a reasonable indication of when the mill was built. John Vaudrey was listed as a cotton spinner

FIGURE 97. VIEW OF ONE OF THE OLD COTTON MILL BUILDINGS AND ASSOCIATED COTTAGES AT BATH VALE.

at Bath Vale until his death in 1828. In his will he mentions *"cotton and silk factories, steam engine, engine house, gas works, apparatus and buildings"*.[4] Obviously, by the time of his death, the waterpower at Bath Vale had already been augmented by steam power. The silk factories mentioned in the will were located in Macclesfield. John Vaudrey was succeeded by his two sons William & Charles Vaudrey who continued in partnership as cotton spinners until 1838. As Bath Vale Mill lies on the boundary between Buglawton and Congleton it was mentioned in a Boundary Commissioner's report of 1830 as having four buildings and a large pool.[5]

When the newly built Macclesfield Canal was opened in 1831 it passed across the Daneinshaw Valley just over ½ mile east of Bath Vale. In 1835 Charles Vaudrey leased some land from Lewis Porter for a coal wharf north of what was, then, bridge 76 on the canal at a rent of £1-0s-0d per year for 99 years. The Vaudreys were also able to charge ½d rent from Brookhouse Farm for the farm's use of the road that the Vaudreys constructed to the wharf.[6] Obviously steam was by now the dominant power at the mill requiring these arrangements to ensure a cheap delivery of coal by this new means (for Congleton) of transport. After William Vaudrey's death in 1838 Charles continued cotton spinning at Bath Vale until about 1850 when he leased the mill to Joseph Berresford.

Charles Vaudrey died in 1858[7] but Joseph Berresford maintained the cotton spinning business until 1865 when he was declared bankrupt. As part of the winding up of his business interests the machinery at Bath Vale was auctioned off as follows:-

"To be sold by auction by the mortgagees of Mr. Joseph Berrisford [sic] on the premises of Bath Mills in Buglawton Vale, Congleton, Cheshire.

The excellent machinery for preparing and spinning cotton comprising 36 inch opener by Platt; ditto; two beater ditto with lap by Sydall & Grime; 28 inch two beater ditto, ditto, by Kay; 17 carding engines, 48 inch on the wire; drawing frame four heads each four deliveries; three ditto, three heads, each six deliveries; three slubbing frames containing 120 spindles, 10 inch lift; one intermediate frame, 96 spindles: 10 roving frames containing 976 spindles; 13 pairs of mules containing 12,104 spindles; 15 throstles containing 3,184 spindles; winding frames 19 reels; two warping mills, skins, cans, bobbins, driving straps, mechanic's tools, lathe, etc. sundry stores and other effects."[8]

The mill was to be let separately. Three years later in 1868, on the death of a Mr. Bagshaw who must have taken over after Joseph Berresford's bankruptcy, the Vaudrey family decided to sell the Bath Vale Mill, giving a description of the premises as follows:-

"To be sold by private treaty the valuable spinning mills known as Bath Vale Mills situated in Buglawton, near Congleton, in the County of Chester latterly occupied by Mr. John Bagshaw deceased, comprising two mills (one of which is a recently erected stone building) engine house, boiler house, blowing room, office, warehouse, storehouse, mechanic's shop, gas house, and large yards.

There are two excellent steam engines, mill gearing and shafting, steam, water and gas piping and other fixtures throughout. The premises are well supplied with water."[9]

In spite of most of the attention being given to the steam engines it was still considered necessary to mention the water supply available.

It is quite possible that the mills were impossible to sell as they were recorded as disused in 1876 and also in 1882.[10] The large scale Ordnance Survey map of 1875 shows that the pool had badly silted by this time. However, by the middle of the 1880s new tenants in the form of Conder & Company took over the occupation of Bath Vale Mill bringing with them the trade of silk spinning, which prevailed until midway through the first decade of the 20th century.[11] In the 20th century the mill was used for various branches of the textile trade such as cotton weaving and shirt making but was eventually acquired by an engineering company called Polarcold. During the 20th century the

99

mill buildings were altered and added to. The area of the mill pond was filled in and built over as well. The premises are still in business today as an engineering works and one of the early stone buildings from the cotton spinning era could be seen until recently. On the canal the remains of "Vaudrey's Wharf", with its short canal arm, can still be seen just to the south of what is now Brookhouse Lane Bridge.

References

1. Congleton Survey, 1770, held by the editor of the Congleton Chronicle.
2. Cowdray's Directory of Chester, 1789.
3. CRO, Richard Martin's Notebook, DCB/1595/7/11.
4. CRO, Will of John Vaudrey, 1828.
5. Boundary Commissioner's Report, 1830, Congleton Public Library.
6. CRO, Lease of Wharf 1835, DCB/1595/33/3.
7. Commemorative stained glass window, Astbury Parish Church.
8. *Macclesfield Courier*, 1st April 1865.
9. *Macclesfield Courier*, 5th October 1868.
10. Ordnance Suurvey maps, 1876 & 1882.
11. Kelly's Directories of Cheshire, 1890 - 1906.

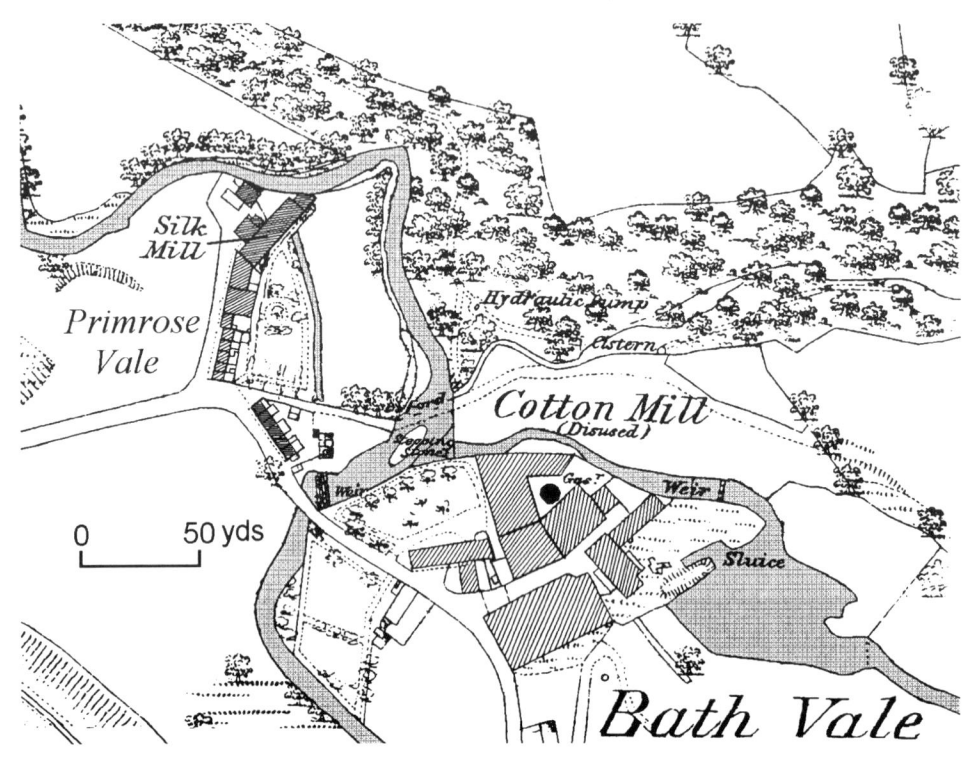

FIGURE 98. PART OF THE 1875 ORDNANCE SURVEY MAP OF BATH VALE WITH THE MANY BUILDINGS OF THE COTTON MILL WITH ITS MILL POOL FORMED ON THE TIMBERSBROOK. JUST DOWNSTREAM OF THE CONFLUENCE OF THE TIMBERSBROOK AND THE DANEINSHAW BROOK A SILK MILL IS INDICATED AT PRIMROSE VALE.

---

34. PRIMROSE VALE MILL   SJ 872634

Just to the west of Bath Vale, about 50 yards downstream of the confluence of the Timbersbrook and Daneinshaw Brook is the area known as Primrose Vale. There was a mill at Primrose Vale much earlier than at Bath Vale. It was recorded in the Land Tax returns from 1785 to 1796 as a paper mill owned by William Drakeford and occupied by Elizabeth Fearnley.[1] However, in the survey of Congleton made in 1775 the land owned by John Vaudrey at Bath Vale was described as *"land at the paper mill"*.[2] It was also probably indicated on Burdett's map of Cheshire published in the same year.

Between 1796 and 1806 William Drakeford was described as both the owner and the occupier of the paper mill. After 1806 William Drakeford was listed as the owner but the occupiers were noted as himself and Goodwin, and from 1810 to 1819, as himself and Krinks.[3] It is not clear what the relationship was between Drakeford and the other listed occupiers, possibly they were employees managing the paper mill with Drakeford or possibly the function of the mill was split between paper making and some other trade. It is known that Krinks went on to be a silk throwster at Lower Washford Mill(*q.v.*) and the following entry

FIGURE 99. A HOLLANDER SIMILAR TO THAT PROBABLY USED IN THE PAPER MAKING PROCESS AT PRIMROSE VALE MILL IN THE 18TH CENTURY.

appeared in the Macclesfield Courier in 1811, namely:-

"*Births - At the paper mill near Congleton, Mrs Barlow wife of Mr. Charles Barlow, silk manufacturer, of two boys who with their mother are likely to do well. It is not more than two years since she had two girls at birth who are both living.*"[4]

This was at a time when both Drakeford and Krinks are listed as the occupiers by the Land Tax returns, so where Charles Barlow practised silk manufacturing is unknown. It is possible that the mill was shared between paper making and silk throwing, with Drakeford continuing to manufacture paper and Krinks leasing a part of the mill for silk manufacture. Charles Barlow was possibly a sub-tenant or employee of Krinks. This scenario is purely conjectural but fits those facts that are known.

That some of the mill was still being used to make paper is confirmed by the allocation of excise number 251 to William Drakeford in 1816.[5] (This was so that the tax on paper could be collected at the point of manufacture.) Also paper makers continued to appear in local parish registers up to 1820. William Drakeford appeared in a Trade Directory as a paper maker in 1818,[6] and William Corry, writing about Congleton in 1817, mentioned that there was "*one paper mill in Congleton*".[7] In fact it was only in 1821 that the General Excise Letter recorded that William Drakeford had ceased making paper and excise number 251 was re-assigned.[8]

From this date, although the mill was still owned by William Drakeford, Charles Barlow is recorded as the occupier with the premises described as a silk mill.[9] Charles Barlow continued the silk trade up until his death in 1840 when his brother John succeeded him at Primrose Vale.[10] Five years later there was an attempt to lease the mill which gives a description of both the mill and its machinery as follows:-

"*To Silkmen & Throwsters - To Be Let*
All that excellent silk mill with machinery, a commodious dwelling house, outhouses, and land situated at Primrose Vale, Congleton, lately occupied by John Barlow.
The silk mill is four stories high, 90 feet long by 25 feet wide and contains about 1300 swifts, 700 cleaners, 400 doublers, 350 dozen of spinning and 110 dozen of throwing mill spindles turned by a waterwheel supplied by the River Daneinshaw and an excellent steam engine of 15 horse power. The dwelling house is near the mill with gardens, pleasure grounds, carriage house, stables, etc. also about five statute acres of superior meadowland near the premises."[11]

It is unlikely that the building described is that used from before 1775 for paper making which would not have required such extensive premises, presumably it had been rebuilt since papermaking ceased in 1821. The advertisement makes clear that the mill was engaged in silk spinning as well as throwing. The steam engine used to supplement the waterwheel was quite powerful at 15 h.p. for such a relatively small silk mill.

After this attempted leasing, Charles Barlow's son, Charles William, continued to operate the family silk business at Primrose Vale. The Barlow family's involvement in the silk trade in Congleton grew with Charles William operating a mill in Bridge Street (*q.v.*), Congleton, as well as that at Primrose Vale, and his brother George was also in the silk trade on his own account. This involvement came to an end in 1866 with the imprisonment of George Barlow on an accusation of theft.[12] The disappearance of the Barlow family from Primrose Vale did not end its use as a silk mill as the Barlows were superseded by Eli Hunt and then by his son, John Hunt, who continued silk throwing until the mid 1880s.[13]

Conditions in the silk trade were such that throwing at Primrose Vale came to an end by the late 1880s and the mill was taken over by Walter Hoyle for manufacturing towels. In 1890 the firm became known as Burberry Bros. (late of W. Hoyle & Co.) who made turkish and fancy towels at Primrose Vale.[14] They were followed in the mid 1890s by Pears, Moffett & Co., also towel makers, who were also at Lower Washford Mill.[15] In 1898 the following item appeared in the Congleton Chronicle:-

"It is good to see that Mr. James Parkes of the Olde Kings Arms has purchased the commodious Primrose Vale Mills which he is fitting with up-to-date machinery, and will soon find employment for a large number of hands in the manufacturing of towels."[16]

However, it would appear that this was not a successful venture so the mill eventually ceased to be of any use to the textile industry and was finally demolished leaving virtually no trace today of its location. The only buildings remaining are the adjacent dwelling house, now called Primrose House, and the coach house which has been converted into living accommodation. The weir on the Daneinshaw Brook can still be seen from the road bridge near Bath Vale. It is slightly concave, made of stone blocks about one foot square, and still in excellent condition. The leat to the mill is now completely overgrown.

References

1. CRO, Buglawton Land Tax Returns.
2. Survey of Congleton, 1770, in the possession of the editor of the Congleton Chronicle.
3. CRO, Buglawton Land Tax Returns.
4. *Macclesfield Courier*, 8th June 1811.
5. PRO, Excise General Letter, 1816.
6. Commercial Directory, 1818.
7. Corry, W., *History of Macclesfield*, 1817.
8. PRO, Excise General Letter, 1821.
9. CRO, Buglawton Land Tax Returns.
10. CRO, Will of Charles Barlow.
11. *Macclesfield Courier*, 30th August 1845.
12. Henry Madden's Diary, *Congleton Chronicle*, July 1974 - August 1975.
13. Morris's Directory of Cheshire, 1874.
14. Slater's Directory of Cheshire, 1890.
15. Kelly's Directory of Cheshire, 1896.
12. *Congleton Chronicle*, 15th September 1898.

---

35. BANK MILL   SJ 869635

Downstream of Primrose Vale Mill, about half way from that mill to the main Congleton to Buxton Road, alongside a meander in the stream, is the site of Bank Mill. It has been said that Bank Mill was built in 1810 as a silk mill by Thomas Johnson,[1] but he would only have been aged 19 at that time, making it unlikely that he would have had the capital or experience for such a venture on his own account. The Johnson family was very active in the silk industry in the Buglawton area and an entry in the Buglawton Enclosure document of 1813 might be relevant. It states:-

"I, Francis George Glynne Johnson of Congleton in the County of Chester claim to be entitled to an allotment of common in Buglawton under the Buglawton Inclosure Act in respect of messuages and a silk mill in the occupation of Mr. Butler Johnson."

Unfortunately the site of the silk mill mentioned is not specific enough to ensure that it was indeed Bank Mill.[2]

Certainly, from around 1820, Bank Mill was always associated with Thomas Johnson who lived at Bank House which was situated on top of the valley side to the north, overlooking the site of the mill. A sale advertisement for the mill in 1832 reveals a few details as follows:-

"The power is water with a steam engine attached as auxiliary power capable of throwing 500 lbs to 1000 lbs of organzine per week, according to the character of the silk. The quantity of spindles is about 700 dozen."[3]

Although by the time it reaches Bank Mill the Daneinshaw Brook is a moderate sized river it was still considered necessary to supplement the waterwheel with a steam engine. The number of spindles employed only indicates a moderate size of operation.

Five years later when another attempt to sell the mill took place rather more information was made available when it was described as:-

"One of the most complete silk throwing establishments in England standing five stories high, 27 yards long and 9 yards wide replete with every convenience. The machinery is nearly new and of the best construction. The premises are walled as a garden for the convenience of perambulation by the watchman. The power is chiefly water with a steam engine of five horse power, boiler of twelve horse power as an auxiliary power and to steam the factory. There is one room empty which may be filled with power looms.

The present is one of those rare instances where premises so pleasantly situated can be obtained, standing alone in a pleasant valley surrounded with picturesque scenery, so much so, that the superintendence is but like an agreeable recreation.

Apply to Thomas Johnson, Bank House, Buglawton."[4]

The prose used makes one wonder why he wished to sell the mill at all and, if he had been successful, no doubt he would have been able to start a new career as an estate agent! Although the mill would have been about 13 bays long with five stories, the small size of steam engine may seem surprising but was in fact fairly typical as waterpower would have been used whenever possible. Even with such a small amount of power available Thomas Johnson was employing as many as 180 hands in 1840.[5]

Thomas Johnson was unsuccessful in selling the mill in 1837, in fact, he never was able to sell it in spite of another attempt in 1847[6] and he continued with silk throwing at the mill up until his death in 1867. Having no direct heir to take over the business, the mill was offered for sale at auction. An interesting statement in the auction advertisement was that there were "*two iron waterwheels attached (the principal power being water) with steam engine, a boiler as an auxiliary power or to warm the mill.*"[7] This is the only mention of two waterwheels at Bank Mill and is quite puzzling. The large scale Ordnance Survey map shows the mill with only the one, short leat running from its weir into the centre of the mill. If there had been two waterwheels positioned in the centre of the mill they would both have been mentioned in the various attempts to sell the mill. Closer examination of the weir shows that there is a position, about four feet wide, for a waterwheel in the weir itself, separate from the mill. The purpose of this second wheel is unknown but may have been used to pump water up to Bank House on the top of the valley side either for a potable supply or even just for the pleasure gardens that surrounded Bank House.

The timing of the auction in 1867 was disastrous as no-one was interested in entering or expanding in the silk trade at that time, trade being very bad with many mills shut and hands out of work.[8] Consequently Bank Mill was just left to decay slowly. In 1923 interest in the mill was stirred by an item in the local press which reported on a proposal to set up a doubling and bleaching works at the mill eventually employing up to 300 workers and that the waterwheel, still on site, was to be used to generate electricity for lighting. Like so many times in this mill's history this proposal came to nothing and not long after the mill was demolished.[9]

Today the only remain visible is the weir in the Daneinshaw Brook and some of the wall around which, presumably, the watchman perambulated. At some stage in the later 20th century someone had fitted a metal fabricated waterwheel, about 8 feet in diameter and 1 feet 9 inches wide in the position in the weir, presumably in an attempt to generate electricity. The remains of a sprocket and chain drive from this waterwheel were also visible. The mill site is now part of a garden but the ground is still thickly covered with bricks from the demolished mill. The tailrace from the mill crosses the adjacent field in a culvert and the stone lined inspection shaft half-way along the tailrace can still be found in the middle of the field. Bank House, Thomas Johnson's residence, went the same way as the mill and the site was occupied by large 20th century factory premises. However, permission was granted in the year 2000 to demolish this factory and the site of Bank House is now covered with modern housing. In spite of all this recent building the mill site is still "*...situated in a pleasant valley surrounded with picturesque scenery*" that Thomas Johnson so enjoyed and the footpath along the valley is known as "Tommy's Lane".

References

1. Stevens, W. B., *History of Congleton*, University of Manchester, 1970.
2. CRO, Buglawton Inclosure Act 1813, DCB/1716/7/2.
3. *Macclesfield Courier*, 3rd November 1832.
4. *Macclesfield Courier*, 19th August 1837.
5. Head, R., *Congleton Past & Present*, 1887.
6. CRO, Will of Thomas Johnson.
7. *Macclesfield Courier*, 2nd may 1867.
8. Henry Madden's Diary, *Congleton Chronicle*, July 1974 - August 1975.
9. *Congleton Chronicle*, 11th November 1983.

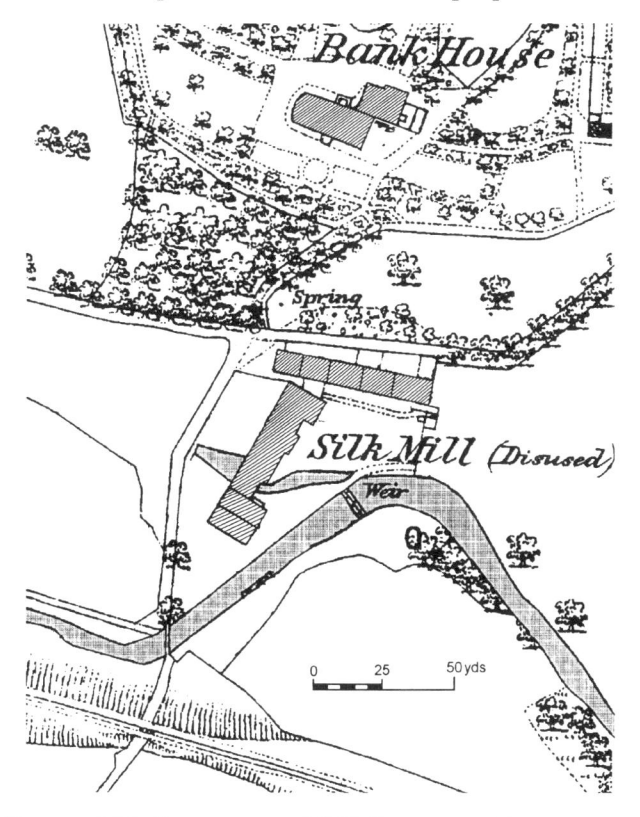

FIGURE 100. PART OF THE 1875 ORDNANCE SURVEY MAP SHOWING THE POSITION OF BANK MILL AND BANK HOUSE.

36. DAVENSHAW MILL SJ 866635

As the Daneinshaw Brook approaches its confluence with the River Dane in Buglawton, it makes a very tight "U" bend just before running under the Congleton to Buxton road, now the A54. It is on the land enclosed by this "U" bend that Davenshaw Mill, originally called Nearer Daneinshaw Mill, was built.

This land was leased from Jonathan Harding by Philip Antrobus in 1761 when there was a only a dwelling and croft occupied by Thomas Morris on the site. However, Philip Antrobus was granted *"...the power and liberty to erect a mill or any other building in such a manner as therein particularly mentioned but also to make cut or drive and arch and cover over a sough sluice or tunnel from the eastern side of the said croft and the said garden or orchard to the mill or other building so to be erected."*[1] The mill was no doubt built soon after the date of this lease, possibly as a flint mill.[2] Certainly a mill was present on the site before 1777 because it was indicated then on Burdett's map of Cheshire at this date. Some time in the 1770s the mill was converted into a cotton spinning mill.

However, at this time Richard Arkwright had a monopoly on the use of the water frame and carding engine through patents that he had taken out in 1769 and 1775 respectively. These two machines were essential for the machine spinning of cotton but Arkwright would only allow other cotton manufacturers to use these inventions on payment of a heavy licence fee. There were some businessmen who were not impressed with the idea of paying Richard Arkwright a share of their profits and so a number of "pirate" cotton spinning enterprises appeared. Eventually Richard Arkwright sued ten cotton spinners for breach of his 1775 patent on the carding engine. One of the defendant firms in this legal action was a John Barnes of Congleton, but no information was given concerning the location of his mill.[3] In 1781 the case came to court with Arkwright losing due to his patent being *"obscure and unintelligible"*[4]

In 1783 the firm of John Barnes advertised their cotton mill for sale, probably the costs and disruptions of the patent case with Arkwright had caused them to run out of money. The advertisement appeared in the Manchester Mercury as follows:-

"To be Sold at Auction

The fee simple and inheritance of a cotton mill with all the machines as they now stand together with a small meadow of near a Cheshire acre, and also the leasehold for 1000 years of a cottage, garden and other premises sufficient to build in an extensive manner situated and being on Danenshaw [sic] Brook in the townships of Buglawton and Congleton.

The above premises are very desirably situated for the manufacture of cotton twist being in a county where labour is cheap, and plenty of hands may readily be got. The mill never wants for water in the driest of seasons and is never in back water but in the highest of floods and even then, not in flood above an hour or two. There is a very good turnpike road from Manchester to Congleton, lately made, and regular carriers upon the road. The above premises did belong to the firm of Thomas Morriss of

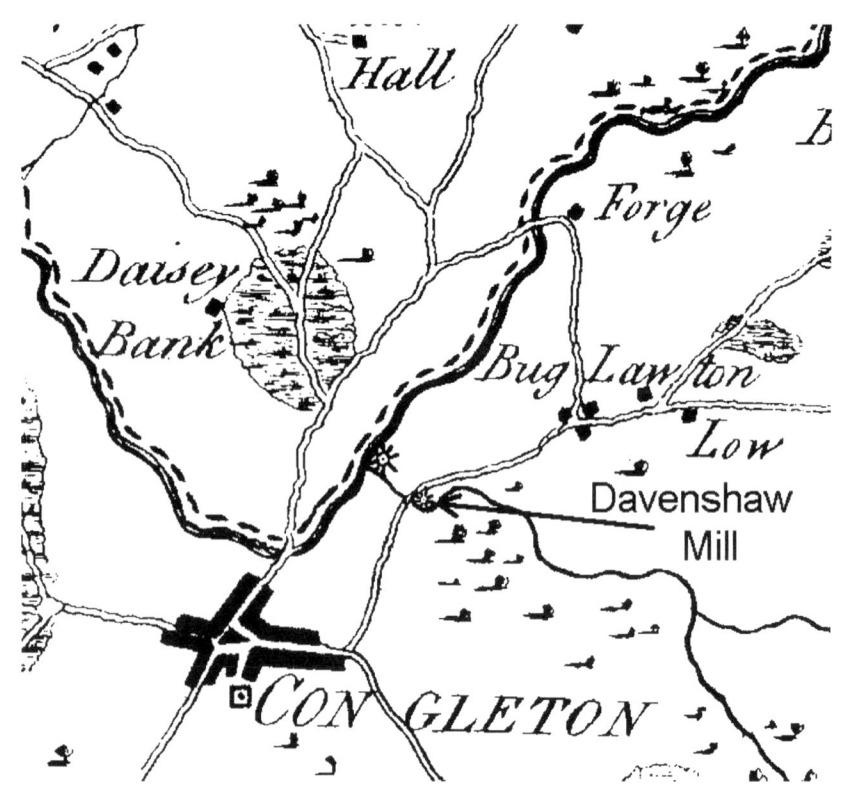

FIGURE 101. PART OF BURDETT'S MAP OF CHESHIRE, 1777, WITH A WATERWHEEL SYMBOL ON THE DANEINSHAW BROOK ALONGSIDE THE TURNPIKE ROAD FROM CONGLETON TO BUXTON. (CCALS, PM12/16)

Congleton, James Robinson, Henry Mather and John Barnes all of Warrington. The said partnership being now totally dissolved all persons who may have any demands against the said partnership prior to the above date are desired to send in their accounts with all convenient dispatch to the said James Robinson, Henry Mather and John Barnes in order that the accounts may be examined and discharged with all convenient speed."[5]

The location of the mill given in this advertisement as being on the Daneinshaw Brook in the townships of Buglawton and Congleton shows that the mill in question is, in fact, Davenshaw (or Nearer Daneinshaw) Mill. The later tithe maps and apportionments show that this is the only mill on the Daneinshaw Brook where part of the building is in Congleton and part in Buglawton. This location is confirmed by the involvement of Thomas Morris(s), who was the original occupier of the land when it was first leased to become a mill.

In 1785 the patent dispute with Richard Arkwright came to a head. In February 1785 Arkwright managed to get the 1781 decision overturned which caused uproar in the cotton trade. So much so that a writ of *scire facias* was taken out to try the validity of all Arkwright's patents once and for all. In the end, the carding patent was annulled and Arkwright even lost the water frame patent because it was proved that he was not the original inventor of this machine and that it was based on the work of Thomas Highs from Leigh in Lancashire. It is perhaps significant that three members of the partnership at the Davenshaw Mill came from Warrington, not far from Leigh, and were, in all probability, aware of the true circumstance concerning the invention of the water frame. This would no doubt have been influencial in decision to ignore Arkwright's patents.[6]

The ownership and use of the mill after this sale is not clear, but the Land Tax returns suggests that in 1786 it might have been owned by Richard Martin. In 1788 a partnership was drawn up between Richard Martin and two brothers, John and Thomas Whitfield, that stated that they would work *"together in the art trade and business of spinning cotton by water into twist and selling such twist henceforth during the term of 21 years. The firm to be called Thomas Whitfield & Company with the manufacturing part carried on at the factory belonging to partners on the River Daneinshaw called Near Daneinshaw and Further Daneinshaw and all sums of money shall be advanced by the partners in equal shares."* Richard Martin & John Whitfield were to receive £100 per year on top of their share of the profits for running Further Daneinshaw Mill and Nearer Daneinshaw Mill respectively, and Thomas Whitfield was to sell the twist in Manchester at 2.5% commission. Nearer Daneinshaw Mill eventually

FIGURE 102. A REPRESENTATION OF A WORKER OF THE LATE 18TH CENTURY OPERATING ONE OF THE WATER FRAMES USED IN THE EARLY MECHANISED SPINNING OF COTTON. (*Helmshore Textile Museum*)

became known as Davenshaw Mill and Further Daneinshaw Mill can be identified as the textile mill just north of the Congleton to Leek road near Daneinshaw Corn Mill.[7]

However, Richard Martin was in business on his own account as a cotton spinner, before this partnership, with customers in Glasgow and elsewhere.[8] Although Further Daneinshaw Mill is mentioned in the partnership of 1788, it is quite possible that it was not built until that year and that Richard Martin was using Nearer Daneinshaw Mill in 1786/7 (see Daneinshaw Silk Mill). There is an interesting letter from Richard Martin to a Mr. Barnes complaining of Barnes's nephew taking some raw Pernambuca cotton (of the highest quality) belonging to Martin and turning it into yarn. Martin wanted compensation for this cotton.[9] It is possible that this is the same Barnes who was involved in the patent case with Arkwright and who was possibly still in business at the mill, perhaps sharing it with Martin. Otherwise it is difficult to explain how the dispute could have occurred. A further note indicates that in 1788 Barnes was paying interest of

£14, presumably on a mortgage or other loan, to either Antrobus or Whitfield (Whitfield seems to have acted as an agent for Antrobus at times).[10]

After the partnership of 1788 between Richard Martin and the Whitfield brothers came into being, the mill was used by John Whitfield to spin cotton. The partnership prospered until 1795 when Richard Martin died. However, there is a curious entry in a local survey that states that the mill at Davonenshaw [sic] Bridge, with the meadow, was sold to Mr. Barnes for about £500 by Mr. Antrobus in 1794.[11] This raises the question as to whether Barnes was still occupying part of the mill at this time. The death of Richard Martin meant that the Whitfield brothers had to purchase Martin's share of the partnership which involved raising a mortgage to buy Further Daneinshaw Mill.[12] They were noted as the owners at Nearer Daneinshaw Mill in the Land Tax return of 1799, although the note in the Congleton Survey states that it was in 1800 that "*Barnes sold it to Whitfield & Co. for £1300 tho' it cost him upwards of £1900.*"[13]

Also in 1800 Thomas Whitfield died and his will makes it quite clear that he was the owner of both Nearer and Further Daneinshaw cotton mills. His whole estate included lands recently purchased in Rushton Spencer, Rushton James, and Congleton as well as the estate at Further Daneinshaw, including many cottages and the corn mill. After payments of specific small requests the residue was to be used by the executors to provide an income of £300 per year for his widow provided she did not remarry (on remarriage this was to drop to £50 per year). Thomas Whitfield's family were resident at Henshall Hall at this time. The size and diversity of Thomas Whitfield's estate shows the level to which a relatively obscure person could rise in just 12 years in the cotton trade at that time.[14]

After Thomas Whitfield's death the Nearer Daneinshaw Mill is listed in the Land Tax returns as being owned and occupied by Henry Whitfield. As Thomas Whitfield's only child was his daughter Sarah, then it is possible that Henry Whitfield was the son of John Whitfield, the last surviving member of the partnership of 1788. Although the mill was occupied by a tenant called Foden in 1805 and 1806, from 1807 until 1812 the records show that Henry Whitfield was not only the owner but the occupier.[15] During this period he tried to sell the mill and its machinery in both 1811 and 1812 when he advertised it for sale as follows:-

"*For Sale by Henry Whitfield*

A capital cotton mill with 17 feet of head and fall on the River Daveninshaw [sic], to be sold by auction by Mr. Twemlow at the Roe Buck Inn in Congleton in the County of Chester on Thursday the 2nd day of July between the hours of four and six o'clock in the afternoon subject to conditions to be produced.

All that excellent cotton mill called the Nearer Dane Mill (with or without the machinery therein contained) being four stories high with a foundation laid for considerable enlarging the same, with nearly a new waterwheel, 16 feet diameter and 8 feet 6 inches broad with three sets of arms and wrought iron buckets, with two dwelling houses contiguous thereto, and a warehouse, counting house, batting room and smith's shop.

Also a piece of land now used as a garden containing about ½ acre of land of statute measure, more or less, adjoining thereto with an excellent piece of meadowland contiguous and which also adjoins the River Daveninshaw, and contains one acre of land of the like measure, or thereabouts, more or less. Late in the occupation of Mr. Henry Whitfield.

The machinery consists of 600 spindles of water frame spinning, one 36 inch carding engine, ten 18 inch ditto, one grinding cylinder, one devil with a 20 inch cylinder, one drawing frame of four boxes, one skeleton frame of four heads, one stretcher of 90 spindles, 106 small cans, 75 large ditto, one feeding frame, one lathe and spindle with vices, a pair of smith's bellows and anvil, lathe & small screw plates with taps, writing desk, skips, straps, and all other requirements necessary to carry on the cotton spinning business.

The above mill is eligibly situated on the turnpike road leading from Congleton to Buxton within a short distance from the former place, and the same can at moderate expense be converted either to a silk mill or any other kind of mill where a considerable power of water is wanted and there is plenty of hands in the neighbourhood.

To view the premises and for further particulars apply at the office of Mr. Lockett, solicitor, Congleton."[16]

This advertisement gives a good description of the waterwheel, and the machinery described, being 600 water frame spindles, would appear to be the original machinery from the 1770s/80s. They were certainly not a very competitive proposition in 1812 when most cotton spinners were using Crompton's mule rather than "Arkwright's" water frame.

According to the Land Tax returns the mill stayed in the ownership of Henry Whitfield and the Whitfields' heirs until 1825. However, the note in the Congleton Survey states that on 20th June 1812 "*Whitfield sold it to Goodale for £1950 for tho' he had laid out on a new wheel £500.*"[17] The reference to the new wheel would explain why so much information about the waterwheel was given in the advertisement of that year. Although Goodale does not appear in any Congleton records, he is entered in the Crompton Spindle list of 1811 as working 13,680 mule spindles and 1,440 throstle spindles in the Cheshire area,[18] a much greater number than the 600 old-fashioned water frame spindles at Davenshaw Mill. With that quantity of mule spindles Goodale could only have been interested in the mill building, not its old water frame machinery. It is perhaps no surprise that in November of 1812 all the machinery at Davenshaw Mill was again advertised for sale.[19]

In 1818 and the early 1820s Henry Hogg is listed as a silk throwster in Buglawton in a number of directories.[20] In 1825 both the deeds of the mill and the land tax returns agree that Henry Hogg became the owner of Davenshaw Mill, so presumably he had previously been a tenant there since at least 1818. The firm of Henry Hogg was to remain at Davenshaw mill until the end of the century, with two generations of the family operating the mill. Both Henry Hogg himself, and his son and successor Capel Wilson Hogg, had large families who lived at the nearby Davenshaw House. They were pillars of local society, both of them serving as magistrates, aldermen and mayors. Unlike many silk mill owners they did not close the mill when times were difficult but attempted to smooth out the effects of periods of depression in the silk trade, keeping their workers in employment, albeit on reduced wages. They introduced gas lighting to the mill as early as 1835 and employed 250 hands in 1840.[21] Henry Hogg died in 1864 at a time when the silk trade was very badly hit by the Free Trade Act of 1860, but his successor Capel Wilson Hogg managed to keep the business afloat and to continue in the silk trade when many mills were closing or switching to the fustian trade. Fortunately, the diary of the manager at the mill in the 1860s, Henry Madden, has survived to give a glimpse of life in Congleton in the 1860s. Trade picked up slightly in 1868 such that Hoggs invested in a new waterwheel that was fitted by Edwin Scragg, a local engineer. Henry Madden states that this was to replace the old wheel which was 28 years old. So Davenshaw Mill had new waterwheels in 1812, 1840 and 1868, each one lasting only 28 years in spite of having wrought iron buckets. Although steam was used to heat the mill from an early time, a steam engine was not installed to drive the mill until just before 1875.[22]

Towards the end of the century the Hoggs had to give up the silk throwing business, one of the last firms to do so in Congleton, and the waterwheel was no longer used. They tried their hand at shirt making which was a popular activity in the town in the 1890s but gave this up early in the 20th century. In 1897 the firm of Messrs Hogg & Son were prosecuted for employing under age children, one of whom was only 11 years of age, for making up shirts. The firm pleaded guilty and was fined 10s for each under age girl employed.[23] This

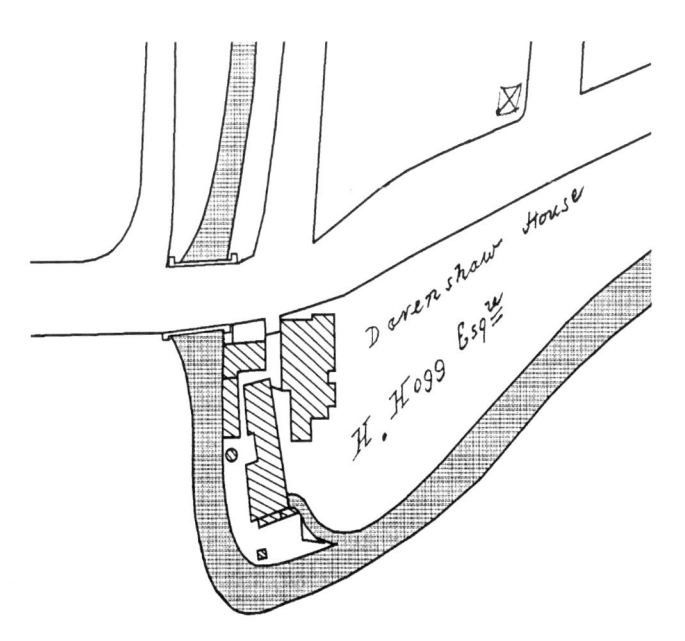

FIGURE 103. PART OF A MAP DRAWN UP IN 1844 FOR THE CONVEYANCING OF WASHFORD MILL (*q.v.*), SHOWING THE LAYOUT OF THE BUILDINGS AT DAVENSHAW MILL. ALTHOUGH NOT VERY ACCURATE, THIS MAP DOES INDICATE THE LIKELY POSITION OF THE WATERWHEEL AT THE SOUTH-EASTERN END OF THE CENTRAL BUILDING.
(*Private deposit on loan to Congleton Town Museum*)

was not an isolated event in the making up trade at this time as costs had to be minimised in the face of extreme competition. Capel Wilson Hogg died in 1909 and then the mill was used for fustian cutting and by the Granby Shirt Manufacturing Company.[24] At the end of the First World War the mill was sold to Jonathan Hopkins who also bought Washford Mill (*q.v.*). He was listed as a paper bag manufacturer and flint grinder in the trade directories.[25] In 1979 the premises were bought by Congleton Engineering Company who occupy the building today.

Today the mill is a hotchpotch of buildings which have been erected at different times. With such a long ownership by one family there are no sale notices after 1812 to describe the state of the buildings so the only evidence for the development of the site has to be interpreted from maps of the area. The survey sketch map of 1803, shows a rectangular building running at right angles to the Congleton to Buxton turnpike road i.e. lying in an approximate north-west to south-east orientation.[26] However, this sketch is by no means an accurate, scaled, representation. The map produced for the conveyancing of Washford Mill in 1844 (see Figure 103) shows a similarly oriented main building having a short leat running to the south-eastern end of the mill with associated buildings to the north and north-west.[27] Unfortunately, this map is not entirely accurate either, being in reality a rough sketch which even has a northern indicator pointing to the west!

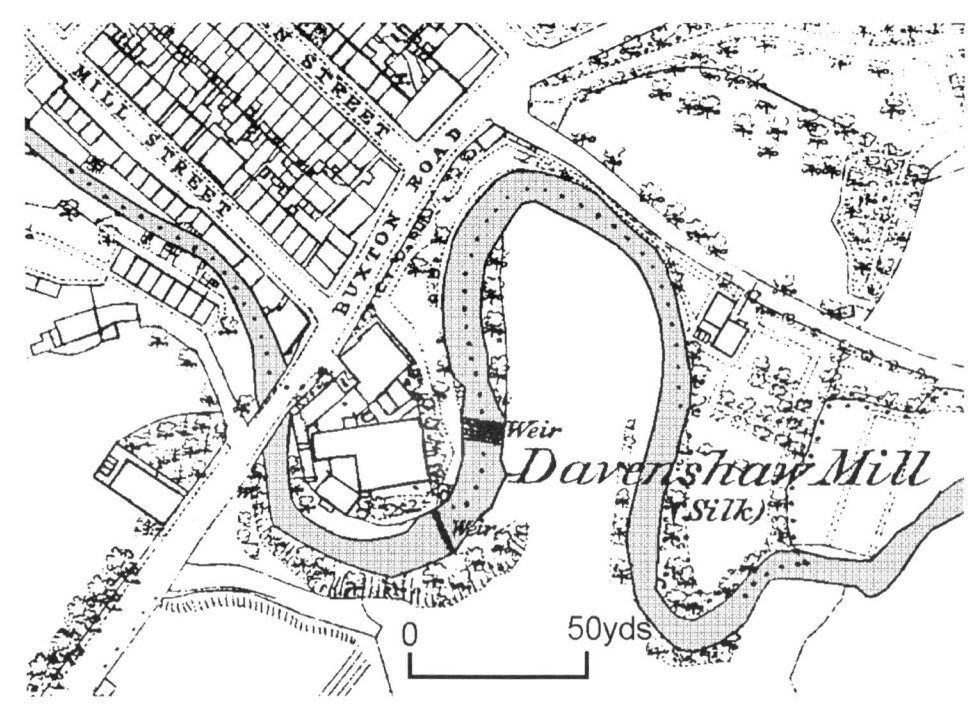

FIGURE 104. PART OF THE 1875 ORDNANCE SURVEY MAP SHOWING THE LATER BUILDINGS AT DAVENSHAW MILL AND THE TWO WEIRS.

A year later the tithe map shows the same buildings in greater accuracy (see Figure 60). By the time of the large scale Ordnance Survey map, in 1875, the main building has become a different shape, being much wider for most of its length, possibly indicating that it had been rebuilt at some stage. Later, in the 20th century the spaces between and around the buildings have been filled in with further constructions, joining all the different buildings to each other to form a single structure.[28] It is entirely possible that some remains of the pre-1780 mill still survives amid the present building. The 19th century maps show two weirs on the Daneinshaw Brook indicating that the head available was raised at some time. The larger of these two weirs is still to be seen behind the mill to the south-east. It is thought that there are some remains of the waterwheel or possibly a turbine in a sealed area of the mill, but this has not been confirmed.

References

1. Deeds held by Congleton Engineering Co.
2. Note in Congleton Survey book, held by the editor of the Congleton Chronicle.
3. Fitton, R. S., *The Arkwrights, Spinners of Fortune*, 1989.
4. Baines, E., *The History of the Cotton Manufacture in Great Britain*, Cass, 1966 (Reprint of the 1835 edition).
5. *Manchester Mercury*, 6th May 1783.
6. Baines, E., 1966 (1835), *op. cit.* 4.
7. CRO, Partnership Agreement, 1788, DCB/1595/27/14.
8. CRO, R. Martin's Notebook, DCB/1595/7/11.
9. *Ibid.*
10. *Ibid.*
11. Note in Congleton Survey book.
12. CRO, Indenture, 1797, DCB/1595/2/5; CRO, Lease, 1797, DCB/1595/8/6.
13. Note in Congleton Survey book.
14. CRO, Will of Thomas Whitfield, 1800.
15. CRO, Land Tax Returns, Congleton.
16. *Macclesfield Courier*, 14th November 1811 & 6th June 1812.
17. Note in Congleton Survey book.
18. Crompton Spindle List 1811, Bolton Local History Library.
19. *Macclesfield Courier*, 14th November 1812.
20. Commercial Directory, 1818; Pigot's Directory of Cheshire, 1822.
21. Head, R., *Congleton Past & Present*, 1887.
22. Henry Madden's Diary, *Congleton Chronicle*, July 1974 - August 1975.
23. *Congleton Chronicle*, 19th December 1997.
24. Kelly's Directory of Cheshire, 1910.
25. East Cheshire Textile Mill Survey, Macclesfield Silk Museum.
26. Survey of Congleton, 1803, held by the editor of the Congleton Chronicle.
27. Conveyancing map of Washford Mill to Mr. John Johnson, 1844, held by Congleton Town Museum.
28. CRO, Sale notice, 1965, D4744/2.

Section 4. The Town of Congleton

Just after leaving Buglawton, where the Daneinshaw Brook meets the River Dane, the river changes from flowing in a southerly direction to a westerly one through the town of Congleton. The name "Congleton" supposedly means "the town at the bend in the river". Prior to the great flood of 1451 the river ran somewhat further to the south of its present course before passing under the town bridge. To the south of the town, on higher ground, lies an area of marshy ground called Congleton Moss. A number of small streams drain the Moss, running northwards towards the River Dane in the town. The largest of these small streams is known as The Howty. The Moss acts as a large sponge, soaking up the rainfall, consequently the Howty provides a regular and sustained flow even in drought conditions. The course of all these streams have been culverted in the town and two of them have been diverted into the Howty at some time, instead of following their original course directly into the river.

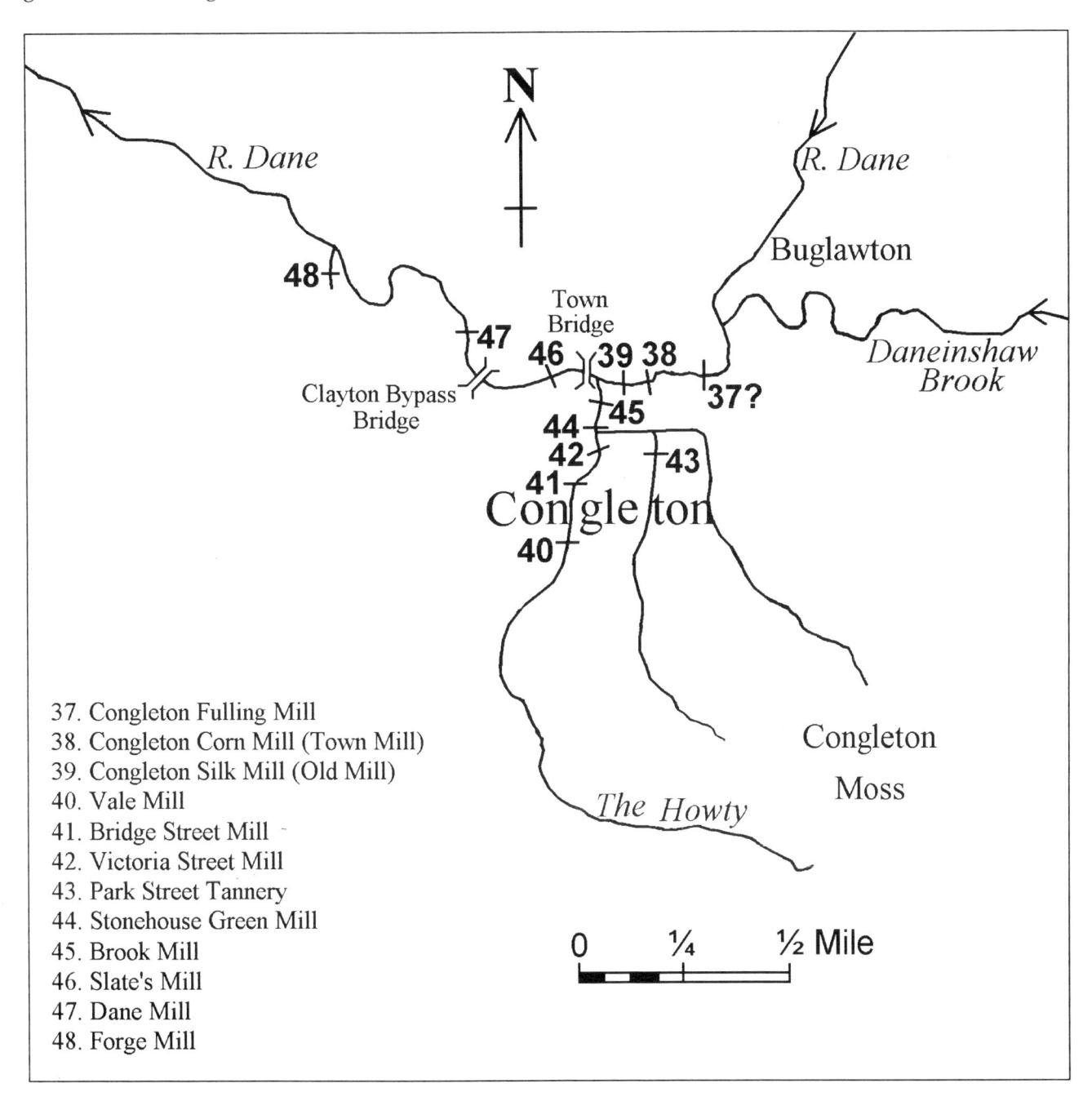

37. Congleton Fulling Mill
38. Congleton Corn Mill (Town Mill)
39. Congleton Silk Mill (Old Mill)
40. Vale Mill
41. Bridge Street Mill
42. Victoria Street Mill
43. Park Street Tannery
44. Stonehouse Green Mill
45. Brook Mill
46. Slate's Mill
47. Dane Mill
48. Forge Mill

FIGURE 105. MAP OF THE RIVER DANE IN THE TOWN OF CONGLETON WITH ITS TRIBUTARIES RISING IN CONGLETON MOSS TO THE SOUTH OF THE TOWN.

37. CONGLETON FULLING MILL   SJ ??????

During the early medieval period mills provided a large part of the income of a Lord of the Manor and so the landowners of the time were quite enthusiastic about maximising their revenue from their water rights. In places where there was sufficient capacity in the existing corn mills and there was a surplus of available water power, it was tempting for the landowners to build fulling mills to supplement their income revenues. Such was the case in Congleton in the early part of the 14th century. In 1353 there is reference to a *"Molendinum fullonicum de novo leuata"*[1] and by 1361 the accounts show that a rent of £1-12s-2d was being paid for the fulling mill in Congleton.[2] Even when Roger and William Walker were paying an increased rent of £2-6s-8d the following year, the income to the manor was much less than the £12 being paid as the rent for the corn mill at that time.[3] It is interesting to note that the lessees' surname reflects their trade, "walker" being another name for a fuller.

With the coming of the economic downturn after the Black Death, many of the fulling mills built in the earlier prosperity failed to survive as the cost of repairs and maintenance became greater than the rents that they could attract. Congleton's fulling mill was no exception to this trend and after spending £2-15s-3d on repairs in 1369 no further entries are recorded in the manor accounts directly concerning the fulling mill.[4] However, the manor continued to collect a rent of 2s per annum up to 1453 on "*a cottage next to the fulling mill*"[5] and references occurred in various documents to a "*Walkynmylnhous*". The last known reference to this property by this name occurred in 1514.[6] It is not known whether these references were to the original fuller's house, which would have been adjacent to the fulling mill, or to a conversion of the fulling mill itself into a dwelling.

The location of Congleton's fulling mill is also a mystery. The best site would have been near Foundry Bank with the mill powered by the stream that used to join the River Dane at that point. The steep bank would have provided the necessary buttressing to absorb the continuous impacts of the fulling hammers. This use of an earthen bank to provide a certain amount of shock absorption was characteristic of many medieval fulling mills. However, the low level of the rental that could be charged may have prevented the capital investment needed for the building of a unique mill site with all its necessary water supply engineering. It is possible that the fulling mill was built alongside the town corn mill (*q.v.*) in order to take advantage of existing water supply arrangements. This could have been on the River Dane itself or possibly on The Howty. Unfortunately, no evidence has been forthcoming in support of any particular location for Congleton's fulling mill.

References

1. Dodgson, J. McN., *The Place Names of Cheshire, Part 2*, Cambridge University Press, 1970.
2. Stevens, W. B., *History of Congleton*, University of Manchester, 1970.
3. *Ibid.*
4. *Ibid.*
5. *Ibid.*
6. Dodgson, J. McN., 1970, *op. cit.* 1.

---

38. CONGLETON CORN MILL   SJ 860633

The earliest mention of a corn mill in Congleton occurs in the charter granted to the town in 1272 by Henry de Lacy. Although Henry de Lacy was the Earl of Lincoln and Constable of Chester, it was as the Lord of Halton that he owned the manor of Congleton. His charter of 1272 granted that the people of Congleton should "*grind their corn at our mill of Congleton, on payment of the twentieth grain, so long as the mill is sufficient*".[1] The implication of this charter is that the mill was already in existence and was emphasising the level of toll to be paid to the mill at a twentieth of the customer's grain. This was not a particularly generous rate, but on the other hand was not excessive for the time either. The charter also enshrines the principle of mill soke whereby all the inhabitants of the manor were constrained to use their lord's mill and no other, unless there was some reason why the mill could not keep up with demand. In reality the charter probably only provided a confirmation of the existing custom and practice rather than granting the people of Congleton any special advantages, but it did legitimise the lord of the manor's privileges of mill suite and toll charge.

The medieval records do not give any details of the mill but it appears always to have been "farmed", or rented, rather than operated by the lord of the manor himself. In 1295 the mill was worth £8 per annum, rising to £10-6s-8d by 1356, when it was leased to Roger Mortimer, and to £12 in 1361.[2] However, this was to prove to be the highest rent in the medieval period, reflecting the increase in population and prosperity being experienced generally in the country between 1250 and 1350. In the middle of the 14th century the county was ravaged by the Black Death which caused a substantial decrease in the population and a general recession in economic activity that lasted until towards the end of the 15th century. This recession was reflected in the rent obtainable by the mill in Congleton, which fell to £6-13s-4d in 1378 when it was leased to Thomas Whelok and William Nesfield for twenty years. The rent was still at this level when leased to Hugh Moreton for 30 years in 1397 and also in 1428 when it was leased to John Yates for 20 years.[3] After the death of Henry de Lacy, the manor of Congleton became the property of the Dukes of Lancaster and in 1399 became the property of the crown on the accession of Henry Bolinbroke as King Henry IV. Thereafter the mill in Congleton was known as the King's Mills.[4]

Some idea of the terms of the mill leases can be gained from items of expenditure on the mill that occur in the manor records, indicating which items the lord was responsible for and hence the items that do not occur were probably the responsibility of the lessee. In 1365 the manor spent £1-10s-0d on the repair of the wheel and associated carpenter's wages and in the following year spent money on the mill dam, the waterwheel and a cog wheel. Only three years later they spent £3-7s-0d for labourers felling timber for the mill in Whatley Bank and for digging a new mill pond and in 1372 over £4 was needed for further repairs to the dam.[5] These accounts indicate that the manor was spending about a third of the income from the mill on maintenance. The split in responsibilities between lessor and lessee were spelt out in the lease of 1369 when the mill was rented by Richard de Moreton, Adam Atkinson and Richard Eygnon for six years; the lord of manor was to provide the wood and iron for maintenance while the lessee supplied the millstones. Similarly, in the lease of 1397 Hugh Moreton was granted wood by the lord of the manor to maintain the mill.[6]

It would seem from these leases that the lessees were local businessmen rather than the miller himself. From around 1378 the mill was usually let together with the meldhouse and bakehouse as well as the town fair and market, which shows that the lessees were substantial businessmen able to control the means of production from the raw material to finished product, i.e. from toll corn to flour through to the baked bread, including the retail outlets of the fair and weekly market. As such they would have employed a number of workers of which the miller was just another example.

Unfortunately there are no details available describing this medieval mill. It is possible to speculate that it was probably built of wattle and daub with a thatched roof, as were most of the secular buildings of this time. The technology of the time was such that a single vertical waterwheel would have been used to drive a single pair of millstones through one pair of simple gears. The building was probably single storeyed as there was no need for much storage area. Customers would bring small amounts of grain, as little as one sack, to the mill to be ground while they waited. Even the town bakehouse would send grain to the mill a sack at a time as they would not want to keep flour in storage since it would quickly deteriorate. If the mill was out of action or the business was so great that the customers had to wait more than about a day then they had the right to take their grain to any other mill. Reference to the King's Mills (plural) indicates that there was more than one waterwheel and corresponding pairs of millstones housed in the mill building.[7]

In the same way as it is possible to speculate over the details of the medieval mill itself it is also possible to speculate on its location. There are a number of possible sites for the mill that have been proposed; firstly, on the old course of the River Dane itself; secondly, on one of the tributaries to the Dane that flow through the town from the south, either at what is now Foundry Bank, or on the stream that now runs under the buildings alongside Kinsey Street, or on the Howty. A site on the River Dane itself would have had a number of disadvantages. The capital cost of taming such a large stream would be a disincentive and the ongoing cost of repairs due to the nature of the river, with its proneness to flooding which would cause continual damage, would be far higher than on one of the smaller tributaries.

Also, mention in the accounts of mill dams and mill ponds is not the terminology consistent with the position of an undershot waterwheel on a big river, even if it had a weir. However, this conclusion is dependent on the ability of the translators of the manorial accounts to distinguish "dam" from "weir" and "pond" from " pound" in medieval Latin (and also in modern English!). The site at Foundry Bank is possibly the most advantageous mill site on the small tributaries in the town as the fall to the river at this point was considerable and consequently suitable for an overshot waterwheel installation. The best clue to the mill's site is the mention in a document of 1405 of Mill Street.[10] This name would indicate that the mill was either on the River Dane at the end of Mill Street or else on the Howty which runs alongside Mill Street (providing of course that the present Mill Street is the same as the one mentioned in 1405). The balance of evidence so far available probably indicates a location on the Howty near its confluence with the River Dane.

In 1451 there was a great flood on the River Dane causing considerable damage to the town of Congleton, so much so that the town petitioned the king for relief. King Henry VI issued letters patent on 29th June 1451 which stated:-

> "...by the impetuous and fluctuating flow of the stream of the river Dane, a great part of the town has become devastated and the King's Mills within his lordship are dilapidated, etc. to the great hurt and perpetual destruction of the said town. The king....concedes and licenses that the said mayor, etc. may change the course from its ancient place as far as and above another course called Biflete so that the water may for ever be able to run there, and to construct one, two, or three water, wind, or horse mills. Also that the said mayor, etc. may hold the said mills so constructed for ten years without rent. After which time they may hold the same for ever at a rent of £1-6s-8d annually."[8]

The mills could have been damaged by the flood itself or could have been in a poor state of repair anyway. Certainly Yates, writing in 1821, indicates that the town took the opportunity of the flood to replace the old King's Mills which were already in a state of decay.[9] The new watermill was built on the Byfleet, a position that it retained until 1966.

In accordance with the letters patent of 1451 the new mill was built by the corporation, and leased by them *in perpetua* from the crown, thereby becoming known as Town Mill. To ensure the smooth running of the mill and its finances two mill reeves and a mill clerk were appointed by the corporation every year. The new mill, built on the Byflat or Byfleet, was on the north side of the River Dane about 100 yards upstream of the town bridge. The year after the new mill was built the king discovered other mills were being erected without his authority, so he instructed the Duke of Buckingham, as steward of Congleton, to stop any new mills being built and forbade anyone to grind corn at these new mills.[11] Possibly with the demolition of the King's Mills the townsfolk felt they had also abolished the duty of mill suite, but the king felt obliged to re-assert his rights in favour of the Town Mill. This was to be a recurring problem as lord's rights weakened generally and the old manorial customs became no longer enforceable. Unfortunately, as it was the king who was lord of the manor at Congleton, the townspeople were unable to break the manorial monopoly as happened in many other manors. In 1521, Henry VIII forbade anyone to erect any other mill *"within the Lordship of Congleton, nor in any wise suffered to grind any part of their stream, and that the inhabitants should grind their grain at the town's mill, and at none other."*[12]

Little is known of the operation of the new mill immediately after 1451 but the Congleton Corporation accounts have survived from 1587 and include many entries concerning the town mill. Analysis of these accounts provides a picture of the mill and its operation during the 17th and 18th centuries.[13]

The weir for the mill, on the River Dane, consisted of poles stuck into the river bed holding back loose rocks and stones. The gaps between the stones were filled with a mixture of clay and straw, and any residual leaks were stopped by applying moss or straw into the fissures. This type of construction would only be suitable for providing a low head of water, any greater head was provided by digging out the bed of the river downstream of the mill. Also, this type of construction was very prone to flood damage and appears to have had to be rebuilt most years. The details of the construction of the weir are derived from the supply of materials that coincide with the labour of repairing flood damage to the weir.

The mill building itself was probably built of timber, standing on a stone base, with a thatched roof and a brick chimney. The building would have been either one to two storeys high. Although it seems odd to have a chimney in a mill, an entry in the accounts for 1652 states:-

"Pd Mr. Moreton for bricks for the mill chymnry....£3-0s-0d"

Other entries for the delivery of timber, thatching work and entries such as:-

"Spent on workmen who were working at the stone wall at the mill.....6d"

confirm the other details of the mill building itself.

There were two wooden waterwheels with their shafts resting in brass bearings. These wheels had wooden starts and "ladles" (paddles?) made of simple planks. Due to the head of water available the waterwheels would have been low breastshot. The two waterwheels each drove a single pair of millstones, which were known as the "wheat mill" and the "rye mill". The configuration of the gearing would have been strictly "Vitruvian" with a wooden pitwheel meshing with a lantern pinion wallower mounted on the millstone spindle. These millstone spindles were made of iron and the tentering of the millstones was achieved using bridgetrees. Although the mill would not need any great storage area there were arks which were probably used to collect the finished meal. Nowhere in the accounts is there any mention of the purchase of cloths for a boulter so it is assumed that there was no effort made at all to separate the meal at the mill. However, hand sieves were being purchased from the beginning of the 18th century at the rate of one every two years

In 1680 there is mention of two millstones for *"the new mill"*. This is not the phrase that was used when one of the existing pairs of millstones was replaced

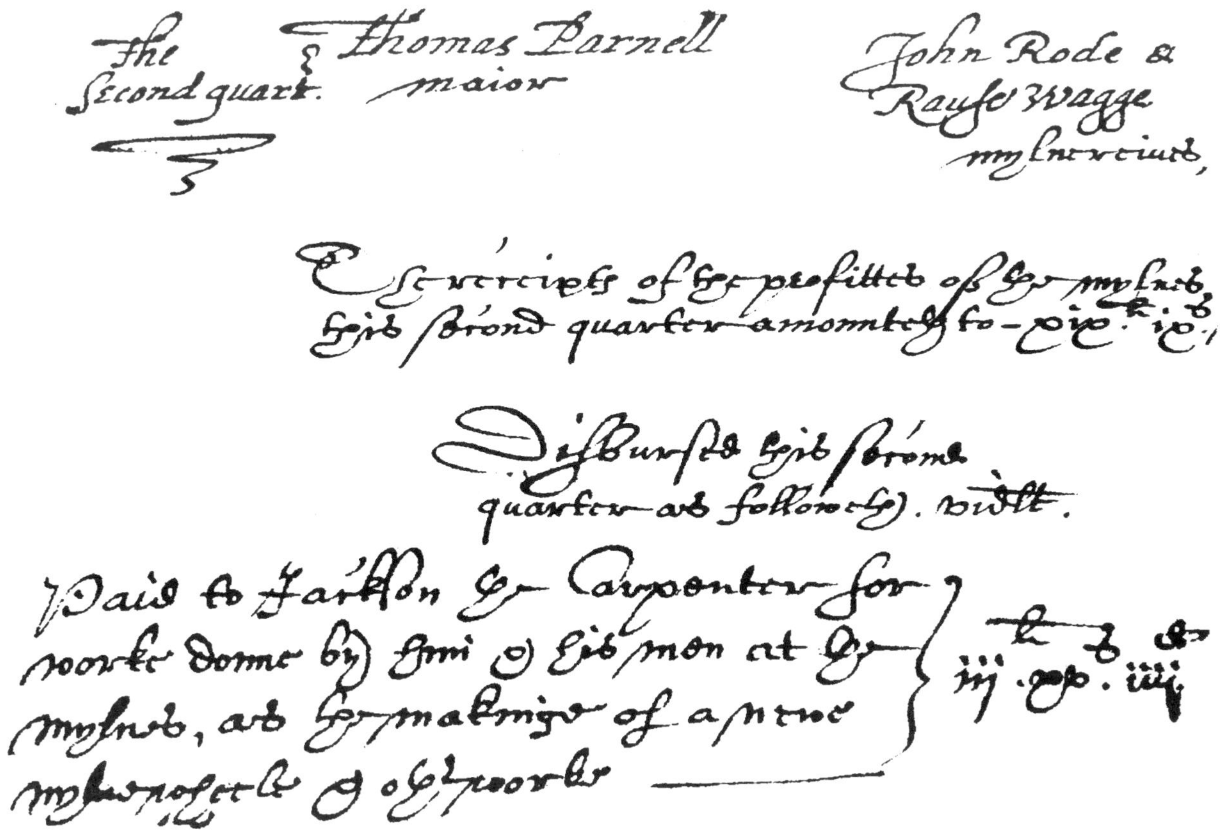

FIGURE 106. AN EXTRACT FROM THE CONGLETON CORPORATION ACCOUNTS FOR THE SECOND QUARTER OF 1620-1 WHEN THOMAS PARNELL WAS MAYOR. THE ENTRY SHOWN IS FOR PAYMENT TO A CARPENTER AND HIS MAN FOR THEIR WORK ON A NEW MILL WHEEL. (*Congleton Town Council*)

or refurbished and so suggests that a third pair of millstones was added to the mill at this time. There is no mention of any third waterwheel being built so it is possible that this "*new mill*" was driven in some way from the pitwheel of one of the existing mills. In the same year references to a malt mill start to occur. Although some malt mills in the 18th century consisted of a pair of rollers, an entry in the accounts for 1722 records a payment

"*To Matthew Lawton for two stones for the malt mill*"
As Matthew Lawton was a millstone maker based at Mow Cop this confirms that the malt mill did consist of two millstones rather than rollers. This evidence for a second pair of millstones driven from a single waterwheel at the mill as early as 1680, although circumstantial, is quite strong. The first illustration of this type of configuration published later in the 18th century represents a mill dated at 1723. A kiln is first mentioned at the town mill in 1712 with the purchase of rods and barrs which might indicate that it was being built (or possibly rebuilt) at that time.

Maintenance of the corn mill required considerable effort and expenditure. Most years the weir suffered from storm and flood damage and the river bed and the mill fleams needed clearing out or "spurring". This would have been particularly cold and wet work and the workers were traditionally bought ale to drink after completing the job. The waterwheels also seemed to suffer from damage with paddles going missing quite often. In the winter of 1730 three men were paid 2s-0d for "*cutting ice of the mill wheels*". The brass bearings on the waterwheels seemed to need replacing on average every three years (but as there were two wheels this could represent a six year period per wheel). The lantern pinion style of gearing was by no means a piece of precision engineering and new sets of cogs and "rownds" were needed virtually every year. The iron mill spindles operated under a considerable twisting force due to the torque applied. In the period up to the late 1690s, due to the brittle nature of the iron available, the mill spindles needed repair approximately every two years (again, as there were two mills they could have been lasting for about four years each). From the late 1690s onward a better type of iron was available, such as that produced by the blast furnace at Lawton (*q.v.*), which greatly reduced the number of repairs needed to the mill spindles with the average time increasing to six (or twelve) years between repairs.

The millstones themselves were the most expensive single item to replace, costing between £3-10s and £4 each. However, in 1694 the town paid £5-4s-0d for a millstone and when this was installed in the mill 3s-0d was paid "*For raising the great millstone*". The millstones were acquired from Mow Cop, which is about three miles south of Congleton, and appear to have lasted for about two (or four) years before needing to be replaced. The work involved in setting up a new millstone, which required some heavy lifting, was another activity traditionally concluded by supplying ale to the workmen. The mill chisels used to dress the millstones required steeling every couple of years. Each year there were a number of minor items that needed replacement and small jobs needed to maintain the building. The thatching of the mill roof needed attention roughly once every three or four years. During the early part of the 18th century the mill was purchasing "*two wiskitts*" about every two years which were some form of basket, but their exact use has not been determined.

During the 17th and 18th century the mill was still operated on a toll basis with customers bringing their grain to the mill for grinding and the mill receiving one twentieth of the grain as payment. Fortunately the accounts for 1676 show that of the twentieth that was kept by the mill, one quarter went to the millers and three quarters went to the town corporation. Quite significantly, an item in the accounts of 1682 shows that Thomas Oakes was fined 2s-0d for breaking the mill suite, indicating that the townspeople were still forced to mill their grain only at the Town Mill at a time when this condition had been abolished or was ignored in many communities.[14] For one quarter of the year 1742 there is a breakdown of the various types and quantities of grain that made up the toll. This shows that malt was ground in the greatest quantity, at 39% of the total throughput of the mill, with wheat being the next largest total at 34% of the throughput. The rest of the work done by the mill was split between barley, 18%, the shulling of oats, 6%, and rye, 3%. It is not surprising that wheat and malt make up by far the largest amount as they provided the population with their main food and drink, i.e. bread and beer.

The amount of wheat taken in toll by the corporation in one quarter in 1742 was 215 pecks. If the amount taken by the millers is added to this then the total toll wheat was 286 pecks. As this toll represents one twentieth of the total then the full amount of wheat ground by the mill was 5720 pecks or 1430 bushels per quarter. It has been long recognised that with the diet prevalent in the 18th century each person needed two bushels per quarter to maintain life, so the mill was only sustaining approximately 700 people.[15] The amount of malt toll taken by the corporation in this quarter was 246 pecks which represents a total processed for the year of 6560 bushels. If a work rate of five bushels per hour is assumed and a six day working week[16] then there would have been a total of about four hours continual grinding per day on the malt mill, similarly the wheat mill would have been in operation for about three and a half hours per day and the rye mill for just over two hours per day. This, of course, is assuming the four quarters had an equal throughput which was not necessarily the case due to the cyclic nature of growing grain. Unfortunately, it is not known which quarter these figures represents but it gives some idea of the amount of work needed for a mill to support a small population of 700 people.

It has been recorded that the millers received a quarter of the total toll grain, so in the quarter analysed above they would have received a total of 212 pecks of grain or about 16 pecks per week. Fortunately, the accounts for this quarter also show the price per peck for the various grains, with malt and wheat at 12d per peck and barley and shullings at 7d per peck. The millers' share of the toll works out to 16s in total or between 5s and 8s per week depending on whether there were three or two millers employed. Nothing much is known about the millers themselves. It is possible that William Burges, who was listed as a "*milner*" paying 2s-0d in the Poll Tax records of 1660, was the miller employed at Town Mills at that time.[17] In 1728/9 there were two millers, James Beech and Samuel Baxter, who were joined by Daniel Chell in 1730. During these few years the millers were paid a weekly wage, directly by the corporation, of 6s-0d each which corresponds to the likely amount received when the millers were paid by the toll. To put these wages in perspective, the accounts show that unskilled labourers were paid 9d per day i.e. 4s-6d per week, and more skilled workers in a particular trade, or a foreman, were paid 1s-0d per day (6s-0d per week), whereas workers who had a higher professional status, such as millwrights, were paid 1s-6d per day (9s-0d per week) and were usually rewarded with free ale on completion of any job. These figures give some insight into the status and rewards obtainable by millers in the 17th and 18th centuries which was akin to that of time-served, skilled labourers. Certainly, at this time, they were not the rich capitalists of myth and legend. After 1740 the corporation paid a kiln man, employed separately from the millers, at the rate of 8d for drying a kilnful of oats. However, there were sometimes ways for the millers to supplement their income; in 1743 the corporation was paying a bounty of 1d for every rat caught which enabled the millers to earn an extra 8s-3d for 99 rats, but by 1748 when they caught 180 rats they were only allowed ½d per rat. The number of rats gives some idea of the conditions in and around the corn mill.

One problem of being paid in kind, as the millers and the corporation were, is that the derivable income fluctuated with the price of grain, which was high in times of scarcity but not so good in times of glut when prices were low. The income derived by the corporation from the mill in the second half of the 17th century averaged out at £115 per year (figures available for 104 quarters), the figures varying from £47 to £16 per quarter over the period. The best year was in 1661 with a total income for the corporation of £156 and the worst was in 1677 at around £76. It is not possible to calculate the total cost of maintenance on the mill because, in any given year, many entries in the accounts do not indicate where an item of work took place. However, from those items that can be identified as being at the mill, it can be seen that in many years the costs were a significant percentage of the profits. The town corporation viewed the money received from the sale of the toll grain as "profit" without considering the level of expenditure on the mill.

There were also a series of standard charges that the town corporation had to pay with respect to the mill. The tithe charge was 13s-4d per annum and the rent to the crown continued to be paid at £1-6s-8d per year as fixed in 1451. Other charges such as the Poor Lay, the Highway Lay and the Land Tax were paid by the corporation for the mill and the town lands combined. Other standard costs were the payments to two mill reeves at 1s-0d each per quarter and to the mill clerk at 6s-8d per quarter.[18]

In 1752 the corporation leased land at the Byfleet, near to the corn mill, to a consortium of businessmen who wished to build a silk mill. In this new venture they employed James Brindley, who was later to become a famous canal engineer, as a millwright.[19] It is interesting that an entry in the town accounts for 1752 shows a payment of £2-3s-6d to James Brindley, millwright, for work on the corn mill. Was it this work on the corn mill that brought him to the notice of the silk mill builders or was he already on site working on the silk mill when he was paid for his services at the corn mill? This was one of the last entries in the accounts of payments on behalf of the corn mill because in 1754 the mill and its water rights were leased to the silk mill consortium. The rental agreement set the corn mill's rent at £150 per annum plus forty bushels of malt, seventy two pecks of barley, and 30s-0d in lieu of malt dust. They were allowed to grind for the townspeople at a toll of a sixteenth for corn and a twenty fourth for malt. The length of the lease issued by Congleton Corporation was for the duration of the lives of John Clayton, Samuel Pattison and Nathaniel Pattison, three of the men involved in building the silk mill, and for a further twenty one years thereafter.[20] Having acquired the lease to the corn mill, Pattison and Clayton were allowed to take water from the corn mill pool, through a culvert which was ten inches square, to supply the silk mill. In return for taking water from the mill pool John Clayton was instructed to alter "...*the wheels and soughs of the present corn mills at his own expense, so that they may be worked as well as usual with less water.*" Even so the town officials were concerned that this modification would not be successful and so authorised Clayton to raise the weir by six inches if necessary.[21]

The corn mill at this time was much as built in 1451 with the probable addition of a third pair of millstones. It must have been obvious that altering the wheels and soughs would not satisfy the requirement of the lease to maintain the mill's output even if the weir was raised by six inches.[22] Eventually in 1758 a decision was taken to rebuild the corn mill entirely,[23] providing a more modern configuration of machinery. This meant more than one pair of stones could be driven from one waterwheel using the latest gearing and transmission techniques to make the mill more efficient. These changes resulted in the new mill being built with three storeys, providing a garner floor for grain storage above the stone floor and machinery floor of the mill. At this time James Brindley was still involved with the silk mill, in fact he was still making alterations to the silk mill in the 1760s.[24] It is highly probable that when Clayton and Pattison needed to rebuild the corn mill they would have employed James Brindley to act as the millwright, as he had been working for them for six years on the silk mill with great success. Unfortunately, no documentation has yet come to light to confirm his involvement at the corn mill other than the single entry in the corporation accounts of 1752.

The exact layout of the machinery installed in the mill in 1758 is not known, but as the building survived until 1966 it is likely to have had some similarity to the configuration present in the 20th century. Certainly the layout of the watercourses indicates that the mill had two waterwheels, a configuration that was inherited from the earlier mill. In spite of this rebuilding in 1758, the rent claimed by the Corporation remained the same as that set in 1754. Then in 1797 the rent was raised from £150 to £187 per annum, a rise probably only reflecting the level of inflation that occurred during the Napoleonic Wars at the end of the 18th century.[25] It is not known whether the silk mill owners operated the mill themselves, employing a miller, or sublet the property.

In spite of the turbulent times the country experienced around the turn of the 18th century the mill continued to work with only local difficulties to contend with. One such was in 1811 when Joseph Potts,

FIGURE 107. THE FRONT OF TOWN MILL, BUILT IN 1758. THE TOP FLOOR WAS ADDED AT THE END OF THE 19TH CENTURY. (*Congleton Town Museum Trust*)

a local carter, was declared bankrupt and as well as having to sell all his wagons and horses there was the small matter of *"a quantity of flour now at Congleton Corn Mills and a quantity of flour bags"* that had to be disposed of.[26] Of greater significance was the destruction of the mill weir on the River Dane in 1814. This was a wooden structure and had appeared time and time again in the Corporation's accounts as being in need of repair. This time the weir was totally reconstructed out of stone, with such quality of construction that this weir is still in place today.[27]

Just a year after the rebuilding of the weir, a new miller called George Watson came to the town to operate the mill, placing an advertisement in the local newspaper to proclaim his good intentions, as follows:-

"Having just entered these complete and able mills and hoping that my character in the neighbourhood of Cheadle in this county will bear the most particular enquiry, where I have been employed at Cheadle Mills for these two or three years past. I beg to assure the public in general that nothing shall be wanting on my part to give those who may favour me with their custom entire satisfaction"[28]

George Watson stayed at the mill until 1826 when the owner of the silk mill leased the corn mill to two local men, Roger Broadhurst, a victualler, and George Cookson, a corn dealer and miller.[29]

Broadhurst and Cookson continued at the mill while running a bread shop in the town until their partnership was dissolved in 1832,[30] in spite of an attempt to sell both the silk mill and the corn mill in 1828 on the death of the owner of the silk mill, Nathaniel Maxey Pattison. The 1828 sale advertisement gives the first tantalising glimpse of the machinery in the corn mill at this time, as follows:-

"Also capital water corn mills containing five pairs of stones near the above [Congleton Silk Mill] *with shulling mill and fan, two dressing machines, drying kilns, etc. The corn mill is held for 120 years determinable at the end of 21 years from the death of the survivor of two lives now in being."*[31]

Although five pairs of millstones are mentioned in this advertisement, another pair of millstones would also have been used in the shulling mill which produced oatmeal. Oatmeal production was common in Cheshire mills and the corn mill in Congleton was no exception with its drying kiln, shulling mill, a fan which was used to separate the husks from the chaff and a dressing machine for separating the different qualities of oatmeal. It is noticeable that even today oatcakes are still a firm favourite with the people of Congleton.

Shortly after the break-up of the partnership between Broadhurst and Cookson, George Cookson was declared bankrupt[32] but Roger Broadhurst continued the business at the mill until sometime in the 1850s.[33] Again he had to contend with attempts to sell the mill whilst he was operating it, this time by the Corporation itself when in 1839 they offered for sale,

"The reversionary interest of the said mayor aldermen and burgesses of and in the water corn mill called Town Mill, with the cottage, gardens and appurtenances occupied therewith and the use of part of the waters of the River Dane, in the occupation of Samuel Pearson, esquire [owner of Congleton Silk Mill] *or Mr Roger Broadhurst his undertenant."*[34]

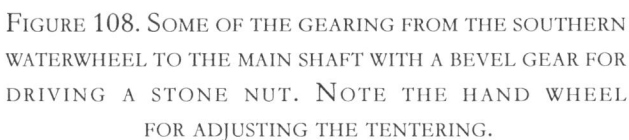

FIGURE 108. SOME OF THE GEARING FROM THE SOUTHERN WATERWHEEL TO THE MAIN SHAFT WITH A BEVEL GEAR FOR DRIVING A STONE NUT. NOTE THE HAND WHEEL FOR ADJUSTING THE TENTERING.
(Rex Wailes, 1964, for the National Monuments Record)

FIGURE 109. ONE PAIR OF MILLSTONES DRIVEN FROM THE SOUTHERN WATERWHEEL AND ITS ASSOCIATED FURNITURE. NOTE THE SMALL CIRCULAR HOPPER ON ITS CURVED TRIPOD SUPPORT AND THE SILENT FEED ARRANGEMENT.
(Rex Wailes, 1964, for the National Monuments Record)

By 1860 the miller at the Town Mill was Thomas Hall who occupied the mill for about twenty years. His business at the mill flourished and by the late 1870s not only was he milling at the Town Mill but also at Forge Mill in Odd Rode(*q.v.*) and at the new steam mill, also in Odd Rode. His milling business provided a level of profits that enabled him to become a gentleman farmer living at Dukes Oak Farm in nearby Brereton.[35] After Thomas Hall relinquished the mill, James Snelson became the miller for a few years until, in 1884, Samuel Pearson, the owner of the silk mill, declined to renew the lease and hence the mill stood idle.[36] As they were unable to renew the lease at the Town Mill the Snelson family of millers became involved in Swettenham Mill (*q.v.*) a few miles west of Congleton.

It is possible that Samuel Pearson declined to renew the lease because his tenure held from the Corporation may well not have had long to run. In 1885 the Corporation itself advertised the mill to let, as follows:-

"To be let by tender, all that excellent and substantial old water corn mill with appurtenances called the Town Mill situated near to the Dane Bridge in the Borough of Congleton with the weir, weir-pool, and right of poundage of the river Dane. The premises are in good order and repair and comprise mill and six pairs of millstones, stabling and all the requisite outbuildings and conveniences, yard, gardens, etc. and are the property of the Mayor, Aldermen, and Burgesses of the Borough of Congleton, under their ancient charter."[37]

This advertisement was not successful as the mill was reported to be still idle in 1887. These were difficult years for the traditional type of corn mill using waterpower and employing millstones for grinding the grain at a time when an entirely new type of milling using rollers was being introduced into the country together with the exploitation of overseas sources of grain. It was not until about 1890 that a new miller, John Mellor, was found to take on the Town Mill. He stayed until 1897 when John Orme took over, working

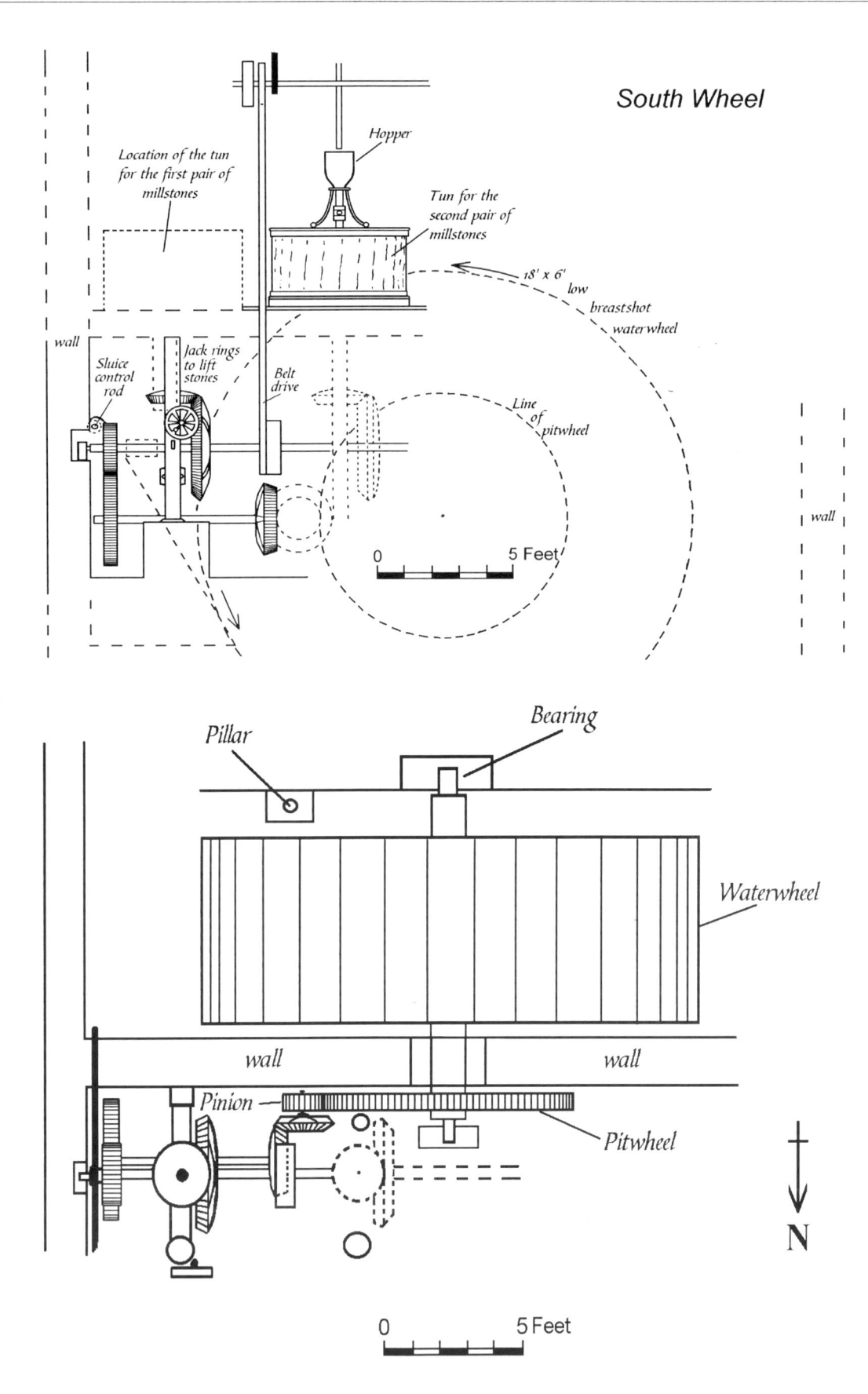

FIGURE 110. A SECTION (*top*) AND PLAN (*bottom*) OF THE POWER TRANSMISSION SYSTEM ASSOCIATED WITH THE SOUTHERN WATERWHEEL IN THE TOWN MILL. GEARING ETC. IS SHOWN FOR TWO SETS OF MILLSTONES ONLY, FURTHER GEARING, ETC. DROVE A THIRD SET. FIGURES 110 & 111 HAVE BEEN CONSTRUCTED USING INFORMATION FROM PHOTOGRAPHS AND A DESCRIPTION OF THE MILL JUST BEFORE ITS DEMOLITION. THE DOTTED LINES REPRESENT THE DEDUCTION OF THE LIKELY CONFIGURATION. (*Tim Booth*)

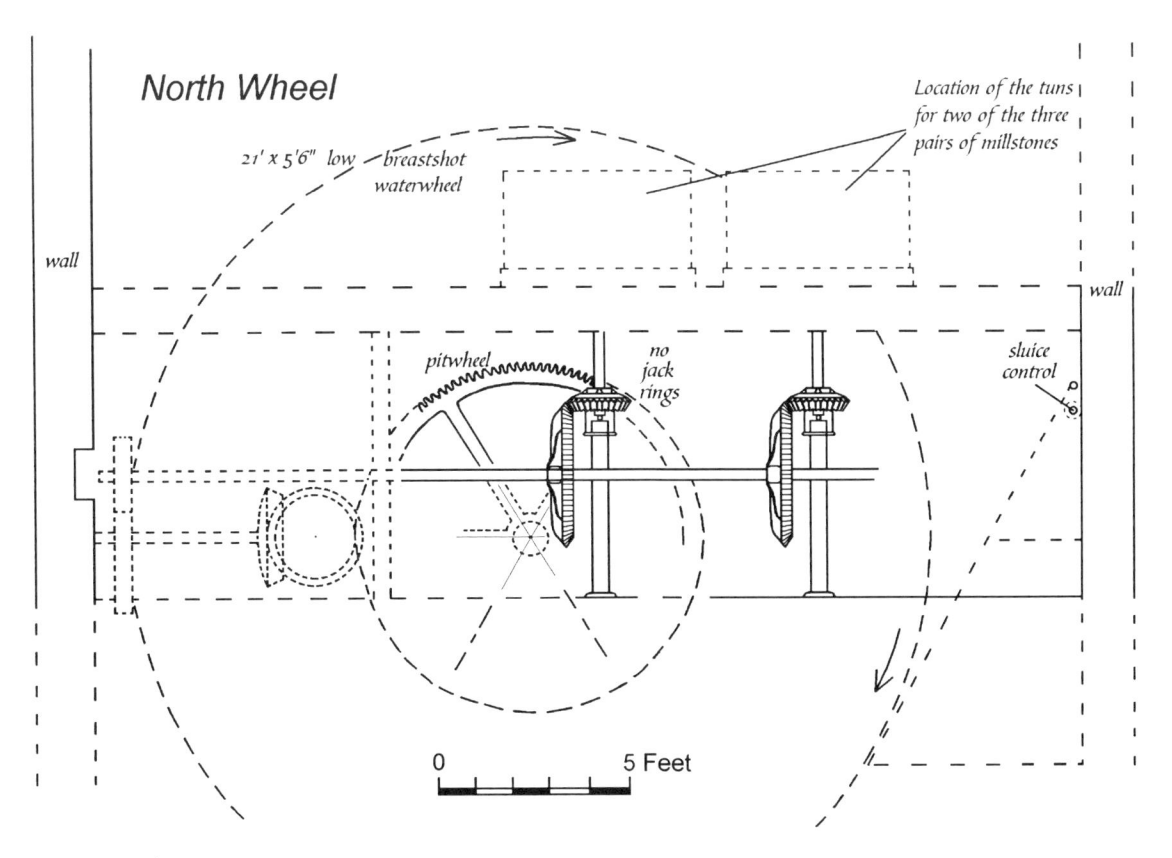

FIGURE 111. A SECTIONAL DRAWING OF THE POWER TRANSMISSION SYSTEM ASSOCIATED WITH THE NORTHERN WATERWHEEL. THERE WOULD HAVE BEEN GEARING, ETC. FOR A THIRD PAIR OF MILLSTONES. (*Tim Booth*)

with his brother Ernest. The Ormes had previously been involved briefly at Somerford Booths Mill (*q.v.*), Havannah Mill (*q.v.*) and also at Royle Street, a steam powered mill, in Congleton.[38]

By the end of the 19th century the mill had, together with its predecessors, been engaged in toll milling for the townspeople and local farmers since the 13th century. Although merchant milling, with the miller himself trading in grain and flour, etc., would have largely taken the place of the toll system in the 19th century, the advent of the large steam driven roller mills caused a significant change in the Town Mill's business. The production of flour, etc. for human consumption was abandoned in favour of provender milling for the production of animal feeds and sub-contract work for other millers. This type of business required the miller to store much larger quantities of grain than previously and so in 1899 the granary storage at the mill was increased by the addition of an extra storey.[39]

Fortunately the mill accounts have survived from the first quarter of the 20th century which show that the Ormes were milling a wide range of products such as wheat, rye, barley, corn (maize), bran, beans, peas and oil cake. However, they had a relatively small number of customers as they mainly acted as subcontractors for such as W. Harrison & Co (Daneinshaw Mill), F. R. Thompstone & Co (Bosley Mill), Hovis Bread Co. (at Macclesfield and later Trafford Park); other regular customers locally included A. Wallworth and A. Lancaster. There are also many entries in the account for various repair work and materials. In 1897 there are many items for building materials, presumably required for the extension of an additional storey, which culminated in a payment of £255 in 1900 "*to Matthew Cooke to contract on corn mill and extras*". Also in that year 7s-0d was paid for a joiner to raise a French stone, and £2 to Jonathon Booth, the local millwright, for a "*coggin and nut*", but lengthening a "*sharp and worm*" cost £10-11s-0d. In the following year, raising two pairs of stones cost 18s-5d, paid to A. J. & J. Worall, and 17s-9d was paid to Baxendales & Co.(a Manchester hardware company) for a new hoist chain. After 1916 the Orme brothers insured the mill with the Prudential Assurance Co. at a yearly premium £23-12s-6d. To offset some of these on-going costs they did receive £5 per quarter from the owners of the silk mill as water rent.

An analysis of the accounts from 1900 to 1924 shows that the average annual profit generated by the mill was £150. However, there was considerable fluctuation in these profits with the maximum amount being recorded at £446 for 1919 and the minimum being a loss of £87 in 1923. Closer inspection of the accounts shows that every week the two Orme brothers paid themselves £5 each under the heading "expenses",

FIGURE 112. THE NORTHERN WATERWHEEL.
(*Rex Wailes, 1964, for the National Monuments Record*)

FIGURE 113. THE SOUTHERN WATERWHEEL.
(*Rex Wailes, 1964, for the National Monuments Record*)

this it would seem was their personal "wages" in addition to the mill business profits. This amount went up to £6 each per week after the First World War. These amounts show that the Ormes were making a very comfortable income from the mill at a time when many small mills went out of business altogether. This must be attributed to their concentration on subcontracting and their provender business rather than providing the flour for Congleton's daily bread.[40]

Sometime early in the 20th century the Ormes had an electric generator installed to be driven by one of the waterwheels to provide lighting for the mill, there being no public supply of electricity in Congleton until the 1930s. Life at the mill did not always run smoothly because in 1929 R. H. Lowe & Co., the owners of the silk mill at this time, complained to Congleton Council about the decay of one of the waterwheels which was letting water leak through. Consequently, the silk mill was only able to get two hours use of the water supply. This leak also seriously affected the hire of rowing boats on the river in the adjacent park. The Congleton millwrights, Booth & Sons, estimated that a replacement wheel would cost £600.[41] Whether a new waterwheel was fitted or the old one repaired is not known but the mill had two working waterwheels in later years.

The Orme family continued to mill at the Town Mill up to and throughout the Second World War and on into the 1950s when they were able to buy the mill from the local Council in 1958. Only five years later, in 1963, the Council gave J. Orme & Son fourteen days notice of compulsory purchase in order to demolish the mill as they deemed it to be unsafe.[42] The reason given by the Council was the presence of a large crack in one of the mill walls which was said to have been caused by a steam lorry reversing into the wall in 1914 (if only all local authorities showed such alacrity!). The Ormes protested and were able to delay the compulsory purchase for another three years but finally they were given compensation of £159 for the waterwheels and gearing, with £446 for the other machinery in spite of the scrap value of the metal being worth a total £900.[43] After this shabby treatment the Ormes moved their business to Siddington Mill, about five miles north of Congleton, and the Council proceeded to demolish the mill and to fill in the mill pool.

FIGURE 114. MR. ORME, THE LAST MILLER AT THE TOWN MILL IN CONGLETON, WITH A MODERN MILLING MACHINE. (*Congleton Chronicle*)

FIGURE 115. MR. ORME WITH THE SOUTHERN WATERWHEEL. (*Congleton Chronicle*)

Fortunately, just prior to its demolition the mill was visited by a couple of enthusiasts who were able to describe the mill with its machinery and take some photographs. These have enabled a layout of the internal power transmission systems to be "reconstructed" from an eyewitness account and photographs (see Figures 110 & 111). The mill itself was a brick building with a slate roof, 50 feet wide by 30 feet deep. The two waterwheels were situated internally but located inside two side wings on the north and south sides of the main mill block. The waterwheels were both made of iron and were of the low breastshot type. The northern wheel was 21 feet diameter by 5½ feet wide and the southern wheel, which was said to have been made in Etruria, was 18 feet diameter by 6 feet wide. The power from the wheels, said to be about 40 h.p. each, was transmitted into the mill by their six sided central shafts which were also made of iron. The wheelshafts were 10 inches and 15 inches in diameter respectively.

The gearing driven by both the waterwheels had almost identical layouts. The wheels drove spur geared pitwheels which meshed with a small pinion mounted on a short horizontal shaft whose axis was parallel with the wheelshaft. Also mounted on this small shaft was a bevel gear which meshed with its counterpart mounted on a horizontal shaft running at right angles to the wheelshaft. At the end of this shaft was a spur gear which meshed with a pinion directly above but mounted on another horizontal shaft. These two gears reversed the direction of the drive back along a shaft which ran across the inside face of the pitwheel but at a higher level than the main wheelshaft. This shaft carried the three sets of bevel gears that took the drive upwards to three pairs of millstones, giving a total of six pairs of stones in the mill. All the millstones were either 54 or 52 inches diameter. The disconnection of the drive to the millstones was by levers on the northern set and by ring and screw on the southern set. There was a sack hoist located in the centre of the mill which was driven by a slack belt system using a jockey pulley to engage the drive.[44] Although the iron waterwheels and gearing were probably fitted in the second half of the 19th century they would, in all probability, be straight replacements for their wooden antecedents installed in 1758.

FIGURE 116. THE DEMOLITION OF THE TOWN MILL IN 1966. (*Congleton Chronicle*)

The demolition of the mil was a great loss to the town of Congleton. At the time of its demolition in 1966 the mill had two working waterwheels driving four pairs of millstones which were still in use. The position of this working watermill at the entrance to Congleton Park from Mill Green should have fired the elected representatives of the town with enthusiasm for the potential of such a building as a tourist attraction. At the time there were rumours that the Council had plans for redeveloping the site. However, since 1966 nothing has been forthcoming and the site is still an area of derelict land, testimony to the lost opportunity, with the only physical evidence that the mill ever existed being the nearby weir on the River Dane. In 2002 the site of Congleton Corn Mill was included in an application for redevelopment that was approved by the Borough Council ensuring that site of the mill will be obliterated under the car park for modern housing.

References

1. Head, R., *Congleton Past and Present*, 1887, (Reprinted by Old Vicarage Publications,1987).
2. Stevens, W. B., *History of Congleton*, University of Manchester, 1970.
3. *Ibid.*
4. Head, R., *1887, op. cit.* 1.
5. Stevens, W. B., *1970, op. cit.* 2.
6. Stevens, W. B., *1970, op. cit.* 2.
7. Holt, R., *Medieval Mills*, 1990.
8. Head, R., *1887, op. cit.* 1.
9. Yates, S., *A History of the Ancient Town and Borough of Congleton*, 1821, (Reprinted by The Silk Press, 2000).
10. Dodgson, J. McN., *The Place Names of Cheshire, Part 2*, Cambridge University Press, 1970.
11. Stevens, W. B., *1970, op. cit.* 2.
12. Head, R., *1887, op. cit.* 1.
13. Congleton Corporation Accounts, held by Congleton Town Council.
14. Stevens, W. B., 1970, *op. cit.* 2.
15. Davies, D. Ll., *Watermill, Life Story of a Welsh Cornmill*, Ceiriog Press, 1997.
16. *Ibid.*
17. Lawton, G. O., ed., *Northwich Hundred Poll Tax 1660*, Record Society of Lancashire & Cheshire, Vol. 119, 1979.
18. Congleton Corporation Accounts, held by Congleton Town Council.
19. Phillips, J., *Inland Navigation*, 1805, (reprinted by David & Charles, 1970)
20. East Cheshire Textile Mill Survey Records, The Silk Museum, Macclesfield.
21. Corry, W., *History of Macclesfield*, 1817.
22. Deeds of Congleton Silk Mill quoted in the East Cheshire Textile Mill Survey Records, The Silk Museum, Macclesfield.
23. Rathbone, C, *Dane Valley Story*, 1954.
24. Smiles, S., *Lives of the Engineers, Vol. I*, 1862.
25. Congleton Corporation Accounts, held by Congleton Town Council.
26. *Macclesfield Courier*, 28th December, 1811.
27. Head, R., *1887, op. cit* 1.
28. *Macclesfield Courier*, 11th March 1815.
29. Stevens, W. B., *1970, op. cit.* 2.
30. *London Gazette*, 27th March 1832.
31. *Macclesfield Courier*, 10th May 1828.
32. *London Gazette*, 12th February 1833.
33. Bagshaw's Directory of Cheshire, 1850.
34. *Macclesfield Courier*, 20th April 1839.
35. Various Directories of Cheshire, 1860 to 1878.
36. Head, R., *1887, op. cit.* 1.
37. *The Miller*, 6th April 1885.
38. Kelly's Directory of Cheshire, 1890 to 1900.

39. *Congleton Chronicle*, 6th January 1985.
40. John Orme's Ledger Book, held by the editor of the Congleton Chronicle.
41. East Cheshire Textile Mill Survey Records, The Silk Museum, Macclesfield.
42. *Congleton Chronicle*, 2nd August & 6th September 1963.
43. *Congleton Chronicle*, 7th January & 28th October 1966.
44. Norris, J. H., "The Water-Powered Corn Mills of Cheshire", *Transactions of the Lancashire & Cheshire Antiquarian Society*, Vol. 75, 1965.

---

39. CONGLETON SILK MILL SJ 859633

In 1749 the government passed an Act of Parliament reducing the duty on imported raw silk from China compared to that charged on Italian silk. This Act encouraged three gentlemen, Nathaniel Pattison, a London merchant, John Clayton, a silk throwster from Stockport, and the Rev. Joseph Dale, also from Stockport, to join together as partners and "...*joint dealers in the art and mystery of manufacturing of raw silk...*".[1] That these three men should form this partnership was no coincidence since they had connections to the three silk mills built prior to this date to exploit the Italian method of throwing silk. Nathaniel Pattison had lived with Sir Thomas Lombe prior to 1724 when Lombe was commissioning the first, waterpowered silk mill to be built in this country at Derby and Nathaniel's son Nathaniel Maxey Pattison was later apprenticed to the manager of the Derby silk mill.[2] Obviously the Pattison family brought considerable technical "know how" concerning the new Italian process of powered silk throwing, introduced by Lombe into this country, to the new partnership. The second partner, John Clayton, had been mayor of Stockport when the town mill there had been leased by John Guardivaglio, one of Lombe's ex-workers, to become the second silk mill to be established in the country in 1733.[3] The Rev. Joseph Dale was a silent investor in the Congleton enterprise but had some insight into the profits to be made in this new industry as his son-in-law was Samuel Lankford who was a business partner of Charles Roe, the builder of the third of these new silk mills at Macclesfield in 1744.[4] Possibly reflecting their respective contributions, Nathaniel Pattison had a half share in the partnership, Clayton a quarter and the Rev. Dale a quarter which he shared with his son.[5] The setting up of this partnership was to prove of immense significance to the development of the town of Congleton. The new methods and techniques that were introduced in the four earliest textile mills and the construction of the mills themselves, especially the Congleton silk mill, were seminal to the whole process of industrialisation that was to occur in the country in the late 18th and 19th centuries.

In order to build the silk mill, John Clayton opened negotiations with the Corporation of Congleton to secure the water rights to the town corn mill. The acquisition of these water rights would allow the partnership access to a controlled water supply for their proposed silk mill without having to invest heavily in a new weir and water management scheme. As the ex-mayor of Stockport who had been involved in leasing Stockport's town mill for silk throwing he was the ideal member of the partnership to persuade Congleton Corporation of the benefits of a similar arrangement in their town. Eventually, in 1752, an agreement was signed that allowed John Clayton

"*the liberty to erect in such part of the said garden or Byflat as he should think fit any building or buildings to be used as a silk mill with engines of various kinds for the winding doubling twisting and working of silk or other goods with a water wheel for the turning or working of such mill and engines...*"[6]

The Byflat, where the mill was to be erected, was on the northern side of the River Dane just upstream of the town bridge. At this time the land was being used as a garden for the town workhouse and the new mill was to be built adjacent to the town corn mill. Although the agreement allowed Clayton to take an amount of water from the corn mill pool that could be delivered by a ten inch square aperture, the corporation insisted that the performance of the town mill must not be compromised.[7] Another worry of the corporation was that the silk mill would attract workers and apprentices from other parts of the country and that they could eventually become a burden on the corporation. Clayton had to give security that the mill workers would not gain settlement in the town by the fact of working at the mill.[8] For this far reaching agreement Clayton was to have paid £80 consideration money to the corporation and the annual rent was to be one shilling for the next 300 years. However, there was some dispute over whether the Lord of the Manor of Congleton, who at this time was Mr Shakerley of Somerford Hall, had any rights concerning the Byflat. This question was eventually settled in favour of the Corporation but Clayton withheld his payment of £80 until the resolution had been reached.[9] Some months

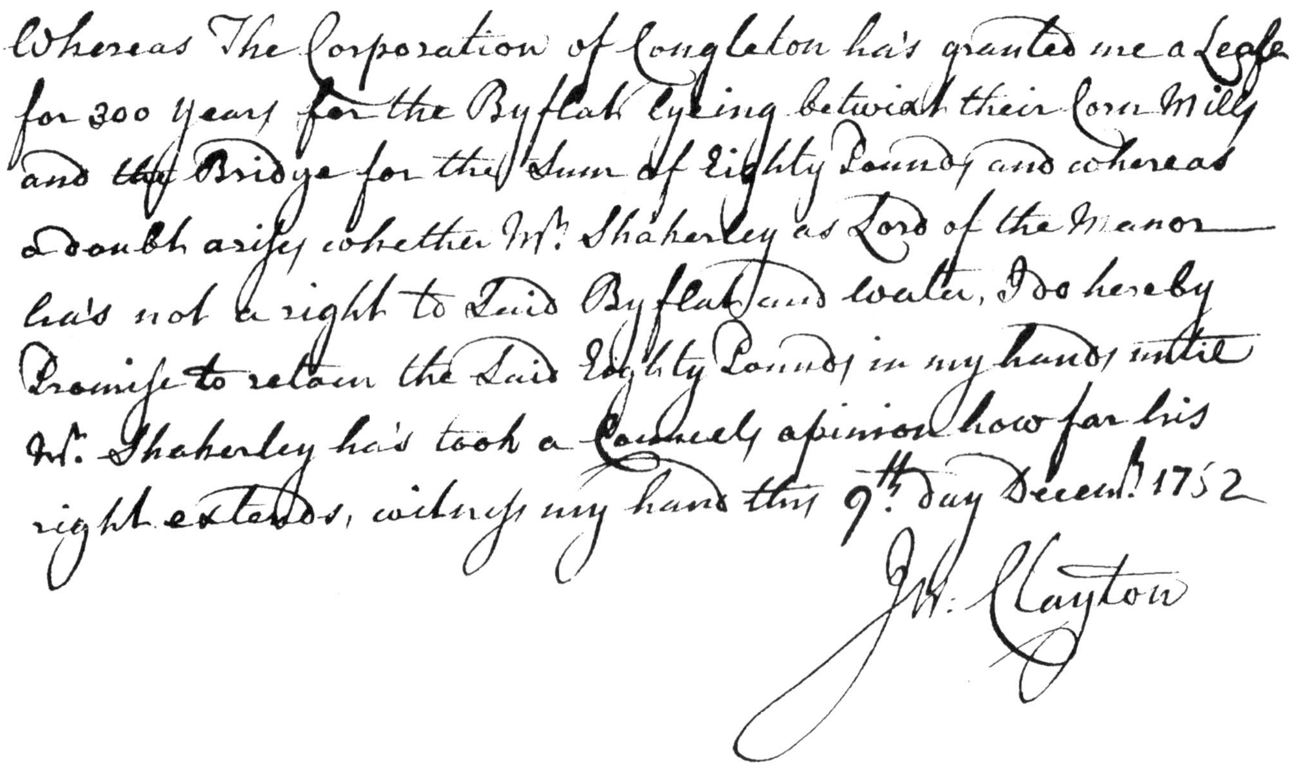
Whereas The Corporation of Congleton ha's granted me a Lease for 300 years for the Byflatt lying betwixt their Corn Milly and the Bridge for the Sum of Eighty Pounds and whereas a doubt arises whether Mr. Shakerley as Lord of the Manor ha's not a right to said Byflat and water, I do hereby Promise to retain the said Eighty Pounds in my hands untill Mr. Shakerley ha's took a Councels opinion how far his right extends, witness my hand this 9th day Decem. 1752

Jn. Clayton

FIGURE 117. THE NOTE FROM JOHN CLAYTON TO CONGLETON CORPORATION AT THE END OF 1752 CONCERNING HOLDING UP PAYMENT FOR THE LEASE ON THE BYFLAT DUE TO DOUBT OVER THE LORD OF THE MANOR RIGHTS. (CCALS, DSS1/6/82)

later, early in 1753, Pattison reimbursed Clayton for a half share of the consideration money.[10]

Once the agreement was signed the partners proceeded to build their silk mill. The mill they were to build was to be larger than any of those built previously at Derby, Stockport or Macclesfield. As such the building was certainly visionary in size, being far larger even than most of the mills that were to be built later in the 18th century, including those mills to be built for the cotton industry. The partners obviously believed in "thinking big" and saw the benefits of large scale production with its economies of scale. However, trying to build the largest mill to date was not without its problems. Fortunately, one of those involved was James Brindley who was a local millwright based in Leek but was later to become one of the most famous of the canal engineers. After Brindley's death the story of the building of the silk mill in Congleton was told, as follows:-

"*James Brindley was employed by Nathaniel Patterson* [sic], *Esq. of London, and some other gentlemen to execute the larger wheels of a new silk mill at Congleton in Cheshire. The execution of the smaller wheels, and of the more complex part of the machinery was committed to another person,* [a Mr. Johnson according to Samuel Smiles, writing later in the 19th century] *who had the suprintendency of the whole, but who was not equal to the undertaking, and confessed himself unable to complete it. The proprietors were greatly alarmed, and called in the assistance of Mr. Brindley, but still left the chief management to their former engineer, who refused to let him see the whole model, and affected to treat him as a common mechanic. Mr. Brindley felt his own superiority to the man who thus assumed consequence, and would not submit to such unworthy treatment, and he told the proprietors if they would let him know what was the effect they would wish to have produced, and would permit him to perform the business in his own way, he would engage to finish the mill to their satisfaction. This assurance, joined to the knowledge they had of his ability and integrity, induced them to intrust the completion of the mill solely to his care; and he accomplished that very curious and very complex piece of machinery in a manner far superior to their expectation, by constructions of many new and useful improvements, particularly one for winding the silk upon the bobbins equally, and not in wreaths; and another for stopping, in an instant, not only the whole of this extensive system throughout its various apartments, but any part of it individually. He invented likewise machines for making all the tooth and pinion wheels of different engines. These wheels had hitherto been made or cut by hand with great labour. But by means of Mr. Brindley's machines as much work could be performed in one day as before as required fourteen.*"[11]

The introduction of the system of lifts for winding the threads onto bobbins, although a very simple idea, was to become standard practice for all textile spinning machinery throughout the industry. The ability to remove power from each separate item of the machinery in the mill was obviously seen as a complete novelty and was copied extensively in future textile mills. Its advantage was twofold, it allowed a certain degree of "unit operation" by making the various machines somewhat independent of each other, but secondly, it would be essential for the machinery in the mill to be started sequentially so as to reduce the inertia that would be present when starting the waterwheel.

The description of the manufacture of the mill gearing is quite startling as it suggests the use of some form of machine tools. This possibility has previously been rejected by other historians in favour of the use of some special kind of planes and jigs.[12] However, the use of hand planes and jigs would have had the effect of de-skilling the job, allowing the use of common labourers for the task thereby reducing costs, but it would not have speeded up the task significantly. An increase in output by a factor of fourteen could only have been achieved by some form of automatic and powered machinery, in other words some form of machine tool, however simple. Due to the size of the mill and the number of gears required, the employment of some kind of machinery for the production of gears was undoubtedly a great benefit. Unfortunately this type of machinery does not appear to have been used in the years after the building of the silk mill in Congleton, or if it was, it raised no comment. It is likely that the requirement for gears during the rest of the 18th century could have been satisfied by the skilled labour available without recourse to such machinery. It is generally accepted that it was not until the massive requirement for rigging blocks by the Royal Navy in the earliest years of the 19th century that machine tools were introduced.[13] However, it is possible that some rudimentary form of powered tools were used in Congleton as early as the 1750s. The gearing in the mill would have used lantern pinions, etc. which could have been produced in quantity using powered versions of a saw, a lathe and a drill. It is quite feasible that James Brindley could have used a waterwheel driven by the River Dane to power these basic tools.

The mill building itself was certainly impressive. At 240 feet long by 24 feet wide and 48 feet high it consisted of five storeys. It was 29 bays long with a five bay wide central section which projected forward of the main building line and was surmounted by a pediment. This central section was flanked by ten bay long ranges which terminated in recessed, two bay long ends. A cupola surmounted the centre of the building and the end bays had pyramidal roofs.[14] At the rear an almost centrally placed wing, at right angles to the main building, was five bays long and rose to the full five storey height. The brickwork of the building was all laid in Flemish bond and the gable end walls, which were three bays wide, were built with alternative dark stretchers and light headers.[15] Obviously no expense was spared in the mill's construction which resulted in a fineness in the quality of its appearance that was not to be matched by much of the later mill construction. The headrace, which was constructed from the corn mill

FIGURE 118. A SKETCH MAP OF "MILLS: WOOD: ETC", 1803, (*North is towards the bottom of the page*) SHOWING THE WEIR ON THE RIVER DANE, TOWN MILL, AND THE SILK MILL BUILT IN 1752 ON THE BYFLEET. (*Congleton Chronicle*)

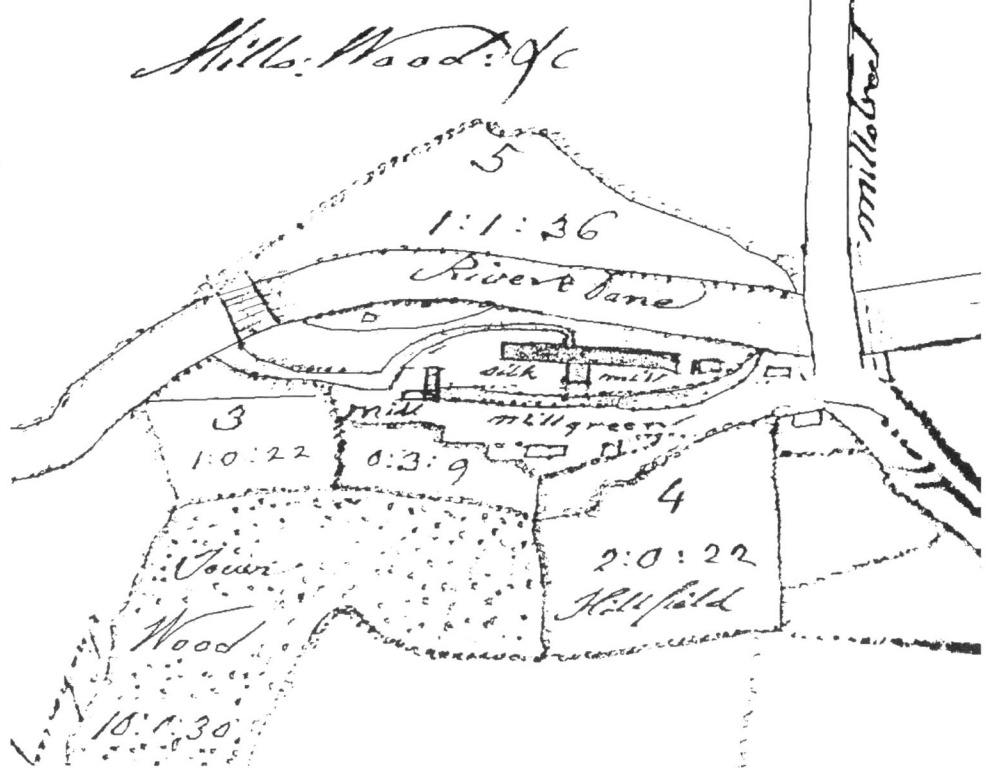

pool, ran around the south face of the silk mill, between the mill itself and the River Dane, and entered the mill at the eastern end of the central pedimented section of the building. The waterwheel was breastshot with a diameter of 19 feet 6 inches and a width of 5 feet 6 inches.

The throwing machines were circular in form, turned by vertical shafts passing through their centres, and were nearly twenty feet high with a diameter of twelve to thirteen feet.[16] These machines passed through the first floor which had cut-outs to accommodate the machines. In effect the first floor was a platform for the operatives to stand on while working on the top part of the throwing machines. At a later date Nathaniel Pattison indicated that there were eleven Italian throwing machines in total.[17] The position of the waterwheel, being somewhat offset from the centre of the building, implies that there were six throwing mills to the west of the waterwheel and five to the east. The three upper floors contained machinery for winding the silk onto bobbins and for cleaning the lengths of silk.

Unfortunately, there is no evidence that shows how the gearing and transmission was arranged in Congleton Silk Mill to distribute the power from the waterwheel to all the machinery. It has been suggested that the waterwheel directly drove horizontal shafting on both sides of the waterwheel that ran under the throwing machines where sets of gears allowed the power to be distributed upwards, powering the respective throwing machine, and continuing upwards to power some of the winding machines on the top three floors.[18] Unfortunately this configuration would not have been compatible with the statement that it was possible to stop each machine individually because stopping one of the throwing machines would have stopped the corresponding winding machines on the floors above.

It is more probable that James Brindley followed the sucessful system installed at the first silk mill built at Derby in 1721 by Thomas Lombe. In this type of configuration the waterwheel immediately drove a vertical shaft through a pair of gears. From this upright shaft other gearing distributed the power horizontally

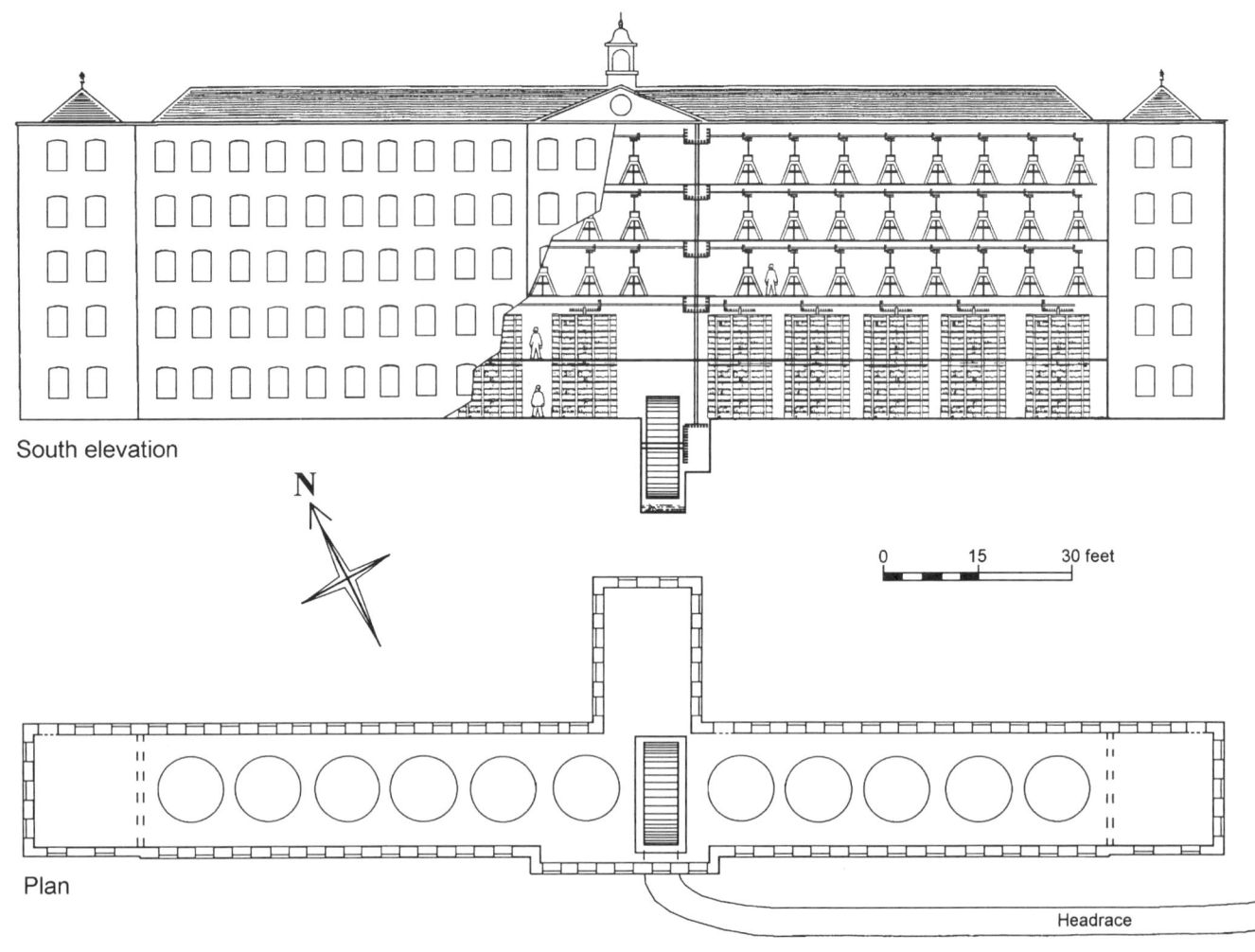

FIGURE 119. A CUT-AWAY VIEW AND PLAN OF THE OLD MILL SHOWING THE ELEVEN ITALIAN THROWING MACHINES OCCUPYING THE TWO LOWER FLOORS WITH THE WINDING ENGINES ETC. ON THE TOP THREE FLOORS. THE POWER FROM THE WATERWHEEL WAS PROBABLY DISTRIBUTED VIA ONE UPRIGHT SHAFT DIRECTLY FROM THE WATERWHEEL.

along shafts running above the throwing machines each of which was connected to the horizontal shaft by a pair of gears (see Figure 119).[19] The vertical shaft would also have delivered the power to the top three storeys for use by the winding engines, etc. Each machine would then have been capable of being disconnected individually from the power distribution system.

The question then arises as to how many upright shafts were used. Two upright shafts, one each side of the waterwheel, would have had the benefit of tending to equalise the twisting forces imparted on the waterwheel and certainly the extra cost involved would not have deterred the owners. However, the statement that it was possible to stop, in an instant, the whole of this extensive system[20] confirms that the power for the whole building probably passed through one set of gears that could be disengaged. This implies that it is most likely that there was only a single upright shaft to distribute the power to each floor.

In the 1750s the size of this building, the complexity of the machinery and the methods of working would have appeared to be a quantum leap forward compared to the small waterpowered installations such as the nearby corn mill with which the people of the time were familiar. Likewise its effect on society and the daily lives of the people who were to work in the new mill was immense. These first four silk mills to be built, culminating with the Congleton Silk Mill in 1752, were to set the pattern for the start of what became known as the Industrial Revolution.

To avoid problems over the supply of water, John Clayton negotiated the lease of the town corn mill from the corporation in 1754 for the sum of £150 during three lives and 21 years. This also allowed him to take Nathaniel Pattison as his partner in the venture.[21] It has been claimed that Samuel Pattison, Nathaniel's brother, took over the responsibility for building the silk mill from John Clayton in 1754. The evidence for this is a memorial in St. Peter's Church, dated 27th May 1756, which states :-

"Samuel Pattison... resided during a year before his death in the town as director of the silk mills: whereby his great abilitys and unwearied application, he render'd the most important services: and enjoyed the satisfaction of living to see all the works compleated, and the manufacture brought to perfection"[22]
Unfortunately, it is not known how much the mill cost to build and equip but £5000 was still owing in 1756 when the account was settled by Nathaniel Pattison.[23]

Having acquired the lease of the town corn mill Clayton and Pattison then had to fulfil the condition to ensure that the mill ground the same amount of grain as previously. This possibly proved to be harder to achieve than they first thought for in 1758 they decided to rebuild the corn mill in its entirety utilising the latest ideas on corn mill configuration.[24] Also in this year John Clayton died, followed six weeks later by his wife, Mary.[25] This unfortunate occurrence left the share of the partnership owned by Clayton in the hands of his children who wished to realise their inheritance. Consequently, Nathaniel Pattison purchased the lease for five shillings and bought out his old partner's share in the silk mill and the corn mill for £2500.[26] No doubt this and the rebuilding of the corn mill left Nathaniel Pattison short of cash, after all the silk throwing business had only been in operation for about two years and had not had time to generate large profits. Pattison's need to raise money probably explains why he appeared to give the silk mill in Congleton to Samuel Smith and John Bell of London, presumably as a surety. This agreement made Smith and Bell Pattison's executors with the right to

FIGURE 120. MEMORIAL TO SAMUEL PATTISON IN ST. PETER'S CHURCH, CONGLETON.

purchase the silk mill outright in the three months after Pattison's death, for the sum of £5000.[27]

After this unplanned set-back Pattison's silk throwing business at the mill thrived for many years. As his business prospered minor tuning took place in the mill with James Brindley making improvements in 1761 and then in the following year *"settling the gearing"*.[28] When called upon to give evidence on the state of the silk industry to the House of Commons in 1765 Nathaniel Pattison was seen as a leading member of his profession.[29] By the 1770s he was employing around 600 people at the mill and was reckoned to be the fifth largest landowner in the Congleton area when he was assessed for land tax.[30]

On Nathaniel Pattison's death in 1784, Smith and Bell did not execute their option of buying the silk mill but transferred the property to Christina Pattison, Nathaniel's eldest child, with Nathaniel Maxey Pattison, James Pattison, and Mary Pattison, all children of Nathaniel, remaining as tenants in partnership with Smith and Bell. It was further agreed that James & Nathaniel Maxey would become owners of the silk mill on Christina's death.[31] During the rest of the 18th century the Pattison business dominated employment in the town and the Pattison family were leading members of the community with Nathaniel Maxey Pattison being mayor of the corporation on four separate occasions.[32] On the death of Christina Pattison in 1805 Congleton Silk Mill duly became the property of Nathaniel Maxey Pattison, James Pattison and James Pattison Jnr.[33] However, by the end of the first decade of the 19th century the machinery in the silk mill was definitely obsolete and becoming uncompetitive. In 1811 the Pattisons insured Congleton silk mill with the Royal Exchange Fire Insurance as follows:-

"On the building of their silk mill turned by water situated at Congleton in the County of Chester £1000. On stock in trade in the same £1000. On the machinery therein £1000. All brick built, tiled & slated. Warranted no steam engine."

The following year the insurance entry stated that the mill was heated by steam.[34] These values do seem to be quite a low valuation for a mill that cost well over £5000 and was still the largest silk mill in the country.

In 1814 a flood washed away the weir on the River Dane and so a new stone weir was built at a cost of £2000.[35] In 1820, Yates published a long description of the Congleton Silk Mill and its machinery. The building was still the same as it was constructed in 1754, as probably was the machinery for winding and cleaning. However, the eleven large Italian throwing machines had been removed a few years before and replaced with more modern machinery which had doubled the output capability of the mill. This new throwing machinery was described as follows:-

"There are thirty eight spinning mills, which perform 39,520 movements. The spindles run in steps of glass, and the threads are conducted over glass rods from one bobbin to another. There are twenty two doubling engines, which have 6820 movements. There are likewise upwards of sixty women employed in doubling silk on single wheels. The throwing mills, which wind the silk from bobbins upon reels, and form it again into skeins ready for the dyers, perform 11,076 movements."[36]

Another interesting item is mentioned with respect to the waterwheel which states that:-

"The wheel is regulated to go and keep time with a clock, by a contrivance that admits more or less water upon the wheel, so as always to make the motion uniform. This timepiece, worked by means of the wheel, does not vary a minute, in the course of the day from the common clock."[37]

This certainly describes a device for controlling the speed of the waterwheel and not a mechanism for determining piecework payments, as is the more usual arrangement where a clock was geared to a waterwheel. Whether this mechanism was installed by James Brindley in 1754 is not known but if it was a speed regulator it is more likely to have been fitted much nearer to its 1820 description, possibly when the new throwing machinery was installed.

During the 1820s the silk industry suffered from a great depression after the Napoleonic wars. In 1820 there were only about 400 people employed at the silk mill and by 1826 the Pattisons had found it necessary to close the mill altogether causing much distress in the town's population.[38] Then, in 1827, Nathaniel Maxey Pattison died and his share in the business passed to his four daughters. The following year James Pattison and his son dissolved their partnership and the mill became the sole property of James Pattison Jnr. and was offered for sale together with the town corn mill. It is in this sale advertisement that the name "Old Silk Mill" was first used.[39]

In the prevailing business climate this offer for sale was unsuccessful and James Pattison Jnr. had to wait until 1830 when he was able to sell the mill to Samuel Pearson for a total of £8200, £5800 for the mill and £2400 for the contents.[40] Up until this time the Old Mill was still basically how it had been built in 1754 but Samuel Pearson decided to take advantage of the slow recovery from depression in the industry to invest in developing the mill by extending it and introducing steam power to augment the existing water power. Fortunately for the Old Mill, Samuel Pearson carried out this extension in a way sympathetic to the original construction using the same Flemish bond for the brickwork. The mill was lengthened by 142 feet which added another seventeen bays in total with five bays providing a pedimented section that balanced the original pediment. This extension was built over the headrace at the south east end of the mill. The completed new building had the original configuration of two bays at each end with pyramidal roofs and the cupola was moved to the central position of the new building. A steam engine house and boiler house were built onto the south east face of the short side wing that stood in the middle of the original building at the rear. The engine house had a semi-circular headed window, the arch of which was surrounded by an imposing moulded sandstone architrave with a prominent keystone and corbels. The parapet was surmounted by elaborate moulded stonework in the form of a scalloped shell flanked by two S-shaped scrolls with moulded leaf decoration and at the corners of the parapet were two stone urns. It is obvious that Samuel Pearson lavished a good deal of money and effort on upgrading the mill to regain in the 19th century the prominent position in the silk trade that the mill had held in the previous century. Certainly the size of the extended mill was even more impressive than the original building. It dominated the prospect of the town and was the largest silk mill to be constructed in this country.

The arrival of Samuel Pearson at the mill coincided with an upturn in the fortunes of the silk industry and trade revived. Riots in Coventry in 1831 persuaded some businesses to switch their trade to Congleton and the Old Mill started making plain black sarsenets for ribbons used in the hatting trade.[41] Samuel Pearson became the mayor of Congleton in 1835 and by 1840 the number of people employed at the mill had risen back up to 600.[42] Samuel was joined in the business by his son James in 1841 which no doubt allowed Samuel more time for his wife Fanny and his seven young children and nine servants all living on his estate at Buglawton Hall.[43]

There was a hiatus in this steady march of progress the following year when the Old Mill was besieged by chartists who stopped the steam engine and extinguished the boilers so bringing the whole Pearson enterprise to a halt.[44] However, soon after the riots business returned to normal. In the 1850s the prospects for the silk industry had improved so in 1855 Samuel Pearson bought some land for expansion. This parcel of land lay to the north of the road alongside the Old Mill that led to the Corn mill.[45] However, he was fated not to use this land as the opportunity arose to acquire Dane Mill (*q.v.*) about ¼ mile downstream on the River Dane which the Pearsons then operated in conjunction with the Old Mill. By this time both Samuel and James Pearson were taking a more "hands off" approach to manufacturing and were employing managers to run their two mills.[46]

In the forty years from 1830 when he had purchased the Old Mill Samuel Pearson kept the mill open regardless of the state of business. No doubt the Old Mill was better placed than some mills to ride out the vagaries of trade due its sheer size, which would keep unit costs low, and the ability in times of reduced output to operate with just the waterwheel so saving fuel costs. There was also probably a degree of paternalism and charity on the part of the Pearsons that caused them not to lay off all their employees in times of hardship and depression. In 1871, just before

FIGURE 121. THE OLD MILL AS IT APPEARED IN THE 19TH CENTURY AFTER BEING LENGTHENED BY SAMUEL PEARSON. NOTE THE REAR OF THE TOWN MILL ON THE EXTREME RIGHT HAND SIDE. (*Head*, 1904)

FIGURE 122. AN AERIAL VIEW OF THE OLD MILL AS IT APPEARED IN ITS PRIME PRIOR TO 1939. (*Reproduced by kind permission of Simmons Aerofilms*)

his death, the employees at the Old Mill presented Samuel Pearson with an illuminated address which suggests that they held their employer is some esteem.[47]

When Samuel Pearson died, leaving £90,000, the Old Mill was shared between James Pearson and his three younger brothers.[48] After James's death in 1881 there does not seem to have been the same level of commitment to the business by the surviving brothers. Even though they were still employing 564 people in 1885 they had decided by 1887 to hand all the business over to its creditors led by Charles Durant. This resulted in the mill being put up for auction at no reserve. The auction advertisement gives a picture of the Old Mill and its machinery at that time, as follows:-

> "To be Auctioned with no reserve, All that valuable silk mill known as the Old Mill situated in Congleton in the County of Chester with the yards, woodturners and joiners shops, smithy, washhouses, engine and boilerhouse and other outbuildings, waterwheel and water rights thereunto belonging, for many years occupied by Messrs S. Pearson & Son.
>
> The factory which is five stories high, is 120 yards long, 80 of which are 9 yards wide and the remainder 12 yards wide. There is also an extension, two stories high, 20 yards long by 12 yards wide. Together with the silk throwing machinery, utensils, and effects in and about the said mill and premises comprising 88 winding, 66 cleaning, and 15 doubling frames; 64 spinning and 28 throwing mills; compound steam engine, 20 h.p., and two steam boilers capable of working to 40 p.s.i. pressure."[49]

The Old Mill was purchased at the auction by Dennis Bradwell, a Congleton silkman, for £2050, about a quarter of what Samuel Pearson paid in 1830, which gives some indication of the state of trade in the silk industry in the late 1880s.[50] Dennis Bradwell continued operating at the mill under the name of Samuel Pearson & Son until the late 1890s when the mill was empty.[51]

In 1899 Dennis Bradwell's executors leased the Old Mill to Messrs. Robert H. Lowe, a hosiery manufacturer, for five years at £15 per year with an option to purchase the mill after five years for £2700. One odd stipulation was that R. H. Lowe were prohibited from subletting the mill for fustian cutting. At the end of this lease in 1904 R. H. Lowe did not buy the mill but extended the lease for another five years, a practice that they repeated in 1909. In fact Messrs. Robert H. Lowe, in the guise of William Charles Sigismund Albanus Higginson Lowe, did not buy the Old Mill until 1918.[52] During this period extending through the years of the Great War the Old Mill was still being partly operated by water power. One of the first jobs given to a young lad called Charles Taylor, who started work at the Old Mill in 1916 at the age of thirteen, was to clear leaves during the autumn from the waterwheel intake, a sure sign that it was still being used.[53] Also, in 1929 it was reported by Councillor Lowe, when complaining to the council about the state of the corn mill, that the Old Mill "*...could not get more water from the river than would drive the machinery for more than two hours a day*" due to problems with the waterwheel on the corn mill.[54] It was probable at this time that the wheel and the steam beam engine were removed and replaced by a twin cylinder Crossley heavy oil engine with a six rope power train. The oil engine put so much strain on the building that buttresses had to be fixed to the walls.[55]

In 1935 a new mill, called Roldane Mill, was built to the north of the Old Mill on the land purchased by

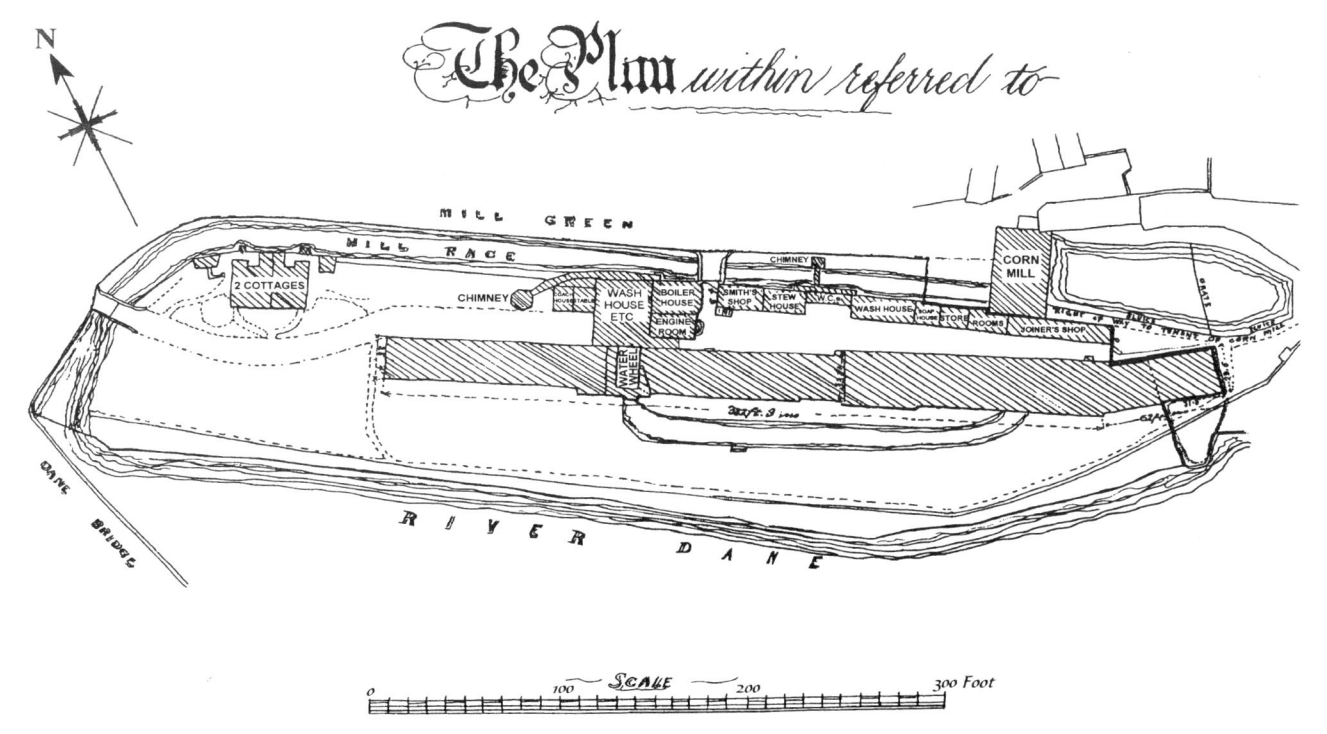

FIGURE 123. PLAN PRODUCED FOR THE SALE OF THE OLD MILL IN 1887, INCLUDING TOWN MILL.
(CCALS, DCB/2114/75)

Samuel Pearson in 1855 for £120. On completion, the main production work of R. H. Lowe & Co. moved to this new mill. By 1939 the state of the fabric of the building of the Old Mill was causing some concern and the top three storeys were removed, reducing the building down to two storeys with a flat roof.[56] R. H. Lowe continued to use the building for making-up and dyeing until about 1984, except for a period during World War II when the mill was used for making ammunition.[57] In the 1990s, when the main business of the site was the manufacture of replica football shirts, the Old Mill was used as a factory shop. In 1999 the business of the R. H. Lowe group fell from a profit of £4.1M to a loss of £3.4M.[58] It was regrettable after being in continual use in the textile industry for approximately 250 years, and having a fine record of maintaining employment in even in the most difficult circumstances, that the apparent dependence on a single customer resulted in the closure of the whole site almost instantaneously.

In the period between 1939 and 2003, the two storey remains of the Old Mill were difficult to see as they were surrounded by utilitarian additions built during the 20th century. The easiest parts to be seen were the north west gable wall of three bays which showed off the two shaded Flemish bond brickwork and a section in the middle of the building facing the River Dane. The engine house and boiler room, probably built around 1830, were still present, the engine house having lost some of its stone pretensions. The stone weir, built in 1814, which was also used by the town corn mill, is still in position in the River Dane.

The bottom two storeys of the Old Mill remained until 2003 when they were demolished to make way for housing. During the demolition the wheel pit was exposed and investigated. The most obvious feature to be seen was the curved breast at the front of the wheelpit which indicates that the waterwheel was a true breastshot wheel driven by the weight of water in the buckets rather than by the force of impact. The stonework of this breast was obviously constructed in two phases, originally being suitable for Brindley's 5 feet 6 inches wide waterwheel, but later widened by about 2 feet 6 inches to accommodate a wider wheel of about 8 feet width. When this widening took place is not documented. It could possibly have occurred when the Italian throwing machines were removed and replaced with throstles which would have needed more power. However, the most likely time for the installation of a wider waterwheel would be when Samuel Pearson extended the mill and introduced steam power in 1830. There is no evidence of lengthening the wheelpit so subsequent waterwheels would have been limited in diameter to a size similar to Brindley's original of around 20 feet. In fact in later years the waterwheel did not utilise the whole length of the wheelpit being only 15 feet in diameter by 8 feet wide.[59] Examination of the ground floor of the mill shows no evidence to support the idea of horizontal shafts being used at that level to drive the throwing machines. In fact, there is evidence for a pit wheel and upright shaft on the south-east side of the wheelpit. If similar

FIGURE 124. MARK FLETCHER OF MATRIX ARCHAEOLOGY INSPECTING THE EXCAVATED WHEELPIT AT THE OLD MILL IN FEBRUARY 2003. NOTE THE CURVED BREAST AND THE DIFFERENCE IN THE STONE WORK NEAR MARK'S LEFT FOOT INDICATING WHERE THE WHEELPIT HAD BEEN WIDENED.

evidence existed on the north-west side, it was removed by the widening of the wheelpit. The arrangement of large stone blocks in the floor could well indicate the bases of possibly two upright shafts used to distribute the power up through the mill.[60]

Because the mill was so inconspicuous after 1939 it is perhaps not surprising that this mill's position in the history of the development of industry has largely been forgotten. However, the entrepreneurs who built the Old Mill, John Clayton and Nathaniel Pattison, were amongst the first exponents of the factory system and rank among the initiators of the first factory and industrially based economy in the world. Richard Arkwright, who was to develop the cotton spinning industry some twenty years or so after the building of the Old Mill, is often credited as the creator of the factory system, however, not only had a factory system been established for at least fifty years in the silk industry, but the building type needed to house those processes was already in existence in the form of the first four silk mills to be built, but especially the Old Mill in Congleton.[61] Lately it has become fashionable to refer to the style of the early cotton mills as "Arkwright mills". It is perhaps ironic that the Old Mill in Congleton was the archetype "Arkwright mill" in spite of being in existence some twenty years before Arkwright built his mills. Since the first three silk mills to be built at Derby, Stockport and Macclesfield had all been demolished in earlier centuries then the two remaining storeys of the Old Mill, built in 1753, were the earliest remains of a textile mill in Britain, and hence in the world. While the mills at Cromford and Lanark are feted as World Heritage Sites, it is much to be regretted that the status of the Old Mill is largely ignored and the building's history is hardly acknowledged. So much so that the demolition of this mill was approved by the planning authority to make way for housing in spite of the protest of various individuals and groups that recognised the important contribution that this mill had made to the development of the industrial revolution.

References

1. Deeds of the Old Mill as quoted in the East Cheshire Textile Mill Survey.
2. Warner, Sir F., *The Silk Industry of the United Kingdom*, Dranes, 1921.
3. Hutton, W., *History of Derby*, 1791, quoted in Yates, S., *An History of The Ancient Town & Borough of Congleton*, 1821 (Reprinted by The Silk Press, 2000).
4. Chaloner, W. H., "Charles Roe of Macclesfield", *Transactions of the Lancashire & Cheshire Antiquarian Society*, LXII, 1951/2.
5. Deeds, *op. cit*. 1.
6. Deeds, *op. cit*. 1.
7. Corry, W., *History of Macclesfield*, 1817.
8. *Ibid*.
9. CRO, Note of J. W. Clayton, 1752, DSS 1/6/82.
10. Deeds, *op. cit*. 1.
11. Phillips, J., *Inland Navigation*, 1792 (Reprinted by David & Charles, 1970).
12. Boucher, C. T. G., *James Brindley - Engineer*, Goose, 1968.
13. Cossons, N., *The BP Book of Industrial Archaeology*, David & Charles, 1975.
14. Yates, S., *An History of The Ancient Town & Borough of Congleton*, 1821 (Reprinted by The Silk Press, 2000).
15. East Cheshire Textile Mills Survey, held at the Macclesfield Silk Museum.
16. English, W., "A Study of the Driving Mechanisms in the Early Circular Throwing Machines", *Textile*

History, 2, 1971.
17. Journal of the House of Commons, XXX, 1765, p213.
18. Calladine, A., & Fricker, J., *East Cheshire Textile Mills*, Royal Commission on the Historical Monuments of England, 1993.
19. Williamson, F., "George Sorocold of Derby", *Journal of the Derbyshire Archaeological and Natural History Society*, 57, 1936.
20. Yates, S., 1821 (2000), *op. cit.* 14.
21. Corry, W., 1817, *op. cit.* 7.
22. Memorial in St Peter's Church, Congleton.
23. East Cheshire Textile Mills Survey, *op. cit.* 15.
24. Stevens, W. B., *History of Congleton*, Manchester University Press, 1972.
25. CRO, Wills of John & Mary Clayton.
26. Deeds, *op. cit.* 1.
27. *Ibid.*
28. Smiles, S., *Lives of the Engineers, Vol. 1*, 1862.
29. Journal of the House of Commons, XXX, 1765, p209/220.
30. Survey of Congleton, 1775, held at the offices of the Congleton Chronicle.
31. Deeds, *op. cit.* 1.
32. Head, R., *Congleton Past & Present*, 1887 (Reprinted Old Vicarage Publications, 1987).
33. Deeds, *op. cit.* 1.
34. Royal Exchange Fire Insurance Policy No 261759, 1811.
35. Head, R., 1887 (1987), *op. cit.* 32.
36. Yates, S., 1821 (2000), *op. cit.* 14.
37. *Ibid.*
38. Head, R., 1887 (1987), *op. cit. 32*; Yates, S., 1821 (2000), *op. cit.* 14.
39. *Macclesfield Courier*, 10th May 1828.
40. Deeds, *op. cit.* 1.
41. Stevens, W. B., 1972, *op. cit.* 24.
42. Head, R., 1887 (1987), *op. cit.* 32.
43. CRO, Census return, Buglawton, 1841.
44. Pigot & Slater, Directories of Cheshire, 1842.
45. Deeds, *op. cit.* 1.
46. Kelly's Directory of Cheshire, 1857, 1865.
47. Illuminated Address, Congleton Museum Trust.
48. East Cheshire Textile Mills Survey, *op. cit.* 15.
49. *Macclesfield Courier*, 6th August 1887.
50. *Congleton Chronicle*, 21st December 1995.
51. Kelly's Directories of Cheshire, 1892, 1896.
52. Deeds, *op. cit.* 1.
53. *Congleton Chronicle*, 30th October 1998.
54. East Cheshire Textile Mills Survey, *op. cit.* 15.
55. *Congleton Chronicle*, 10th April 1986.
56. East Cheshire Textile Mills Survey, *op. cit.* 15.
57. *Congleton Chronicle*, 25th April 1986.
58. *Congleton Chronicle*, 26th March 1999.
59. Gibson, P. et al., "Knitting and Making-up", in Alcock, J. ed., *The Industrial Scene*, Congleton History Society, 1972.
60. Visit to the Old Mill wheelpit excavation, February 2003.
61. Calladine, A., & Fricker, J., 1993, *op. cit.* 18.

---

40. VALE MILL   SJ 857627

Following the Howty from its source, running northwards into Congleton, the first mill encountered is known as Vale Mill. This mill is situated in the valley between Moody Street and Waggs Road in Congleton where it was convenient to dam the Howty to provide a large mill pool. Yates, in his *History of Congleton* published in 1821,[1] stated that *"Southwards of Moody Street, two modern silk manufacturers terminate the buildings in a very pleasing manner as they afford a demonstration of the commercial prosperity and spirit of improvement."* This reference, which can only refer to Vale Mill and to Moody Street Mill, a nearby steam powered silk mill, shows that Vale Mill was built prior to this date. Consultation of the Land Tax Returns for the late 18th and early 19th century shows that the land on which the mill eventually stood was owned by Isaac Faulkner from 1810 to 1831. He was also the occupier of the land up to 1816 but from then until 1823 others shared the occupancy. It is conceivable that this indicates that the mill could have been built in 1816.[2] However, the ownership ascribed to Isaac Faulkner from 1810 onwards is contradicted by a note on the sketch plan of the site made in 1803 which has had a note added in 1818 stating that the property was bought by Falkner [*sic*] for £4000 from someone called Hodgson in that year.[3] Consequently it is not possible to say who actually built Vale Mill as it could have been either Hodgson or Isaac Faulkner, although Faulkner is the more likely candidate.

One of the early trade directories, published in 1822, listed the firm of Gent & Norbury, silk throwsters and ribbon manufacturers, as being at Vale Mill[4] and the Land Tax confirms Charles Gent as the occupier from 1824 to 1827. Later in the 1820s the firm is referred to as Gent & Smeeton who also appear as the occupiers of Vale Mill in the Land Tax returns from 1828 to 1831 when the returns cease.[5] This state of affairs continued until 1838 when Charles Gent died and the

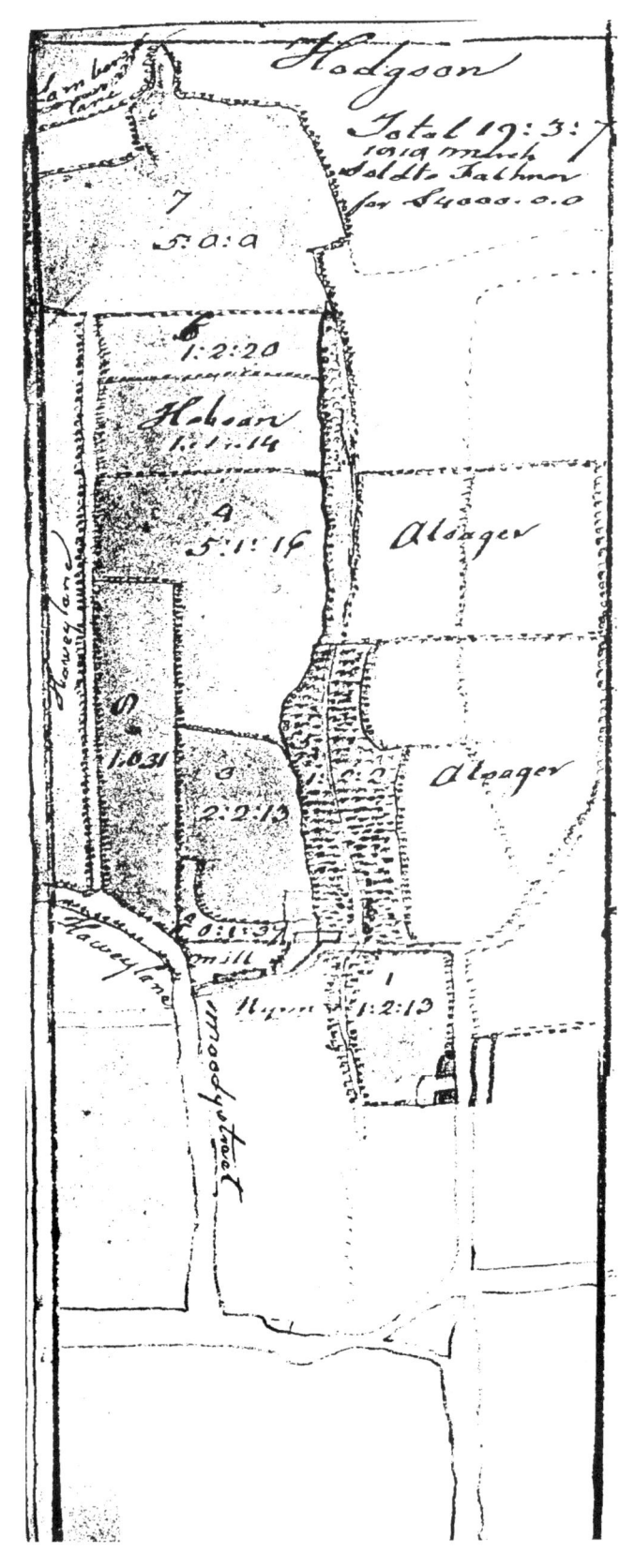

FIGURE 125. A SKETCH MAP OF THE MOODY STREET AND HOWEY LANE AREA OF CONGLETON, 1803, SHOWING HODGSON'S LAND (*North is towards the bottom of the page.*). THE NOTE AT THE TOP OF THE MAP WAS ADDED LATER AND STATES "1818 MARCH SOLD TO FALKNER FOR £4000.0.0" AND A SMALL RECTANGLE HAS BEEN ADDED TO REPRESENT VALE MILL.(*Congleton Chronicle*)

mill was advertised for sale as follows:-

"*All that desirable silk manufacture and throwing concern belonging to the late occupation of Mr. Charles Gent deceased, in the Vale, Congleton comprising a silk mill, engine house and other outbuildings, capacious gentleman's residence for the principal and four neat cottages.*

The mill has an excellent fall of water with a waterwheel 18 feet diameter, 3 feet wide and efficient and complete gearing. The condensing engine of 8 horse power is also attached together with the excellent and modern machinery, looms and gearing. The proposed site of the station for the Manchester-Birmingham Railway is nearby.

Apply to Mrs. Gent, The Vale, Congleton."[6]

This shows quite clearly that the mill was powered by both a waterwheel and a steam engine. The fall of water which is still visible today shows that the waterwheel was of the low breastshot type. The geography of the site would indicate that the waterwheel was situated at the western end of the mill. Unfortunately, the railway station referred to did not materialise as the railway was eventually built on a different alignment, passing further south of Congleton at Hightown.

This sale was not successful and the mill was advertised to let two years later giving details of the mill which was four storeys high and 120 feet long. The machinery used in the mill for silk throwing and ribbon weaving consisted of

"*broad silk looms, spinning mills, throwing mills, winding and cleaning engines, warping mills, filling engines, rubbing machines and numerous other requisite for carrying on an extensive business; warehouse fixtures, a large quantity of gas and steam piping, tallow and oil, late the property of the said Charles Gent, deceased.*"[7]

This advertisement was once again unsuccessful and the property was re-advertised in 1841 to be sold at auction over a "*Friday and Saturday and the following days (Sunday excepted) until the whole is disposed of.*"[8] Presumably Mrs. Gent was getting slightly impatient at trying to realise her late husband's estate. This auction was probably successful because when the Vale Mill was advertised for sale five years later in 1846 it was described as "*late in the occupation of Mr. W. S. Reade*" who presumably operated at the mill from 1841. The sale advertisement in 1846 gives even greater detail of the machinery then in use at the mill although this may have changed since the mill was originally equipped. There were "*...73 power looms, four hand looms, two throwing mills, winding, cleaning and doubling engines, and five filling engines*" which shows that by this time the ribbon weaving side of the business greatly outweighed the silk throwing business. This was a reflection of the state of health of these respective branches of the textile industry in the late 1830s and early 1840s. There was also a large amount of steam piping used to heat the mill and a considerable amount of wrought iron shafting of 1 inch, 1¼ inch and 1½ inch diameter, culminating in 53 feet of 12 inch shafting which was, presumably, the main drive from the power sources to the four floors of the mill. It is interesting to note that the mill was also lit by sixty gas lights using a metered supply from the town gas works.[9]

After this sale in 1846 the mill was occupied by Peter Hunt until 1864 when the mill was again offered for sale. Peter Hunt also traded as a silk throwster at West Street Mills as well as at Vale Mill. The 1846 advertisement gave the dimensions of the mill as four storeys high and 108 feet by 25 feet (inside measure). The mill was driven by both water and steam power giving a total of 18 h.p. Providing that the steam engine is the same as mentioned in earlier advertisements then it would seem that the waterwheel (18 feet diameter by 3 feet wide) could generate around 10 h.p. The machinery mentioned at this time is quite different from that quoted in 1846 representing a change from ribbon weaving back to silk throwing and spinning. The winding engines had lancewood swifts with "*two extra swifts for Taysaan and Tsatlee silk*" and the twelve spinning mills and four throwing mills had "*steel spindles, brass bolsters and steps, and iron sledging.*"[10]

In spite of the 1860s being a depressing time for the silk industry in general, Vale Mill continued to throw silk under Jacob Bradwell & Co. for the rest of the decade.[11] They were followed by Arthur Bull until about 1878.[12] Charles Pedley and Son continued until the end of the 1880s, but when they gave up the business, silk throwing at Vale Mill came to an end.[13] After the end of the silk business the mill was used for fustian cutting by James Cook & Son until after the turn of the century and then by John Shepherd & Son up to the Second World War.[14] This trade, however, did not use machinery and at some time the waterwheel and steam engine were removed. After the Second World War Vale Mill was used for hosiery production for a period of time.

Vale Mill is still in existence today, a four storey brick building with a slate roof, 14 bays long by two bays wide, used as industrial units including the manufacture of office equipment. Unfortunately, the pool on the Howty has been filled in to provide a car park with the Howty culverted underneath. Some indication of the power once derived from this stream can still be appreciated by walking along the footpath, known as Vale Walk, along the north side of the mill and looking over the wall at where the Howty emerges from a large underground pipe. Here the stream falls about five or six feet with considerable force.

References

1. Yates, S., *History of Congleton*, 1822, (reprinted by The Silk Press, 2000.)
2. CRO, Congleton Land Tax Returns.
3. Survey of Congleton, 1803. Held by the editor of the Congleton Chronicle.
4. Pigott's Directory of Cheshire, 1822.
5. CRO, Congleton Land Tax Returns.
6. *Macclesfield Courier*, 17th November, 1838.
7. *Macclesfield Courier*, 15th March, 1840.
8. *Macclesfield Courier*, 13th March, 1841.
9. *Macclesfield Courier*, 4th October, 1846.
10. *Macclesfield Courier*, 3rd September, 1864.
11. Slater's Directory of Cheshire, 1869.
12. Morris's Directory of Cheshire, 1874.
13. Slater's Directory of Cheshire, 1883.
14. Kelly's Directories of Cheshire, 1892, 1906, & 1939.

FIGURE 126. THE THREE BAYS AT THE WESTERN END OF VALE MILL VIEWED FROM WHERE THE MILL POOL USED TO BE. THE WATERWHEEL WOULD HAVE BEEN POSITIONED ON THIS END GABLE WALL.

41. BRIDGE STREET MILL　　　　SJ 857629

About 100 yards downstream of Vale Mill on the Howty Brook the stream disappears under the main street of Congleton, called Bridge Street at this point. Late 19th century Ordnance Survey maps show a silk mill just south of Bridge Street where the Howty also appears to be impounded. When this mill was first established is not accurately known. Certainly there is no mill shown on Moorhouse's Map of Congleton dated 1818. In 1822 when Yates was describing the industry of Congleton he listed a number of silk throwsters in Bridge Street, one of which was a James Johnson. This same James Johnson was bankrupt in 1828 and he was described as previously holding a silk mill in Bridge Street when the property was advertised for auction.

This 1828 auction advert gives details of the machinery in the mill at that time which shows that James Johnson's business was very much on the small side as it lists only two doubling engines, four spinning mills and two throwing mills with a total of 174 dozen spindles. The advertisement concludes with details on the motive power of the mill which was described as *"an excellent steam engine complete with cast iron pump."* There is no mention of a waterwheel at this time.[1]

Eight years later, in 1836, a mill belonging to Paul Barlow, and stated to be situated near Bridge Street, was offered for let. The mill building itself was described as being *"four stories high, 31 yards long by 8 yards wide, heated by steam pipes and worked by a steam engine of sufficient power"*. Again there was no mention of any waterwheel but the statement was made that *"Appertaining to the mill is a stream of water pounded up sufficient to work the mill several months of the year."*[2] which implies that the use of waterpower was a possibility. Unfortunately, it is impossible to determine whether or not water power was ever used at this mill. The 1836 advertisement raises the possibility that water power may have been used but the evidence is inconclusive.

After 1836 the mill is listed until the middle of the 1860s as being operated by Charles William Barlow who was also operating Primrose Vale silk mill (q.v.).[3] Charles William Barlow had to give up the silk business in 1866 when his brother George was imprisoned for theft.[4] After this date it appears that silk throwing ceased and the mill was later used for other purposes, such as fustian cutting, which did not require a prime mover. The building survived until 1978 when it was demolished.

References

1. *Macclesfield Courier*, 2nd September, 1828.
2. *Macclesfield Courier*, 8th October, 1836.
3. Morris's Directory of Cheshire, 1864.
4. Henry Madden's Diary, *Congleton Chronicle*, July 1974 to August 1975.

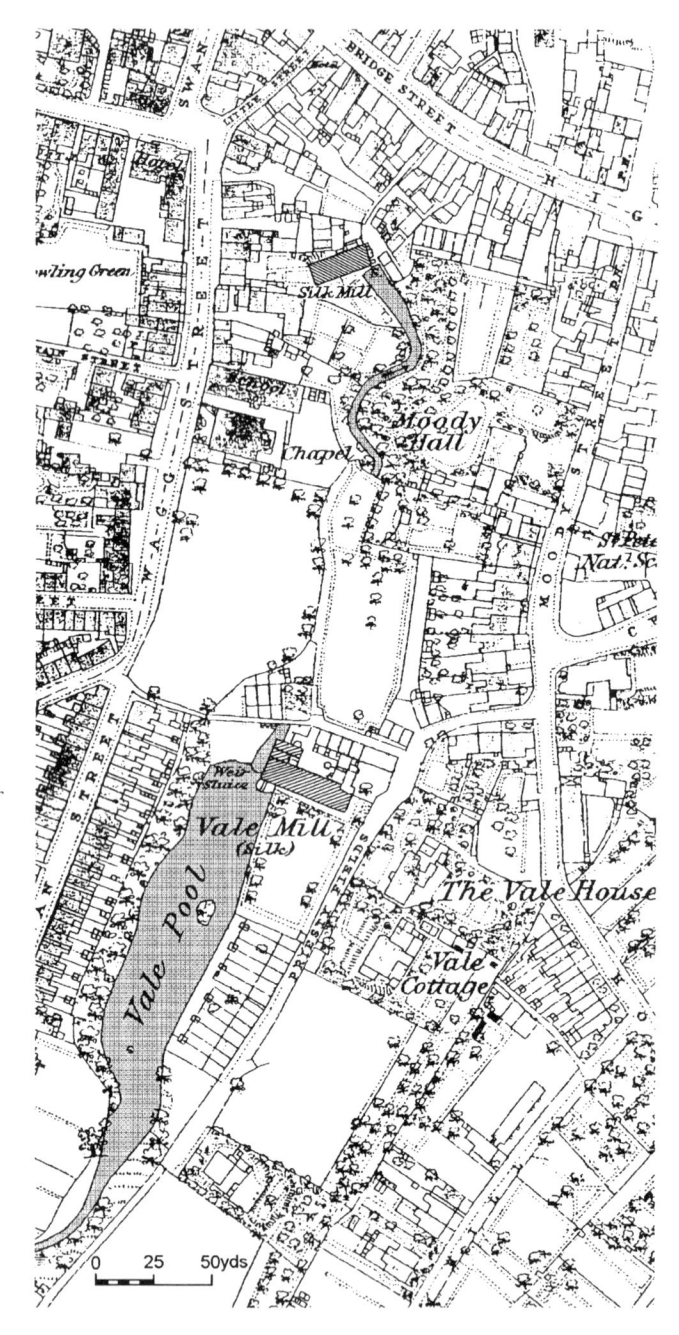

FIGURE 127. PART OF THE ORDNANCE SURVEY MAP OF 1875 SHOWING VALE MILL AND POOL WITH BRIDGE STREET MILL SOME 200 YARDS DOWNSTREAM WHERE THE STREAM HAS OBVIOUSLY BEEN IMPOUNDED.

42. VICTORIA STREET MILL   SJ 858630

On the northern side of Bridge Street, in the centre of Congleton, the Howty Brook emerges from its culvert under the main street at a point some way down Victoria Street, which runs northwards from Bridge Street. This is the site of another small silk mill known as Victoria Street Mill. Again the origins of this mill are unknown. It is first noted in a sale advert in 1840 when it was owned by Bradford & Gent who traded as ribbon manufacturers. It is possible that Charles Gent who was a ribbon manufacturer at Vale Mill (*q.v.*) was involved in this partnership. At this time the mill building was described as being three stories high, 27 yards long by 9 yards wide and containing "*40 steam ribbon looms and winding engines, presses, scales & weights, doubles and all other apparatus for manufacturing ribbons, nearly new*". The power source for the mill was stated to be a steam engine and water power combined.[1]

In 1848, the tenancy of Victoria Street Mill, held by Peter Hewit, a silk throwster, expired and was advertised in the local press stating that the three storey building was 40 yards long by 9 yards wide. It would seem that the mill had been extended sometime in the eight years since 1840. The power supply arrangements at the mill were described as "*by a steam engine of 4 h.p. and a water wheel of 3 h.p.*"[2] These figures highlight the tiny amount of power required by the small silk mills of the middle of the 19th century. In spite of this small amount of power needed, the mill was purported to employ 70 hands in 1850 when it was operated by Mr. Hunt who also threw silk at West Street and Vale Mills.[3]

A curious detail was mentioned in 1857 when the then tenant, a Mr. Hill, offered for sale at the mill "*A valuable patent twisting machine containing 144 spindles with tubes, change wheels, and the patent right*" along with the other silk throwing machinery when he was said to be declining the business. The patent machine was further extolled by the statement that

> "*The lots may be viewed on application to Mr. Hill and any experiments of the capabilities of the patent twist machine tried so that the parties may fully satisfy themselves of its great powers of producing, as well as seeing the beauty of the work produced and its adaptation for sewing purposes. It cannot be equalled by any principle of machine known for the production of sewing silk and twist used in the sewing machine or for tailors' purposes. It will likewise produce a better article and greater quantities of embroidery silk, lace cord, hosiery, fish lines, etc. In fact every kind of silk seems to be improved by the tension it gives the threads. A boy can produce daily upon this principle from 60 lbs. to 100 lbs. of sewing, according to the size of the article.*"[4]

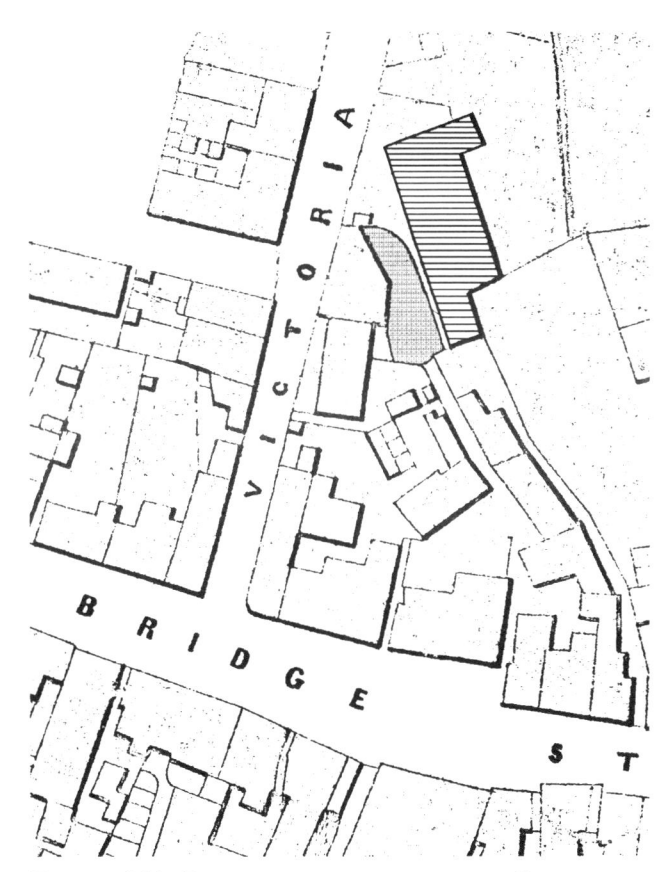

FIGURE 128. PART OF THE TITHE MAP OF CONGLETON SHOWING THE JUNCTION OF VICTORIA STREET AND BRIDGE STREET (THE MAIN THOROUGHFARE IN THE TOWN). THE ORIENTATION OF VICTORIA STREET MILL APPEARS TO HAVE BEEN DICTATED BY THE DIRECTION TAKEN BY THE HOWTY, BEING OFFSET TO THOSE BUILDINGS THAT ADHERED TO THE STREET PATTERN. A SMALL POOL IS INDICATED ALONGSIDE THE MILL TO SUPPLY THE SMALL AMOUNT OF WATERPOWER USED BY THE MILL. (CCALS)

This machine was an adaptation of the spinning mule to work silk.[5]

After Mr. Hill, the tenant became David Green who reverted to ribbon manufacturing at Victoria Street Mill and also Moody Street Mill.[6] In the 1860s both Levi Brassington and William Woolnough were listed as silk throwsters at Victoria Street Mill but after their appearance the mill is not mentioned again.[7] It would seem that the mill ceased work in the 1860s and was probably demolished as Victoria Street was developed in the following years.

References

1. *Macclesfield Courier*, 17th October, 1840.
2. *Macclesfield Courier*, 17th June 1848.
3. Head, R., *Congleton Past & Present*, 1887.
4. *Macclesfield Courier*, 22nd August, 1857.
5. Patent number 1930, 1854.
6. Kelly's Directory of Cheshire, 1857.
7. Kelly's Directory of Cheshire, 1865.

43. PARK STREET TANNERY SJ 862629

One of the small streams that flow through Congleton rises at a spring nearly ¼ mile to the south of the town, close to Canal Road. It flows northward under Lawton Street heading to the River Dane. Just to the north of Lawton Street was the site of the tannery which is remembered in the street names of the area today, such as Tanner Street and Bark Street.

The origins of the tannery are obscure but it was offered for sale in 1846 with the following newspaper advertisement:-

"To be sold at Auction. All that established tannery in Park Street, Congleton, now in the occupation of Mr. Lownds. The tannery is distinguished by the names Old Tan Yard and New Tan Yard.

The Old Tan Yard has 44 pits namely 16 handlers, 16 layers, four spenders, four limes, two water pits, and two Marstruis with adequate drying sheds. The New Tan Yard contains 94 pits namely 35 handlers, 31 layers, 28 spenders, with adequate drying sheds. Also two newly erected bark sheds capable of holding 200 tons of bark. Also a bark mill which with the pumps are turned by water power of which there is a plentiful supply."[1]

This advertisement conveys the size of the operation. Bearing in mind that it was situated in the middle of the town it no doubt exuded an all pervading odour. The stream providing water to the tannery is quite small but would appear to have been sufficient to drive the bark mill and pumps, which would have been used intermittently, and to supply the various treatment pits.

This sale was not successful as the tannery was advertised again early in 1847.[2] It is quite likely that the premises were then purchased by a Mr. Broadhurst because two years later in 1849 the tannery was advertised to let as being *"formerly belonging to Mr. Lownds, afterwards to Mr. Broadhurst now deceased."*[3]

Unfortunately the advertisement of 1849 is the lastest information so far discovered concerning the tannery. It obviously had a much longer life span than just the three years between 1846 and 1849 as is testified to by the names of nearby streets but no further information has come to light about the owners or the water powered machinery used at the tannery. At some stage the premises were converted to a wood yard which is still in business on the site today.

References

1. *Macclesfield Courier*, 8th August 1846.
2. *Macclesfield Courier*, 30th January 1847.
3. *Macclesfield Courier*, 9th June 1849.

---

44 & 45. STONEHOUSE GREEN MILL and BROOK MILL SJ 858631-2

Stonehouse Green and Brook Mills are situated on the Howty between Victoria Street Mill and the stream's confluence with the River Dane. These two mills will be treated together as for most of their life they were under the same ownership.

In the third quarter of the 18th century the land ownership at Stonehouse Green, which had been held by a variety of owners, all came under the control of Thomas Garside, a Congleton sieve maker. In 1766 he purchased a house close to Mill Street which had been converted from a barn by John Willis. In 1783 he leased land on the south side of Stonehouse Green from the trustees of the late Thomas Hall. The following year he also purchased from John Willis the triangle of land between the Howty and its small tributary stream which flowed into it from the east (This tributary stream is the one that powered the Park Street Tannery (q.v.)). In the same year he purchased the rest of Stonehouse Green from Philip Antrobus. In total the whole parcel of land finally owned and leased by Thomas Garside was about 4,300 square yards on which he built four dwelling houses with a long room or workshop over them.[1]

In May 1785 Thomas Garside leased all this property to George and William Reade for 999 years at £30 per annum with the agreement that *"it shall be lawful for the said George and William Reade... to erect any buildings, wears [sic], dams, shafts, wheels, waterworks, basons [sic], reservoirs, soughs, gutters, and other things upon through and under the said premises hereby demised."* For their part the Reades agreed to take out £400 of insurance against *"loss or savage fire"*. Less than a month later, in June, the Reades leased a plot of land from the Burgesses of Congleton in order to construct a sough *"from the workshop or dwelling house and premises lately leased by them from Thomas Garside through a part of the Byflat and which sough empties itself into the River Dane."*[2] This sough was obviously going to be the mill wheel's tailrace. The Reade family had close ties with Richard Martin who had taken over at Davenshaw Mill (q.v.), spinning cotton, in the aftermath of the Arkwright patent case. Possibly the lure of the fortunes that seemed to be obtainable in the cotton spinning trade at this time and their relationship with existing cotton manufacturers gave the necessary inspiration to the Reades to acquire the

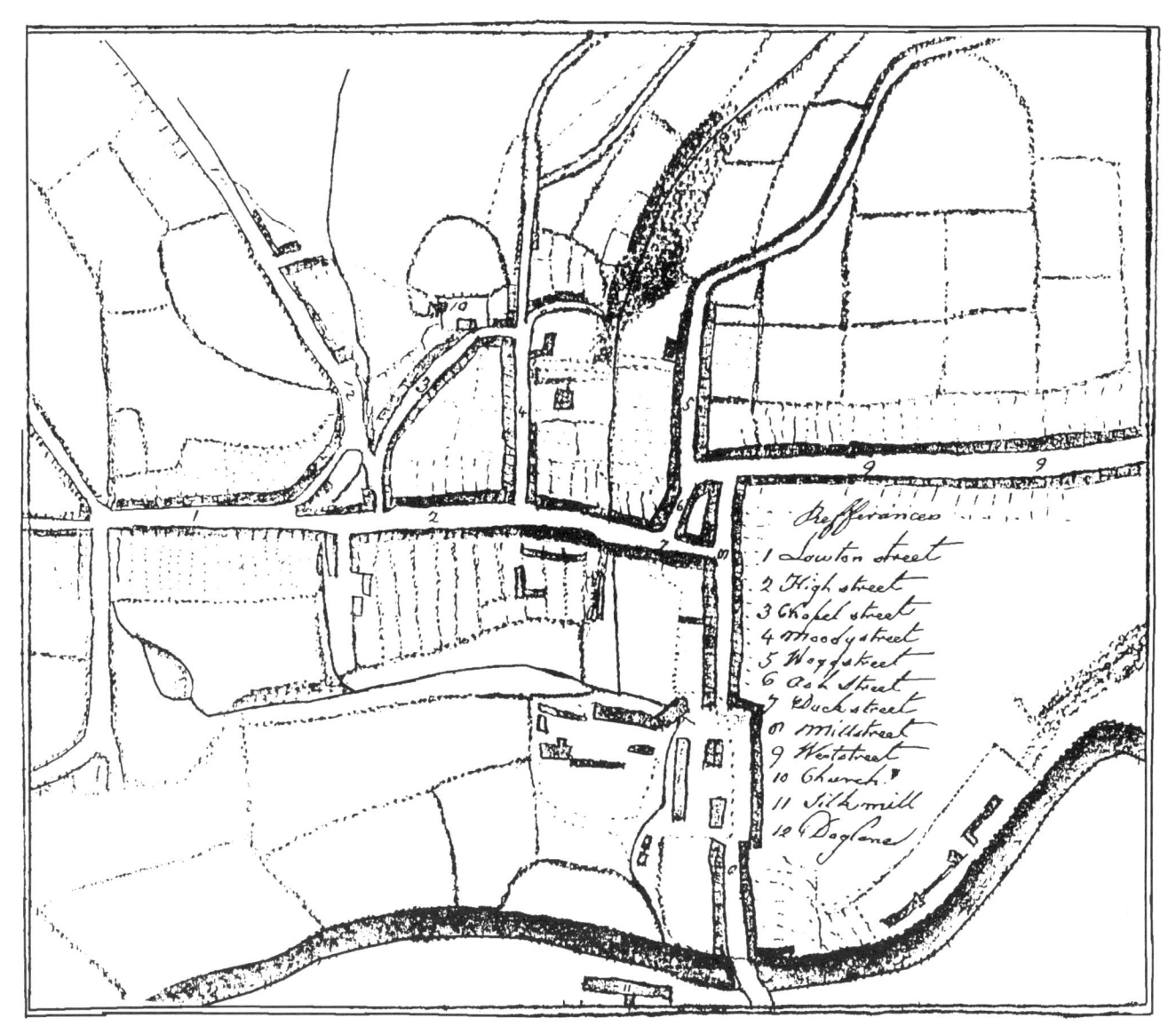

FIGURE 129. A SKETCH MAP OF CONGLETON, 1803. (*North is towards the bottom of the page.*) (*Congleton Chronicle*)

Stonehouse Green property for the spinning of cotton using waterpower.

In 1795, shortly after the death of Richard Martin (see Daneinshaw Silk Mill), George Reade married Richard Martin's widow. Maybe the Reades' cotton spinning business had been profitable or George Reade's marriage brought him new capital but in 1796 the Reades purchased an orchard from Thomas Knight which was just to the north-east of their property at Stonehouse Green. Then in 1802 they bought land from John Bayley next to the orchard and contiguous with their mill, in order to expand their cotton spinning factory.[3] Unfortunately, the Land Tax returns do not reflect this complicated land ownership, in fact they only list George & William Reade as the occupiers of property owned by Richard Hawar, described as a "ribbon shop" from 1790 to 1803, and then described as a "cotton shop" in 1804.[4]

The first indication of any details of the mill at Stonehouse Green is to be found on the survey plan of 1803 produced for taxation purposes. This shows that the Stonehouse Green Mill was of an open "V" shape running east to west across the path of the Howty Brook. It is not known if this was a new factory erected on the site of Garside's original building or if it was an adaptation of that original building. Its use of water power is indicated by the fact that the Howty Brook flows through the western part of the factory where the waterwheel would have been situated. Spinning cotton takes considerably more power than that needed for throwing silk and the plan indicates that the Reades had to combine the flow of three streams to provide enough water supply for their factory. The most easterly of these streams flows northwards from Congleton Moss, entering the town at its eastern end near the bottom of Wallworth Bank where a pool is indicated on the 1803 plan. This pool may possibly have had a connection with the original Town Mill (*q.v.*) or the medieval fulling mill. After leaving the pool this stream runs westwards, joining with the small stream that powered the Park Street Tannery, before joining the Howty at Stonehouse Green.[5]

Also shown on this 1803 plan is a long building just to the north of the Stonehouse Green Mill running along the western side of the Howty. This could be the first indication of the position of the earliest part of Brook Mill. This building, which still exists, has the appearance of a late 18th century textile mill. However, the building is shown on the plan to the west of the Howty when in fact today it is situated on the east of the stream, which leaves some doubt as to what is being represented on the 1803 plan.

George & William Reade continued to spin cotton at Stonehouse Green up until 1817 when the partnership was dissolved with George buying out William's share of the business. In the indenture drawn up at the time the assets are described as *"all those three cotton mills or factories with the three warehouses, counting house, steam engine house, stable, carthouse and stovehouse which have been lately erected built and made by the said George Reade and William Reade"*. This indenture is concerned only with the Stonehouse Green Mill. Of the three mills referred to, two form the arms of the "V" shaped building on the 1803 map, plus the third is an extension to the west of the original mill built on the land acquired in 1802 from Thomas Bayley. The existence of a waterwheel at Stonehouse Green is confirmed in the indenture of 1817 which mentioned *"and also half part of all...waters, watercourses, waterwheel, ...etc"*.[6] That the waterwheel was in use is confirmed by an agreement George Reade made with his neighbour James Twemlow that allowed him to pen the stream *"to work and turn the said factory"* providing that the resulting pond did not expand beyond a certain marker stone that was placed on James Twemlow's land.[7] The land tax returns and the early trade directories continue to show George Reade as a cotton spinner up to 1825. However in 1821 the following advertisement appeared:

"To Let: A substantial & newly erected silk mill or factory situated at or near Stonehouse Green in Congleton with the engine house and excellent steam engine of six horse power recently erected lately in the occupation of Messrs Brown & Bracegirdle but now by Mr. John Brown.

The factory is a substantial and handsome brick building, four stories high and 16 yards long by 9 yards wide. The machinery is worked during the day by the steam engine and the stream of water together (the latter sufficient to turn a waterwheel of two horse power) and the same is worked during the night by the stream of water alone. The factory is in the neighbourhood of colliers and coals are 6d/cwt and another mill might be erected on adjoining vacant land.

The aforementioned brook and liberties and appurtenances are held for the remainder of the several terms of 196 years (commencing lst May 1785) & 999 years (commencing 12th May 1785) & 197 years (commencing 3rd June 1785) subject to a yearly rents of £25-2s-6d and 1s. The factory is completely fitted with new machinery of the best and most modern construction containing upward of 200 doz. spindles for throwing & spinning with a proportionate quantity of winding engines, doubling machines etc. and with an excellent waterwheel with new gearing complete."[8]

The dates attached to the rents leave no doubt that this is the Stonehouse Green Mill of George Reade but it is described as a silk mill (not cotton). It is possible that one of the three mills, mentioned at the dissolution of the partnership in 1817, had been leased to Messrs Brown & Bracegirdle at some time. The advertisement does show that only eight horse power was needed, of which the waterwheel provided a quarter of the power, to drive 200 silk throwing spindles and associated machinery. At the height of a depression in the silk industry, in the mid 1820s, the Reades turned their attention to silk throwing in conjunction with cotton spinning and are shown as owning and occupying two factories, one of which was described as a silk factory. This double entry in the land tax returns probably indicates that George Reade was operating from Brook Mill as well as Stonehouse Green Mill from 1825. Certainly the entry for 1830 lists him as a cotton spinner at Stonehouse Green and George Reade & Sons as silk throwsters at Brookside, as Brook Mill was sometimes known.[9]

This state of affairs continued until 1834 when the businesses were totally reorganised by George Reade who leased Brook Mill to his sons John Fielder Reade, Thomas Reade and James Reade while keeping direct control of Stonehouse Green Mill in partnership with his son Thomas. These two concerns traded as Reade Bros. and George Reade & Son respectively. John Fielder Reade was named after John Fielder, a merchant resident in Bath who was the executor of Richard Martin's estate (see above) when he died in 1795. Whatever transpired with regard to this estate when George Reade married Richard Martin's widow he obviously felt sufficiently grateful that John Fielder was commemorated in the naming of his eldest son. The evidence that Brook Mill was also operated to some extent by water power is provided in the indenture at the time which stated that *"...the paddles sometime since put down and now standing at the bottom of the bank and adjoining or near to the south west corner of the said premises of the said John Fielder Reade, Thomas Reade and James Reade whereby the water of the brook is divided and turned into the wheelrace of the said silk mill."*[10]

This injection of a new generation into the business

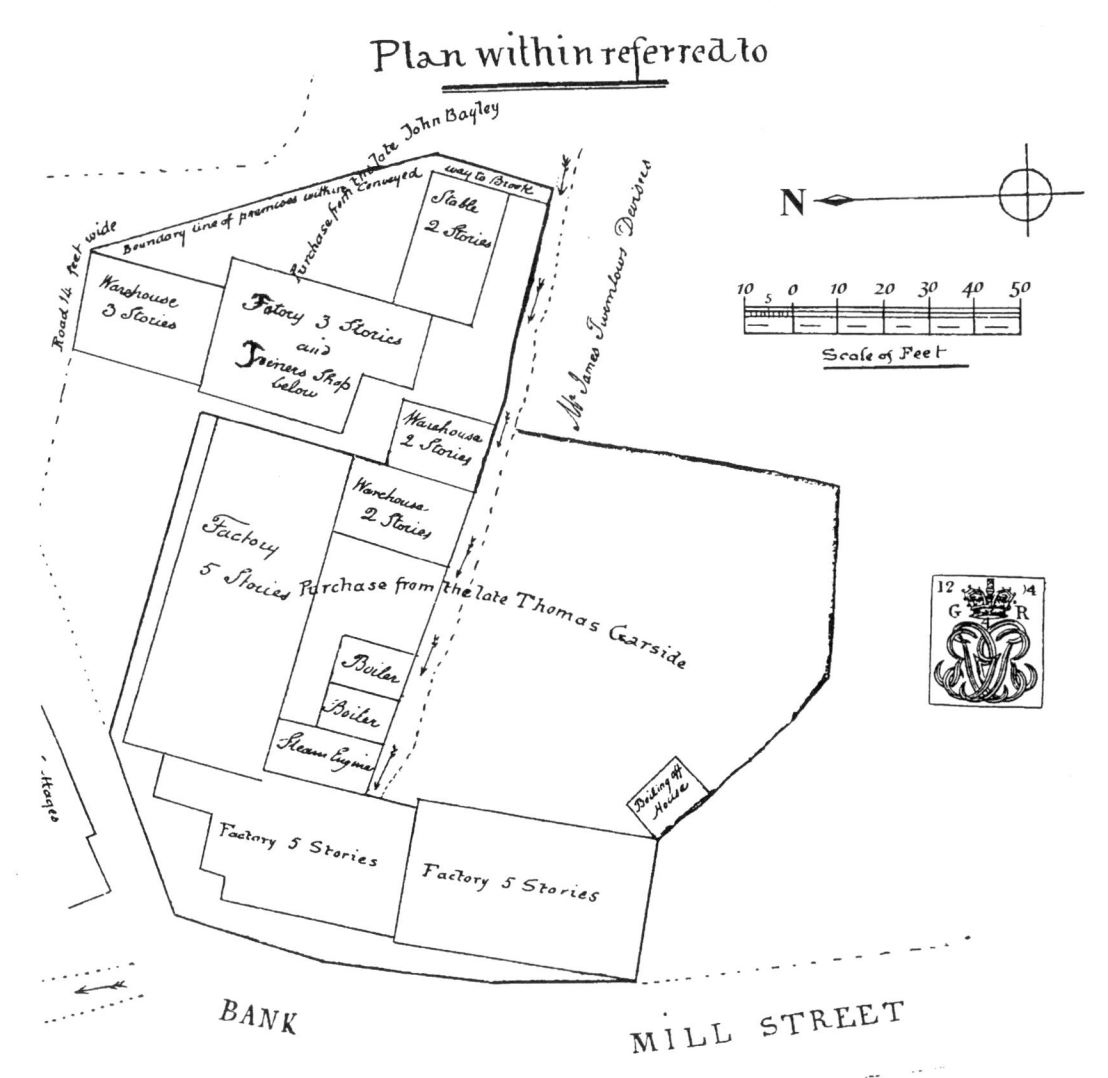

FIGURE 130. PLAN OF STONEHOUSE GREEN MILL, 1839, SHOWING THE LINE OF THE WATERCOURSE RUNNING WESTWARDS (*indicated by arrows*) INTO THE FIVE STOREY BUILDING AT THE WESTERNMOST END OF THE RANGE OF BUILDINGS. (CCALS)

also came at a time when the Reades decided to concentrate on silk spinning at the expense of the cotton spinning side of the business. A new six storey mill was added at the site of Brook Mill, driven entirely by a steam powered beam engine called "The Lady Mayoress". In 1835 John Fielder Reade was Mayor of Congleton and this beam engine was named after his wife who officially opened the mill in that year.[11] The design of this mill suggests that it was built to accommodate powered silk weaving. Also the main part of Stonehouse Green Mill was demolished and rebuilt as a five storey block for silk spinning. George Reade died in 1838 at which point his son Thomas purchased the Stonehouse Green Mill for £2705. As part of this sale, a plan of the mill was drawn which shows the water course flowing through the property together with a steam engine house and two boiler houses (one for the engine and the other to heat the mill). As well as these buildings, the Stonehouse Green complex consisted at this date of four factories, three warehouses, a boiling-off house and a stable.[12]

The early 1840s were a difficult time in the silk industry with many mills closing and employees laid off. Reades continued operating in this difficult climate but from October 1841 to February 1842 the work force of around 200 workers were "turned out" and the mills closed. Later in 1842, some of the men returned to work on reduced wages with a lot of bitterness. Feelings were running high especially as some men were in work but others laid off. Further reductions in the wages caused disturbances in the town and a mob of union workers assembled at Reades' mills and stopped those who were working from getting to the mills. The workers were physically manhandled and escorted to the Union House where they were harangued and intimidated into joining the strikers. Although the strikers eventually went back to work on reduced terms some of the ringleaders at the riot outside Reades were prosecuted and went to gaol.[13] At the same time the business was under pressure from its customers. In a letter from one of their French customers dated 6th July, which was the same day as

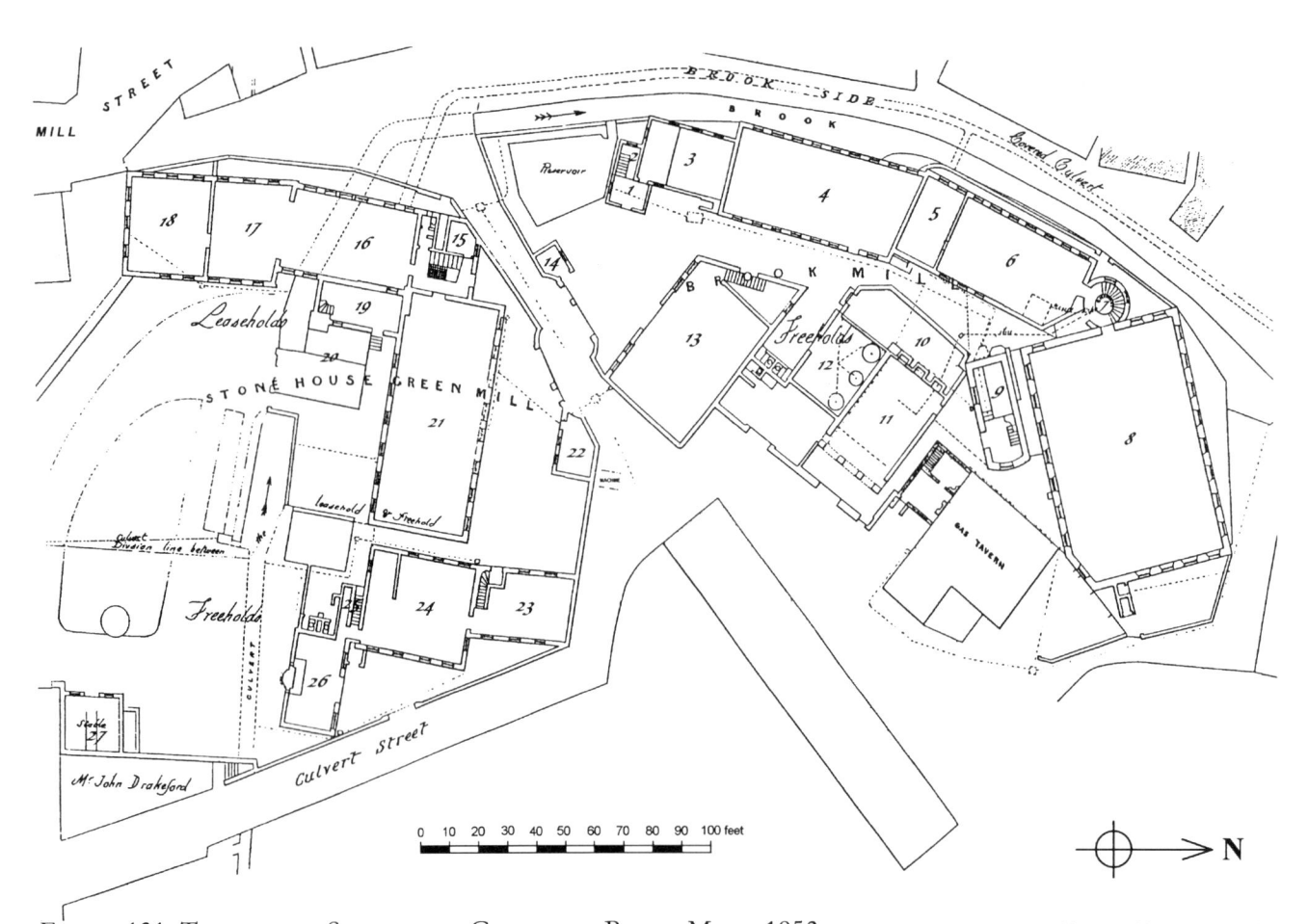

FIGURE 131. THE PLAN OF STONEHOUSE GREEN AND BROOK MILLS, 1853, SHOWING THE CULVERTED WATERCOURSES. THE MILL BUILDINGS HAD VARIOUS USES AS FOLLOWS:-
1. PORCH, 2. STAIRCASE, 3. WAREHOUSE, 4, 5 & 6. MILLS WITH DRESSING, CARDING AND TWO SPINNING FLOORS, 7. CHIMNEY, 8. MILL WITH DRESSING, CARDING, DOUBLING AND TWO SPINNING FLOORS, 9. ENGINE HOUSE, 10. DRYING ROOM, 11. BOILER HOUSE, 12. BOILING HOUSE, 13. WAREHOUSE, 14. LODGE, 15. WAREHOUSE, 16, 17 & 18. MILLS WITH DRESSING, CARDING AND THREE SPINNING FLOORS, 19. ENGINE HOUSE, 20. BOILER HOUSE, 21. MILL WITH DRESSING, CARDING, DOUBLING AND TWO SPINNING FLOORS, 22. LODGE, 23. JOINER'S SHOP, 24. MECHANIC'S SHOP, 25. STAIRCASE, 26. BLACKSMITH'S SHOP, 27. STABLE.

the riot at the mills, they commented *"Mr. Reade promised to send that bale very quickly as we have not yet received it, we beg of you to inform us if we can rely upon you"*. This was followed in August by a letter containing the following *" ...We certainly confirm our engagement with you to take 300lbs of your yarn, but you must really go with the times and reduce your prices a little, or rather if you do as others have done reduce them a good deal."* Later letters make it quite clear that Reades just could not supply the quality of thread that they had previously shown to the French representatives on an earlier visit to London.[14]

No doubt all these labour troubles and their inability to satisfy their customers weighed heavily on Thomas Reade and it is perhaps no surprise that he advertised the Stonehouse Green Mill for sale in March 1846 when it was described as having *"a steam engine of 20 horse power and two boilers and a plentiful supply of water"*.[15] However, the state of the trade was so bad that a sale was not possible before Thomas Reade eventually died in 1848. His death caused the mills to be put on the market again in 1849 and it is perhaps significant that water power is no longer mentioned in the advertisements.[16] Although the silk industry was recovering from the recession of the earlier 1840s it was not possible to sell the mills until 1852 when they were purchased by a consortium made up of Edward Harrison Solly, Arthur Isaac Solly, Richard Clegg, and John Fielder Hall who were to continue trading under the name of Reade & Co. as silk spinners at both Brook Mill and Stonehouse Green Mill. This is perhaps not surprising as all the gentlemen were related in some way with the Reade family. This sale generated a plan of both mills that shows some of the watercourses in more detail. The path of the Howty ran westwards through Stonehouse Green Mill and then northerly along the western side of Brook Mill. Another culvert is shown running from Stonehouse Green Mill which crosses the Howty and then runs parallel to the stream. Part way along, a branch is shown from this culvert where the water is *"turned into the wheelrace of the said*

FIGURE 132. THE WESTERN SIDE OF THE EARLIEST RANGE OF MILL BUILDINGS AT BROOK MILL DATING FROM THE END OF THE 18TH CENTURY. THEY STAND ALONGSIDE THE HOWTY STREAM.

silk mill." that provided power for Brook Mill. The plan also shows the position of the steam engine and boiler house at Brook Mill.[17]

In 1866 the Sollys were sued by John Drakeford who had purchased the properties belonging to James Twemlow just upstream from Stonehouse Green. He claimed that the Sollys had penned the stream such as to cause it to overflow and flood his property, ignoring the agreement made in 1817.[18] The fact that the Sollys had started to dam the stream again seems to suggest that they might have been using the waterwheel once more. This might have come about in an attempt to reduce costs in the difficult trading conditions of the 1860s after the passing of the Free Trade Act and the resulting increased foreign competition.

The business survived the crisis of the 1860s and the Sollys were still employing about 100 hands in 1886.[19] In 1891 Arthur Solly bought out his brother William's share of the business for £750 which shows how much its value had declined.[20] In fact, by this time part of the mills were occupied by fustian cutters. Nevertheless, the silk spinning business of the Sollys continued until it went into receivership in 1929. The mills were then sold to Messrs George & J. H. Berk, a coal & metal merchant.[21] In 1946 they were taken over by the De Luxe Manufacturing Co. who specialised in the making up of garments for the clothing trade and who remained at the mills until 1985. After the closure of this business the Stonehouse Green Mill was demolished to become a car park. In the meantime the Brook Mill had been acquired by Bossons, a firm that specialised in the manufacture of painted pottery wall plaques. This business ceased in 1996.

The surviving Brook Mills consist of two original four storey brick buildings possibly dating from the very late 18th century with the square six storey steam mill of 1835 at the northern end of the site. The older range of buildings are 22 bays long and less than 30 feet wide with both king and queen post roofs possibly indicating separate building dates. Some of the cast iron pillars on the ground floor have shafting housings built into the top of the columns. Of the watercourses only the Howty Brook can be seen running alongside these older parts of Brook Mill but where the stream emerges from the culvert at the south end of the building there are some remains of one of the sluice gates. These older buildings used to be joined to the 1835 building by a round stair tower but this has now been demolished. The large, eight bay by five bay, brick built, mill of 1835 has dated rainwater hoppers at the top of the drainpipes and at its eastern end has a dry chute privy tower with small openings to each floor and an open ashlar base where the removal of the soil took place. The rest of the ancillary buildings are as shown on the 1853 plan except that the engine and boiler houses have been replaced with more modern buildings. The whole of the Brook Mill site has been designated a Grade 2 listed building, however, planning permission has been granted to convert the buildings into appartments. The plans include the demolition of the smaller ancilliary buildings and the attachment of flying roofs and external, cladded stairways to the main buildings

There is nothing to be seen of Stonehouse Green Mill except that the war memorial commemorating those workers from the mill who fell during the First World War has been preserved in one of the walls of the new Bridestones shopping facility that was built around 1990. There is a flight of steps leading from Mill Street near the site of Stonehouse Green Mill that are still known as "Solly's Steps" in recognition of the owner of the mills in the late 19th century.

References

1. East Cheshire Textile Mills Survey, Macclesfield Silk Museum.
2. *Ibid.*
3. *Ibid.*
4. CRO, Congleton Land Tax Return.
5. Survey Book of Congleton, 1803, held by the editor of the Congleton Chronicle.
6. East Cheshire Textile Mills Survey, Macclesfield Silk Museum.
7. CRO, Solicitor's brief 1867, DCB/2114/137.
8. *Macclesfield Courier*, 21st April, 1821.
9. CRO, Congleton Land Tax Returns.
10. East Cheshire Textile Mills Survey, Macclesfield Silk Museum.
11. *Congleton Chronicle*, 6th June 1897.
12. East Cheshire Textile Mills Survey, Macclesfield Silk Museum.
13. CRO, Prosecutor's brief, Luddite Riots 1843, DCB/2114/138 - 139.
14. Manchester Central Library, Geo Reade Letters, MUSC 93/1 - 93/6.
15. *Macclesfield Courier*, 21st March, 1846.
16. *Macclesfield Courier*, 11th August, 1849.
17. East Cheshire Textile Mills Survey, Macclesfield Silk Museum.
18. CRO, Solicitor's brief 1867, DCB/2114/137.
19. Head, R., *Congleton Past and Present*, 1887.
20. East Cheshire Textile Mills Survey, Macclesfield Silk Museum.
21. *Ibid.*

---

46. DANE MILL (SLATE'S) SJ 856632

Not long after 1785, William Slate decided to build a textile mill on the south side of the River Dane just downstream of Congleton's Town Bridge. Access to this mill was via the road alongside the rope walk situated at the south-west corner of the Town Bridge. Slate's negotiations for a lease to the land from the owner, Ann Whittaker, were not successful until 1788, when, for an annual rent of £12, William Slate acquired

"All those two closes, pieces or parcels of land or ground the inheritance of her the said Ann Whittaker lying and being on each side of, and adjoining to the River Dane....and also that silk mill, bleaching trench, sluice, watercourses, and other works then lately erected, cut and made by and at the expense of the said William Slate the elder."[1]

This latter statement indicates that by the time the lease was concluded the building of the mill had already been completed.

Although the lease, agreed in 1788, mentions that the mill was a silk mill, when the firm of William Slate and Son first advertised their business they described themselves as both silk and cotton manufacturers.[2] The reference to watercourses in the lease confirms that the mill was powered by a waterwheel, but the mill's position just downstream of the Town Bridge would have limited the configuration of the waterwheel to being undershot. In the years following the establishment of the mill, the business prospered as the start of the Napoleonic wars raised the demand for home produced cloth. In 1796 William Slate sublet some of the land leased from Ann Whittaker to Isaac Pyot and John Ball to build another mill next to his original building.[3]

In 1799 Thomas Slate took over the business from his father and immediately invested in a steam engine from the firm of Boulton & Watt of the Soho Works in Birmingham.[4] This six horse power rotative crank engine was initially purchased to power the new mill next to William Slate's original mill. However, Thomas Slate eventually arranged for the engine to be integrated with the waterwheel of the original mill. Thomas Slate continued manufacturing silk thread until, in August 1811, he put the mill up for sale, describing the machinery and the mill building as follows:-

"All that extensive factory for throwing silk with machinery fixed to the building, and 136 doz. of spinning spindles, 119 doz. of tram, and 2500 swifts, near Congleton Bridge on the River Dane, and worked thereby, in the possession of Mr. Thomas Slate, the same four stories high, each of the rooms or stories 84 feet in length by 27 feet in width, the mill room 11 feet in height, first engine room 7 feet 2 inches in height, the second engine room 7 feet 4 inches in height and the uppermost or doublers room 8 feet in height; with a good piece of building adjoining, which has been used for two weaving rooms."[5]

There was no rush of prospective purchasers for the mill which continued to be regularly advertised for over six months. These continuing advertisements show that there was a *"large waterwheel to turn the machinery...(and) there is a large steam engine, this work which has been lately erected on Messrs Boulton & Watt's Patent for another factory intended to be built."* Eventually, Thomas Slate was able to lease the mill to Charles Roe in 1812 for twenty one years. The terms of this lease were surprisingly advantageous for Thomas Slate giving him £200 per year for the first fourteen years and then £260 for the remaining seven years.[6]

During the next ten years Charles Roe continued

FIGURE 133. A SKETCH MAP OF WHITTAKER'S LAND DOWNSTREAM OF CONGLETON BRIDGE, 1803 (*Note that north is towards the bottom of the page.*). THE MAP SHOWS SLATE'S DANE MILL IN PLOT 2. THE ENTRY HAS BEEN ALTERED AT A LATER DATE WITH THE ADDITION OF "NOW SLATE" TO THE HEADING. (*Congleton Chronicle*)

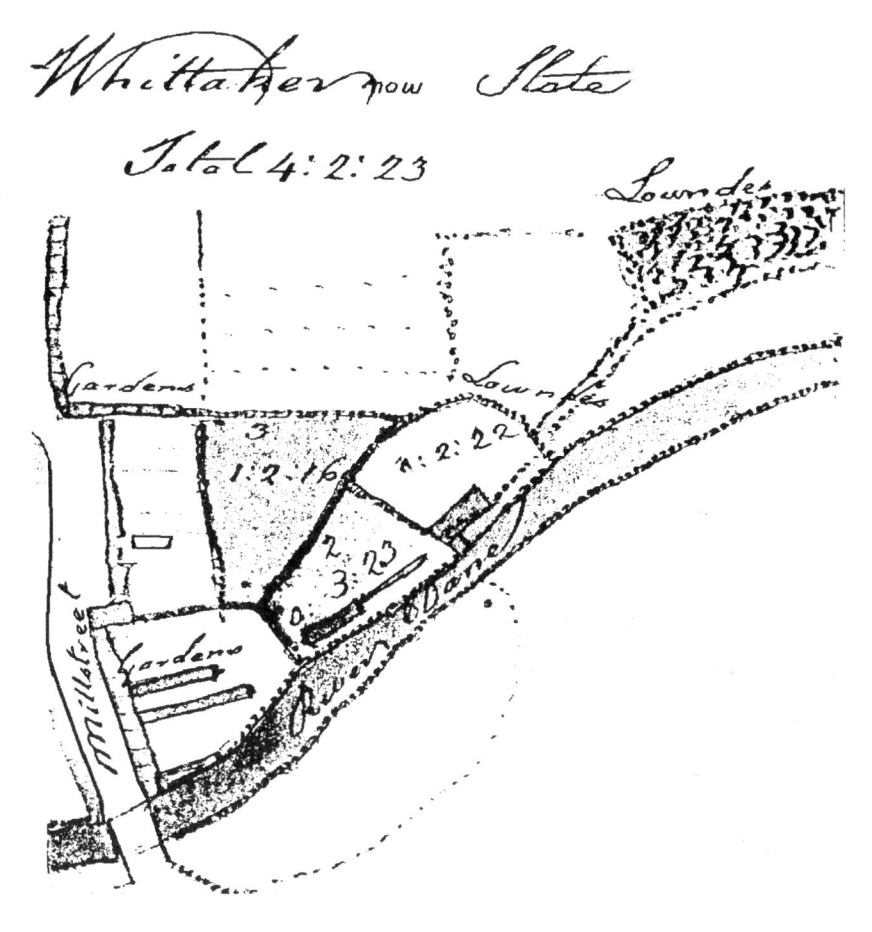

the silk throwing business at the mill but sublet some of the mill to Rowlinson & Co.[7] The expensive terms of Roe's lease were acceptable during the boom times during the war years. After the war, with the competition of imported foreign goods, recession began to bite causing the terms of the lease to become too great a burden. Consequently, by 1822, Charles Roe was looking to sell the lease. The terms of the sale notice described the mill's prime movers as "*a steam engine capable of turning 500 doz. spindles. The waterwheel is capable of driving 400 doz. spindles. The waterwheel and steam engine are adapted to work together.*"[8]

Charles Roe managed to sublet parts of the mill to various people but eventually a Mrs. Allman became the sole tenant. Mrs. Allman was not able to prosper in the silk trade and by 1826 the mill machinery was being advertised for sale together with all Mrs. Allman's household furniture and chattels.[9] It would appear that after the failure of Mrs. Allman's business the mill was untenanted and not used. Certainly prospective sale notices in the years up to 1831 described the mill as being empty.[10] These unsuccessful attempts at selling the mill in 1826, 1828 and again in 1831 do give a bit more detail about the mill. As part of the mill's estate there was "*a piece of land on the north west side of river, opposite the silk mill, for purposes of repairing or altering the weir across the river belonging to the silk mill.*" Any tenant of the mill had also to pay annual rents of £2-12s-6d and 5s for access to the mill over the road alongside the rope walk and another rent of £5-5s-0d was payable for the water right to Messrs Pattison (later Pearson) of the Congleton Silk Mill.[11]

Charles Roe's agreement came to an end in 1833 after which it is possible that the mill acquired a new tenant because when it was to be sold at auction in 1835 it was described as a "*silk mill on the River Dane at Congleton with steam engine, etc. lately occupied by Mr. Thomas Slate but part now used as a brewery by Messrs Bayley & Nixon and other part unoccupied.*"[12] Mr. Bayley was involved in the rope walk near the mill and possibly saw converting the mill into a brewery as a financial investment rather than a potential new career because he offered the new brewery to let in 1836.[13] Although it is often thought that running a brewery cannot fail it is sad to note that by the time the Congleton Tithe map was drawn up in 1845 the mill/brewery had disappeared completely. Presumably the building had been demolished sometime in the ten years between 1836 and 1845. This means that today there is absolutely no sign of a mill on the site, or even any remains of the weir in the river. The site has been built over since the middle of the 19th century and these buildings have also been demolished leaving an open space. The only indication of the industries that once occupied this area is the access road that is still called Rope Walk.

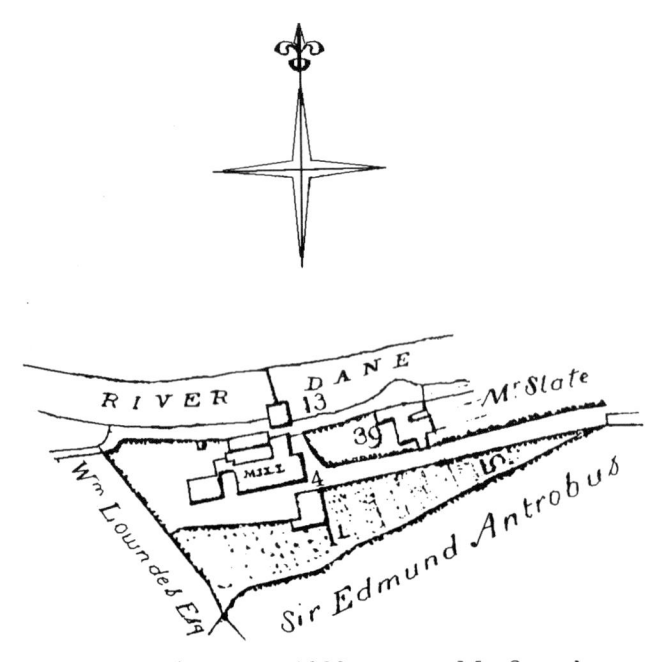

FIGURE 134. A PLAN OF 1820 SHOWING MR. SLATE'S MILL ON THE SOUTH SIDE OF THE RIVER DANE. (CCALS)

References

1. CRO, Lease, 1788, D/2612/2.
2. *Chester Guide*, 1789.
3. CRO, Lease, 1796, D/2612/2.
4. Chaloner; W. H., "Cheshire activities of Matthew Boulton & James Watt, of Soho, near Birmingham, 1776-1817", *Transactions of the Lancashire & Cheshire Antiquarian Society*, Vol. 61, 1949.
5. *Macclesfield Courier*, 17th August 1811.
6. *Macclesfield Courier*, 30 November 1822.
7. CRO, Congleton Land Tax Returns.
8. *Macclesfield Courier*, 30 November 1822.
9. *Macclesfield Courier*, 4th November 1826.
10. CRO, Congleton Land Tax Returns.
11. *Macclesfield Courier*, 19th April 1828, 22nd January 1831.
12. *Macclesfield Courier*, 18th July 1835.
13. *Macclesfield Courier*, 29th October 1836.

---

47. DANE MILL   SJ 854634

After passing under the Town Bridge the River Dane leaves the town of Congleton flowing in a generally westerly direction. At a point just less than a half mile downstream of the town bridge the river makes a tight loop forming a peninsular surrounded on its east, north and west sides by the river. It is on this piece of land, ideally suited for utilising the available waterpower, being almost encircled by the River Dane, that the building known as Dane Mill is located.

Unfortunately nothing is currently known about the construction and early days of Dane Mill. The earliest indication of its existence is provided by a mill symbol on Greenwood's Map of Cheshire published in 1819. Although the mill was indicated on later maps of the county, it is only in 1845 that evidence shows that the mill was at that time owned by James Pearson, who was also owner of the Congleton Silk Mill (*q.v.*), by now known as the Old Mill.[1] Dane Mill remained in the possession of the Pearson family spinning silk until around the end of the 19th century[2] but no evidence has been found as to the types and numbers of the machines used in the mill. Apart from cartographic evidence there is only indirect information that implies that the mill was, in fact, water powered. This occurred in 1865 when Congleton town council passed a resolution that resulted in an official letter being sent to a Mr. H. Newton-Eldershaw at Dane Mill instructing him to remove a six inch high addition to the weir at the mill.[3] Mr. H. Newton-Eldershaw was married to one of the daughters of Samuel Pearson, owner of the Old Mill, and sister of James Pearson, owner of Dane Mill. Presumably Mr. Newton-Eldershaw was managing Dane Mill at this time for the Pearsons although eight years earlier the manager had been George Beech.[4]

During the latter half of the 19th century silk manufacturing suffered serious decline which affected even the Pearson businesses and in 1867 James Pearson mortgaged Dane Mill for £2000 from the Reverend Jonathon Wilson and Christopher Moorhouse.[5] After James Pearson's death in 1881 Dane Mill passed to his sister Victoire Marie Louse Newton-Eldershaw. The silk trade was still in decline at this time and Mrs. Newton-Eldershaw raised mortgages of £2000 and £500 from Edmund Clemence (the mortgage of £2000 might have been a re-mortgaging of the 1867 loan to James Pearson) and a further £500 from Edward Montague-Brown.[6] Four years later, in 1885, Mrs Newton-Eldershaw sold Dane Mill to her husband for £5.[7] In spite of the Old Mill being sold by the Pearson family in 1888, silk manufacturing continued at Dane Mill under the name of James Pearson & Co until just prior to the First World War, although they sub-let part of the building, firstly to Denis Bradwell (purchaser of the Old Mill) and then to Peter Wild & Co. Ltd. who were silk spinners.[8] Peter Wild & Co Ltd continued silk spinning at Dane Mill until 1939 when they were replaced by Conlowe Ltd. who manufactured ladies underwear.[9]

Dane Mill, being built in a loop in the river, is

FIGUER 135. PART OP THE ORDNANCE SURVEY MAP OF 1875 SHOWING THE LOCATION OF DANE MILL ON VIRTUALLY AN ISLAND. IT IS SURROUNDED ON THREE SIDES BY THE RIVER DANE AND ON THE OTHER SIDE BY THE MILL'S BYWASH.

surrounded on three sides by the River Dane. This river loop is cut through by the by-wash so Dane Mill is, in effect, situated on an island. The mill building is L-shaped with the main wing of the building running north-east to south-west, and the smaller wing running north-west to south-east. The Ordnance Survey maps show the headrace for the waterwheel entering the middle of the main wing but this area is now covered by a single storied modern building lying parallel with the main wing. Within the L-shape of the building there was a boiler house with an octagonal chimney. The main wing of the building is four stories high made of brick with a slate roof. It is 24 bays long by three bays wide. The central four bays on the south-east face are surmounted by a pediment that has a clock face in its apex. The north-west (or rear) face of the main wing shows 21 bays (three bays taken up by the other wing) consequently the central pediment is five bays wide but its central section flat rather than reaching an apex. Due to the different lengths of the two faces of the main wing these central pediments are offset from each other. The other wing is similar in style but only 10 bays long without any pediments. The two wings of Dane Mill have survived intact but are split into a number of units which are used today by a variety of businesses.

FIGURE 136. THE GEORGIAN FACADE OF DANE MILL SEEN FROM THE SOUTH LOOKING DOWNSTREAM. THE DIVERSION OF THE RIVER INTO THE BYWASH CAN CLEARLY BE SEEN.

References

1. CRO, Congleton Tithe Apportionments.
2. Various Trade Directories of Cheshire.
3. CRO, Letter from Congleton Town Council, DCB/2114/134.
4. Kelly's Directory of Cheshire, 1857.
5. CRO, Conveyance of Dane Mill, 1885, DCB/1595/8/7.
6. Ibid.
7. Ibid.
8. Kelly's Directories of Cheshire, 1902, 1906, 1910, 1914.
9. Kelly's Directory of Cheshire, 1939.

---

48. FORGE MILL   SJ 849636

About three quarters of a mile downstream from the Town Bridge over the River Dane in Congleton lies the site of Crossledge Forge. Although the forge was situated quite close to the River Dane it did not use the river for power but instead used a small spring that originated in the field next to the river. The spring was used to feed a small pool.[1] It is not known when the forge was built but it could have been in existence from the Middle Ages. The earliest records of the forge to have been found so far show that it was owned by Hugh Ford from before 1786 until his death in 1814.[2] He was succeeded as owner of the forge by his son James who continued to operate it to support his wife Hannah and his two young children Lavinia and George.[3]

As the industrial revolution expanded and trade with other parts of the country became easier, James Ford could possibly have found that it was too difficult to compete with factory-made goods. For whatever reason, he decided in 1822 to offer to let the water powered site of the forge to those industries that appeared to be prospering at the time, as follows:-

"To let - A very desirable situation near the town of Congleton admirably calculated for a throwing mill or paper mill known by the name of Crosslidge [sic] Forge. The above situation is well worth the attention of silk dyers and throwsters being well watered by the purest springs and possesses considerable head and fall."[4]

This attempt was unsuccessful, possibly because by this time steam powered mills were much preferred to water powered ones, as they offered greater flexibility in location. Two years later an advertisement was published as follows:-

"For Sale - Two fields situated near the forge in Congleton called Carr Field and the Crosslach [sic] meadow containing three acre of land in the tenure and occupation of Mr. Joseph Davenport, cordwainer.

The fields lie contiguous to Congleton and the River Dane flows at the bottom of Crosslach Meadow where is a considerable fall of water. There are also streams of water on each side of the same meadow so situated very excellently for a factory to be erected. A yearly sum of 2s has been paid to the owner of the forge for the benefit of the stream of water which turns the same and which is one of the streams alluded to above."[5]

Again the availability of water power was stressed but the sum paid to James Ford for the water rights was miserly in the extreme.

It would appear that this latest sale notice was also unsuccessful as the site was indicated only as a forge on maps published in the years 1830 and 1831.[6] But at some stage the site must have been sold (or leased) for development because in 1840, when Thomas and Archibald Templeton were bankrupt, the sale of all their assets included a silk spinning mill at Crosslach described as follows:-

"For Sale - By order of the assignees of Messrs Thomas & Archibald Templeton, bankrupts.

LOT 1. The capital silk spinning concern situated at the Forge otherwise Crosslach in Congleton comprising of two recently built spinning mills, six dwelling houses, smith's & joiner's shops, and other offices, two steam engines, one of 26 horse power and the other of 10 horse power, boilers, heavy gearing, & fixtures.

The principal mill is four stories high, 96 feet long by 30 feet wide and 46 feet high and being well supplied by a constant stream of excellent water (the properties of which are particularly adapted to the process which silk has to undergo).

The machinery at Crosslach Mill comprises two 30 inch blowers, pickers, with double scutcher, 36 carding engines of 30, 36, and 40 inches in the wire, and four double engines of 48 and 54 inches each, drawing, stubbing, & roving frames, cans & skips; twelve doubling frames, from 180 to 240 spindles each; hydraulic screw and making up presses; three power looms, turning & drilling, lathes and tools, two copper boilers with furnaces, etc. The whole of the gearing through the mill except the main shaft and a general and excellent assortment of the necessary implements including mechanics & joiner's shops."[7]

Obviously a silk spinning mill(s) had been built on the site in the ten years between about 1830 and 1840. It is noticeable that the mill described was driven by two steam engines signifying that the use of water power on the site came to an end with the demise of the forge. It is not known whether the mill was built by the Templetons, who also owned Pool Bank (Woodhouse) Mill (*q.v.*) at Timbersbrook. The list of machinery indicates that Forge Mill was a relatively small concern which is confirmed by the number of employees, which was around seventy hands.[8]

After the bankruptcy of the Templetons, Forge Mill was owned and occupied by Robert Thompson & Co. but they had been superseded by James & Joseph Holdforth by 1848.[9] The Holdforths together with their manager, John Taylor, continued operating the mill as silk spinners up to around 1870 when they gave way to Wild and Bradwell.[10] By the turn of the century the operator of the mill was Peter Wild & Co. who also operated the nearby Dane Mill (*q.v.*).[11] They were the last silk spinners in Congleton but ceased spinning some time before they left Forge Mill in 1953.[12]

Fortunately the description of the mill in the 1840 advertisement can be augmented by various drawings and a plan of the mill made in 1865 which shows a typical 19th century steam powered textile mill.[13] The site is currently occupied by Grahams Pumps Ltd.

References

1. CRO, Moorhouse's Map of Congleton, 1818, D 4552/1.
2. CRO, Land Tax Returns for Congleton.
3. St. Peter's Parish Record, Congleton.
4. *Macclesfield Courier*, 30th November 1822.
5. *Macclesfield Courier*, 17th July 1824.
6. CRO, Swires & Hutchin's Map of Cheshire, 1830; Bryant's Map of Cheshire, 1831.
7. *Macclesfield Courier*, 19th December 1840.
8. Head, R., *Congleton Past and Present*, 1887. (Reprinted Old Vicarage Publications, 1987).
9. CRO, Congleton Tithe Apportionments, 1845.
10. Worrall's Directory of Cheshire, 1872.
11. Kelly's Directories of Cheshire, 1896, 1902, 1906, 1910, 1934, 1939.
12. Congleton Chronicle, 7th March 2003.
13. CRO, Plan of Forge Mill, 1865, DCB/2114/MP5.

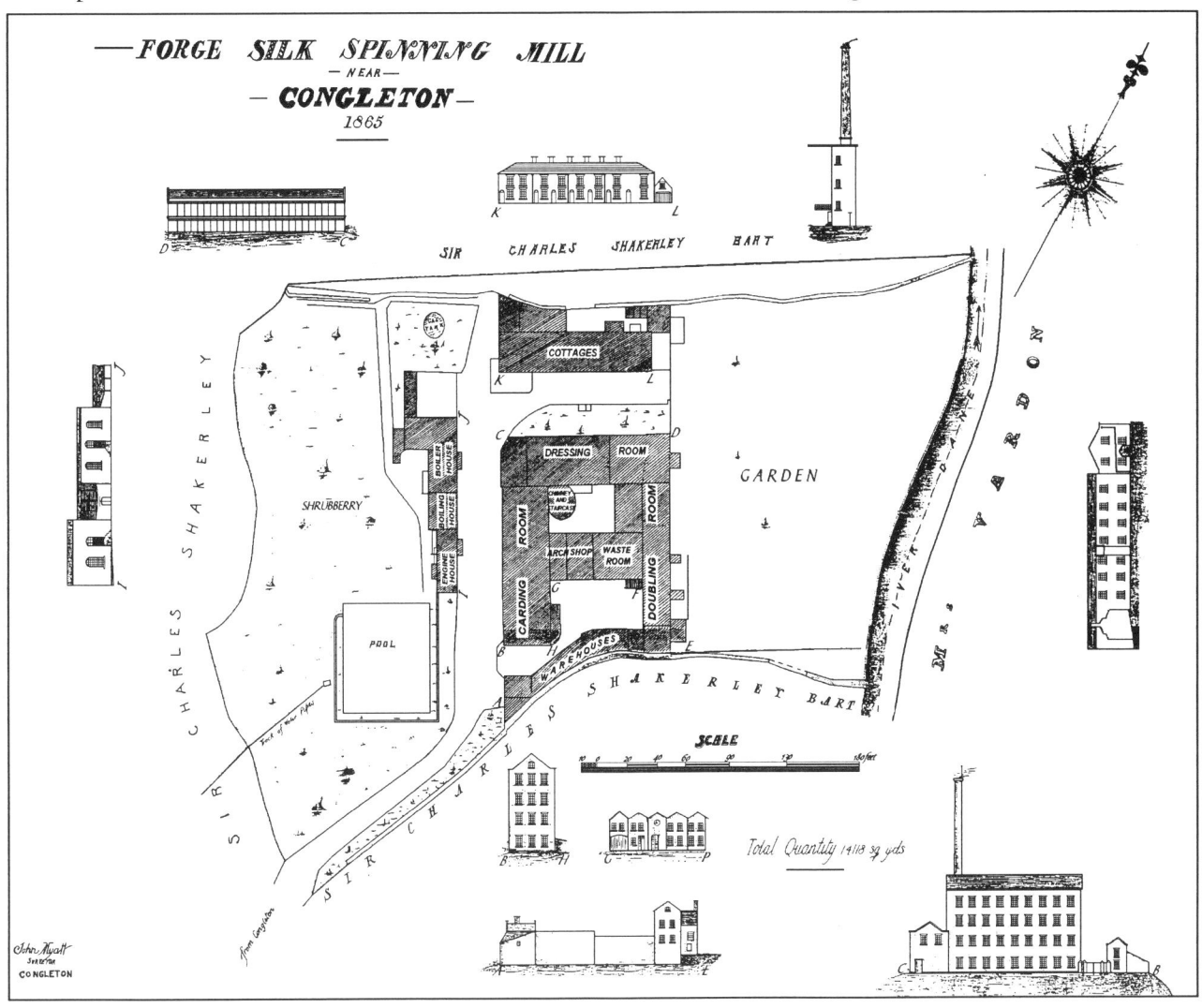

FIGURE 137. A PLAN OF FORGE MILL, 1865. (CCALS, DCB/2114/83)

Section 5. Congleton to the Confluence with the River Wheelock

After flowing through the town of Congleton the nature of the River Dane changes as the hills are left behind and the river meanders north-westerly through a rich, flat, agricultural landscape. During the ten miles from Congleton until the river meets its largest tributary, the River Wheelock near Middlewich, it is joined by three smaller tributaries. On the south side of the river, the Loach Brook, which rises on Congleton Edge not far from Nick i' th' Hill, runs for approximately six miles. It joins the River Dane some 3½ miles downstream from Congleton. On the north side of the River Dane, about eight miles west of Congleton the Swettenham Brook joins the river. This stream rises in the slightly higher ground between Marton and Gawsworth, north of Congleton, running about six miles to its confluence with the River Dane. This stream has a number of names being known also as Chapel Brook in Marton parish and Midge Brook in Somerford Booths parish. The River Croco rises about two and half miles south west of Congleton. This river is hardly bigger than the local streams and brooks, running for only about nine miles in a more or less north-westerly direction, south of the village of Holmes Chapel, and then through Middlewich to join the River Dane near Croxton.

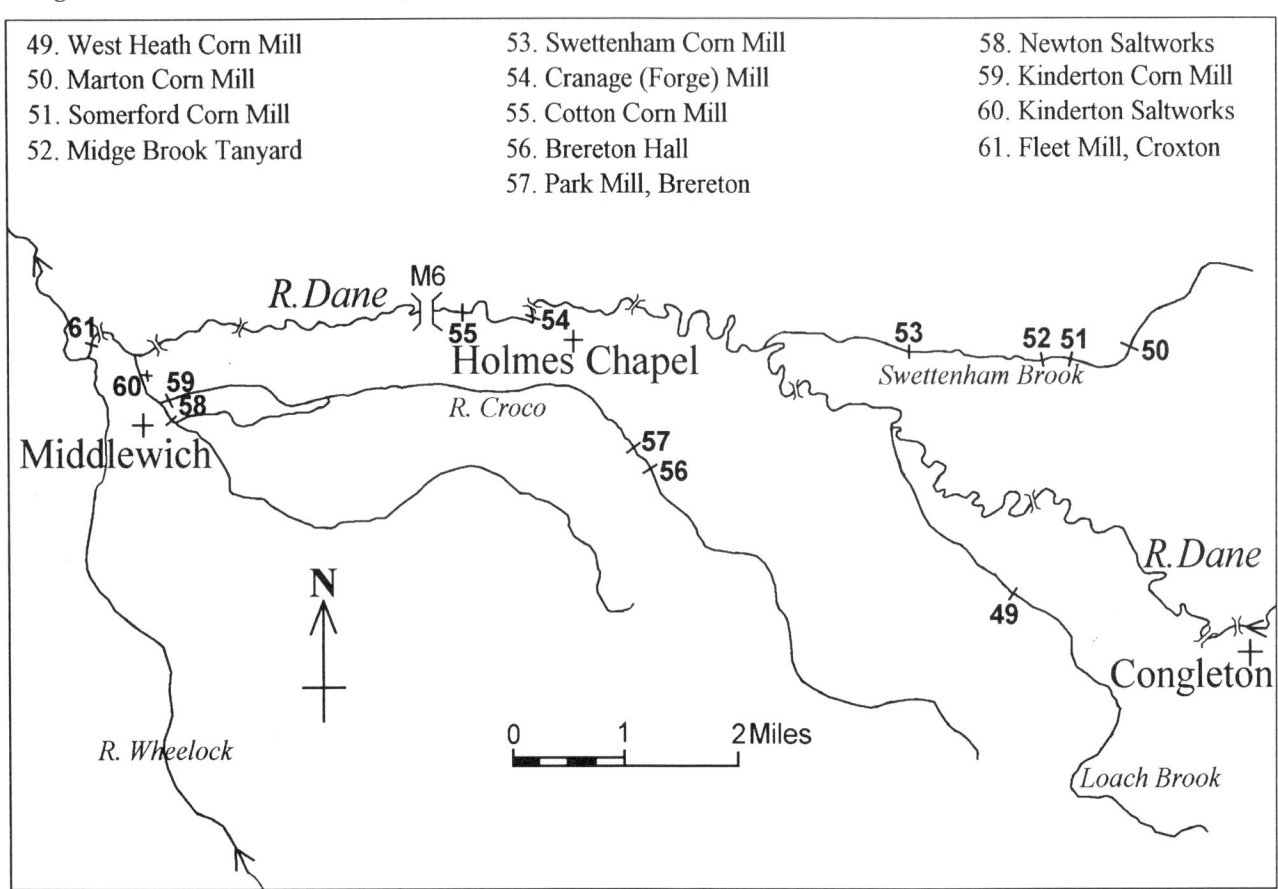

49. West Heath Corn Mill
50. Marton Corn Mill
51. Somerford Corn Mill
52. Midge Brook Tanyard
53. Swettenham Corn Mill
54. Cranage (Forge) Mill
55. Cotton Corn Mill
56. Brereton Hall
57. Park Mill, Brereton
58. Newton Saltworks
59. Kinderton Corn Mill
60. Kinderton Saltworks
61. Fleet Mill, Croxton

FIGURE 138. MAP OF THE RIVER DANE AND ITS TRIBUTARIES FROM CONGLETON TO ITS CONFLUENCE WITH THE RIVER WHEELOCK NEAR MIDDLEWICH.

49. WEST HEATH CORN MILL   SJ 830637

The upper reach of the Loach Brook is called Dairy Brook. Shortly after crossing under the Macclesfield Canal the stream runs past Mill House Farm which is possibly the site of the medieval manorial mill of Astbury, however there is no trace of a mill at this location apart from the farm's name.

West Heath Corn Mill was situated about two miles west of Congleton alongside the Congleton to Holmes Chapel road (now the A54), close to the junction of Chelford Road that leads to Radnor Bridge over the River Dane. There is a reference in 1599 to the building of a dam at West Heath Mill.[1] This might indicate that the mill was first built at this date but equally could be referring to a repair of the dam. Over a hundred years later there is a lengthy correspondence in 1713 between Mr. Gorst, the steward of the Somerford Estate, and Mr. Shakerley concerning the state of the dam at West Heath Mill.[2] The mill dam appears to have been constructed with timber and there was some disagreement on the best way to proceed. In the middle of June Mr. Gorst's regular report to Mr. Shakerley contained the following statement:

"As to West Heath Mill I am quite at a loss you do not aprove of the method proposed for the other way will certainly above double the charge. And this having a butress of earth behind. therefore I think I must only repair the breasting with stone or wod as formerly at this time and leave the Great Works to your own management and consideration."[3]

Nine days later he was recommending some workmen for the task of repairing the dam to Mr. Shakerley with the comment *"As to West Heath Mill dam I will commend you to Joseph Henshaw and Thomas Bobington .... both will not fail to give you a full account"*[4] After examining the problem, Thomas Bobington wrote direct to Mr. Shakerley with his opinion of the necessary work as follows:

"Sir,
According to your orders William Dale, Joseph Henshaw and I went to West Heath Mill finding the dam out of order, it is plashed with wod[sic] and it will not last above three years and it must be done with stone or plashed for the width it is three score and ten yards and it must be done two yards high the timber work never have taken care of....and I think it will be the best for to do it with stone for it will last for ever.

Thomas Bobington"[5]

At this time the miller was Hugh Marton who was paying a rent of £25 per year and the poor ley was 4s-4d per year.[6] The mill itself was described as being *"in good repair having a new shaft and two new stones and will go with little change...the West Heath Mill is the most able mill."*[7] Presumably the repairs to the mill dam were carried out successfully because the mill continued to operate for a number of years after 1713.

In March 1750 Peter Shakerley received a quotation from William Forster for work on the mill which included a new shaft, two gudgeon, a wallower, a bridgetree, a mill brass, a pair of foot stones, a new cog wheel, and a pair of mill irons, etc. The quotation ended with the statement *"As for millwright work I never had less then six pounds for putting up a trebble [sic] mill"*.[8] This reference to a "trebble mill" refers to the practice of fitting an extra pair of millstones so as to be driven by two-step gearing from the pitwheel in addition to the original pair of stones driven directly from the pitwheel via one-step gearing. Although the first publication of a drawing of this type of arrangement referred to a mill built in 1728 there is some evidence that the practice had been introduced some time earlier (see Congleton Corn Mill).[9] Around this time millwrights started to introduce the multi-millstone configuration mill based on the design using an upright shaft and great spur wheel which superseded the design used in "treble mills".

Although West Heath Mill is clearly shown on Burdett's map of Cheshire published in 1777, it is possible that West Heath Mill went out of use when the nearby Somerford Booths Mill (*q.v.*) was rebuilt around 1818. This rebuilt mill then contained four pairs

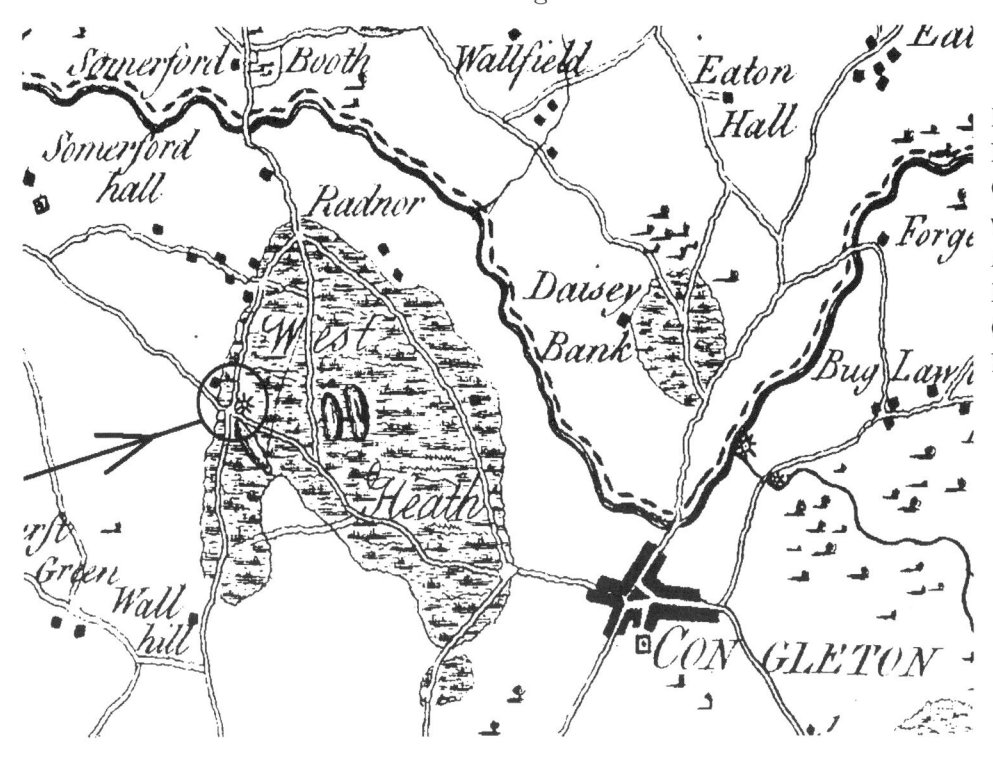

FIGURE 139. PART OF BURDETT'S MAP OF CHESHIRE, 1777, SHOWING A WATERMILL SYMBOL (*arrowed*) AT THE LOCATION OF WEST HEATH MILL, WEST OF CONGLETON. (CCALS, PM12/16)

of millstones and presumably could cater for all the milling needs of the area. Certainly, by the time that Bryant's map of Cheshire was published in 1831, the site is only identified by the words "West Heath Mill Farm". A few years later, on the Ordnance Survey first edition, the site is only indicated by the designation "mill pool". If the mill had ceased to operate by the early 19th century then it is not clear where Joseph Wood, a miller at West Heath mentioned in Kelly's Directory of Cheshire for 1857, was active. After this date there is no further evidence that can be connected with a mill at this location.

Today the mill pool referred to in 1840 on the Ordnance Survey map can still be seen. It is located on the northern side of the A54 just east of the junction with Chelford Road which leads to Radnor Bridge. It is now a favourite haunt of the local fishermen. Possibly by sheer coincidence, just downstream from the pool are the premises of Young's animal feed business.

References

1. CRO, Letters, 1599, DSS 1/7/26.
2. CRO, Letters, Gorst to Shakerley, 1713, DSS/1/3/261.
3. CRO, Letter, 20th June, 1713, DSS/1/3/261/18.
4. CRO, Letter, 29th June, 1713, DSS/1/3/261/20.
5. CRO, Letter, Bobington to Shakerley, 1713, DSS/1/3/262.
6. CRO, Steward's report, 1713, DSS/1/3/261/3.
7. CRO, Letter, 18th May, 1713, DSS/1/3/261/6.
8. CRO, Estimate from Wm. Forster, March 1750, DSS/1/6/82.
9. Watts, M., *The Archaeology of Mills & Milling*, Tempus, 2002.

---

50. MARTON CORN MILL SJ 844672

The first mill downstream from the source of the Swettenham Brook, or Chapel Brook as it was known in Marton, was situated about one mile south-west of the village of Marton on the south side of Marton Hall Lane near Marton Hall which is about a third of a mile west of the A34.

The location of the mill strongly suggests it was a medieval manorial corn mill, a proposition supported by a reference to the "*molendinum de Merton*" in 1307.[1] However, the earliest detailed evidence relating to the mill occurs in 1761 when it was part of the Davenport family's Capesthorne Estate. In that year the mill was leased to Josiah Browne and William Lawton for 21 years at £30 per year "*if David Davenport should so long live*". The terms of the lease stated that all "*...taxes, edifices, buildings, wares*[sic]*, dams, paddles, floodgates, wheels, millstones, troughs, arks, measures, sieves, bolting mill, clothes, crows, chizels, picks, malls, hammers, and all other utensils are the tenants responsibilities*".[2] William Lawton held the lease of the mill at this time and was also probably the miller.

The profession of miller ran in families and certainly William Lawton's son, John, was later identified as the miller at Marton between 1813 and 1831 when the mill was recorded as owned by Davies Davenport, Esq.[3] Also in 1831 John Lawton was noted as the tenant of Marton Mill and just over 29 acres of farmland.[4] Unfortunately, the following year John Lawton was proclaimed a bankrupt.[5] However, some five years later he was still the tenant of the mill and associated farm.[6] By 1841 John Lawton had been superseded at the mill by John Lewis who was recorded as a farmer and miller,[7] but his stay at the mill was to be short, being replaced after only a few years by Peter Lucas.[8]

In 1851 Peter Lucas, then aged 59, described himself as a farmer and it was his son Edwin (or Edward) who was the miller at Marton Mill. Edwin lived at the mill house with his parents and four siblings.[9] By 1857, around the time of Peter Lucas's death, the farm had been reduced in size to just over 18 acres.[10] It would seem that the miller's job at Marton was always a part time affair as he was often listed not just as a corn miller but also as a farmer. Certainly this was the case in 1861 when Edwin Lucas was described as a miller and farmer living with his wife and two small children at the mill house.[11]

By the 1880s the mill seems to have ceased to grind as the tenants of the mill house were thereafter merely described as farmers.[12] The part-time nature of the miller's work might indicate that Marton Mill may have been a small mill, only having one pair of stones driven by the waterwheel. If that were the case the mill may never have been upgraded from its medieval configuration. The level of business would perhaps have never encouraged any modernisation that would only have produced a greater capacity that was not required. The large scale Ordnance Survey map of 1875 shows that the mill and mill house occupied a range of buildings about 100 feet long by about 20 feet wide

FIGURE 140. PART OF THE ORDNANCE SURVEY MAP, 1875, SHOWING THE POSITION AND LAYOUT OF MARTON MILL.

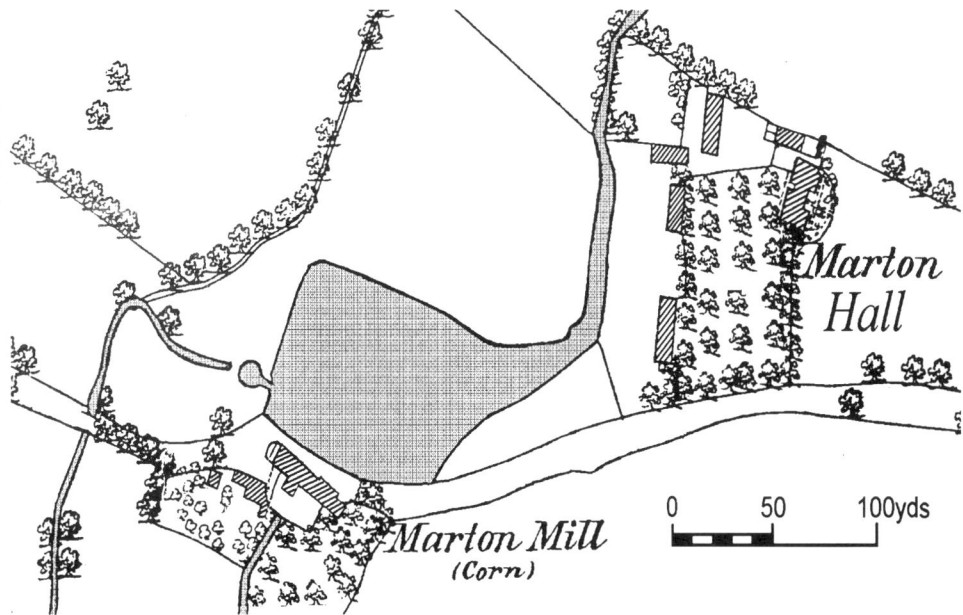

with some ancillary farm buildings or outhouses to its rear. The mill pool was about 250 feet by 200 feet with the mill building orientated along the pool's longest side. The shape of the tailrace is indicative of a single waterwheel although it could also be consistent with two waterwheels mounted side by side in the centre of the mill.[13] The ending of trade at Marton Mill coincides with the demise of many small country mills due to the introduction of roller milling into the trade in the late 19th century.

Today there is very little sign that Marton Mill ever existed. There are just a few small hillocks in the field near to Marton Hall Lane that indicate where the mill once stood. The area that was once the mill pool, which was north of the lane, has reverted to being just part of the field.

References

1. Dodgson, J. McN., *The Place Names of Cheshire, Part 2*, Cambridge University Press, 1970.
2. Giles, K., *The Bromley Davenport Papers*, 1999.
3. CRO, Marton Land Tax Returns.
4. Giles, K., *op. cit.* 2.
5. *London Gazette*, 9th November, 1832.
6. Giles, K., *op. cit.* 2.
7. CRO, Marton Census Returns, 1841
8. CRO, Marton Tithe Apportionments, 1845.
9. CRO, Marton Census Returns, 1851.
10. Dodgson, J. McN., *op. cit.* 1.
11. CRO, Marton Census Returns, 1861.
12. Various Trade Directories of Cheshire, 1864 - 1883.
13. Ordnance Survey, 1875.

51. SOMERFORD CORN MILL   SJ 836668

It is only about ½ mile downstream from Marton Mill, where the Chapel Brook becomes known as the Midge Brook, that the stream is dammed to form a pool at the western side of which stands Somerford Corn Mill (later called Somerford Booths Mill). The earliest record found to date of a mill in Somerford occurred in 1630 in the inquisition post mortem of Edmund Swetenham which listed amongst his holdings a water mill and land in Somerford.[1]

The only other information that might be connected with Somerford Corn Mill in the remaining part of the 17th century and the whole of the 18th century is the recorded deaths of two millers at Somerford, namely John Deane in 1680 and Zachary Shepley in 1773, although it is possible that they had worked elsewhere, such as at West Heath Mill.[2] However, the mill was represented on Burdett's map of Cheshire in 1777.

The earliest information in the 19th century concerning Somerford Mill occurs when the mill was advertised to let in 1818, as follows:-

"All that long established flour mill on the Midge Brook in Somerford Booths near Congleton which the proprietors propose to rebuild on a plan that may be fixed upon between parties."[3]

Although this advertisement gives no details of the mill it indicates that a medieval mill may well have been demolished and a more modern mill built around this time. Almost certainly this new mill would have been capable of operating a number of pairs of millstones. Even so the mill at Somerford Booths does not appear to have been able to provide a full-time occupation for a miller. In the tithe apportionments of 1839 the owner is given as Clement Swetenham, while the occupier was William Blackshaw. However, two years later, William Blackshaw is described as a farmer and it is his fifteen year old son George who was the miller.[4] George obviously did not see this arrangement as providing a worthwhile future and soon departed from Somerford Booths leaving his father William, who lived at Mill Bank Farm, to act as both farmer and miller until the 1850s.[5] George Blackshaw's ambition led him to take over the steam mill alongside the banks of the Macclesfield Canal in Hightown, Congleton. By 1857 William Blackshaw had been superseded at Somerford Booths Mill by Thomas Lockett who was himself replaced by William Hall by 1874; all of whom described themselves as both millers and farmers.[6]

Arthur Ward was 25 years old, married with two small children, when he took the lease on Somerford Mill on the 25th March, 1880.[7] At that time the schedule in the lease shows that the mill was powered by a six horse power steam engine and its associated boiler, as well as a high breast shot waterwheel with a diameter of 12 feet and a width of 5 feet. The waterwheel had iron buckets and a wooden drum (sole boards).[8] It is possible that when the mill was rebuilt around 1818 it was designed to use steam as well as waterpower.

From a number of leases that have survived from this time it has been possible to "reconstruct" the machinery of the mill. The main machinery inside the mill centred on an upright shaft running up through the first floor. At the ground floor level this shaft carried a crown gearwheel which could be driven from the steam engine via a shaft and pinion, and a wallower (also described as a crown wheel) was driven from the waterwheel through a pitwheel. Turned from one of these crown wheels was a shaft with pulleys and belts for driving a "shooding" or shulling mill, which was located on the stone floor. However, the shooding stage and its fan used to separate the oatmeal from the husks was accommodated on the gound floor. This shaft also provided power, via pulleys and belts, for a dust sieve and its associated fan, which were used to clean the grain prior to grinding, and a jigger. Underneath the position of the millstones there was a meal trough for collecting and bagging the output from the whole milling process.

On the first, or stone floor, the upright shaft turned a spur wheel that drove three sets of millstones via stone nuts and spindles arranged in an overdrift configuration i.e. the spur wheel and stone nuts were located above the millstones. There was one pair of French stones, 4 feet 4 inches in diameter, with its associated furniture, known as the "wheat mill". The other two pairs of millstones were made of millstone grit from Mow Cop and were each 4 feet in diameter. These two pairs of millstones were known as the "offal (or indian meal) mill" and the "meal mill". There was a pulley fixed to the drive spindle to the meal mill which operated an elevator by means of a belt. The fourth pair of millstones, known as the "shooding mill", also consisted of 4 feet diameter stones from Mow Cop. Because this pair of millstones were driven by a belt on the ground floor, they were underdriven. Also on the stone floor there was iron shafting and bevel gears to drive a smutting machine for cleaning the mould off the grain on its arrival at the mill, and a cylindrical silk flour dresser for separating the meal after grinding. Pulleys and belts attached to this shaft also provided power for another elevator and the sack hoist. The drive to the sack hoist, which was located on the top floor, was via a slack belt operated by a lever.

In 1880 the cylindrical silk dresser was brand new having just replaced a flour machine that had a cylinder with six brushes. The mill also had a kiln attached that would have been used mainly for the roasting of oats which would then be processed in the shooding mill. There was also all the necessary equipment needed to dress the millstones, consisting of a crowbar, wrenches and a set of lewises for lifting the top stones (note that there was not a stone crane), twelve steel picks of various shapes for the actual dressing and two staffs to check the work. There was also a variety of miscellaneous items such as a weighing machine with weights, a truck, a spade, a kiln poker, etc.[9]

All the machinery mentioned, including the millstones, was, of course, the property of the landlord. At the beginning of a tenancy it was normal practice at this mill to measure the thickness of the stones and assign a value to each pair. At the end of the tenancy the millstones' thickness would be measured again and any diminution in thickness and hence in value was charged to the outgoing tenant.

In December of 1880 there was a dispute between the miller, Arthur Ward, and the mill owner, Clement William Swetenham, concerning the state of the sluice and paddles at the mill.[10] These were in such a poor state of repair, a condition that the owner admitted was the case at the start of Arthur Ward's tenancy, that any flood on the stream in the coming winter would very likely wash the sluices away. Arthur Ward thought that it was the owner's responsibility to repair the sluice but the owner considered that the tenant should carry out the repair if the owner supplied the necessary timber. According to the terms of the lease the owner was strictly correct that the tenant should do the repair but the sluices should have been in good repair at the start of the tenancy. This was a good bargaining point for the tenant when allied with the clause in the lease that allowed him to claim relief from his rent if the mill was "*damaged by tempest*". The owner was obviously penny pinching and had been expecting this problem to arise as in his letter to his solicitor, George Reade, he states "*You will recollect I some time since told you I apprehended trouble with Ward on account of the mill dam and the paddles & sluices*".[11] The outcome seems to have been reached amicably in the end.

Possibly this problem with the sluices prompted the owner to improve the condition of the mill, or perhaps the competitive nature of the flour trade at this time forced his hand. In 1883 he approached the local millwright in Congleton, Jonathon Booth, to make some modifications and improvements to the mill. These improvements included the fitting of a worm and two elevators together with a new sieve to divide the offal. There was also to be a new silk reel, 14 feet 6 inches long and 3 feet 3 inches in diameter, with one conveyor to feed this new reel and two conveyors to handle the outputs from the reel (one half-way along and the other at the bottom of the reel). There was to be some re-arrangement of shafting and pulleys and the addition of a set of chilled iron rolls (10 inches diameter and 12 inches long) with all the necessary mechanics. The total package of work was quoted at £154 with the millwright taking all the old shafting and fixings etc.[12] This quotation was accepted and Jonathon Booth carried out the work during 1884. However, problems arose due to conflicting claims to the old material no longer being used. One of the gear wheels removed from the mill at this time, and so claimed by the millwright, was said by the miller to belong to him and not the owner and therefore could not be claimed by the millwright. Also the stairs to the middle floor were replaced by the millwright but before he could claim the old wood the miller used it to make the frame of an exhaust chamber. It seems that the work on the screen was paid for by the miller not the owner which complicated the terms and conditions applicable. The owner and the miller then complained that the millwright had not fixed some brickwork which he had verbally agreed to do.[13] Possibly business was not getting any better and tempers were getting short but the matter was still outstanding in August 1885. Hopefully Jonathon Booth finally got his money!

Arthur Ward stayed at Somerford Booths Mill until 1890 when he is listed as operating the mill with William Barber.[14] From 1890 until 1894 William Barber worked the mill alone, but seems to have been an undertenant of Arthur Ward who remained at the mill house and ran the farm. This meant that William Barber was unable to occupy the traditional miller's accommodation at the mill house and had to reside at Tanyard Farm nearby.[15]

In 1894, when the mill was leased to John Orme at a rent of £50 per year, the contents in the schedule of the lease reflected the changes that were taking place in the flour industry. The steam engine was removed either for sale or for use elsewhere on the estate as the throughput of the mill no longer justified its expense. Also removed from the mill at this time was the smutting machine and the silk dresser, but an oat crushing machine had been recently acquired. These changes indicate that the mill was no longer intended to produce flour for human consumption but was concentrating mainly on processing animal feed. The disappearance of all the millstone picks, chisels and weighing machine with its weights indicates either a need to raise cash or possibly misappropriation. Since

FIGURE 141. PART OF THE ORDNANCE SURVEY MAP, 1875, SHOWING THE LAYOUT OF SOMERFORD BOOTHS MILL AND ITS MILL POOL.

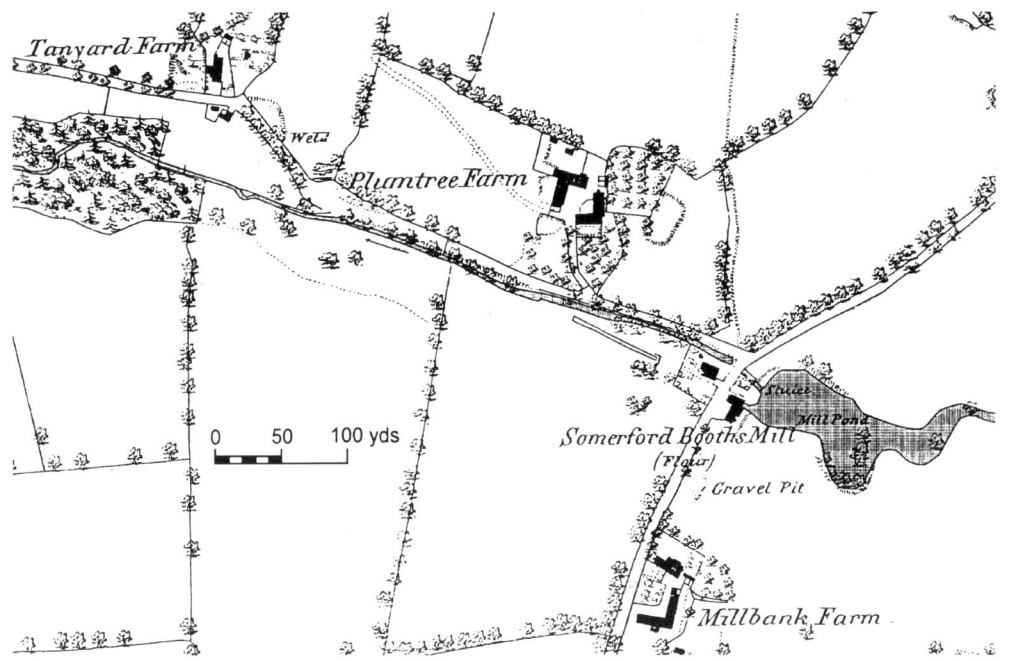

FIGURE 142. A RECENT VIEW OF SOMERFORD BOOTHS MILL. THE WATERWHEEL IS A MODERN CONSTRUCTION. THE ORIGINAL BREASTSHOT WHEEL HAD ITS SHAFT AT A LOWER POSITION, ABOUT LEVEL WITH THE GROUND.

1880 the pair of French millstones had been renewed, as had the pairs of stones in the offal mill and the meal mill, although the Mow Cop stones of the offal mill had been replaced with a pair of French stones. The millstones of the shooding mill appeared to be the same as those present in 1880. In the case of the shooding mill the top stone was 1¾ inches thinner, and the bedstone one inch thinner, than they were fourteen years earlier in 1880.[16] It has been estimated that some millstones wear by about ¼ inch per year when used to their full capacity, although this would vary, dependant on the hardness of the millstones and of the grain milled.[17] Applying this estimate to the wear at Somerford Booths Mill in the fourteen years between 1880 and 1894 shows that about seven years work was done, i.e the mill was working at half its capacity.[18] Further indication of the loss of status by the mill is evident in the values assigned to the various pairs of millstones in 1880 and 1894. In 1880 the pair of French stones, which were used to produce flour, were valued at £14 whereas in 1894 the two pairs of French stones were valued at £5 and £6-10s respectively in spite of being thicker than the pair in 1880.[19] This reflects the impact of the large roller mills on traditional flour production using millstones, the value of French millstones being greatly reduced due to the surplus in second-hand stones as mills closed. It is noticeable that the pairs of Mow Cop stones, used primarily for producing animal feed, retained their value over the same period except for those used for shulling oats which had reached the end of their life by 1894.

John Orme did not renew his lease on the mill in 1895, deciding to move to a much larger and more modern establishment at the Havannah Mill (q.v.), and then to the Town Mill in Congleton (q.v.). His place at Somerford Booths Mill was taken by Richard Clarke.[20] It is, perhaps, indicative of the difficulties being experienced by the small country miller at this time that a receiving order against Richard Clarke was issued in March 1896.[21] This state of affairs caused the assignation of the lease for 1896 to be hurriedly changed from Richard Clarke to John Condliffe who was the miller at Smallwood Mill (q.v.) at this time. This last minute change of plans was exploited by John Condliffe who was able to negotiate a reduction in rent to £40 per year in these circumstances.[22] John Condliffe had the contract for supplying flour to the workhouse at Arclid at this time[23] and the acquisition of Somerford Booths Mill gave him the extra milling capacity he needed to satisfy this contract.

This work came to an end by 1902 when the miller at Somerford Booths became Richard Goodwin who was also operating Siddington Mill about three miles away, some five miles north of Congleton.[24] He was replaced by R. Walley in 1910 and then by J. Green by 1914. Green continued milling at the mill until at least 1939.[25] After the Second World War the mill building was used as a store until it was sold for house conversion in 1966 for £3500.[26] Today the mill is still used as a residence and has a large wooden waterwheel attached to the side of the building which is not a replica of the original mill wheel. Inside the mill there are the remains of the sack hoist and also a recess cut into the brickwork on the first floor that used to accommodate the great spur wheel. At the rear of the mill building there are three millstones laid into the patio, two of which are French burr stones and the other is of millstone grit made up of two equal pieces and surrounded by iron bands. This type of millstone is typical of those from Mow Cop nearby, and probably is all that remains of the Mow Cop millstones mentioned in the late 19th century lease schedules. Adjacent to the building, the mill pool and its associated sluices still exist.

References

1. CRO, Inquisition post mortem, 9th April 1630.
2. CRO, Wills of John Deane, 1680, and Zachary Shepley, 1773.
3. *Macclesfield Courier*, 3rd October 1818.
4. CRO, Somerford Booths Census Return, 1841.
5. Bagshaw's Directory of Cheshire, 1850; CRO, Somerford Booths Census Return, 1851.
6. Various Directories, 1857 - 1874.
7. CRO, Lease of 1880, DCB/1716/4/1/1; Somerford Booths Census Return, 1881.
8. Norris, J. H., "The Corn Mills of Cheshire", *Transactions of the Lancashire & Cheshire Antiquarian Society*, Vol. 75, 1965.
9. CRO, Lease of 1880, DCB/1716/4/1/1.
10. CRO, Letter from Arthur Ward to C. W. Swetenham, 2nd December 1880, DCB/1716/4/1/7.
11. CRO, Letter from C. W. Swetenham to George Reade, 1st December 1880, DCB/1716/4/1/7.
12. CRO, Quotation from Jonathon Booth, 28th August 1883, DCB/1716/4/1/5.
13. CRO, Particulars of work, August 1885, DCB/1716/4/1/4.
14. Slater's Directory of Cheshire, 1890.
15. CRO, Somerford Booth Census Return, 1891; CRO, Lease of 1894, DCB/1716/4/1/3.
16. CRO, Lease of 1894, DCB/1716/4/1/5.
17. Davies, D. L., *Watermill, Life Story of a Welsh Cornmill*, Ceiriog Press, 1997.
18. Russell, J., "Millstones in Wind & Water Mills", *The Engineer*, 31st March 1944.
19. CRO, Lease of 1880, DCB/1716/4/1/1; Lease of 1894, DCB/1716/4/1/5.
20. Kelly's Directory of Cheshire, 1896.
21. *The Miller*, 30 March 1896.
22. CRO, Lease of 1896, DCB/1716/4/1/2.
23. Information from A. J. Condliffe, retired editor of the Congleton Chronicle, and great-nephew of John Condliffe, miller.
24. Kelly's Directory of Cheshire, 1902.
25. Kelly's Directories of Cheshire, 1910, 1914, 1934, 1939.
26. Norris, J. H., 1965, *op. cit.* 8.

---

52. MIDGE BROOK TANYARD   SJ 833671

Just under ¼ mile downstream from Somerford Booths Mill on the Midge Brook is the site of a former tanyard at Midge Brook Farm. The date at which this tannery was first established is not known but when it was advertised to let in 1824 it was described as "*A commodious and conveniently situated tan yard which being long established with pits and other conveniences for carrying on the tanning concern.*" This suggests that it could well have been built in the 18th century or even earlier. At the time of this advertisement it was being operated by Mr. James Lowndes.[1]

The tanyard appears on Swires and Hutchins map of 1830 and also on Bryant's map of 1831 complete with a watermill symbol.[2] The use of water power at this tannery was confirmed by an advertisement in 1834 when the nearby Midge Brook Farm was to let, giving the following details of the adjacent tannery.

"*Also a valuable tan yard adjoining the bark house, warehouse, bark mill and bone mill and other convenient offices and fixtures. The bark and bone mills are worked*

FIGURE 143. BRYANT'S MAP OF CHESHIRE, 1831, SHOWING THE LOCATION OF THE FOUR MILLS ON THE SWETTENHAM BROOK, NAMELY (*from right to left*) MARTON MILL, SOMERFORD MILL, THE TANYARD AT MIDGE BROOK FARM AND SWETTENHAM MILL. (CCALS, M5.2)

by water of which there is an abundant supply. The premises possess every requisite for carrying on an extensive and lucrative business."[3]

Not only was water power being used to grind up bark to provide a source of tannin for the tannery but animal bones were also being ground to provide a fertiliser for use on nearby farms. In the first half of the 19th century, before the introduction of chemical fertilisers, the use of bonemeal was a popular way of improving the quality of soils.

How long the tanyard continued to operate is also not known and there is now no trace to be found of the tan yard except in the name of the nearby Tan Yard Farm.

References

1. *Macclesfield Courier*, 18th August 1824.
2. CRO, Bryant's Map of Cheshire, 1831.
3. *Macclesfield Courier*, 18th January 1834.

---

53. SWETTENHAM CORN MILL SJ 811673

About two thirds of a mile east of Swettenham church can be found the remaining buildings of Swettenham Mill. Although the mill building and its machinery survived intact almost until the end of the 20th century there is not much documentary evidence of its history and very little to illuminate the major changes that took place at the mill during its lifetime. Especially for the mill's early history, information is largely dependent on oral tradition and the evidence provided by the building itself. Details of the mill's later history, in the 20th century, reflects the memories of the last miller's family.

The existing evidence suggests a mill site of considerable antiquity although the mill is not mentioned in the Domesday Survey of 1087. The earliest reference to be found is in a document of 1345 which mentions a *"molendinum de Swetenam"*.[1] No other detail is given so although the mill was most likely to have been on the current mill site it is also possible that it was sited elsewhere in the parish.

Oral tradition claims that the mill was built on its present site in 1675 by Thomas de Swettenham with William Lea as miller. Certainly there is an inscription carved on one of the half timbered beams at the front of the mill that reads "TS 1675" and the Lea family history has a tradition of a long family involvement with the mill from its building in 1675 up to the middle of the 19th century.[2] There is another inscribed date on the building, namely "WS 1714". This is on a stone which is positioned low down in the wall between the mill and the pool, opposite the kiln building. Its significance in the history of the mill is not known.

That there was a watermill at Swettenham during the early 18th century is confirmed by two indentures. The first is dated 26th November 1724 between Thomas Swettenham and Thomas Norbury of Lower Withington who is instructed to "... *grind all his corn at the mill of the said Thomas Swettenham, in Swettenham or pay 2s-6d for every bushel of corn (Winchester measure) which he shall grind from the said mill, and also, if any extraordinary breach shall happen in the weir or dam at Swettenham Mill pool, shall send sufficient workman with a shovel to assist to help repair the same for so many days (not exceeding four) as shall be necessary or pay 3s-4d to the said Thomas Swettenham in lieu thereof*". Similarly an indenture of 13th January 1736 between Thomas Swettenham and William Lockett who is charged with "...*providing a workman with a shovel to repair Swettenham Mill dam in case of any extraordinary breach therein....all corn to be ground at Swettenham Mill*".[3] It is perhaps no coincidence that the Arms of the Swettenham family contains three shovels!

Tradition has it that the current mill was rebuilt after a fire in 1765, which is confirmed by the datestone built into the gable of the mill which is inscribed "Rebuilt by TS 1765". The type of machinery installed at this time probably used an upright shaft with great spurwheel driving two or more pairs of stones, similar to the present machinery but made almost entirely out of wood. This could have replaced the previous system where each pair of stones was driven by a separate waterwheel. In the stone wall separating the ground floor of the mill from the waterwheel there are two holes that have been bricked up at some time. These are the remains of two shaft holes for two waterwheels that would have driven one pair of stones each. Also on the ground floor, situated in the base of the stone wall that forms the wall of the dam, there is a bricked up arch. The original purpose of this arch is not known, perhaps it was for the original by-wash, or for a very early undershot waterwheel used before the pool was formed.

Evidence for the layout of the site just prior to the rebuilding of 1765 is provided by a map of the Swettenham Estate from 1762.[4] This clearly shows the water supply system as it is now, with pictograms of the mill and kiln buildings showing that they were in exactly their same positions as today. As for the millers during the 17th and 18th century, there is only the Lea family history to rely on. This claims that the original miller in 1675 was William Lea, followed in 1690 by

John and Mary Lea. In 1710 James & Ann Lea lived at the mill and in 1745 the mill was worked by another John and Mary Lea.[5] Unfortunately, the Swettenham parish register does not include any Leas prior to 1736 and even then does not record their occupations. Similarly, the Land Tax of 1780 to 1792 lists a John Lea at Swettenham paying 19s-8d per annum but does not give any information as to the nature of the property involved. In fact, the mill is not mentioned as such in the Land Tax returns at all in the period 1780 to 1830.

There is evidence that John Lea left Swettenham in 1792 to go to Bosley, a village a few miles to the east of Congleton, where he was the miller at Bosley Corn Mill (q.v.).[6] In 1796 he returned to Swettenham with his wife Mary, and his son John and his wife Sarah.[7] It is possible that he went to Bosley to have his own mill and returned to Swettenham when it was possible for him or his son to become the miller there. In 1818 William Lea is mentioned as a miller of Swettenham in the parish register together with his wife Anne Lea (née Newton), but almost immediately after this date they moved to Park Mill, Brereton (q.v.), in the next parish, eventually moving to Lower Peover Mill, which is not far away in Cheshire. After they had moved to Brereton, James (the son of John & Sarah Lea) took over as miller at Swettenham with his wife Catherine. They appear quite regularly in the parish register at the baptisms of their many children.[8] Their long tenancy of over thirty years continued in spite of the fact that Swettenham Mill was offered to let in 1829 with an advert in the local newspaper that stated *"To be let for a term, a well accustomed water corn mill and farm in Swettenham."*[9] They and their family were enumerated in the 1841 and 1851 censuses and it is interesting that William Newton (presumably a relation by marriage to the Lea family) was listed as a miller and servant to James Lea in 1841.

The Lea family history claims that James & Catherine left Swettenham in 1852 to go to Park Mill at Brereton due to their landowner's insistence that they convert to Roman Catholicism. However, some of the family were still at Swettenham up to 1856 because Henry Lea is mentioned in the parish register as a miller, residing in Swettenham, when he married Sarah Collins in 1854 and at the baptism of their first child in 1855. After 1855 Henry Lea is recorded as being a miller living in Brereton.

After the Lea family left Swettenham the miller there is listed in 1857 as Francis Hine but by 1860 it was Arthur Malkin.[10] Arthur Malkin and his wife Hannah came from local Swettenham families and Arthur is listed in the 1861 census as being 26 years old with one house servant, aged 13, one miller's servant, aged 15, and two children. By 1871 this had increased to six children.

In 1873 the parish register records William Newton as a miller of Swettenham with his wife Jane. The Newton family had been connected by marriage to the Lea family since 1818 and they operated the mill with their son Theophilus until the mid 1880s, although during this period the Trade Directories listed Dakin & Co. as millers under the heading of Swettenham so presumably the Newtons were employees of this company.

By 1888, the mill had been taken over by James Snelson and his son Henry.[11] In the 1891 census James Snelson is listed as a corn dealer living in Kermincham (the next parish to Swettenham) with his son Henry, whereas a Price Englefield is actually living at the mill as the employed miller. In 1893 the Snelsons were paying a rent of £80 per year for the mill.[12] Eventually,

FIGURE 144. SWETTENHAM MILL SEEN FROM THE WEST IN 1999, WITH THE KILN BUILDING ON THE RIGHT.

the historical circle was completed when the Lea family once again provided the miller at Swettenham, in the guise of Herbert James Lea, grandson of James and Catherine, who operated the mill between 1902 and 1916.[13]

At some time during the latter half of the 19th century, the main machinery and waterwheel were replaced with cast iron versions, as were the kiln tiles. Although oral tradition assigns Jonathon Booth as the millwright probably responsible for this work, this is has not been confirmed and there is no indication of a date when it took place.

In 1924, the mill was taken over by Wilf Lancaster after his marriage. It had been purchased from the Swettenham Estate by his uncle just after the end of the First World War. Wilf was born at Tan House Farm, Buglawton, near Congleton, in 1899 and left school at thirteen years of age to work on the farm. By sixteen he was driving steam rollers and traction engines, although he needed a box to stand on to see over the front end (Wilf could never have been described as tall even as an adult!). One of his stories from these early years was about driving a steamer in Buglawton when it ran away down the steep hill there and all he could do was blow the whistle continuously, hang on for dear life and hope for the best. Fortunately, he survived to become the miller at Swettenham until his death in 1986.

By the 1920s it was obvious that there was no future for the small country watermills as just producers of flour for human consumption. Consequently Wilf Lancaster had to diversify in order to make a living from Swettenham Mill. To do this he installed extra line shafting to provide the necessary power to drive ancillary machinery such as a suite of wood saws (a flat bed plank saw, a circular saw and a band saw) with which he was able to make all manner of wooden items such as farm carts, farm machinery, etc. As an adjunct to these activities he became a wheelwright and also used the mill's waterpower to turn a lathe and produce wooden bobbins which were sold to the silk mills as far away as Leek.

Shortly after arriving at the mill, Wilf installed a direct current (d.c.) electric generator which he acquired from Jonathon Booth, the Congleton millwright, in exchange for some timber. It has been said that this generator started life as the motor for an anchor winch on a tug working on the Manchester Ship Canal! When coupled to the waterwheel via belting this generator enabled him to provide lighting in his farmhouse and to drive some household appliances. It also opened up another developing market to be exploited, namely that of radio (or the "wireless" as it was then called). In rural England at this time there was no mains electricity power supply, nor even in some of the smaller towns (Congleton received its first mains electricity in the 1930s), so wireless sets, which were becoming very popular, were powered by batteries (these were quite large and heavy, containing liquid acid). With his d.c. generator, Wilf was able to provide a battery charging service which, in its heyday, covered the area of South Manchester and East Cheshire.

At about the same time, Wilf purchased a second hand refrigeration plant which he installed between the mill and the pool, again driven by belting off the lineshafts. Without a widespread electricity supply, ice was not readily available. Large houses had used ice houses in the past to store ice taken during the winter from local sources but, by the 1920s, the level of demand resulted in most ice being imported from Scandinavia. The ice produced at Swettenham Mill was sold to hotels and restaurants throughout the same area covered by the battery charging service, even as far afield as Derby. It was while supplying ice and recharged batteries that one of his customers who owned a plant nursery, being unable to pay his account, settled his debt to Wilf by giving him a large quantity of daffodil bulbs. These were planted in the valley by

FIGURE 145. WILF LANCASTER, THE LAST MILLER AT SWETTENHAM MILL, AGED 84, AT A STEAM RALLY HELD AT THE MILL IN 1983. (*Dr. Barry Job*)

FIGURE 146. A DRAWING OF SWETTENHAM MILL. THE MAIN MILL BUILDING, WITH ITS LEAN-TO COVERING THE WOODWORKING SAWS, WAS ONLY GENUINELY HALF-TIMBERED ON ITS UPPER STOREYS. THE GROUND FLOOR OF THE MILL AND THE KILN BUILDING ON THE RIGHT WERE PAINTED TO GIVE A HALF-TIMBERED APPEARANCE.

the mill and so started Daffodil Dell which was beloved by many people in South Cheshire and in the Potteries, as a place to visit in the springtime.

With his early experience with steam power, operating a successful set of businesses based on the mill, working as a blacksmith and running a farm, it is not surprising that Wilf had a long line of various steam engines to assist him. At one time the mill was even the base for road making gangs using steam rollers. Wilf was a great aficionado of steam wagons, having an Atkinson steam lorry and also one by Foden (he claimed to have been present for the trial run of the first Foden steam lorry). He used an early Wellington steam tractor on the farm and for road transport, as well as having a portable steam engine driving the saw bench at the mill.

With the coming of the internal combustion engine, he had an early Titan tractor which he could also belt up to the mill to drive the saws and the ice machine. On 20th June 1928 Wilf purchased a brand new semi-diesel engine from Crossley Brothers of Manchester. This was installed to provide a drive capability for the d.c. generator that was independent of the water supply and hence provide a less fluctuating supply of electricity. This seems somewhat out of character as Wilf normally procured most of his equipment second hand or through a barter system.

After the Second World War, it became increasingly difficult to get electrical equipment that would work on a d.c. system and so Wilf installed an alternating current (a.c.) generator (second hand, of course). It is not known were this generator came from but the regulator used in conjunction with it was from a surplus R.A.F. radar system. This worked very well and enabled the farmhouse to switch over to an a.c. supply. Many years later, when the electricity board were installing the mains supply in Swettenham, Wilf claimed he would not have it unless they could supply at less than 1d per day, which is what he estimated it cost in grease to keep the mill working and generating.

Eventually, the need for battery charging, ice, and even self-generated electricity subsided and the

FIGURE 147. THE COMPRESSOR USED FOR MAKING ICE. ON THE LEFT IS THE BRINE TANK WITH COOLING PIPES AND CONTAINERS IN WHICH THE WATER WAS FROZEN. NOTE THE HOLE IN THE MILL WALL THAT ALLOWED THE DRIVE BELT TO CONNECT TO THE FAST AND LOOSE PULLEYS ON THE COMPRESSOR.

FIGURE 148. THE STONE FLOOR WITH SOME OF THE MILLSONES AND THEIR FURNITURE. IN THE BACKGROUND ARE THE BENCHES THAT USED TO HAVE VARIOUS LATHES, ETC. DRIVEN FROM AN OVERHEAD LINESHAFT. NOTE THE COMPLETE LACK OF SAFETY GUARDS AROUND THE MAIN BELT DRIVE.

associated machinery in the mill stopped work and was allowed to slowly decay. However, there was still a market for wooden furniture which Wilf continued to make almost up until his death using the water powered sawing equipment of the mill. In his later years, he used to specialise in furniture such as coffee tables made from yew wood. He even had an arrangement with the Church Commissioners for the first choice of any yew trees that were felled on church lands around Cheshire and North Staffordshire.[14]

After Wilf Lancaster's death in 1986, his son Alf sold the mill to a local businessman who had plans to restore the mill and open a heritage centre. Some work was done on the mill, not very proficiently (according to Alf), but the project came to grief when the local authority, Congleton Borough Council, refused planning permission. They insisted that as a listed building dating from the 18th century, the mill should be restored using 18th century materials only and that any extra buildings for the heritage centre would not be in keeping with the mill's setting. After that the mill passed through various hands with the mill buildings getting more and more derelict. Once the integrity of the roof failed, rain was able to enter the building and eventually the floors rotted and the walls cracked and bulged alarmingly. It was this desperate state that persuaded Congleton Borough Council to finally grant planning permission in the late 1990s for the mill to be converted into living accommodation as the only way of providing the investment of the very large amount of money needed to save the building. Although there was a legally binding agreement with the owner and his successors to keep and preserve the waterwheel, main milling machinery and the kiln floor, this protection did not cover the ancillary machinery, most of which was scrapped prior to the mill being convered into a house in 1999/2000.

The mill building is "L" shaped with three storeys. The east-west arm of the mill is built of brick with a stone tiled roof whereas the upper floor of the north-south wing is timber framed with brick infill. The waterwheel is completely enclosed in its own chamber within the building along its northern side. The area enclosed by the two wings used to house much of the woodworking machinery and was protected from the elements by a timber roof at first floor level. The plank saw was housed in a covered area just to the west of the mill and the ice making machine was accommodated on the eastern side between the mill and the pool. The kiln was a separate building to the

FIGURE 149. THE LOWER FACE OF ONE OF THE CAST IRON KILN TILES.

south of the mill with a two storey store at its western end. The cast iron kiln floor was housed in the eastern end of the building at an intermediate level between ground level and the first floor of the store. The furnace was under the kiln floor with its stoke holes accessible from the southern side of the building.

The headrace from the mill pond enters the mill at the north-east corner of the first floor. The pentrough consists of a metal tank half of which is inside the mill. The sluice which fed the water onto the waterwheel is controlled by a rotary handle on the ground floor. The all iron waterwheel is pitchback, measuring 14 feet diameter by 4 feet 6 inches wide. It is an elegant piece of casting, with scalloped naves and two sets of six slightly tapering arms, indented on the outside and with a strengthening rib on the inside. The shrouds, jointed in six places, are 10 inches deep and there are 54 angled buckets. The wheel is keyed onto a 7 inches diameter iron shaft which passes through a small opening into the mill.

An iron pitwheel, 6 feet 6 inches diameter, is fitted on the inside end of the waterwheel shaft. The pitwheel is cast in two parts, bolted together, having eight arms and an octagonal hub. The space between the hub and the wheel shaft is filled by a casting 23 inches across. This suggests that there may previously have been a wooden wheel shaft and that the waterwheel and present shaft are late replacements. The pitwheel meshes with a cast iron wallower mounted on the wooden upright shaft. The footstep bearing of the upright shaft rests on a substantial cast iron arch which lifts the shaft clear of the plummer block housing the bearing for the inside end of the waterwheel shaft. Above the wallower is the cast iron great spur wheel which has eight arms, is cast in two parts, and measures 8ft in diameter.

The hurst consists of four pairs of circular cast iron columns rising from the stone block floor to individual stone pans above. The pairs of columns are linked by cast iron bridgetrees which incorporate the centre-lift tentering mechanism, the means of adjusting the gap between the millstones. Each stone nut casting incorporates a groove and collar on its upper face to hold the journals on the ends of the forked arm of the disengaging lever. The fulcrum of the lever is on one of the hurst columns and, as with the bridgetrees, is adjustable over about 6 inches The hurst columns, bridgetrees and disengaging levers are very well engineered to an excellent design. The stone pans above were certainly made for this specific location as they are cast with flanges which allow them to be bolted together and to the wheelpit wall. They each have three screw pins set in them to enable the bedstones to be accurately levelled. As usual, the whole of this machinery is boxed in with a timber frame and boards.

On the stone floor are to be found the four pairs of

FIGURE 150. A SCHEMATIC VIEW OF THE MAIN GEARING BETWEEN THE WATERWHEEL AND THE MILLSTONES. THE DRIVE FROM THE CROWN WHEEL OPERATED A COMPLEX SET OF LINESHAFTS AND PULLEYS.

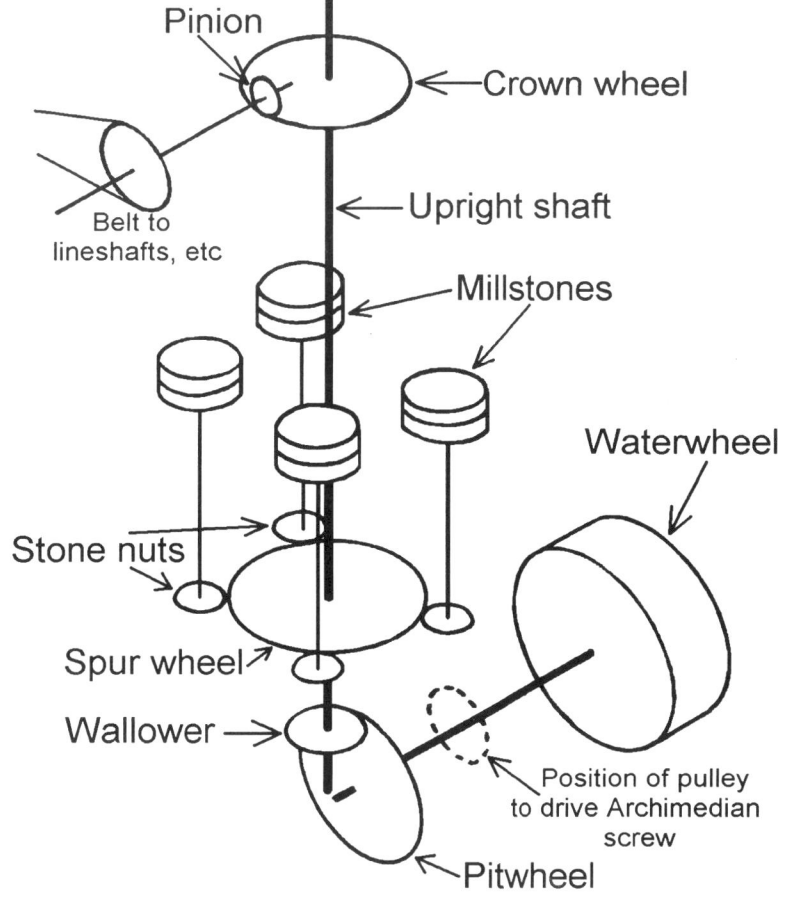

millstones, complete with all their furniture; circular tuns, plain box-framed horses and large hoppers. There are two pairs of French burrs, both 4 feet diameter. The runners both have four balance boxes and one is named at the eye "Davies & Sneade, Charters Street, Liverpool" which is thought to date it to the 1890s. The other two pairs are millstone grit, almost certainly from Mow Cop which is only about seven miles south-east of the mill. One pair is 4feet 2 inches diameter and the other 4 feet diameter. All four runners are mounted on simple balance rynds. The two pairs of stones thought to be from Mow Cop are all made up from four pieces of millstone grit using the same configuration, the segments being cemented together and banded with iron.. The pair of millstones used for grinding oatmeal, which were removed during the conversion work, were also probably from Mow Cop and had been manufactured to the same configuration, as had an old stone buried in the approach path to the mill. The significance of this configuration of pieces is not known, nor is it known if the pattern was specified by the millstone maker or the purchaser.

The upright shaft reaches up into the first floor where it is surmounted by a crown wheel of 7 feet 4 inches diameter. From this crown wheel a pinion takes the power to a number of lineshafts in the mill all interconnected by belting. These lineshafts and pulleys were installed by Wilf Lancaster and were used to drive all the ancillary machinery. The sack hoist was operated on the "slack belt" principle using a jockey pulley to tighten the drive belt when required. The ice machine compressor, situated on the dam between the mill and the pool was driven by a belt passing through a crudely made hole in the mill wall. Alongside the compressor was the brine tank which held the 24 buckets of fresh water that was to be frozen. The ammonia cylinder was buried in the dam wall and the cooling tubes were positioned in the headrace.

On the same floor as the millstones was a bean splitter which worked in conjunction with the adjacent elevator. The oatmeal stones also passed their product into a jigger on the ground floor before its output was passed into a fan-driven winnower. All this equipment was driven by an assortment of pulleys and belts, some of which ran in confined spaces at almost impossible angles.

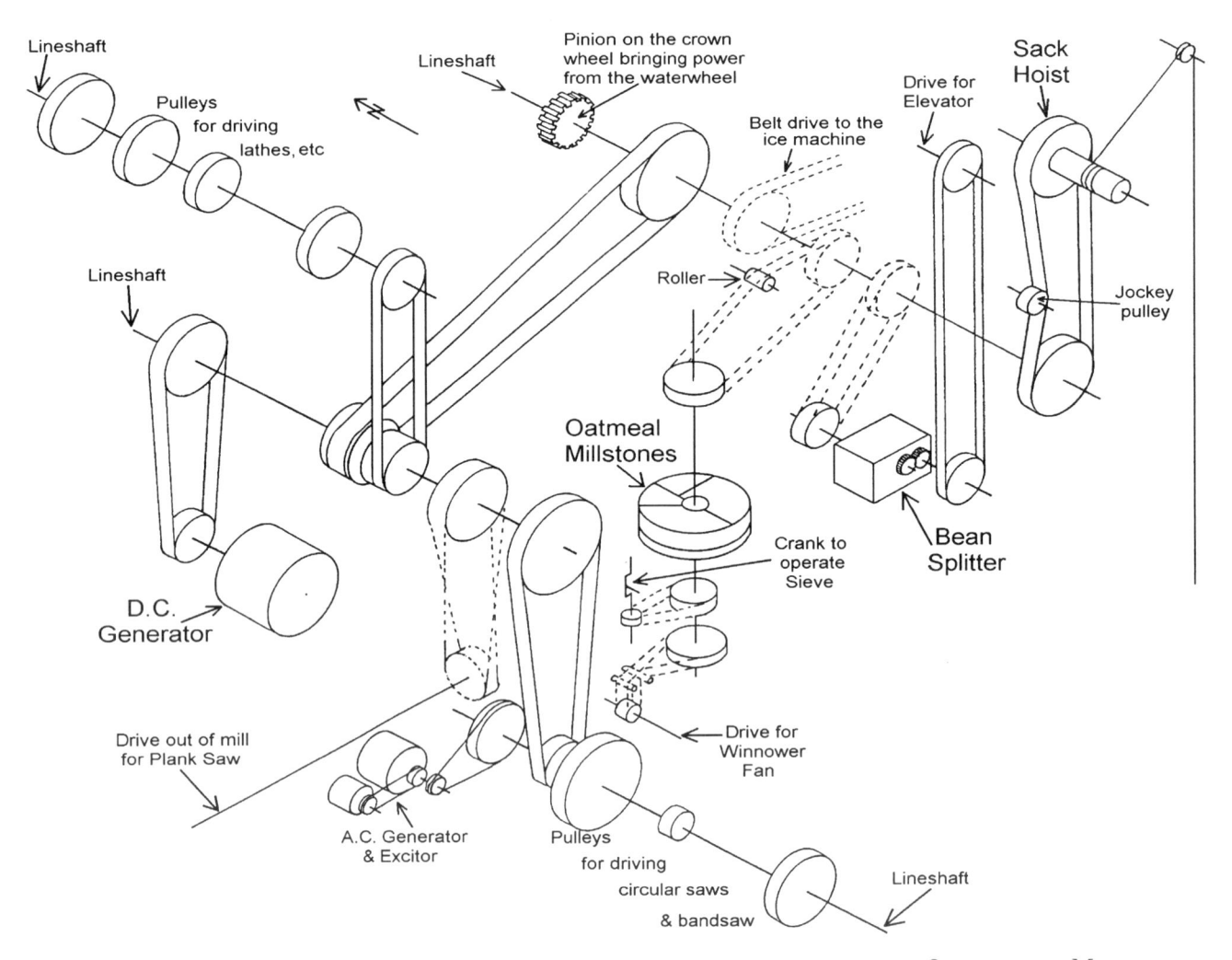

FIGURE 151. AN ISOMETRIC VIEW OF ALL THE LINESHAFTS, PULLEYS AND BELTS IN SWETTENHAM MILL.

FIGURE 152. AN ILLUSTRATION FROM A CATALOGUE OF ROBINSONS OF ROCHDALE, C.1900, SHOWING A PLANK SAW SIMILAR TO THE ONE USED AT SWETTENHAM MILL.

The electrical generation equipment consisted of a very large d.c. dynamo and an a.c. generator and regulator, all of which were belt driven and located on the ground floor.

The wood working machinery was kept outside the mill except for some bobbin lathes which were positioned on the stone floor. The largest wood working machine was a plank saw, made by Thomas Robinson & Son, of Rochdale, which was accommodated in a roofed structure about 10 yards to the west of the mill. Power was required for the carriage which fed the tree trunk to the saw and for the saw blade itself. The other wood working equipment comprised of two circular saws and a band saw which were housed in the lean-to attached on the south-west corner of the mill. These saws were also made by Robinsons of Rochdale.[15]

In the year 2000 the mill was converted into a private house. Most of the machinery was removed although it was intended that the waterwheel, main gearing, and four pairs of millstones and their fittings are restored and retained. Similarly the cast iron kiln tiles were intended to be preserved in situ, although the kiln building became part of the living accommodation. The ground and top floors of the mill were incorporated into the dwelling area and were linked to the kiln building via a covered walkway. According to the planning submission it was not intended that the stone floor become part of the living area. At the same time a new single storey building was erected around the south-western side of the mill pool to provide further living space.

References

1. Dodgson, J. McN., *The Place Names of Cheshire, Part 2*, Cambridge University Press, 1970.
2. Lea, Herbert, *Swettenham Mill*, 1984.
3. Rathbone, C., *The Dane Valley Story*, 1954.
4. CRO, Map of Swettenham Estate, D/4025/8.
5. Lea, Herbert, 1983, *op. cit.* 2.
6. CRO, Bosley Land Tax Returns.
7. CRO, Swettenham Land Tax Returns.
8. Congleton Library, Swettenham Parish Register.
9. *Macclesfield Courier*, 10th Jan 1829.
10. Kelly's Directory of Cheshire, 1857; White's Directory of Cheshire, 1860.
11. Slater's Directory of Cheshire, 1888.
12. CRO, Lease of 1893, D/4025/14.
13. Kelly's Directories of Cheshire, 1902 - 1916.
14. Information provided by Alf Lancaster.
15. Bonson, T., Booth T., & Job, B., "Swettenham Mill, A History and Survey", *Wind and Water Mills*, 18, 1999.

54. CRANAGE (FORGE) MILL   SJ 757677

The village of Holmes Chapel (once known as Church Hulme) is situated about seven miles west of Congleton and four miles east of Middlewich, lying on the high ground just to the south of the River Dane. Just north of the village the main road from the Potteries passes over the River Dane before heading northwards to Knutsford and Warrington. To the east of this bridge over the River Dane are the grounds of the Hermitage estate whilst just to the west of the bridge, lying in a very tight bend of the river, is the site of Cranage Mill.

During the civil war in England much of the iron industry in the south of the country was destroyed. The demand for iron provided an opportunity for the establishment of iron works in other parts of the country, especially where charcoal could be obtained from the local forests. In 1658 John Turner came to Cheshire to build a furnace and associated forge. He chose a site in Lawton (*q.v.*) for his furnace and was expecting to build his forge about eight miles distant from there, probably at Cranage. However, he experienced financial problems and the forge was not built in 1658 as expected. Eight years earlier John Leadbeater had purchased the Hermitage estate in Church Hulme and in 1660 he bought the site of an existing water mill from the trustees of Viscount Kilmorey in order to build a forge.[1] The mill he purchased may well have been Cotton Mill (*q.v.*), which was situated about half a mile downstream from the bridge over the River Dane at Cranage. However, Cotton Mill continued to grind corn and the forge was built upstream near the road bridge over the river in Cranage.

The forge was leased to a William Fletcher of Makeney near Duffield in Derbyshire, who was in partnership with a George Whyte. That they were operating the forge in the early 1660s can be shown from their contracts with local landowners, such as Thomas Cholmondley of Holford, for supplies of wood for making into charcoal for use in the forge. They even purchased some haematite ore from Sir Thomas Preston's Stainton mine in Furness which would have been destined for a nearby furnace, such as that at Lawton.[2] The iron industry in Cheshire during the 1660s was in turmoil with legal actions taking place with respect to Lawton Furnace and there were also severe problems with the quality of the pig iron produced. This probably influenced William Fletcher and George Whyte's decision to sell their lease in Cranage Forge to William Keeling of Odd Rode in June, 1667. Having acquired the forge, William Keeling made over a half share, consisting of *"all buildings and working tools including two hammers and one anvil"* to his brother Francis Keeling of Kinderton for £205 and yearly payments to the original partners in the lease. Francis Keeling then made over half of his share (i.e. a quarter share in the forge) to William Rowley of Church Hulme, who was probably his nephew.[3]

In December of 1667 the Keeling brothers ordered 35 tons of *"best Coltshire[sic] iron pig metall"* at £4-14s-0d per ton, five tons of hammers, anvils and other cast metal at £7-5s-0d per ton for Cranage Forge from Thomas Wright, who was the agent for Sir Thomas Delves's Doddington Furnace, which was located about five miles south-east of Nantwich. They would appear to have been renewing the forge's machinery and acquiring a supply of pig iron for processing. However, this promising start ran into problems when William Keeling became bankrupt in 1668 and Francis Keeling made a payment to Sir Thomas Delves through William Rowley, who was described as the manager of Cranage Forge. The trade with Doddington Furnace seems to have continued as another payment was made in 1670 via Walter Pensall of Lea Forge. However, when Francis Keeling died in 1672 no further payments were made to Doddington although they must have continued to receive supplies for some time. Not surprisingly Sir Thomas Delves brought a case against William Rowley in 1678 for the moneys outstanding.[4]

It is probable that this action caused William Rowley to give up his interest in the forge. At some time in the early 1680s Lawton Furnace and Cranage Forge were acquired by a partnership of William Cotton, Dennis Heyford and Thomas Dickin, with Dennis Heyford having the sole interest in Cranage Forge. Previously these partners had been associated with ironworks in Yorkshire. Around the same time, they leased Abbot's Bromley Forge in Staffordshire from Lord Paget. Later they also worked Cannock Forge and Rugeley Slitting Mill.[5] That Cranage Forge was operating during the early 1680s is shown by the continuing appearance of workmen such as John Butterfield *"ffiner at Cranage forge"* and Richard Bolton, *"clarke of the fforge"* in the parish record of 1679.[6] In 1680 Dennis Heyford was owed £264-16s-0d by Nicholas Withington, a recently deceased nailer in Atherton, Lancashire, for iron probably supplied from Cranage. Also in 1683 the inventory of Thomas Bagnall shows a debt of £50 owed by the forgemasters at Cranage for the supply of charcoal. Confirmation of the partnership's interest in Cranage Forge occurs in 1687 when they were sued by John Crewe's widow for the payment of supplies of cordwood to the forge from the Crewe estate. Another defendant in this case was Thomas Hall who was William Cotton's cousin and manager of the forge at Cranage.[7]

FIGURE 153. A SET OF WATERPOWERED TRIP HAMMERS SUCH AS WOULD HAVE BEEN USED AT CRANAGE FORGE IN THE LATE 17TH AND EARLY 18TH CENTURY. (*Sheffield Chamber of Commerce*)

When Thomas Dickin died in 1692 he was succeeded in the partnership by his son, also called Thomas Dickin. Thomas Dickin and William Cotton both held 25% of the partnership with Dennis Heyford holding the remaining 50%. In the early 1690s Dickin and Cotton were receiving around £1600 per year on average. Between 1692 and 1696 the Staffordshire forges at Bromley, Cannock and Rugeley were sold and Vale Royal Furnace and Warmingham Forge (*q.v.*) in Cheshire were acquired. This grouping of two furnaces, at Lawton and Vale Royal, and two forges, at Cranage and Warmingham became known as the "Cheshire Works". In 1696 the management of the Cheshire Works was organised with Thomas Hall at Cranage Forge and Lawton Furnace; Daniel Cotton, the younger brother of William Cotton, at Vale Royal Furnace; and Edward Hall, the younger brother of Thomas Hall, at Warmingham Forge. At the same time Thomas Hall became a partner and in 1698 he and William Cotton bought out Dennis Heyford's interest in Cranage Forge. Not only did Cranage Forge house a finery and a chafery for decarburizing the iron and hammering it into bars but it was the only site in the Cheshire Works to incorporate a slitting mill for converting bar iron into rods. This facility ensured that Cranage not only received pigs from Vale Royal and Lawton furnaces but also received iron from other forges for slitting. Some idea of the type of business being conducted is illustrated by a shipment of nails to London in 1699 that made a profit of £991 6s 2d.[8]

When William Cotton withdrew from the partnership in 1700 the capital involved stood at £21,426. In this reorganisation, Cranage Forge, together with Lawton Furnace, became the responsibility of Daniel Cotton.[9] In the next few years the partnership continued to expand, becoming involved with Bodfari Forge, in North Wales, and Street Furnace (*q.v.*), in nearby Odd Rode, which was converted into a plating forge. After a very unstable beginning in the 1660s, by the beginning of the 18th century the iron industry in Cheshire had matured into considerably more than a local or even a regional concern with interests in Cumbria, Lancashire and Staffordshire, and with customers throughout the country. The level of profits being made by the partners was commensurate with this widespread activity, so much so that in 1702 Thomas Hall bought the Hermitage estate at Church Hulme from Thomas

Leadbeater and became one of the landed gentry.[10]

In 1706 further reorganisation saw Daniel Cotton taking over the direct management of Cranage Forge and Lawton Furnace from Thomas Hall. A year later Cranage Forge became part of an even larger consortium when the Cheshire Works amalgamated with the "Staffordshire Works", a similar consortium operating in North Staffordshire. The Staffordshire Works consisted of Meir Heath Furnace, Consall, Oakamoor, Chartley, Bromley, Cannock and Tib Green forges together with Rugeley slitting mill. This amalgamation was logical as the Staffordshire Works only had the one furnace and had always relied on supplementing their supply of pig iron from Cheshire. At first sight there appears to be a bias towards the Staffordshire partners as they held four of the six shares in the Cheshire Works (each share valued at £2250) and four of the seven shares in the Staffordshire Works.[11] However, by 1708, due to the deaths of two of the Staffordshire partners, Thomas Hall had acquired two more shares and had become the managing director of the Staffordshire Works. It was in this same year that Thomas Hall presented a double tiered brass candelabra to Holmes Chapel Church, possibly to commemorate his good fortune in the partnership. Not to be outdone, the following year Daniel Cotton presented a set of bells to the same church.[12]

The working of these partnerships was by no means as straight forward as it may seem as a number of the partners had shares in iron making businesses elsewhere, such as in South Staffordshire, Lancashire and in Furness. The partners also sold parts of their shares in the various sites of the main business to other associates to raise money. This kind of arrangement is exemplified by a purchase for £500 by Lawrence Booth of Twemlow of one of the shares in Cranage Forge held by one of the Staffordshire partners.[13] The position is further complicated by alliances formed due to marriages, for instance, Lawrence Booth was also the brother-in-law of Daniel Cotton, one of the Cheshire partners.[14] In 1710, two new partners, Edward Kendall and William Rea, joined the partnership in Staffordshire. Edward Kendall, who had extensive interests in the iron trade in the Forest of Dean and the Stour Valley, married Anne Cotton, one of Daniel Cotton's daughters in 1712. Although Kendall and Rea were mainly concerned at this time with the Staffordshire Works and ironmaking in Lancashire, the family of Daniel Cotton's son-in-law was to be further involved with Cranage Forge in the future.[15]

In 1715 Thomas Hall died and was succeeded at Cranage by his son Edward. Over the next fifteen years or so the business continued to prosper and grow. The Hall family continued to live at the Hermitage near Holmes Chapel and during this time Cranage Forge

FIGURE 154. A TYPICAL FORGE IN ACTION. (*Painted by Mathaus Christoph Hartmann in 1836*)

was one of the main sites involved in the iron manufacturing industry in Cheshire and North Staffordshire. Fortunately a set of accounts has survived for a few years around 1730. Although these are labelled as the Cranage Forge accounts they appear to be more likely the accounts of the Hall's land agent as they cover a number of premises. However, they do include many miscellaneous entries concerning the forge between 1732 and 1736 and a record of stocktaking in 1732. These records provide an insight into many aspects of the business operation at Cranage Forge at this time.[16]

Although the forge building was made of brick it had a thatched roof. However, the windows were glazed and both the finery and chafery hearths had brick chimneys. The forge was driven by two breastshot waterwheels, one of which provided power for one set of hammers and the hearths' bellows, whilst the other wheel powered a slitting mill. In 1732 the waterwheel and gearing, called the "mill engine", was valued at £35. Apart from this main machinery there were a number of ancillary items such as a smith's bellows and anvil, a beam scale and weights, two pairs of cutters, various hammers and tongs, a vice, grindstones, riddles, and various containers. The total value of the stock at the forge in 1732 was £4036-12s-1¼d.

Like any business there were certain fixed overheads that had to be paid. These consisted of the annual rent of £39 for the site, paid to the Leadbeater family, plus various taxes which were levied including the land tax of 19s per year, a poor tax of 10s-8d per year, a window tax of 6s or 7s per year, a church tax of 4s (although other payments were made to the vicar of Sandbach from time to time) and a local highway tax of 6d. There were usually four men working at the forge, three finers and one slitter. In 1732 the finers were George Wilson, William Scott, and John Barthomley but a new finer, Philip Matthews, was taken on near the end of the year, possibly as a replacement. The slitter at this time was John Cope.

Cranage Forge was being supplied with pig iron from Lawton Furnace at the rate of around 100 tons per year and also with bar iron for slitting, mainly from Warmingham Forge. Some bar iron came in smaller quantities from Tib Green and Norton forges and also some blooms (the output of a forge finery) came from Acton Forge. The amount of bar iron from Warmingham Forge was around ten tons per month of which about 25% was coldshort bars made from Staffordshire ores and the remaining 75% was mixed bar, made by adding some Cumberland haematite to the native Staffordshire ore in the furnace. Unfortunately customers are identified by name only

FIGURE 155. A FORGE HEARTH SIMILAR TO THOSE USED FOR THE FINERY AND THE CHAFERY AT CRANAGE FORGE.

so it is difficult to identify where the output from the forge was being sold but much of the output went to Warrington where the partnership leased a warehouse, presumably this was used to supply to nailers, etc. in Lancashire.

There were regular deliveries of charcoal to the forge, which came from up to fifteen miles away, for use in the finery, however the records show that the chafery hearth used pit coal as its fuel. Other items were regularly delivered to the forge such as soap and grease for use in the slitting mill, hides and sackcloth for the bellows and protective clothing, together with lime and glue.

As well as the normal work in the forge there are many entries in the accounts for repairs and maintenance. In 1732 the hammer beam needed repairing at the end of which 2s was spent on "*Ale to the forgemen finishing the hammer beam*". The following year hair was purchased to block up leaks in one of the waterwheels. In 1734 there were expenses for dressing the two finery bellows and the chafery bellows. Just after the bellows had been dressed there are entries of payments to one of the finers, as follows:-

"*To William Scott, ffynor, in full to midsummer* 10s-9d
Paid to him for some labour at the Bellows etc............ 6s-9d
Given him for encouragement to work well.................. 1s-0d"

169

FIGURE 156. WORKING IRON BAR UNDER A TRIP HAMMER. NOTE THE SWINGING SEAT TO ALLOW THE WORKER EASILY TO ADJUST HIS POSITION TO SUIT THE WORK.

The last entry was quite unusual but hints at some situation that needed a small bribe to ensure that the job would be done. Later that year, William Kennerley was paid 11s-0d for mending the slitting mill wheel and other work at the forge. In 1735 four men were paid 16s-8d for four days work mending a breach at the forge plus a bonus of 2s for standing in the water. Later in the year there are entries for "*slutching the pool*" and "*to cutt the anvil*".[17]

These accounts occur at a time when the iron industry in Cheshire was probably at its height, for in 1737/8 the industry went into recession and the total production of bar iron in Cheshire fell from around 460 tons to 290 tons per annum, and many forges especially in North Staffordshire were closed down. This would have seriously affected Cranage Forge but it soldiered on and there was a slow recovery in the output figures throughout the 1740s.

At the end of the 1740s there were major changes in the Cheshire and Staffordshire partnerships with the death of Edward Hall's son Thomas in 1748 and Edward Hall himself in 1750, coupled with the death of Thomas Cotton and his nephew in 1749 and Edward Kendall in 1750. Due to all these deaths a new partnership was established to continue working Cranage Forge and the other parts of the Cheshire Works. The main partners became Jonathon Kendall, son of Edward, who had been running Duddon Furnace in Cumberland, and whose mother was a member of the Cotton family, together with Samuel Hopkins who was the executor of the Cotton's estates. The heir to the Hall estate was a minor, Thomas Bayley Hall, who appears to have played no further part in the partnership.[18]

After 1750 production at Cranage Forge was not the same. Lawton Furnace, its main source of pig iron, had closed and at some stage prior to 1767 the nearby Cotton Corn Mill also closed and the local flour production was moved to the forge.[19] Initially the corn mill probably took the place of the slitting mill inside the forge building. In 1776 the partnership between Kendall and Hopkins was dissolved with the issue of the following notice

"*Partnership between Jonathon Kendall, Henry Kendall, John Hopkins and Thomas Hopkins in the iron trade at Doddington Furnace and at Cranage, Warmingham, and Lea Forges in the county of Chester, at Rugeley Slitting Mill, Rugeley, Cankwood and Winnington Forges and Mearheath Furnace in the county of Stafford and at Dovey Furnace in the county of Cardigan was this day (15th April 1776) dissolved by mutual consent and that the Trade will for the future be carried on by the said John Hopkins and Thomas Hopkins in partnership and except Winnington Forge which is to be worked for one year from Lady Day 1776 by the said John & Thomas Hopkins and afterwards by the said Jonathon & Henry Kendall. All debts to be paid to Jonathon Kendall at Drayton in Shropshire, Henry Kendall at Ulverston in Lancashire, John Hopkins at Rugeley in the county of Stafford or Thomas Hopkins at Cankwood Forge in the same county.*"[20]

The second half of the 18th century saw a great upheaval in the iron industry with the widespread introduction of coal, as opposed to charcoal, as the fuel used by blast furnaces, which dictated that the iron industry moved to the coalfields such as in South Staffordshire. It is likely that all iron working at Cranage came to an end after the partnership was dissolved, with the premises being used solely for flour milling after this time.

FIGURE 157. A TYPICAL CONFIGURATION OF ROLLING AND SLITTING MILL. (*Swedenborg, 1734*)

It is possible that the miller at Cranage up to about 1776 was Richard Dobson because in that year, when he was a prisoner at Chester, he was described as a miller late of Cranage.[21] After this date and until 1792 the miller was William Kennerley who was a descendent (probably the son or grandson) of the William Kennerley who appeared in the Cranage accounts in the 1730s, sometimes doing the work of a millwright. William Kennerley died in 1792 and his son Samuel became the miller, although for administrative reasons his mother, the widow Kennerley, was officially the occupier in 1792.[22] It was also in this year that Samuel Kennerley insured the "utensils and trade" at Cranage for £300. The mill was described in the policy as being brick built with a slate roof and had a drying kiln in the mill building.[23] The mill stayed in the Kennerley family for another generation with John Kennerley taking over from Samuel in 1810. However, by 1812 the mill was occupied by John Swanswick, although it is possible that he sub-let the property to the actual miller.[24]

In 1819 Cranage Mill was purchased by Samuel Pointon, who had been the miller at Colley Mill (*q.v.*), in North Rode.[25] The arrival of Samuel Pointon and his brother Thomas, who were both corn millers, saw the departure of John Swanswick. In operating the mill it would have been obvious to the Pointons that the total power available from the River Dane was much in excess of that needed just for grinding corn for the local market. Consequently, they decided to exploit this potential by building a silk mill which they could then lease out whilst continuing to operate the corn mill themselves. The silk mill was nearing completion by the middle of 1822 when it was described in a newspaper advertisement as follows:-

"*To be let by ticket. A factory now building, and which will be ready by 1st November next, situated at Cranage on the River Dane where there is a powerful and constant stream of water sufficient for turning to any extent.*

The factory will be three stories high and the rooms 108 feet long by 24 feet wide with other conveniences belonging thereto.

For details apply to S. & T. Pointon of Cranage"[26]

However, the Pointon brothers were to find out that the road to riches was not that easy as they were unable to find anyone to lease the new silk mill in spite of repeatedly advertising the property over the next three years.[27] In 1825 they were reduced to trying to let separate rooms in the mill in an effort to generate some income from their investment.[28]

Although they were eventually successful in leasing the silk mill to Richard Burgess in the late 1820s, the experience of tying up their capital in a marginal undertaking took its toll on the Pointon family finances. In 1830, when Richard Burgess's silk business closed, the Pointons tried to recover their money by selling both the corn mill and the silk mill.[29] There were no buyers for either mill at this time so they then tried to sell the silk manufacturing machinery of Richard Burgess at auction as follows:-

"*All that valuable silk machinery and other effects of Richard Burgess & Co comprising one throwing mill containing 28 doz. and eight spindles, four spinning mills containing 26 doz. spindles ea., two doubling engines containing 80 spindles ea., four doubling engines containing 82 swifts ea., three ditto containing 80 swifts ea., six doz. reels, quantity of tram, shafts, bobbins, a double press, gearing belonging to the factory, a quantity of steam pipes, office desk and counter.*"[30]

The details of the machinery highlight just how small

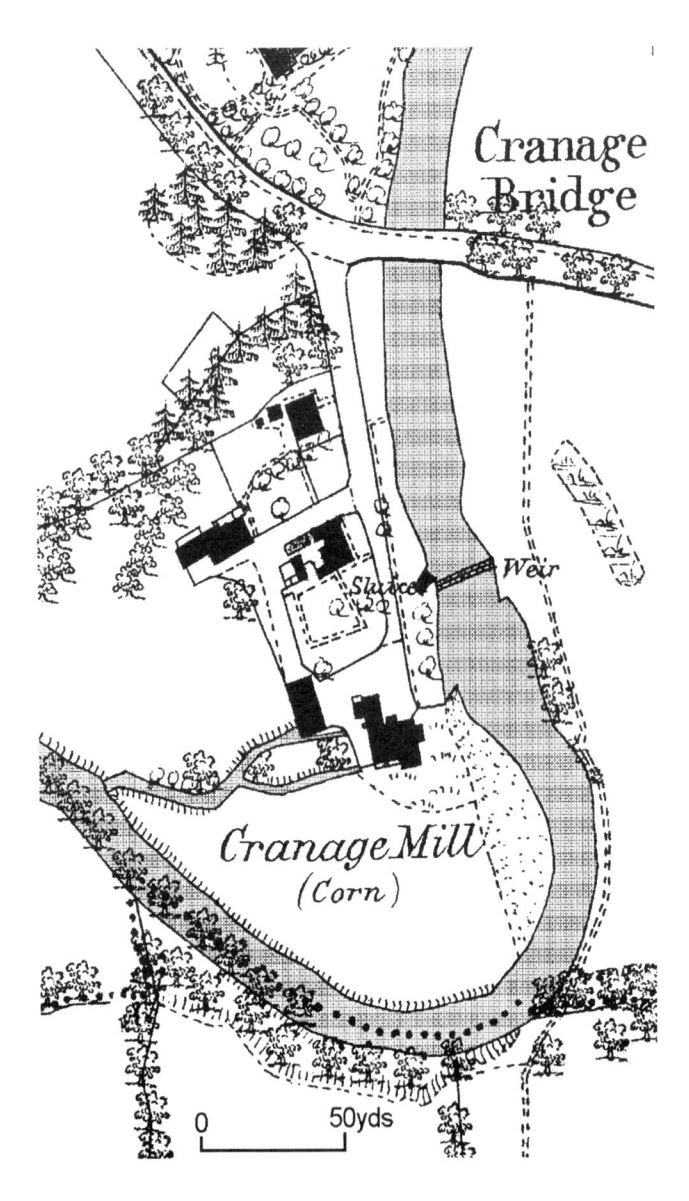

FIGURE 158. PART OF THE ORDNANCE SURVEY MAP, 1872. THE TAILRACES INDICATE THAT THE CORN MILL HAD TWO WATERWHEELS, ONE AT EACH END OF THE BUILDING. THE WESTERNMOST BUILDING WAS THE OLD SILK MILL, BUILT IN 1822, WHICH HAD ONE WHEEL AT ITS SOUTH END.

and hence unprofitable was this silk business. It is probable that they were even unable to sell the machinery, but they did attract another possible client for the silk mill in a Mr. J. Pyott, who realised that to make any profit from such a marginal business would need the costs to be kept to a minimum. This he proposed to do by employing between 20 to 30 apprentices to run the silk mill.[31] Whether this scheme ever got off the ground is not known but two years later the Pointon family were still in financial difficulties and the formal partnership between Samuel and Thomas Pointon was dissolved.[32] Shortly afterwards both Cranage corn mill and the silk mill were sold at auction, the mills being described in the sale notice as:-

"The water corn mill comprises of a warehouse, dressing mill, mill yard and other requisites for grinding corn and is in capital condition and very well accustomed being in a populous district, it stands on the banks, and is turned by, the River Dane and contains five pair of grinding stones of which three are grey stones and two pair French.

The weir has recently and substantially been erected across the river at a great expense and the fall of water is exceedingly powerful and never fails.

The silk factory has been lately erected and is built of brick being 40 yards long by 8 yards wide (inside measurements) and three stories high and a waterwheel supplied from the River Dane and is capable of containing nearly 500 doz. spindles and all other machinery required for an organzine mill."[33]

The description of the corn mill shows that it was a well equipped, up-to-date, and probably profitable enterprise but the building of the silk mill and their inability to lease it had drained the Pointon's reserves. It is probable that the whole of the site was purchased by Lawrence Armitstead who was the current owner of the Hermitage Estate. The corn mill at Cranage was commercially attractive and soon had a new miller called John Blease. However the silk mill had never been a paying proposition and never reopened after the sale of 1834.[34]

Although John Blease's milling business thrived at Cranage, he was unfortunate to lose his first wife just after 1841, but he quickly remarried. In 1850 Lawrence Armitstead sold the Cranage Mill to Benjamin Llewelyn Vawdrey of Kinderton for £4000.[35] At this time John Blease was living at the mill with his second wife and five children, employing two millers to assist in the mill, one of which was his brother William and the other was a young journeyman having just finished his apprenticeship. John and William Blease remained at Cranage Mill until the mid 1870s, sometimes assisted by another brother, Henry Blease.[36]

By 1878 the miller at Cranage was Edward Massey who, together with his brothers, founded a business that still operates at Cranage to this day. Initially they supplemented their income by being involved in some non-milling endeavours. In 1888 they were described as *"corn millers and brick makers"*[37] and then from 1892 to 1914 as *"corn merchants and cheese factors"*.[38] At some stage the waterwheels were replaced by a 90 h.p. turbine which generated electricity for use in the mill. The type of turbine is unknown but the paddle gear is inscribed with the name "Amos Crosta Mills & Co, Heywood". During the Masseys' tenure at Cranage many changes have been made and the business gradually changed into one of modern provender millers producing animal feed which is successfully supplying the agricultural industry in the 21st century.

FIGURE 159. THE WEIR BUILT BY THE POINTONS AT CRANAGE MILL CAN STILL BE SEEN TODAY.

References

1. Awtry, B. G., "Charcoal Ironmasters of Cheshire & Lancashire, 1660-1785", *Transactions of the Lancashire & Cheshire Antiquarian Society*, Vol. 109, 1957.
2. Awtry, B. G., 1957, *op. cit.* 1.
3. Capewell, a, et al., *A Journey through Time*, Intec Publishing Ltd., 1996.
4. Awtry, B. G., 1957, *op. cit.* 1.
5. *Ibid.*
6. Congleton Library, Church Hulme Parish Register.
7. Awtry, B. G., 1957, *op. cit.* 1.
8. PRO, c 9.372/11 as quoted in Awtry, B. G., 1957, *op. cit.* 1.
9. Johnson, B. L. C., "The Foley Partnerships: The Iron Industry at the end of the Charcoal Era", *The Economic History Review*, Second Series, 4, 1952.
10. Earwaker, J. P., *East Cheshire*, 1887/8.
11. Johnson, B. L. C., "The Iron Industry of Cheshire & North Staffordshire: 1688-1712", *Transactions of the North Staffordshire Field Club*, Vol. 88, 1954.
12. Awtry, B. G., 1957, *op. cit.* 1.
13. Johnson, B. L. C., 1952, *op. cit.* 9
14. Awtry, B. G., 1957, *op. cit.* 1.
15. Johnson, B. L. C., 1952, *op. cit.* 9
16. CRO, Cranage Forge Accounts, DTO/3/1-4.
17. *Ibid.*
18. Awtry, B. G., 1957, *op. cit.* 1.
19. CRO, Probert map of Holmes Chaple, CRO DDX329.
20. *London Gazette*, 7th May 1776.
21. *London Gazette*, 3rd September 1776.
22. CRO, Land Tax Returns, Cranage.
23. Royal Exchange Fire Insurance policy 128008, 1792 14th May.
24. CRO, Land Tax Returns, Cranage.
25. CRO, Land Tax Returns, Cranage.
26. *Macclesfield Courier*, 27th July 1822.
27. *Macclesfield Courier*, 14th June 1823; 26th March 1824; 15th May 1824;17th July 1824; 28th August 1824.
28. *Macclesfield Courier*, 9th April 1825.
29. *Macclesfield Courier*, 27th February 1830.
30. *Macclesfield Courier*, 10th April 1830.
31. *Macclesfield Courier*, 7th August 1830.
32. *London Gazette*, 25th May 1832.
33. *Macclesfield Courier*, 23rd Feb 1833; 20th July 1833.
34. CRO, Cranage Tithe Map & Apportionments, 1840.
35. Op. cit. 3.
36. CRO, Census Returns, Cranage, 1851; 1861; 1871.
37. Slater's Directory of Cheshire, 1888.
38. Kelly's Directories of Cheshire, 1892/1914.

55. COTTON CORN MILL   SJ 748677

The site of Cotton Corn Mill is about half a mile downstream of the bridge over the River Dane in Cranage. Access to the mill in the days of the Manor of Cotton was along Mill Lane which ran northwards to the river, past Cotton Hall. When Cotton Mill was built is not known but was probably in medieval times. The earliest information concerning Cotton Mill so far discovered is contained in entries relating to the mill occuring in the accounts of William Hall's land agent at Cranage Forge (*q.v.*) around 1730.

These accounts provide an insight on the mill and its operation in the early 18th century. The mill appears to have been constructed of wood with a thatched roof. It had a single undershot waterwheel, fed by a low weir made of stones, moss, etc. The mill seems to have had only one set of millstones driven from the waterwheel by cog and rung gearing. The millstones used in the mill were purchased from Mow Cop, an entry for 1731 reading

"Mr. Antrobus for a millstone for Cotton Mill....... £4-7s-6d"[1]
Mr. Antrobus is known to have been a supplier of millstones to Congleton Mill (*q.v.*) and is also one of the Mow Cop quarry leaseholders around this time.[2] The shafts in the mill used gudgeons with brass bearings. The meal from the stones was possibly sieved through a boulter, as two sieve cottons were bought during 1731, although it is possible that these were for hand sieves. The system of toll milling was clearly the way that the mill operated in the 1730s. The mill was probably rented by Samuel Moss in 1730 as expenses were claimed for showing him around the mill and letting the mill. However, in 1733 a Thomas Williamson was paying £20 per year rent for the mill.[3]

The ad hoc nature of the mill weir was a cause of considerable labour at Cotton Mill as seasonal floods and storms washed some of the structure away. In 1730 the waterworks at Cotton Mill were receiving attention at a variety of times between the beginning of November and the end of January. This appeared to be an on-going task nearly every year. The custom of giving ale to the men working in the water, as was usual at Congleton Corn Mill, was also practised at Cotton Mill, as shown by the entry in the accounts for 6th July and 5th August, 1734

"Ale and stopping the Dane at Cotton Mill .... 5s-0d"
"Ale at Cotton Mill, men working in the water .... 5s-0d"
The last entries for Cotton Mill in the accounts occur in March 1736 when again work is being done on the waterworks at the mill.[4]

The only evidence of the millers at Cotton Mill, apart from Samuel Moss who is mentioned in the accounts, refers to millers resident in nearby Church Hulme. These were Hugh James in 1660,[5] and William Wharton and John Hall who died in 1732 and 1740 respectively, when their wills described them as millers of Church Hulme.[6] It is likely that these millers all worked at Cotton Mill but there is no direct evidence to support this conjecture.

In the middle of the 18th century Cotton Mill would still have had the appearance of a medieval mill using medieval technology. Its site on the River Dane restricted the mill to using an undershot waterwheel and the lie of the land did not allow for any improvement in this aspect of the mill. At some stage after 1750 corn milling started at Cranage Forge, which was a much more advantageous watermill site with its high weir capable of powering a number of overshot waterwheels. At the same time that Cranage Forge started milling, Cotton Mill went out of use. As both sites belonged to the same estate it is more than likely that Cotton Mill was not considered capable of being improved to produce a greater output and so was closed in order to concentrate corn milling at the better site at the forge. Certainly when Probert's large scale map of Holmes Chapel (i.e. Church Hulme) was published in 1767 the only indication at the location of Cotton Mill was a legend reading "*Here stood Cotton Mill*".[7] The position of the mill was still remembered nearly a hundred years later when the tithe map produced in the 1840s also indicated the "*banks where the mill stood*".[8] One of the millers at Cotton Mill in the last years of its existence was possibly Joseph Yardley whose will, proved in 1768, described him as a miller living in Cotton.[9]

The passage of time since the mill went out of use in the middle of the 18th century has erased all trace that a mill ever existed on this site.

References

1. CRO, Cranage Forge Accounts, DTO 11/4.
2. Congleton Corporation Accounts, held by Congleton Town Council.
3. CRO, Cranage Forge Accounts, DTO 11/4.
4. *Ibid.*
5. Lawton, G. O., ed., *Northwich Hundred Poll Tax 1660 and Hearth Tax 1664*, Record Society of Lancashire & Cheshire, Vol. 119, 1979.
6. CRO, Wills of William Wharton, 1732; and John Hall, 1740.
7. CRO, Probert's Map of Holmes Chapel, 1767, DDX329.
8. CRO, Cranage Tithe map, 1840.
9. CRO, Will of Joseph Yardley, 1768.

56. BRERETON HALL   SJ 780651

In the early years of the 20th century the owners of Brereton Hall installed a small waterwheel inside a brick shed situated at the downstream end of the pool in front of the Hall. At this point there was a small waterfall just before the River Croco entered the millpool for Park Mill (*q.v.*).

This waterwheel drove a d.c. generator to provide electricity for use in the Hall.[1] The electricity was conducted by an overhead cable to the Hall where it was used to charge batteries that supplied a primitive lighting system.[2]

The brick shed was demolished a few years ago and the waterwheel was removed. No remains of this installation can now be seen.

References

1. *Congleton Chronicle*, 2nd August 1985.
2. *Congleton Chronicle*, 24th December 1986.

---

57. PARK MILL, BRERETON   SJ 774659

The whole of the River Croco lies on the Cheshire plain consequently there is very little fall over the length of the river. This fact, coupled with only a moderate flow, caused mill owners to make special arrangements to ensure adequate water supplies at the mills located on the river. About one mile south east of Holmes Chapel, just north of Brereton Hall, the river is dammed to form a pool which provided the water supply to drive a corn mill called Park Mill. This dam, which has a height of approximately twenty feet is quite substantial and considerably larger than most mill dams in the area. The extremely large millpool created by this dam was almost half a mile long, spreading back up the river nearly as far as Brereton Hall.

Park Mill is the only mill site in the area covered by this survey to be mentioned in the Domesday Book. Prior to 1086, Brereton had been held by Wulfgeat but after the Conquest it was held by Gilbert Hunter at a total value of 20s, part of which was made up by the mill, which was listed as being worth 12d.[1] Although there is no indication that the mills enumerated in the Domesday Book were watermills, the village of Brereton situated on the River Croco is ideally placed to support a watermill. The value of 12d assigned to the mill at Brereton is the lowest in Cheshire which probably reflects the small population living in the area at that time. There has been much discussion as to the nature of the mills recorded in the Domesday Book which so far has been inconclusive. In all probability the mill at Brereton would have had a single pair of millstones but the type of waterwheel used at that time could have been either vertical or horizontal.

There are further references to a mill at Brereton in the 15th century and the name *"Parke Millne"* appears in 1619.[2] In the early part of the 18th century Park Mill appears to have been owned or leased by the Hall family who lived at the Hermitage just north of Holmes Chapel. The Halls were partners in the Cheshire iron works and were especially involved with Cranage Forge. Entries occur in the Cranage Forge (*q.v.*) accounts that show payments for such items as land tax, poor law tax, and church leys for Park Mill as well as purchases of boards, thatching and a pump for the mill. This type of expenditure is commensurate with the Halls acting as landlords for the mill.[3]

At the end of the 18th century, from 1784 up until 1821, the owner of the Park Mill was H. Legge with the occupier of the mill listed until 1813 as Abraham Bracebridge but it is not known if he was the miller as well as the tenant.[4] Certainly, Joseph Peak in 1788, Samuel Brindley in 1795 and Thomas Buckley in 1796 are all recorded as millers in the parish register. Samuel Brindley was the younger brother of James Brindley, the famous canal engineer, and of Henry Brindley the owner of Danebridge Mill (*q.v.*). Between at least 1814 and 1817 George Manby was recorded as being a miller living in Brereton.[5] Although the Brereton Hall estate was offered for sale in 1817 the sale advertisement contained no mention of Park Mill, only mentioning *"two fine sheets of water"*, presumably referring to Bagmere (long since drained and reclaimed) and the lake in front of the hall.[6] In 1819 William and Anne Lea came to Park Mill from Swettenham Mill (*q.v.*), staying until around 1825 when they moved to Peover Mill a few miles to the north.[7] After 1821 the owner of the mill was Dr. Stephen Lushington until 1830 when John Howard, a cotton manufacturer with textile mills in Hyde, bought the Brereton Hall estate when by all accounts the mill was in a very poor state of repair.[8]

It is possible that John Howard paid a low price for the Brereton estate due to most of the farms not being profitable because of poor land practices and poor management. Certainly, shortly after purchasing the estate, John Howard started to spend large amounts of his fortune made in the cotton spinning industry on improving his estate through drainage schemes and the liberal addition of marl. One of the improvements he felt necessary was to rebuild Park Mill in line with the latest thinking on mill technology. Being a well respected figure in the cotton industry, with money no object, he approached William Fairbairn of Manchester, the leading millwright of the day, to design a new mill for Brereton based on the improvements that Fairbairn and others had been introducing in the early 19th century. Around this time William Fairbairn was at the height of his influence as a millwright and no doubt John Howard would have been well aware of Fairbairn's skills from his involvement in the cotton industry.

A set of plans dated 30th July 1833 and signed by William Fairbairn demonstrates his response to this challenge.[9] The plans show a four storey building which had a passing resemblance to a small industrial factory (see Figure 160). The drawings show that Fairbairn proposed a suspension type of waterwheel for the mill of 19 feet diameter and 6 feet width. This style of waterwheel was invented in the first decade of the century by Thomas Hewes, who later was to employ William Fairbairn in his works. After Fairbairn set up in business on his own account he specialised in this type of waterwheel, perfecting its design, and being responsible for the erection of many of the famous industrial waterwheels of the time. The design shows the waterwheel was to have ring gear attached to the waterwheel shrouds which meshed with a pinion mounted on a short shaft. A large spur gear mounted on this short shaft drove, via another pinion, the main lineshaft which ran almost the length of the building on the ground floor. Sets of bevel gears were positioned on this lineshaft to drive four pairs of millstones positioned in a row along the front wall of the mill on the first floor. These were intended to have conventional furniture including a damsel and shoe feed mechanism. No doubt in line with John Howard's intention to have only the best possible state of the art machinery the plans also show that there was to be a steam powered beam engine, together with its boiler, capable of driving the mill from the opposite end of the main lineshaft to the waterwheel. There are also two smaller drawings in the same style showing the proposed 14 feet square kiln, but these drawings are not signed.

Whether John Howard did not like the price quoted or had some aesthetic objection, he did not proceed with the building as specified by William Fairbairn but appears to have employed a local builder, in 1835, to construct a three storey mill of less grandiose proportions than the Fairbairn design. According to the working drawings produced when installing the machinery the overall design of the machinery was not

FIGURE 160. A VERTICAL SECTION THROUGH THE FOUR STOREY MILL PROPOSED BY WILLIAM FAIRBAIRN IN 1833. (*Mr. & Mrs. Wood*)

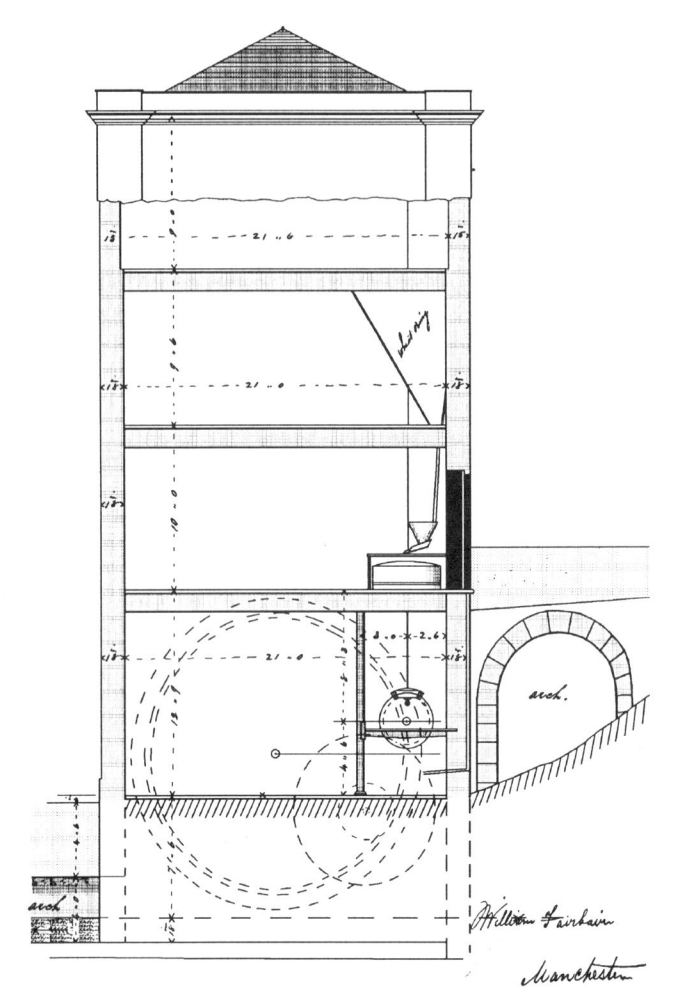

quite as that specified by Fairbairn in 1833. The waterwheel was shown as now being 18 feet in diameter with a ring gear of 16½ feet diameter and there were also minor dimensional changes, particularly with respect to the position of the bevel gears on the main shaft, no doubt caused by the exigencies of fitting the machinery in practice. In spite of these variations, the overwhelming similarity of the installed machinery to the 1833 drawings shows that the mill machinery was indeed purchased from William Fairbairn's engineering company.

Another drawing, produced by a Frank Murch, shows the main milling machinery in elevation. On this drawing the flour millstones are specified as being

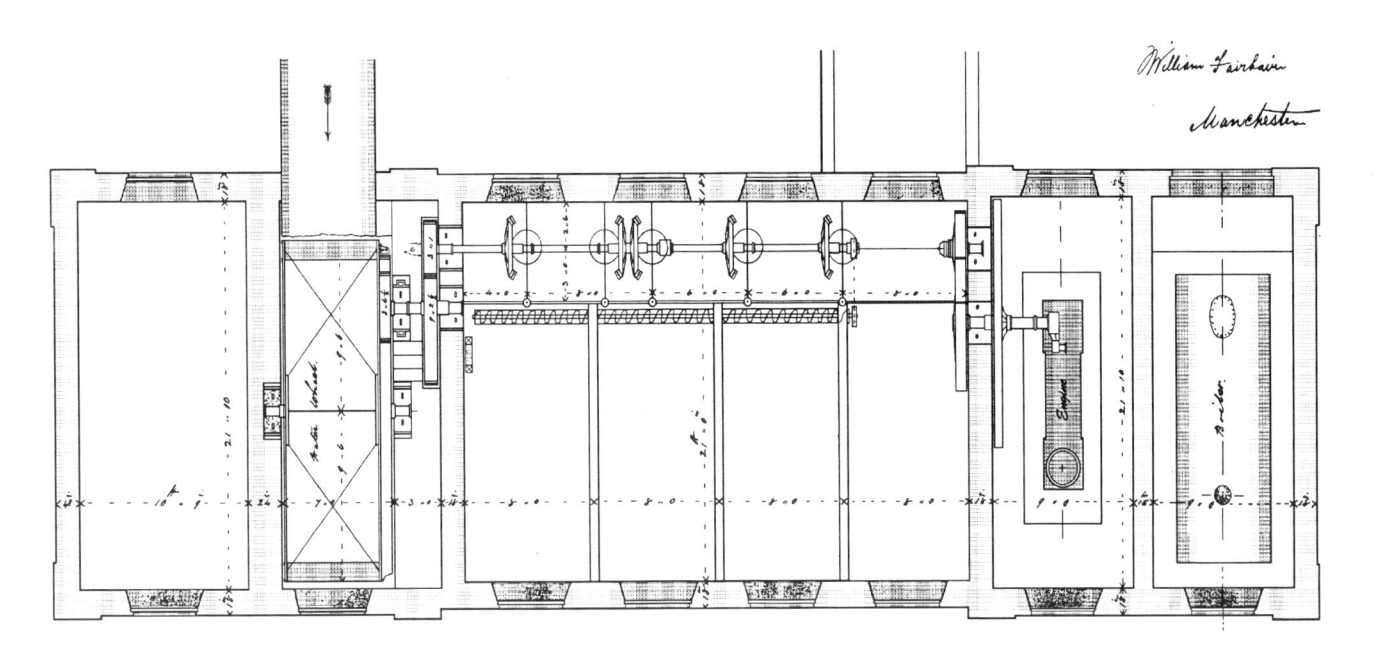

FIGURE 161. A PLAN OF THE PROPOSED PARK MILL DRAWN BY WILLIAM FAIRBAIRN IN 1833. (*Mr. & Mrs. Wood*)

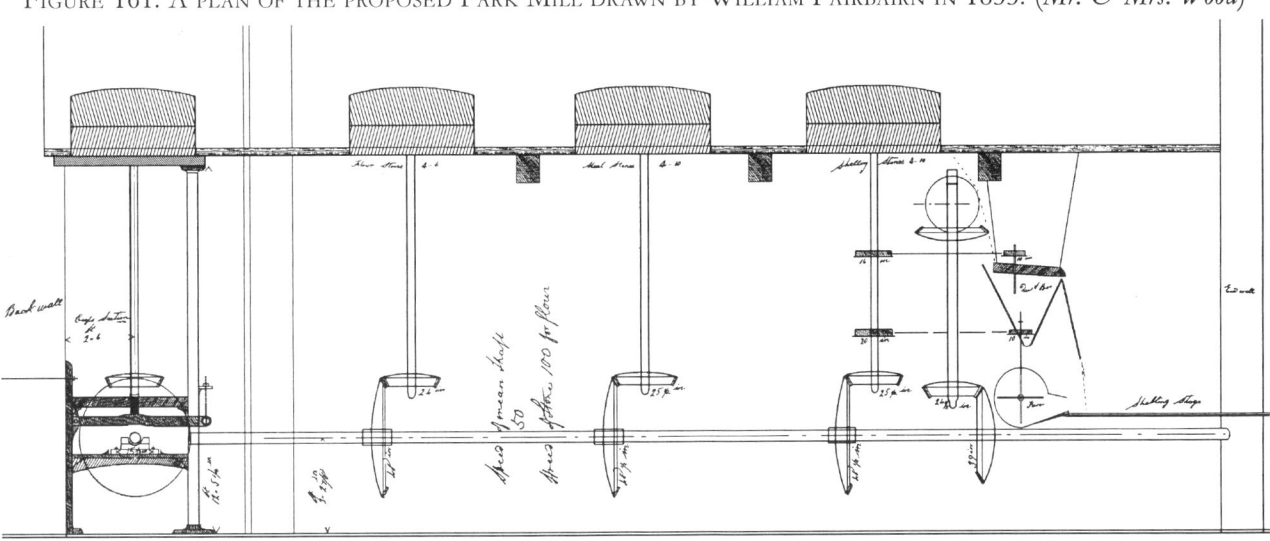

FIGURE 162. A DRAWING BY FRANK MURCH, 1835, SHOWING THE MAIN MACHINERY ON THE GROUND FLOOR AND THE FOUR PAIRS OF MILLSTONES ON THE MIDDLE, OR STONE FLOOR, AT PARK MILL. NOTE THE INSTALLATION OF A SHELLING STAGE (*at the right*) FOR THE PRODUCTION OF OATMEAL AND THE INTENDED SPEED OF THE MAIN SHAFT AND THE MILLSTONES, 50 R.P.M. AND 100 R.P.M. RESPECTIVELY. (*Mr. & Mrs. Wood*)

4 feet 6 inches diameter whereas the meal stones and shelling stones are 4 feet 10 inches diameter. A note on the drawing states that the "*speed of the main shaft 50, speed of the stones 100 for flour*". The tentering shown is by bridgetrees operated by vertical adjustment screws. This drawing also shows equipment for producing oatmeal consisting of a shaking sieve feeding into a fan driven winnower and a shelling stage.

One of the working drawings from 1835 shows that the building was intended to house not only a corn mill but also a bone mill. The bone mill was to be accommodated on the other side of the waterwheel from the corn grinding equipment, with the kiln at the opposite end of the building, probably located beyond the steam engine. No doubt the bone mill was intended to provide fertiliser to improve the soil of the farms on the estate. No drawing has survived showing the bone mill machinery so perhaps it was never fitted but documents indicate that John Howard used prodigious amounts of the material on his farms during the 1840s.[10]

Having built his new mill, John Howard then leased it to Thomas and Henry Hall who occupied the mill until 1850[11] when their partnership as millers and corn merchants at Brereton Mills was dissolved.[12] The fact that they were corn merchants throws light on the type of business they were operating. In the past it is probable that Park Mill operated on a toll basis with

FIGURE 163. THE WATERWHEEL AT PARK MILL WITH ITS CRUCIFORM, FISH-BELLIED, SHAFT IS TYPICAL OF WILLIAM FAIRBAIRN'S DESIGN OF SUSPENSION WHEELS. THE OUTLINE OF THE CURVED BUCKETS CAN JUST BE SEEN ON THE INSIDE OF THE WATERWHEEL RIMS.

FIGURE 164. THE MAIN HORIZONTAL LINESHAFT WITH A PAIR OF LARGE BEVEL GEARS TO DRIVE ONE OF THE MILLSTONES AND A SMALL PAIR OF BEVEL GEARS DRIVING A VERTICAL SHAFT TO THE TOP FLOOR. NOTE THE PIVOTED LEVER FOR DISENGAGING THE STONE NUTS.

the miller taking a percentage of the total output as his payment. It would seem that the building of the new mill in 1835 introduced the merchant system whereby the miller would buy grain to mill, selling the resulting produce on the open market.

In 1851 the miller at Park Mill was Joseph Stockton, but he relinquished the lease in the following year and was replaced by John and Catherine Lea from Swettenham Mill.[13] This was a continuation of a long association between the Lea family and Park Mill which had started with William and Ann Lea earlier in the century. From 1855 to 1867 the miller was James Lea who employed four men at the mill and was able to support two spinster daughters and provide accommodation for two sons, James and Henry, even though Henry was married with four children.[14] John Lea took over and operated the mill from 1867 until his death at a relatively early age of 40 in 1882, but his wife Mary continued the business in her own name with the help of her miller son, Herbert, until about the end of the century.[15] Around 1890 a new steam engine was purchased from Fodens Ltd. of Sandbach which probably replaced the beam engine shown on the Fairbairn drawings of 1833. The new engine and its Lancashire boiler were installed alongside the waterwheel in the area originally designated as the bone mill. This single cylinder, horizontal, engine could be connected to the main gearing via a dog clutch such that it could supplement the power of the waterwheel.[16]

At the start of the 20th century, in 1902, Herbert James Lea took over the tenancy at Park Mill at the age of thirty. He continued to operate the mill until 1918 when he went to live at Wheelock Mill (*q.v.*) which he had purchased in 1911. The Lea family continued in the milling business establishing Morning Foods Ltd in 1941 to make "Mornflake" oats, a breakfast cereal still in production today. Another business set up by the Lea family still exists today as Lea-Oakes (Millers) Ltd.[17]

After the Lea family left Park Mill the tenancy was taken over by Henry Snelson until 1939 when the mill ceased production.[18] After the Second World War the mill was used as an agricultural store until it was sold around 1960. At some stage a private residence was built at the western end of the mill on the area once

FIGURE 165. A TYPICAL SCENE AT PARK MILL IN THE EARLY PART OF THE 20TH CENTURY WITH ITS LOADING BAY (*now removed*) RECEIVING GRAIN FROM A HORSE AND CART. (*Lyndon Murgatroyd*)

FIGURE 166. THE REAR VIEW OF PARK MILL SHOWING ALL THREE STOREYS. NOTE THE CAST IRON WINDOW FRAMES IN THE TOP FLOOR.

possibly occupied by the kiln and original steam engine.

Today the mill is much as it was built in 1835, a three storey brick building built in English garden wall bond with three rows of stretchers between each row of headers with a tile roof. At the front of the mill there is a large porch sheltering the main entrance onto the stone floor and there is a small brick chimney built into the south-east corner of the mill (see Figures 165 and 175). The very large mill pool has been lowered by the water authority and is now silted up and overgrown. The headrace under the lane, which entered the mill under a low brick arch next to the porch, is now blocked off with a concrete plug. From the lane the mill appears to have only two storeys as it is not until the rear of the mill is in view that the ground floor is visible. One attractive feature of the mill is the cast iron window frames on the top floor, all of which have survived.

The ground floor contains the waterwheel, steam engine and main power transmission and gearing. The waterwheel was fed from an iron pentrough via a sluice operated by a large handwheel operating a worm gear and pinion. The 18 feet diameter by 6 feet wide Fairbairn waterwheel, with its fish bellied cruciform shaft, arms and cross braces, has now lost all of its 48 curved buckets due to corrosion (see Figure 163). The ring gear which was originally attached to the wheel has been removed at some stage, probably when the Foden steam engine was installed. However, there is a series of filled-in holes in the waterwheel shroud plates where the ring gear might have been attached. Consequently the pinion that used to mesh with the ring gear now makes contact with a large spur pitwheel on the wheelshaft. Apart from this change the main gearing is as specified by Fairbairn in 1833. However, the far end of the main shaft is perched in a rough

FIGURE 167. THE TENTERING ARRANGEMENT. THERE IS A WORM GEAR BEHIND THE CAST IRON PILLAR.

FIGURE 168. A PAIR OF MILLSTONES IN A WOODEN, OCTAGONAL CASING AT PARK MILL.

FIGURE 169. THE OAT ROLLER AND ITS ASSOCIATED ELEVATOR AND BIN ON THE STONE FLOOR.

FIGURE 170. THE GRAIN CLEANER AND OAT CLIPPER, PATENTED AND SUPPLIED BY BOOTHS OF CONGLETON.

GAZETEER OF MILLS

FIGURE 171. THE PAIR OF MILLSTONES IN A CIRCULAR METAL TUN AT PARK MILL. THIS WAS PROBABLY PART OF THE ORIGINAL MILL FURNISHING SUPPLIED BY WILLIAM FAIRBAIRN. NOTE THE SMALL CONICAL HOPPER AND THE DAMSEL FEED ARRANGEMENT.

FIGURE 172. ONE OF THE CAST IRON STONE PANS SUPPORTING THE MILLSTONES.

FIGURE 173. THE SACK HOIST ON THE GARNER FLOOR. THERE WOULD HAVE BEEN A SLACK ROPE DRIVE AROUND THE SIX SPOKED WHEEL WHICH WAS DRIVEN FROM A PULLEY ON THE MAIN LINESHAFT ON THE BOTTOM FLOOR. THIS WAS TIGHTENED WHEN REQUIRED BY THE LEVER MOVING THE AXLE BEARING SLIGHTLY UPWARDS.

FIGURE 174. THE FODEN HORIZONTAL STEAM ENGINE INSTALLED C.1890. RECENTLY THE GOVERNOR, NAME PLATE AND SOME BRASS FITTINGS HAVE "GONE MISSING".

recess in the end wall brickwork rather than in a bearing. This would indicate that originally the shaft was longer than at present suggesting that the beam engine specified by Fairbairn was indeed originally installed at that end of the mill. The replacement Foden steam engine was in good condition in 1998 (see Figure 174) but has since lost its governor, nameplate and some brass parts, such as oilers.

Over the years the addition of extra equipment and processes to the mill has caused the power distribution system to be extended. On the ground floor, at ceiling height, there is a lineshaft with pulleys, running across the mill, that powered machinery on the upper floors via belts. There is also a vertical shaft from the main horizontal shaft taking power up to the top floor. In the engine room there is another lineshaft running at ceiling height that was driven by a belt around the engine's flywheel. A fast and loose pulley at the end of this shaft shows that it was used to drive machinery housed in the engine room.

There is an oat roller and associated bin on the stone floor with an elevator rising up into the roof space on the top floor (see Figure 169). There are also two jiggers, or dust sieves, one on the stone floor and the other on the ground floor, which were also used in processing oats, but the associated system of worms and chutes is now far from complete. Also on the stone floor is a patent grain cleaner and oat clipper made by Booths, the Congleton millwrights, with a silo above (see Figure 170). These items of machinery were all driven by belts from the ground floor.

There are two pair of French burr and two pair of Peak millstones which, having diameters of 5 feet 3 inches and 4 feet 10 inches respectively, are slightly larger than those specified in 1835. All millstones are tentered using handwheels to adjust the bridgetrees (see Figure 167) and there are also levers attached to the stone nut bevels to lift them out of gear when required. The set of millstone furniture furthest from the waterwheel consists of a metal circular tun, surmounted with a simple metal support for a small conical metal hopper (see Figure 171). The two middle pairs of millstones have octagonal wooden tuns but carry the same metal furniture as the futhest pair (see Figure 168). This would indicate that the wooden tuns are replacements for the original metal ones, which may have been installed by Fairbairn. All four pairs of millstones used damsels to vibrate the feed shoes on the hoppers. The bedstones are positioned on metal stonetrays supported on the outside by cast iron columns but fixed into the brickwork on the inside (see Figure 172).

On the top floor is the sack hoist which was driven by a slack rope drive from the main lineshaft on the ground floor (see Figure 173). It operated two hoist chains, one of which would have been used to lift sacks through a doorway from carts on the roadway outside the mill. This doorway once had a lucam to house the chain pulley (see Figure 165). The other chain would have been used to raise sacks, etc. through the mill from the bottom floor. Power delivered via the vertical shaft from the ground floor was used to operate a system of worms and chutes which have mainly been removed from the mill.[19]

This is a most complete corn mill with an important provenance, being the only watermill (complete or in part) still in existence with machinery supplied by William Fairbairn, the foremost millwright and engineer produced by this country in the 19th century. Its importance has now been recognised by being designated as a listed grade 2* building by English Heritage.

FIGURE 175. A VIEW OF PARK MILL IN THE FIRST HALF OF THE 20TH CENTURY WHEN THE POOL WAS AT ITS FULL HEIGHT. (CCALS)

References

1. Morgan, P., ed., *Domesday Book, Vol. 26, Cheshire*, Phillimore, 1978.
2. Dodgson, J. McN., *The Place Names of Cheshire, Part 2*, Cambridge University Press, 1970.
3. CRO, Cranage Forge Accounts, DTO/3/1-4.
4. CRO, land Tax Returns, Brereton.
5. Congleton Library, Brereton Parish Records.
6. *The Times*, 18th September 1817.
7. Lea, Herbert, *Swettenham Mill*, 1984.
8. Private correspondence held at Brereton Hall.
9. Plans of Park Mill, 1833, held at Brereton Hall.
10. Accounts and letters held at Brereton Hall.
11. CRO, Tithe Apportionments, Brereton Parish; Bagshaw's Directory of Cheshire, 1850.
12. *London Gazette*, 4th October 1850.
13. CRO, Census Return for Brereton ,1851; Brereton Parish Register.
14. CRO, Census Return for Brereton, 1861.
15. CRO, Census Returns for Brereton, 1871 &1891; Slater's Directories of Cheshire, 1883, 1888, & 1890.
16. Norris, J. H., "The Water-Powered Corn Mills of Cheshire", *Transactions of the Lancashire & Cheshire Antiquarian Society*, Vol. 75, 1965.
17. Lea, Herbert, *Swettenham Mill*, 1984
18. Kelly's Directories of Cheshire, 1934 & 1939.
19. Visit to the mill, February 2003.

---

58. NEWTON SALTWORKS   SJ 706661

Newton Saltworks were located about 100 yards south-east of the church of St. Michael and All Angels in Middlewich. Brine had been used to produce salt in Middlewich from as early as Roman times. Being alongside the River Croco it is perhaps inevitable that water power would have been used at some stage to pump the brine at Newton Saltworks. Unfortunately, the only evidence of the use of waterpower occurred in a sale advertisement for the saltworks in the Macclesfield Courier of 27th November 1819 as follows:-

"To be sold at Auction. One undivided third of a part of a share of and in all those valuable freehold saltworks situated for the home trade at Newton near Middlewich adjoining the navigable canal from the Trent to the Mersey now in the holding of Mr John Thomas Brabant consisting of a large range of substantial brick buildings, four pans with newly erected steam engine of 10 horse power, water engines and reservoirs. There is a labourer's cottage, blacksmith's and carpenter's shops and a cheese warehouse."

It is to be assumed that the "water engines" mentioned were in fact waterwheel powered pumps. There are no further mention of the use of waterwheels at Newton Saltworks so presumably they were soon replaced by the use of steam engines. Today Newton Saltworks have been demolished and the area landscaped leaving no trace above ground of this former extensive establishment.

59. KINDERTON CORN MILL   SJ 705664

Kinderton Corn Mill is situated close to the eastern bank of the River Croco on the northern side of the A54 Holmes Chapel to Middlewich road just before the bridge over the Trent & Mersey Canal (see Figure 180). The strategy adopted at Kinderton Mill to overcome the very gradual fall of the River Croco was to use a very long leat. The mill's water supply started about 1¼ miles eastwards of the mill at a small weir on the river which fed almost the whole of the river into a leat. This leat then ran across country, deviating significantly from the course of the river, and eventually fed into a small mill pond adjoining the mill on its eastern side. When the railway came to Middlewich in the 19th century an inverted siphon had to be built to allow the leat to pass under the railway.

The earliest references found mentioning a mill at Kinderton occurred in the 14th century. On 26th March 1330 Hugh de Venables, Lord of Kinderton, granted a messuage in Kinderton to Alexander de Crewe whereby one of the conditions of the grant was that Alexander de Crewe paid suit at Kinderton Mill i.e. he agreed that all the grain from this holding would be ground at Kinderton Mill. Also early in the 14th century there is a conveyance of property from William, son of Hugh of Middlewich, to Robert de Buckley, Felicia his wife and Richard their son, of a plot of land in which the road to Kinderton Mill is mentioned. In a document of the 4th October 1342 Laurence Winnington granted the rent from a piece of land to Hugh de Venables and Katherine his wife. This piece of land had previously belonged to Thomas Sutton who was described as *"of the mill in Kinderton"*.[1]

After the 14th Century little is heard of Kinderton Mill although there is a datestone on the ground floor of the mill just near its main entrance inscribed with the date 1609 which may signify a rebuilding of the mill at that date. However, only ten years later, in 1619, a report of the Constables & Supervisors of Bridges in the Northwich Hundred stated *"Kinderton - one horse bridge over the River Dane at Kinderton built for use of the mill there now in decay."*[2] Later, in the first half of the 18th century, the identity of some of the millers at Kinderton are known from their wills; these are Roger Wood who died in 1707, Rowland Bealey in 1720, and George Waverton whose will was proved in 1745.[3]

In the second half of the 18th century there a few references to Kinderton Mill in the minutes of Kinderton Vestry which in 1745 itemises

"To repairing the mill pools head twice .... 5s-0d"

and then in 1749

"Sinders and spreading them in ye mill lane .... 2s-6d".[4]

The ownership of the mill towards the end of the 18th century and the first thirty years of the 19th century resided with the estate of Lord Vernon and probably had belonged to this estate for many years. From 1782 to 1787 the mill was leased by James Lingum, but whether he was the miller is not known. The tenancy of the mill was taken over by Joseph Lowe in 1788 which he held until 1814 when he was succeeded by William Lowe until 1825.[5]

It was during the tenancy of Joseph Lowe that a tragic accident occurred at Kinderton Mill just before Christmas in 1801. The local newspaper report of the incident is as follows:-

"On Wednesday 18th instant eighteen boys and girls belonging to the cotton mill of Messrs. Henry & John Clarke of Middlewich went upon the ice of the mill pool in Kinderton which suddenly breaking nine of them fell in, seven of whom in a little time were got out, one girl remained in some time longer having stuck in the mud at the bottom, but was drawn out much bruised and cut by the ice, another was not got out until nearly an hour after when she was immediately carried to a house in the neighbourhood and every means used for her recovery by Mr. Taylor and Mr. W. Beckett but ineffectually."[6]

Personal misfortunes seems to have dogged those

FIGURE 176. DATE STONE JUST INSIDE THE DOOR OF KINDERTON MILL.

associated with Kinderton Mill in the early years of the 19th century. In 1804 Ralph Lowe, probably a relative of Joseph Lowe the leaseholder of the mill, was the miller at Kinderton when he was declared bankrupt.[7] This was followed in 1813 when Peter Thorley, who had been a miller in Middlewich, was imprisoned for debt in the Fleet Prison in London although his debts were concerned with his activities as a carrier's agent after he had ceased to be a miller.[8] Joseph Dale, who was described as a maltster and miller in Middlewich, had some bad luck in 1815 when he lost money to a William Gleave, a publican, who was also imprisoned for debt in the castle at Chester.[9] Later, in 1822, the millers at Kinderton were listed as James Dale & Joseph Dale.[10]

At the end of William Lowe's tenancy in 1825 the mill was advertised to let as follows:-

"To be let and entered into upon 25th March 1826, Middlewich Mills consisting of two pairs of French stones, three pairs of grey stones, with two machines and every convenience for carrying on the business either on a large or small scale, as well as two Cheshire acres of land. The mills are situated within 60 yards of the canal from the Potteries to Manchester and Liverpool.

For further details apply to Mr Joseph Dale, at the mill in Kinderton."[11]

As a result of this re-letting Joseph Dale became the direct tenant of Lord Vernon.[12] Although this would have resulted in decreasing the rental paid by Joseph Dale it did not prevent John Dale (Joseph's brother?) who was described as a miller of Kinderton from being declared bankrupt in 1827.[13]

It was in this early part of the 19th century that market conditions encouraged the building of a windmill just to the north of Kinderton Mill to supplement its flour output.[14] It is possible that this might have been the cause of the stream of bankruptcies encountered by the various millers at Kinderton. Possibly the extra output from the windmill could not balance the costs of employing extra workers and the capital tied up in the windmill. The windmill must have been dismantled very soon after its construction as it was never mentioned in any records after the 1830s and was not present on the first Ordnance Survey map which was surveyed in 1839/41.

Although in 1828 the millers at Kinderton were listed as Joseph Goodyear and Thomas Lowe, the tenant of Lord Vernon holding the mill in 1830 was Thomas Arden.[15] Joseph Goodyear (or Goodier as he became) continued at the mill until some time in the 1840s. He was certainly the miller there with his wife Elizabeth and their five children in 1841.[16]

It was in the period between 1835 and 1840 that some calculations were made concerning the financial

FIGURE 177. THE FACADE OF KINDERTON MILL. NOTE THE THREE STONE BUTTRESSES, MULLIONED WINDOWS AND BALL FINIALS AT THE ROOF APEX AND EAVES.

possibilities of Kinderton Mill which give an insight into not only the kind of returns expected of a mill at this time but also some indication of the operation of the mill and the state of its machinery. Of the five pairs of millstones in Kinderton Mill, four pair were considered to be operational for seven months of the year while the other one was only operational for five months per year and then at half time only. This requirement was considered to be attainable using eight to ten horse power. As one horsepower was seen as worth £100 and the return required on the horse power was 8% then the horse power should yield a rent of £64 per annum. The building was constructed *"of brick 18 inches good and strong"* and the slate roof had been recently renewed, so the mill was valued at £425 with £125 deducted for wear and tear. Again a return of 8% was expected from this valuation of £300, yielding a rent of £24 per annum. The machinery was described and valued as follows:-

"9ft-6ins diameter cast iron pit wheel and a 4ft-6ins diameter cast iron crown wheel, both new last year.............. £35-0s-0d
9ft-0ins diameter cast iron spur wheel £4-0s-0d

2 pair of French stones, half done £16-0s-0d
1 dressing machine, worn out .... —
Corn screen, nearly new £24-0s-0d
Sack tackle, good, new £12-0s-0d
5 spindles & nuts at 3£, each £15-0s-0d
Wants a new wheel, cost .... £100-0s-0d"

Although the addition of these figures gives £206, the value of the machinery was actually noted as £263. The machinery was expected to return 10% of its value per year, resulting in a rent of £26 per annum. On top of which the drying kiln and the house were to be rented for £6 per annum. These various estimates of rental values (horse power £64, building £24, machinery £26 and house £6) all added up to a grand total of £120 per annum for the expected mill rent when in sound repair. The recent installation of cast iron gearing and the presence of a nearly new grain cleaner testifies to recent investment in new methods and technology. However, the non-valuation of the other three pairs of millstones (presumably Peak stones) suggested that they needed replacing together with a new waterwheel before a rent of £120 was likely to be achieved.[17]

Although Joseph Goodier continued operating Kinderton Mill well into the 1840s he had been replaced by 1850 by Thomas Walton. However, Thomas Walton was an old man at this time, being 70 years old in 1851, and so it is most likely that the actual milling was undertaken by his sons Thomas, 26, and John, 23 who were both journeyman millers.[18]

Thomas Walton was replaced at Kinderton Mill in 1857 by Samuel Pickering who was already the miller at Sutton Mill (*q.v.*) on the River Wheelock which was about a mile and a half south-east of Kinderton Mill on the other side of Middlewich.[19] Samuel Pickering was no youngster when he came to Kinderton Mill, being 48 years old in 1857, so he employed two journeyman millers and a labourer at the mill.[20] Fourteen years later in 1871 Samuel Pickering was still nominally the miller at Kinderton, sharing the accommodation at the mill with his son Thomas and his family. Thomas worked as a miller as did their lodger Thomas Hulse and room was also found to accommodate a young clerk and his family. After 1871 Thomas Pickering is listed as the miller at Kinderton Mill in his own right and by 1881 he had been joined as a miller by his young son Samuel, aged 18, and an apprenticed miller, aged 17.[21] Thomas Pickering continued to be listed as the miller until 1910 by which time he was 73 years old. Thomas Pickering must have died not long after 1910 because in 1914 his widow Martha Pickering and his son Samuel are listed as the millers at Kinderton Street. Also in both 1910 and 1914 Thomas Rigby & Sons of Frodsham Bridge were listed as millers at Warrington and Kinderton Street.[22] It is not clear what the business relationship was between the Pickerings and Thomas Rigby but the two families were connected by marriage earlier in the 19th century.

After the First World War Frank Hall took over the mill until 1939.[23] After the Second World War milling ceased and the waterwheels were removed in 1963. The mill building has survived to the present day and is now used as part of an automobile repair facility.

The building has four floors and has a floor plan in the form of the letter "T". The arms of the building housed the two waterwheels, one at each side of the building. The mill building is built of brick on a stone base with most of the brickwork laid in English bond with some patches and repairs with no particular bond. There is an additional two storey brick building attached to the north side of the mill which has been executed in English garden wall bond. The mill and its extension now have slate roofs but it is quite likely that they were originally made of stone slabs. The roof of the mill itself is supported by queen post trusses. The brickwork of the front elevation is edged with sandstone quoins and there are three massive stone buttresses supporting the front elevation. The main doorway is to the northern end of the front elevation and has a loading platform, with a timber cover, alongside. The front elevation has three windows, on the first and second floors they are centrally placed with three and four lights respectively and on the ground floor there is one window with three lights offset towards the doorway. These windows are constructed with sandstone lintels, cills, mullions and splayed reveals. There are also sandstone ball finials at the apex of the roof and at the eaves.

Most of the machinery has been removed but the cast iron penstocks can both be seen on the first floor. These are two feet square and their position signifies that the waterwheels were high breastshot in configuration. The top of both wheelpits are about three feet above the level of the first floor. The size of the wheelhouses and a wheel scrape suggest that both of the waterwheels were 18 to 20 feet in diameter and about five feet in width. The floors were originally supported by cast iron pillars, a few of which still survive. On the first floor, there is a short lineshaft, suspended from the ceiling, which can still be seen near the front of the building. On this lineshaft there are three pulleys, ranging from two to three feet in diameter. Directly above this lineshaft, suspended from the second floor ceiling, is the sack hoist mechanism consisting of a pulley, chain drum and control lever. This sack hoist operated with a slack belt from the central pulley on the lineshaft on the first floor and the chain passed from the drum up over a pulley located

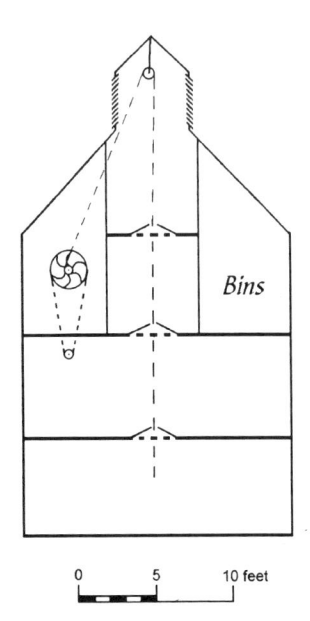

FIGURE 178. A SECTION THROUGH THE FRONT WING OF KINDERTON MILL SHOWING THE POSITION OF THE SACK HOIST WHICH IS THE ONLY MACHINERY TO SURVIVE.

in the lucam in the centre of the roof before passing down through the hatches in the building floors. The second floor was the bin floor which has above it a walkway, or garner floor, running as a raised platform in the shape of a letter "T" in the middle of, but above, the bin floor. Sacks of grain hoisted to this top floor could then be emptied directly into the bins on either side of this raised platform. Although there is a date stone of 1609 just inside the main door and the general appearance of the mill is reminiscent of the 17th century, it is probable that this building dates from much later. The sandstone details of the windows are crisp and sharp, although the stone is relatively soft, and not weathered like some of the stonework in the base of the mill. This lack of weathering is not indicative of a 1609 build date. The size of the mill, which used to hold five pairs of millstones, also suggests a date considerably later than 1609. The most likely explanation is that the current mill was built on the existing base of a previous mill in the style of a much earlier building. No doubt, Lord Vernon and his estate could afford a mill with more architectural pretension than the average country mill of the day.

The site of the windmill is now covered with modern housing but even before this development there was no indication of the site of this mill.

References

1. Varley, J., ed. *A Middlewich Cartulary*, Chetham Society, Vol 1, 1941; Vol 2, 1944.
2. Report of the Constables and Supervisors of Northwich Hundred quoted in Earl, A. L., *Middlewich 900 - 1900*, Ravenscroft Publications, 1990.

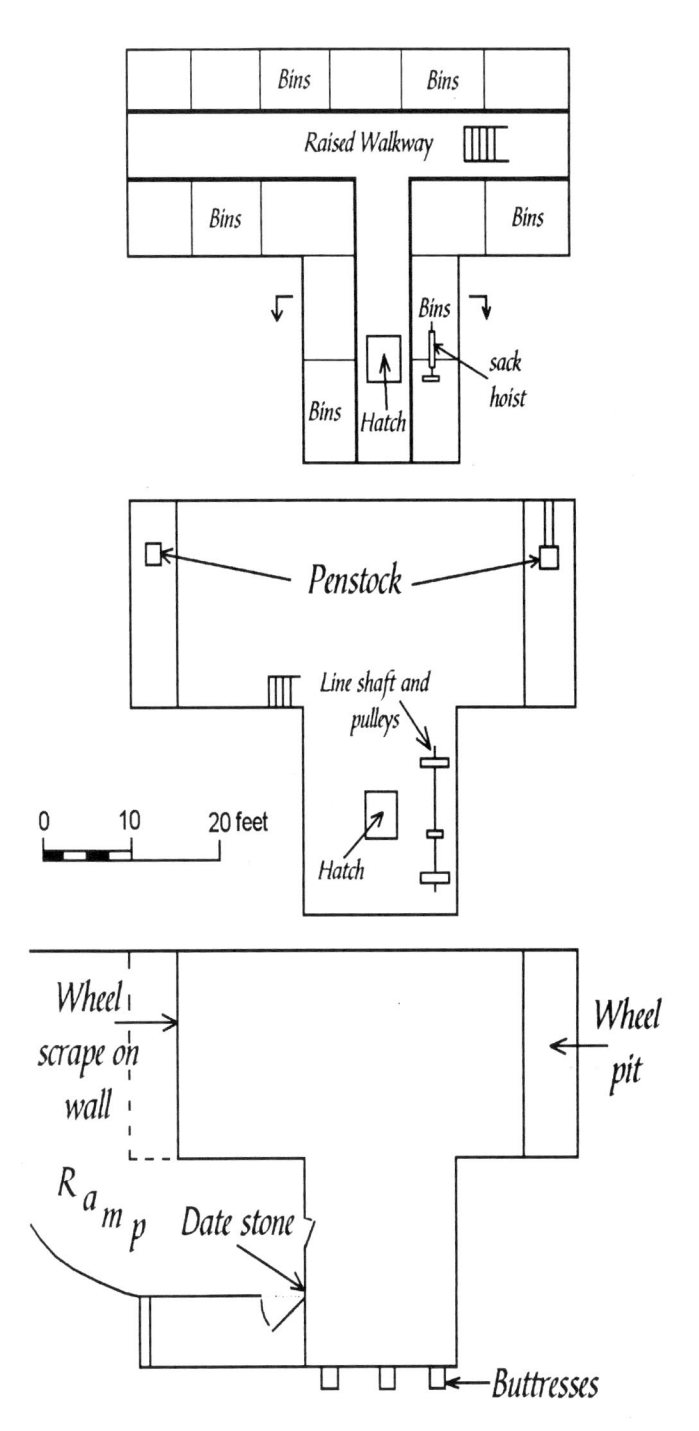

FIGURE 179. FLOOR PLANS OF KINDERTON MILL. GROUND FLOOR (*bottom*), STONE FLOOR (*middle*), AND BIN FLOOR WITH RAISED WALKWAY (*top*).

3. CRO, Wills of Roger Wood, 1707; Rowland Bealey, 1720; George Waverton, 1745.
4. Earl, A. L., *Middlewich 900 - 1900*, Ravenscroft Publications, 1994.
5. CRO, Land Tax Return, Middlewich.
6. The Sun, 22nd December, 1801.
7. *London Gazette*, 31st March 1804.
8. *London Gazette*, 25th December 1813.
9. *London Gazette*, 4th April 1815.
10. Pigot's Directory of Cheshire, 1822.

11. *Macclesfield Courier*, 18th March 1826.
12. CRO, Land Tax Return, Middlewich.
13. *London Gazette*, 8th June 1827.
14. CRO, Bryant's Map of Cheshire, 1831.
15. Pigot's Directory of Cheshire, 1828; CRO, Land Tax Return, Middlewich, 1830.
16. CRO, Census Return, Middlewich, 1841.
17. SRO, Notebook, D 1119.
18. Bagshaw's Directory of Cheshire, 1850; CRO, Census Return, Middlewich, 1851.
19. Kelly's Directory of Cheshire, 1857
20. CRO, Census Returns, Middlewich, 1861.
21. CRO, Census Returns, Middlewich, 1871 & 1881.
22. Kelly's Directory of Cheshire, 1892, 1902, 1906, 1910, 1914.
23. Kelly's Directory of Cheshire, 1934, 1939.

60. KINDERTON SALTWORKS — SJ 703667

Kinderton Saltworks were located on the banks of the River Croco which separates the works from the Trent & Mersey Canal, about a quarter of a mile north of Kinderton Mill in Middlewich. Calvert records that Kinderton had "*two brine seeths or salt pits, great store of salt there is made and vended into parts both near and remote.*"[1] At this time the brine occured as a natural spring without the need for pumping. Kinderton saltworks are noticeable on Bryant's map of Cheshire published in 1831 (see Figure 180). By the 19th century the brine springs were no longer flowing and recourse to pumping was necessary. The use of waterpower for these pumps is confirmed by the advertisement placed when the saltworks were offered to let in 1836 which read as follows:-

"*To be let with immediate possession for a term of 21 years, all those valuable salt works situated at Kinderton, Middlewich and now in the occupation of Mr. Brabant consisting of two pan houses, two stoves, one stove house, sheds and offices, two brine pits worked by water power with pumps, rods, brine pans, pits and premises complete and in good repair on the banks of the Trent & Mersey Canal being advantageously situated for supplying salt to the Manchester and Liverpool markets by canal.*"[2]

It is possible that waterpower was used for many years, even as late as the end of the 19th century. A description of the works published in 1933 read as follows:-

"*Some will remember the two old shafts supplying the Kinderton salt works when the brine was raised by means of a waterwheel working the pumps.*"[3]

This statement was later embellished and became

"*Kinderton Salt Works lies off King Street alongside the footpath across Harvest (Harbutts) Field. Brine from two shafts is drawn up two pipes by means of a waterwheel on a stream that flowed into the River Croco.*"[4]

This intimates that the pump was not operated by the River Croco itself but by a very minor tributary.

After the saltworks ceased production, the site was cleared and it now consists of a piece of derelict waste ground with no sign, above ground, of the saltworks.

References

1. Calvert, A. F., *Salt in Cheshire*, SPON, 1915.
2. *Macclesfield Courier*, 24th September 1836.
3. Lawrence, C. F., "Middlewich and its Salt Works", *Northwich Chronicle*, 25th February 1933.
4. Earl, A. L., *Middlewich 900 - 1900*, Ravenscroft Publications, 1990.

61. FLEET MILL, CROXTON — SJ 693673

The confluence of the Rivers Croco and Dane lies about half a mile north of Middlewich. About 500 yards downstream on the River Dane from this confluence is the site of a weir lying below the bluff on which sits Croxton Hall Farm. The leat from this weir ran about 300 yards, passing under Croxton Bridge, to the site of Fleet Mill on the northern bank of the River Dane just before the river passes under Croxton Aqueduct on the Trent & Mersey Canal.

There is a reference to a corn mill in Croxton as early as 1310 but whether this was at the site of Fleet Mill is uncertain.[1] It is not until the 19th century that details of a flint mill become available when in 1827 George Challoner, who was bankrupt, was described as a flint grinder of Croxton Mill near Middlewich.[2] A little later in the century the mill was being referred to as "Fleet Mill" on a number of maps.[3]

In the middle of the 19th century the mill was owned by Thomas Goodfellow who was described as an earthenware manufacturer but it was the family of Thomas Hoole who managed the mill.[4] Thomas Hoole was recorded as a flint miller and in 1841 he lived at the mill with his wife and five children plus a Thomas Nixon, aged 15 years, who was also listed as a flint miller.[5] In 1851 the Hoole family was still living at Fleet Mill together with the eldest daughter's husband and their infant. By this time the eldest son George was also a flint miller, but the next son William, aged 17, was an apprentice blacksmith, and John, only aged 14 was a farm servant.[6]

By 1860 Thomas Goodfellow had died and the flint mill was being operated by his trustees.[7] Thomas Hoole was still the manager in 1861 with his son William now a fully fledged blacksmith and his other sons Samuel aged 21 and Peter aged 15 described as an engine smith and flint miller respectively. It would seem that Thomas Hoole was trying to breed a complete workforce not only in numbers but also with all the necessary diverse skills. Although Thomas Hoole's eldest son and daughter had left Fleet Mill by this time, Thomas and his wife still lived with five of their children, one grandchild and an 18 year old servant who also worked as a flint miller.[8]

Shortly after 1861 the Trustees of Thomas Goodfellow sold the flint grinding business to Bridgewood & Clark, but Thomas Hoole stayed on as manager.[9] It is not known what happened to Thomas Hoole and his wife but by 1871 the business was being managed and undertaken by Peter Hoole who was only 25 at the time. Peter's older brothers, William the blacksmith and Samuel the engine smith, had left Fleet Mill and Peter was living with his elder sister Annie and his younger brothers Joseph, 21, a millwright, and Thomas, 17, described as a flint grinder.[10] Peter Hoole continued working for Edward Clark at the mill until in 1881 there was only himself and his sister Annie living at the mill.[11]

Soon after 1881 Thomas Smith came to live at Croxton Hall and also acquired the mill which he continued to operate up to about the turn of the century without any of the Hoole family being involved.[12] The business finished and the mill closed sometime around 1900. The position of the mill just alongside the Trent & Mersey Canal made it ideal for supplying crushed flint to the pottery industry in the Stoke-on-Trent area which was directly accessible via the canal. Whether there was a corn mill on the site before the building of the canal is dubious and it is possible that the flint mill was built on a green field site around about 1800. Today the only indication visible of the previous presence of a water powered flint mill is the channel used for the headrace which is traceable, being about twenty feet wide, but it is completely overgrown with nettles and similar vegetation.

References

1. C. F. Lawrence, *Annals of Middlewich*, 1911.
2. *London Gazette*, 13th November 1827.
3. CRO, Bryant's Map of Cheshire, 1831; 1st Edition one inch Ordnance Survey.
4. Bagshaw's Directory of Cheshire, 1850.
5. CRO, Census Return for Croxton, 1841.
6. CRO, Census Return for Croxton, 1851.
7. White's Directory of Cheshire, 1860.
8. CRO, Census Return for Croxton, 1861.
9. Morris's Directory of Cheshire, 1864.
10. CRO, Census Return for Croxton, 1871.
11. CRO, Census Return for Croxton, 1881.
12. Slater's Directories of Cheshire, 1883 & 1888; Kelly's Directories of Cheshire, 1892 & 1902.

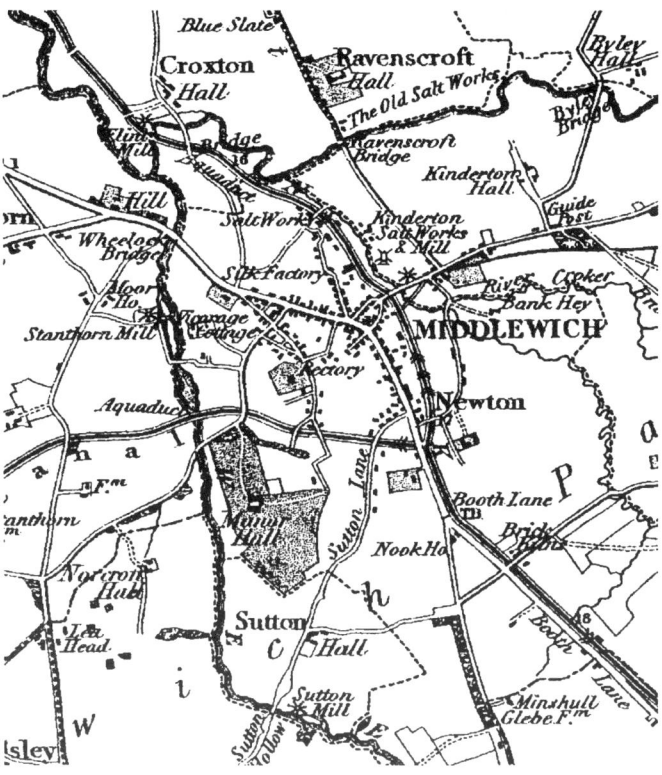

FIGURE 180. PART OF BRYANT'S MAP OF CHESHIRE DEPICTING THE AREA AROUND MIDDLEWICH INCLUDING THE CONFLUENCE OF THE RIVERS DANE AND WHEELOCK. NOTE KINDERTON SALT WORKS AND MILLS (*water and wind powered*) ON THE RIVER CROKER [SIC]. ALSO SHOWN ON THE MAP IS THE FLINT MILL ALONGSIDE THE CANAL AT CROXTON AND THE MILLS AT SUTTON AND STANTHORNE ON THE RIVER WHEELOCK. (CCALS, M5.2)

Section 6. The Upper River Wheelock, Kidsgrove to Wheelock

The River Wheelock with its contributory streams is the largest tributary of the River Dane. The River Wheelock is formed by the joining of four separate streams in the vicinity of the township of Wheelock. The longest of these originates in a number of little streams that rise on the eastern slopes of Mow Cop and passes near Kidsgrove, which used to be a centre of heavy industry, where this stream forms the boundary between Staffordshire and Cheshire. Streams rising on the western side of Mow Cop eventually combine to become the most northerly two streams forming the River Wheelock. The fourth and smallest stream to contribute to the river rises just to the south of township of Wheelock. Although Mow Cop rises to a height of 1000 feet, the main length of the streams runs through the agricultural area of the Cheshire plain.

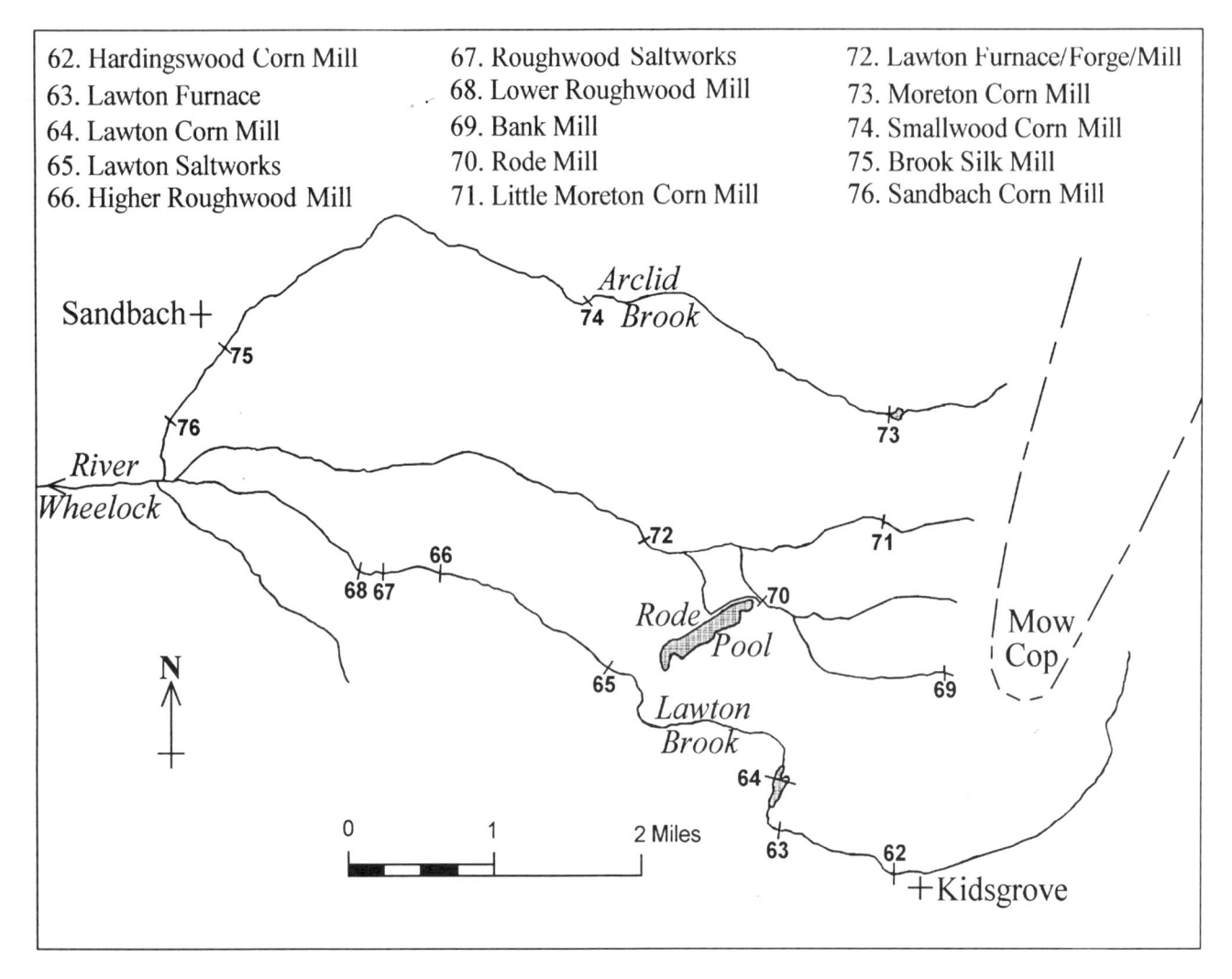

62. Hardingswood Corn Mill
63. Lawton Furnace
64. Lawton Corn Mill
65. Lawton Saltworks
66. Higher Roughwood Mill
67. Roughwood Saltworks
68. Lower Roughwood Mill
69. Bank Mill
70. Rode Mill
71. Little Moreton Corn Mill
72. Lawton Furnace/Forge/Mill
73. Moreton Corn Mill
74. Smallwood Corn Mill
75. Brook Silk Mill
76. Sandbach Corn Mill

FIGURE 181. MAP OF THE STREAMS WHICH RISE ON THE SLOPES OF MOW COP NORTH OF KIDSGROVE AND JOIN TOGETHER JUST TO THE SOUTH OF WHEELOCK BRIDGE TO FORM THE RIVER WHEELOCK.

62. HARDINGSWOOD CORN MILL   SJ 836547

The stream that rises on the eastern slope of Mow Cop in Staffordshire runs southwards before turning to flow west through the low ground between Mow Cop to the north and Harecastle Hill to the south near Kidsgrove. Hardingswood Mill was situated in the town of Kidsgrove close to the top of the Cheshire flight of locks on the Trent and Mersey Canal.

Kidsgrove is an important point on both the canal and railway networks where these transport systems broke through the high ground that separates the Midlands from the Cheshire plain. There was a mill a Hardingswood at least as early as 1747 when a lease was raised for the mill.[1] This was a couple of decades before the coming of the canal and around a hundred years before the railway. Nothing further is known of the mill in the 18th century. It is likely that John Foden at one time operated the mill as his will, which was proved in 1810, named him as a miller resident in Harding's Wood.[2] Later documents show that the mill was part of the estate of John Edensor Heathcote, Esq.[3]

In 1782 John Gilbert, who was the Duke of Bridgwater's agent, bought the Clough Hall estate which covered a large area of land in Kidsgrove. John Gilbert, along with James Brindley, was one of the moving forces behind the building of the Duke of Bridgwater's Canal in the 1760s and early 1770s. The Gilbert family were also closely involved in the building of the Trent & Mersey Canal, whose chief engineer was again James Brindley. It is perhaps not surprising that on completion of the long canal tunnel under Harecastle Hill at Kidsgrove both John Gilbert and James Brindley purchased land in Kidsgrove to exploit the mineral wealth discovered during the building of the tunnel. The commercial possibilities were so great that they encouraged John Gilbert to borrow the money from the Duke of Bridgwater to buy the Clough Hall estate. The mines were then accessed directly from the canal, a technique pioneered by Gilbert and Brindley at the Duke's mines in Worsley on the Bridgwater Canal. On the death of John Gilbert the estate passed to his son, also called John. The younger John Gilbert built a new mansion house at Clough Hall and expanded his commercial interests in the local mines.[4] He died in September 1812 and the whole of the Clough Hall estate was eventually sold in the following year.

One of the lots in this sale was Hardings Wood Mill which was held on a 99 year lease, of which 33 years were unexpired, from Sir John Edensor Heathcote at an annual rent of £8. Any lessee of the mill would also have to pay a £10 annual rental for the use of a stream originating on William Lawton's estate and another one guinea per year to Heathcote for the *"priviledge of conveying the stream to the mill"*. Another payment, necessary for running the mill, was made to Heathcote for a railway and part of a wharf on the canal. Also a payment was required by the canal company for water supplied from the canal reservoir at Bath Pool. The lessor, who had a right to inspect the mill to see that it was kept in repair, was specifically forbidden from tampering with the weirs, dams and pools that could affect the drainage of nearby mines.[5]

Although the conditions of the lease are similar to many such mill leases at this time, there are a couple of surprising items mentioned in the sale particulars. Firstly, in spite of using streams from both Heathcote's and Lawton's land, the mill was also using the Canal Company's water from their reservoir near Harecastle Hill. This is not a situation encouraged by canal companies as any water was needed for their canal. Secondly, John Gilbert had obviously leased the mill from John Edensor Heathcote. These two land owners had been taking part in a running feud since around 1797. Heathcote accused the Gilberts of breaking the conditions of their leases on coal mines in the area. They were served with a writ and eventually lost the ensuing court case, having to pay compensation to Heathcote. If this was not enough to enrage John Gilbert, Heathcote then opened his own colliery nearby in complete contempt of his covenant not to do so. A stalemate continued until 1811 when Heathcote's miners strayed under Gilbert's land. This gave Gilbert the opportunity for revenge as he got his miners to break into Heathcote's level so allowing all of Gilbert's mines to drain into Heathcote's, totally flooding it out. Although this caused the Heathcote colliery to shut permanently it was quite legal. Unfortunately, John Gilbert did not enjoy his triumph for very long as he died the following year.[6]

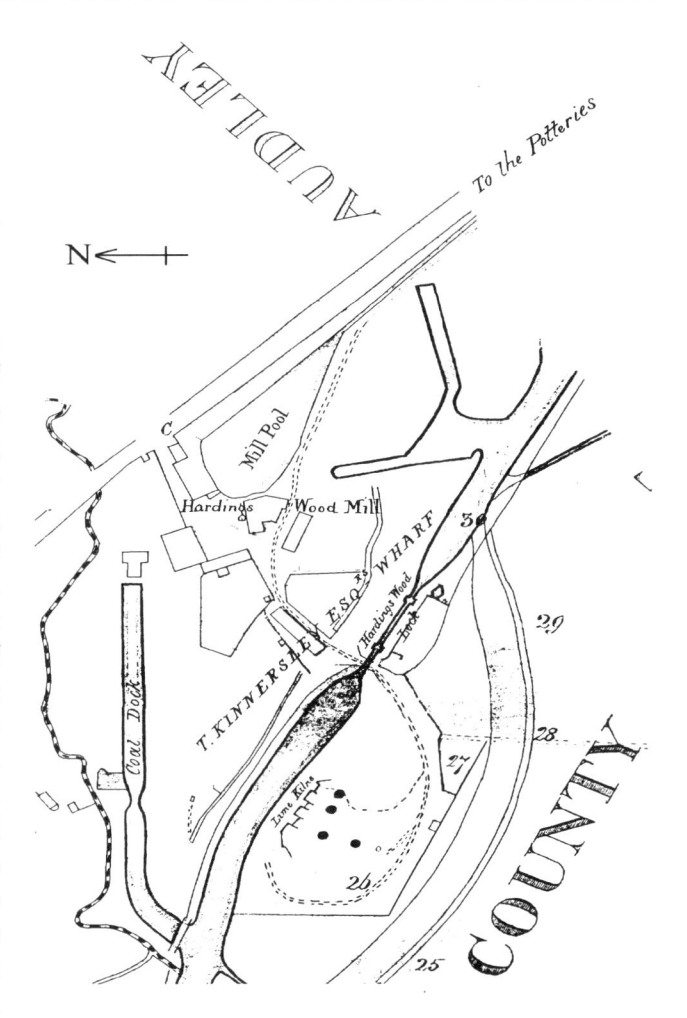

FIGURE 182. PART OF THE PLAN PRODUCED FOR THE CONSTRUCTION OF THE JUNCTION BETWEEN THE TRENT & MERSEY AND MACCLESFIELD CANALS IN 1827. HARDINGS WOOD MILL IS SHOWN TO BE IDEALLY SITUATED TO USE THE CANAL. (CCALS, QDP 76))

Six months after the death of John Gilbert in 1812 notice was given of the dissolution of a partnership between himself and William Smith of Harding's Wood Mill. Undoubtedly William Smith was the miller at Harding's Wood Mill at this time.[7] Later in the century, in 1856, another partnership at the mill was dissolved. This was between William Edwards and Thomas Heath Birks, both of whom were described as millers at Harding's Wood Mill.[8] It was stated at the time that William Edwards would be continuing the business on his own account, but three weeks later he offered the mill for sale when there were still seventeen years to run on the current lease. At the time it was said to *"adjoin the Trent & Mersey Canal and is also in close proximity to the Harecastle Station on the North Staffordshire Railway and is one of the most flourishing districts of North Staffordshire."*[9]

After this date the mill continued to operate and was shown on the Ordnance Survey maps of 1879 and 1890. However, the town of Kidsgrove had changed beyond all recognition in the 19th century, the availability of coal and ironstone having led to the establishment of heavy industry based on these underlying raw materials. These industries continued to expand throughout the 19th century and had a voracious appetite for land. When the corn milling industry began to change towards the end of the 19th century the site of the mill was deemed more valuable for the expansion of the other industries in the town rather than incurring the costs of upgrading the mill to be competitive. By the end of the century the mill had totally disappeared and its site had been redeveloped by other industry.[10]

Today, all the heavy industries of Kidsgrove have disappeared and the mill site is now occupied by the Kidsgrove Working Mens' Club and their car park. Its position, on the main road through Kidsgrove, was only a few yards away from the canal and the railway. The competition brought to traditional mills by the roller mill revolution at the end of the 19th century must have been immense if a mill with such good transport links and expanding local population could not remain in existence. The only possible drawback for the mill could have been its water supply, as it was fed by two quite small streams, but its position alongside the top lock on the Trent & Mersey Canal, and the agreements with the Canal Company, could well have allowed the mill to use water from the summit pound of the canal, returning it below the top lock. Although not a common configuration for a mill's water supply, it would not have been an unknown arrangement for the Trent & Mersey Canal (see Moston Mill).

References

1. SRO, Sale catalogue, D239/M.1703.
2. CRO, Will of John Foden, 1810.
3. SRO, Sale catalogue, D239/M.1703.
4. Lead, P., *Agents of Revolution*, Keele University, 1990.
5. SRO, Sale catalogue, D239/M.1703.
6. Lead, P., 1990, op. cit. 4.
7. *London Gazette*, 3rd March, 1812.
8. *London Gazette*, 11th July, 1856.
9. *Macclesfield Courier*, 2nd August, 1856.
10. Ordnance Survey, 25 inch to 1 mile, 1899.

---

63. LAWTON FURNACE   SJ 828551

The location of Lawton Furnace cannot be determined with any accuracy but would have been near to Lawton Brook as the stream was needed to drive the waterpowered bellows for the furnace. The furnace is thought to have been somewhere close to the present cross roads of the Liverpool and Congleton roads, not far from the Cheshire - Staffordshire border, at Red Bull.

During the civil war in the second quarter of the 17th century a number of established iron works in the south of the country, such as in Sussex, were destroyed. This gave an opportunity to supply the market for iron from new charcoal-fired furnaces, many of which were located in the north of the country near to iron ore deposits. In the 1650s, during the Commonwealth period, John Turner of Stafford was operating forges at Lizard Hill, north-east of Shifnall in Shropshire, under a twelve year lease. However, the rent payable under the lease was excessive which caused John Turner to look for alternative sites to establish a furnace and a forge on better commercial terms. He selected a corn mill in Church Lawton which he leased from William Lawton at an annual rent of £44 per year for twenty one years from April 1658. This lease provided Turner with the necessary waterpower to operate the bellows of his proposed furnace. (For details of the earlier history of this corn mill see under Lawton Mill). It appears to have been John Turner's intention to build a forge within about seven or eight miles of the furnace he was building at Church Lawton. To help finance these plans he reassigned the lease on his forges at Lizard to Rowland Revell and Thomas Fletcher in May 1658. One of his reasons for choosing Cheshire for this development was the easy availability of charcoal from the Cheshire woodlands as a good supply of this raw material was essential to the operation of the furnace. During the period when he was building the furnace, John Turner purchased a supply of timber for £3000, a very considerable sum. He also entered into an agreement to supply 100 tons of pig iron to a Mr. Newbrook at the rate of 20 tons per month, starting in August 1658. Moreover, he also entered into a bond with a person called Higginson, who was the "*farmer of the excise of iron*" in Cheshire, to supply him exclusively with all the output of bar iron during the three years from 1658 to 1660.

Unfortunately John Turner's plans then started to unravel. The leases on the Lizard forges were not taken over until December 1658, which delayed the availability of money needed to build the new furnace, causing its completion to be two months late. This delay meant that the charcoal for the furnace had to be produced in the rainy time of the year and the amount of iron produced was low compared to the amount of charcoal consumed. These problems and their adverse effect on cash flow meant that the planned forge was not built and so the wood collected for it had to be transferred to the furnace at Lawton. Newbrook did not receive any pig iron until November 1658, instead of in August, and the poor quality of the iron meant that he only accepted 12 tons. Without the forge there was no possibility of supplying Higginson with bar iron and so the bond was likely to be forfeited. All in all Turner reckoned to have lost £300 by the end of 1658.[1]

In order to raise further finance Turner mortgaged the furnace to Robert Crompton of Elstow, near Bedford, in February 1659 for £500 to be repaid at the rate of £100 per year for the next seven years. This represents a simple interest rate of 5.7% (the compound rate would be much higher as some of the capital was being repaid every year). At the same time as this loan Turner took John Crompton, an uncle of Robert Crompton, as a partner. It is most likely that this partnership was a condition of the mortgage so that the Cromptons could keep an eye on their investment, but the lack of trust and the degree of coercion inherent in this agreement was bound to lead to future problems. The conditions that Turner had to accept to raise this extra capital shows how desperate he had become in order to keep his business afloat long enough to start making profits.

In July 1661 Turner was invited to a conference at Congleton with John Crompton, ostensibly to discuss the supply of cordwood for the furnace. However, on arrival Turner was arrested and thrown into gaol on a complaint by Robert Crompton for allegedly missing a mortgage repayment, two complaints of John Crompton and a complaint by George Parker, a supplier of wood, for £3000. Whilst Turner was in gaol, Robert Crompton seized the furnace at Lawton and arranged to operate it using his own workers, buying £400 of wood from William Vernon. After three weeks in gaol Turner was released and immediately regained control of the furnace. He persuaded Crompton's workers to change their allegiance and even made use of the wood purchased by Crompton. The result of these practices was that everyone sued each other, ending up in court at Chester in 1668 when Turner was ordered to restore the partnership. At this time Robert Crompton accused John Crompton of siding with Turner which would seem highly unlikely as Turner and John Crompton were very soon in further disputes which led to them both ending up in gaol in 1674. Although they agreed to a settlement whilst in gaol the two men broke their word with further dispute taking place in 1676 just before John Crompton died.[2]

FIGURE 183. A CONTEMPORARY DRAWING OF AN 18TH CENTURY FURNACE WITH WATERPOWERED BELOWS SIMILAR TO LAWTON FURNACE. (R. R. *Angerstein, 1754*)

At some time in the early 1680s Lawton Furnace and Cranage Forge (*q.v.*) were acquired by a partnership of William Cotton, Dennis Heyford and Thomas Dickin. Previously these partners had been associated with ironworks in Yorkshire. Around the same time they leased Abbot's Bromley Forge in Staffordshire from Lord Paget. Later they also worked Cannock Forge and Rugeley Slitting Mill. The pattern of trade established at this time saw pig iron from Lawton Furnace going to the finery at Bromley Forge from where the resultant blooms went to Cannock Forge to become bars which then went to Rugeley for slitting. In Cheshire, Heyford's main concern was Cranage Forge while William Cotton concentrated on Lawton Furnace, but it is likely that Thomas Hall, a cousin of William Cotton, was the actual manager at these sites. In the early days of the partnership the quality of the iron produced could be quite poor, as was pointed out by Roger Wilbraham in 1682 when he is reported to have said "*the woods ...are now cut down and destroyed to make worse iron than we had then from beyond the seas*". In 1692/3 Lawton Furnace supplied small amounts of tough iron, made from Cumberland hematite ores, to Cannock Forge. This was a process started by Turner in 1665 when he bought 71 tons of Cumberland hematite for the Lawton Furnace. Normally the furnace would use the local phosphoric ores from Staffordshire which produced an iron with a hard brittle quality, called coldshear or coldshort iron, that was highly suitable for making nails. It was around this time that the partnership sold Bromley and Cannock Forges and Rugeley Slitting Mill which were then brought together with other furnaces and forges to form the "Staffordshire Works" which Lawton Furnace continued to supply.[3]

In Cheshire it is likely that some reorganisation took place in 1696 with Thomas Hall in charge of Lawton Furnace and Cranage Forge, Daniel Cotton (son of William) at Vale Royal Furnace and Edward Hall (son of Thomas) at Warmingham Forge (*q.v.*) as well as Street and Bodfari forges. This group of ironworks was known as the "Cheshire Works". Under this partnership at the end of the 17th century the output of Lawton Furnace averaged about 700 tons per year.[4] The furnace continued to concentrate on using the phosphoric Staffordshire ore to produce coldshort iron pigs with the occasional mixture of a little Cumberland hematite ore from time to time to produce a mixed pig. As well as supplying the Cheshire forges much of the Lawton Furnace output went to forges in the Midlands with some of it travelling as far as the forges on the River Stour. However, the only delivery to Bodfari forge, at this time, was a shipment of hammers & anvils in 1699. In this year it is possible that William Hall was working at Lawton Furnace because he donated a silver communion paten to Lawton Church inscribed with "William Hall, Iron Master".[5]

Further changes in the partnership occurred in 1698 when Heyford sold Cranage Forge to Cotton and Hall. But in 1700 Cotton withdrew from the partnership altogether causing a realignment of interests in a new partnership with Edward Hall contributing £10251, Daniel Cotton with £8848 and Edward Vernon with £2327.[6] Under this new partnership the output of Lawton Furnace rose to a peak of 900 tons in the winter campaign, or work season, of 1703/4.[7] A further reorganisation of the Cheshire Works in 1706 saw Daniel Cotton taking over from Thomas Hall at Lawton furnace. The following year the Cheshire Works amalgamated with the Staffordshire Works. The new partners in Cheshire were Philip Foley of Prestwood, John Wheeler of Woolaston Hall (managing director of the Staffordshire Works), Obadiah Lane (general manager of the Staffordshire Works), Nathaniel Lane (son of Obadiah), Edward Hall, and Daniel Cotton, all of whom had a share of £2250 in the total stock of £13500. The three Cheshire partners, including Thomas Hall had shares in the Staffordshire Works and over the next few years they strengthened their control over the whole partnership as well as being involved in ironworks in Lancashire and elsewhere. This was achieved by purchasing partner's shares as they died and also by marriage within the families.[8]

A typical example of the production at Lawton furnace can be seen in the figures for the 1709/10 season when the output of the furnace consisted of 437 tons of coldshort pig, 145 tons of mixed pig, and 21 tons of castings. This output was achieved using 1202 doz. of Staffordshire ironstone and 70 tons of Cumberland hematite. The coldshort pig was sold for between £4-15s-0d to £5-2s-6d per ton with the mixed pig being sold at £5-17s-6d per ton giving something in the region of £30,000 in sales for the year.[9] The success of the iron works partnerships saw some of the partners making various gifts to their churches, including a set of bells to Lawton church from "The present ironmasters of Lawton Furnace".

This continued to be the state of affairs up to the 1730s when the Cheshire Works started to contract, leading up to a great depression in the iron trade in 1737/8 when the output of bar iron in Cheshire dropped from 460 tons to 290 tons per year. Considerable quantities of foreign pig iron were imported into the country in the first half of the 18th century and the local iron masters were faced with the problem of a shortage of timber and rising costs. These factors resulted in many ironworks closing, especially in 1737/8 when the full force of the foreign

FIGURE 184. PART OF BURDETT'S MAP OF CHESHIRE, 1777, WITH TWO WATERMILL SYMBOLS SHOWN IN LAWTON. THE SYMBOL SOUTH OF THE ROAD PROBABLY REPRESENTS THE LOCATION OF THE FLINT MILL BUILT ON THE SITE OF LAWTON FURNACE. ALSO NOTE THE LOCATION OF HARDINGS WOOD MILL ALONGSIDE THE "WILDEN FERRY CANAL", BETTER KNOWN AS THE TRENT & MERSEY CANAL. (CCALS, PM12/16)

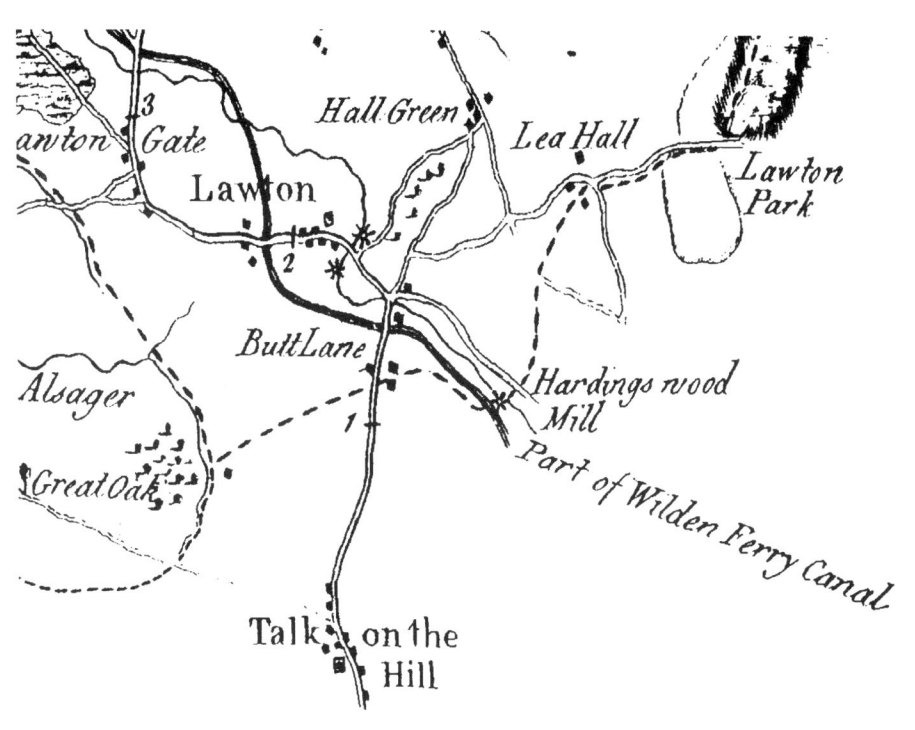

competition took effect.[10] Around 1750 all the principle participants in the Cheshire Works died and the delivery of "Cotton's coldshort" and "Cheshire coldshort" to the Stour forges suddenly ceased. This has been taken as showing the end of Lawton Furnace, however, some years earlier in 1744 Thomas Adams of Ford Green agreed with Stephen Stringer of Little Hassal and Thomas, son of John Wedgwood, "*to take, farme and rent ye old ffurnace at Lawton and convert it to a fflint mill*".[11] If Lawton Furnace closed in 1744 it is not certain where the Cheshire iron supplied after this date was smelted, although Vale Royal Furnace is a possibility. Certainly by 1744 there would be great interest in establishing a flint mill to take advantage of this new technique used by the pottery industry which was establishing itself in North Staffordshire.

The next indication of the use of waterpower in Church Lawton occurred in 1777 with the publication of Burdett's large scale map of Cheshire which shows two mill symbols in the parish. These are shown a short distance north-east of where the Potteries to Liverpool road crosses the Congleton road at Red Bull. One is just north of the Liverpool road, while the other is just south of the road. If Burdett saw fit to show two watermills here then the most likely explanation is that there were in fact two watermills present. It is extremely unlikely that there would be two corn mills so close together. Therefore, the most logical conclusion would be the existence of one corn mill and another type of watermill, possibly the flint mill mentioned as the replacement of the furnace in 1744. Certainly, a few years later in 1781, the land tax returns for Lawton shows John Lawton as owner of a "flint mill" with an annual tax of £1-2s-9d.[12] Further evidence from the land tax returns show that the occupier in 1782 and 1783 was Daniel Cooper, although the premises are only described as a "mill", the valuation was the same as that for the flint mill in 1781. Between 1784 and 1791 the occupier of the mill was William Bate, followed from 1792 to 1809 by Charles Bate. Although the valuation of the property varies it can be traced to being occupied by William Beresford in 1810 and 1811 but after that date there is no occupier recorded, however the owner of the mill, William Lawton, is listed until 1827.[13] After this date the mill does not appear in the land tax returns. This evidence suggests that the mill was in operation until 1811 after which it stood idle until 1827 when it was demolished, fell down or otherwise ceased to exist.

In recent years there has been some speculation as to the location once occupied by Lawton Furnace. Using Burdett's map of the area in 1777 the current theory is that the furnace was located on the western side of the road running northwards from Talk on the Hill to Hall Green, between the canal and the adjacent cross-roads. This location has been favoured because slag was found there and the nearby building is called "Furnace Cottage".[14] However, the cottage is a modern construction and the evidence of its name is doubtful.

The evidence from the documentation suggests a different location for the furnace in so far as the flint mill, which was built on the site of the furnace after 1744, is propoably represented by the southernmost water mill symbol in Lawton on Burdett's map. The map shows this symbol as being just to the south of the road from Hardingswood to Lawton, west of the cross-roads, where the road crosses the stream. Today this position is difficult to locate as the alignment of the modern road (A50) has been altered since the 18th century, as has the line of stream. Of these two sites the latter one looks today much more like a possible waterpowered site, whereas the one by the A34 is considerably higher than the bed of the current stream and it is difficult to see how a leat could have been arranged over the adjacent land to provide the necessary head of water, although the lie of the land has been altered greatly over the years especially with the building of the Macclesfield Canal in the late 1820s.

References

1. Awtry, B. G., "Charcoal Ironmasters of Cheshire & Lancashire, 1660-1785", *Transactions of the Lancashire & Cheshire Antiquarian Society*, Vol. 109, 1957.
2. *Ibid.*
3. *Ibid.*
4. Johnson, B. L. C., "The Iron Industry of Cheshire & North Staffordshire: 1688-1712", *Transactions of the North Staffordshire Field Club*, Vol. 88, 1954.
5. Awtry, B. G., 1957, *op. cit.* 1.
6. Johnson, B. L. C., "The Foley Partnerships: The Iron Industry at the end of the Charcoal Era", *The Economic History Review*, second series, 4, 1952.
7. Johnson, B. L. C., 1954, *op. cit.* 4.
8. Awtry, B. G., 1957, *op. cit.* 1.
9. Johnson, B. L. C., 1954, *op. cit.* 4.
10. Lead, P., "The North Staffordshire Iron Industry 1600-1800", *Journal of the Historical Metallurgy Society*, Vol. 11, 1977.
11. Hardman, B. M., "The Iron Industry of North Staffordshire and South Cheshire in the pre-coke Smelting Era", *North Staffordshire Journal of Field Studies*, Vol. 15, 1975.
12. CRO, Land Tax Returns, Lawton.
13. *Ibid.*
14. Hardman, B. M., 1975, *op. cit.* 11.

---

64. LAWTON CORN MILL   SJ 825556

The earliest mention so far discovered of a watermill in Church Lawton occurs in various quit claims and deeds in the 14th and early 15th century when certain members of the Lawton family transferred lands or other items concerned with a watermill and its accessories at Lawton to the Abbot of St. Werburgh's monastery in Chester. Certainly, by the early 15th century the mill was under the full control of the abbot. The Lawton manor accounts for 1422 to 1425 show that the watermill was farmed (i.e. rented) by Richard Eardley who was paying a rental of 20s per year, but a note states that the rental used to be 40s per year. The accounts also show that a payment of one shilling was made each year to a Beatrice Madyowe in order to keep the mill race in good repair where it crossed her land.[1]

Most of the habitation in Lawton manor was sited near the boundary with Rode manor and it seems that for some inhabitants it was more convenient to use Rode Mill (*q.v.*) than Lawton Mill. As they were bound by the mill soke to use Lawton Mill certain people were always being fined for breaking their obligation by milling at Rode. In 1574 Thomas Rode, gentleman, Margaret Cartwright, widow, with her son Richard, and William Muttchell were each presented and fined 3s-4d for grinding corn away from the lord of the manor's mill in Lawton. In the early years of the 17th century Thomas Rode and Richard Cartwright were fined 10s for the same offence. Undeterred, they continued milling at Rode Mill, no doubt treating the fines in effect as a licence to grind outside their manor. In 1628 the Cartwright family sought to be officially recognised as residents of Rode manor but they must have lost the case because in 1636 Randle Rode and Mary Cartwright were both fined one shilling for grinding away from their lord's mill.[2]

During the Commonwealth period, in 1658, John Turner took over the lease of a corn mill in the parish in order to convert it into a charcoal fired blast furnace using the water supply to power the furnace bellows (see Lawton Furnace).[3] As the main reason that attracted John Turner to the area was the low rent, it is possible that the corn mill was in a poor condition or even not in work at that time.

After the establishment of the furnace, another corn mill was built nearby to satisfy the needs of the local community. When this mill was built is not known but a deed, dated 16th March, 1721, for a parcel of land, mentioned that it was near to Lawton Mill.[4] The mill appeared on Burdett's map of Cheshire, published in 1777, where it is most likely depicted by the more northerly of the two mill symbols at Lawton. This position is just north of the Potteries to Liverpool road, some 500 yards west of where the road crosses the Congleton Road at Red Bull (see Figure 184). Confirmation of the mill's location is to be found on a tramway map of 1806 which shows the position of "Lawton Mill" with respect to the Trent and Mersey Canal.[5] The mill must have ceased to exist soon after 1800 as it does not occur on any of the maps published in the early part of the 19th century. By triangulation on the tramway map of 1806 it is possible to determine that Lawton Mill was located in the area now covered by the lake which lies below the ruins of Lawton Hall. Although the lake was recently drained for maintenance no trace of the site of the mill could be detected due to the large amount of silt present.

References

1. Renaud, F., "Church Lawton Manor Records", *Transactions of the Lancashire & Cheshire Antiquarian Society*, Vol. 5, 1887.
2. Renaud, F., 1887, *op. cit.* 1.
3. Awtry, B. G., "Charcoal Ironmasters of Cheshire & Lancashire, 1660-1785", *Transactions of the Lancashire & Cheshire Antiquarian Society*, Vol. 109, 1957.
4. CRO, Deed ,1721, DBW/M/E/8.
5. SRO, Tramway Map, 1806, Q/RUM/41.

---

65. LAWTON SALTWORKS   SJ 806570

The village of Rode Heath is about 3½ miles south-east of Sandbach and lies alongside the Sandbach to the Potteries road. The Trent & Mersey Canal runs parallel with this road through the village. On the other side of the canal there is a stream running in a valley. This stream forms the boundary between the parishes of Church Lawton to the south and Odd Rode to the north. Lawton Saltworks was located between the canal and the stream just inside the parish of Odd Rode rather than in Church Lawton.

Lawton Saltworks were established in 1693 by John Lawton who spent £9000 on their construction. Originally the works used brine from naturally occurring springs but in 1711 a rock salt pit was dug abutting onto Snape Farm. This farm was in the possession of Thomas Lawton who was a very distant relative of John Lawton. In 1719 there was a dispute between John Lawton and a Samuel Taylor of Nantwich who was found to be wrongfully extracting brine by a tunnel he had dug from Thomas Lawton's land to intersect with John Lawton's brine spring. As a result of this dispute all the saltworks on Thomas Lawton's farm were assigned to John Lawton as settlement of Thomas Lawton's outstanding mortgage.

In 1737 John Lawton agreed to supply all the output of his saltworks exclusively to Thomas Furnival who was also allowed to buy salt from other suppliers if the output from Lawton was not sufficient. Later on in the 18th century, a Joseph Furnival was working Lawton Saltworks.[1] It is claimed that the saltworks were worked by a waterwheel from 1770, presumably because the naturally occurring brine spring was no longer running due to the earlier extraction.[2]

Joseph Furnival was succeeded at the saltworks in 1775 by Edward Salmon, who lived at Hassal Hall, and William Penlington, who was described as a doctor.[3] The building of the Trent & Mersey Canal, which opened in 1777 running through Lawton, made the supply of coal, required in large quantities to evaporate the brine, very cheap and easy to deliver. In fact, the ease of coal supply encouraged the proprietors of Lawton Saltworks to consider using the new emerging technology of a steam engine to pump the brine at Lawton. In November 1776, Mr. N. Henshaw of Knypersley, a friend of the owners of Lawton Saltworks, wrote to Boulton and Watt at the Soho Foundry in Birmingham enquiring about the purchase of one of their engines for the works. In January 1778 James Watt arranged to visit Lawton with one of his workmen, a James Law, to oversee the installation of the engine, at the same time apologising about the delays that has been incurred. By April James Watt was still apologising about the delays due to him being "*so busy*" in Cheshire, but the engine appears to have been put in work early that month.

It would seem that the steam engine was an immediate success which Salmon wished to publicise. He asked James Watt for his approval and received a reply which suggested a mutually acceptable text, as follows:-

"We hear from Lawton Salt Works that Mr. Salmon has now completed his new works on the banks of the Staffordshire Canal, which are esteemed to be the most complete in their way in the country: the reservoir of brine is situated 100 yards above the brine in the pit and higher than the canal. It is filled with brine by a small fire engine constructed by Boulton & Watt of Birmingham which does its business with great ease and a very small consumption of fuel."

The steam engine had an 18 inch diameter cylinder which was cast and bored at John Wilkinson's foundry at Bersham near Wrexham. It had a stroke of 3 feet 2 inch and was rated at 4 h.p. The engine cost a total of £400-15s-3d, of which £100 was for the Boulton and Watt operating licence. The performance of the engine was described in the year after its installation as follows:-

"fire engine constructed by Messrs Boulton and Watt, which, in the waste of three hundredweight of coals, (value nine pence) does in twelve hours throw up to the height of a hundred yards, not less than twenty four thousand gallons of brine."[4]

There can be no doubt that in 1779 the Lawton Salt Works were at the forefront of salt manufacturing

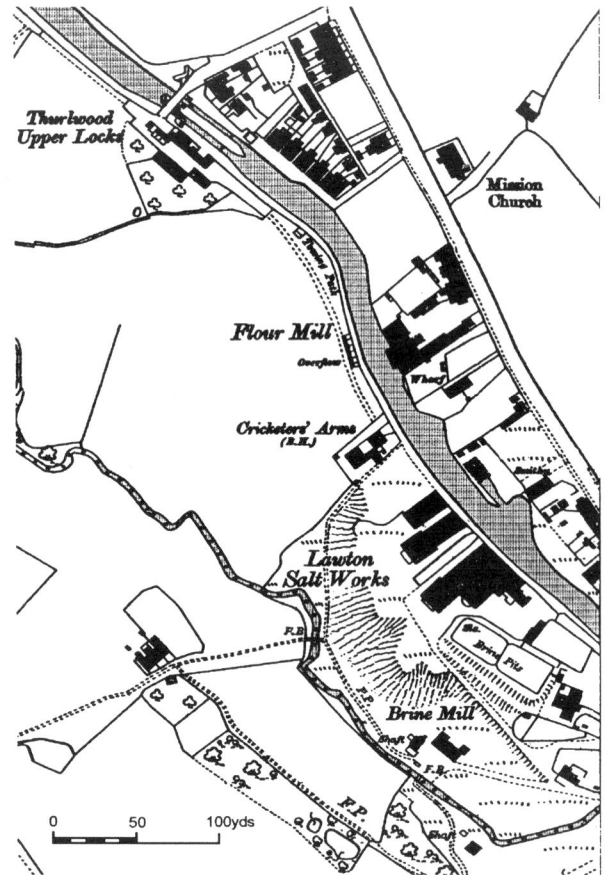

FIGURE 185. ORDNANCE SURVEY MAP SHOWING LAWTON SALTWORKS, 1908. NOTE THE STRUCTURE LABELLED "BRINE MILL" TO THE SOUTH OF THE SITE.

technology, not only being the first to install a steam engine, but it was at Lawton that the first trial borings were made through the normal salt layer, at a depth of 120 feet, that was exploited by the existing salt works in Nantwich, Middlewich and Northwich. This exploratory drilling located two new layers of salt, the first at a depth of 150 feet was only 12 feet thick but the second new layer found at a depth of 207 feet was well over 75 feet thick.[5] This discovery, together with the use of steam power, changed the whole nature of salt extraction in Cheshire. To maintain their leadership of the industry Salmon wrote to James Watt in 1782 to try to persuade him not to supply steam engines to a rival salt company to prevent the possibility of competition. Initially Boulton & Watt replied

"...we look upon ourselves, as so far servants of the public, in consequence of the exclusive privilege granted to us, as not to dare avowedly to refuse to serve any person who offers an adequate compensation for our trouble."

But their attitude was somewhat different eleven months later when they wrote to Salmon as follows:-

"No application has ever been made to us from the quarter you mention, and if it comes, we shall decline it if possible, or at least make them pay properly for it, but we will not do it at all if we can avoid it by any fair means."[6]

This does appear to be a complete reversal of attitude but may just be a form of words to keep an awkward customer satisfied.

Lawton Saltworks appeared to continue to function successfully under the auspices of Salmon and Penlington until the death of Penlington in 1795.[7] After his death the salt works came to be operated by James Robinson until the end of the 18th century.[8]

The agreement between John Lawton and James Robinson leased the saltworks for 21 years at £180 per annum with the possibility of a six year extension at a rent of £230 per annum. Also James Robinson was obliged to spend £1000 taking down the old salt works and erecting new premises which had to include at least four pans.[9] From the available records it would seem that the works enjoyed an ongoing success, however, the innovative use of a steam engine, while initially successful, was bound to fall prey to maintenance problems due to working in such a hostile environment as a saltworks. This was certainly the case at other saltworks using a steam engine, such as Marston, so it was probably also true at Lawton.[10] If the steam engine had any problems then the works would have to revert to somewhat older technology to stay in production, namely the use of water, wind or muscle power. In fact this older technology might well have continued being used on a "belt and braces" principle during the period that Watt's steam engine was also in use. That waterpower was in fact still used at Lawton can be seen when James Robinson was retiring from the business in 1800 and the following advertisement appeared:-

"To be sold at auction all those capital and old established salt works with a small allotment of land situated at Lawton in the said County of Chester now in the possession of Mr. James Robinson who is retiring from the business.

The premises are held under a lease from John Lawton Esq. 19 years whereof are unexpired at Lady day last.

The works are singularly valuable from the advantageous situation lying close to the navigable cut from the Trent to the Mersey. The brine is unquestionably of the first quality in the kingdom of which there is a most plentiful supply raised from two pits contiguous to the pans by a water engine at a very trifling expense. On the premises are also a rock salt pit.

NB. The above lease is renewable for the term of 12 years at a small additional rent."[11]

It is noticeable that there is no mention of the famous steam, or fire engine as it was known, and the works appears to rely entirely on the "water engine" which was how a waterwheel driven pump was usually described at this time.

In response to this advertisement the works were

FIGURE 186. LAWTON SALTWORKS IN 1915. THE STRUCTURE IN THE FOREGROUND APPEARS TO BE LOCATED AT THE PLACE MARKED AS A "BRINE MILL" ON THE 1908 ORDNANCE MAP. (*A. F. Calvert*)

purchased by Sir Thomas Broughton Bt. of Doddington Hall, Staffordshire.[12] One of his first tasks on taking over the saltworks was to request Boulton and Watt to inspect the steam engine.[13] Information published in 1808 shows that the new owners of the works were still using the waterpowered pumps at that date.[14] Later, in 1813, Broughton went into partnership with James Sutton to form the British Eaton Salt Company to run the works. This company eventually became James Sutton & Co.[15] When Messrs. Sutton advertised the saltworks for sale in 1853 there was no mention of the water engine. Perhaps they did not wish to emphasise such old fashioned technology![16] The use of waterpowered brine pumps at Lawton Saltworks may be inferred as late as 1908 when a "Brine Mill" is indicated on a large scale map showing the saltworks (see Figure 184).[17] Earlier maps show a leat running to this buidling. In 1888 the saltworks became part of the Salt Union, but on the expiry of the lease, reverted to the land owner after which the works were acquired by the Commercial Salt Company.[18] The saltworks continued in production until 1926 when they finally ceased working. In 1937 subsidence caused the engine house to collapse causing a steam engine to fall down the brine shaft.[19] Today the site has been completely cleared of any industrial remains and has been landscaped by the Cheshire County Council. The only indication that this was once the site of the most progressive saltworks in the country, that was at one time powered by water, is an interpretation board erected by the County Council.

References

1. Renaud, F., "The Mineral Wealth of Lawton", *Lawton Chronicles*, Vol. 5, Lawton Heritage Society, 2000.
2. Calvert, A. F., *Salt in Cheshire*, SPON, 1915.
3. Chaloner, W. H., "Salt in Cheshire, 1600-1870", *Transactions of the Lancashire & Cheshire Antiquarian Society*, Vol. 71, 1961.
4. Chaloner, W. H., "The Cheshire Activities of Matthew Boulton and James Watt of Soho, near Birmingham, 1776-1817", *Transactions of the Lancashire & Cheshire Antiquarian Society*, Vol. 61, 1949.
5. Chaloner, W. H., 1961, *op. cit.* 3.
6. Chaloner, W. H., 1949, *op. cit.* 4.
7. Cheshire County Council Information Board at the site of Lawton Saltworks.
8. CRO, Land Tax Returns, Lawton.
9. Renaud., F., 2000, *op. cit.* 1.
10. Lead, P., *Agents of Revolution*, Keele University, 1990.
11. *Staffordshire Advertiser*, 12th July 1800.
12. CRO, Land Tax Returns, Lawton.
13. Chaloner, W. H., 1949, *op. cit.* 4.
14. Holland, H., *General View of the Agriculture of Cheshire*, 1808.
15. CRO, Land Tax Returns, Lawton; Bagshaw's Directory of Cheshire, 1850.
16. *Macclesfield Courier*, 2nd April 1853.
17. Ordnance Survey, 25 inch to 1 mile map, 1909.
18. Calvert, A. F., 1915, *op. cit.* 2.
19. Cheshire County Council Information Board at the site of Lawton Saltworks.

---

66. HIGHER ROUGHWOOD MILL   SJ 785579

The Lawton Brook flows westwards, parallel to the Trent & Mersey Canal, through the parish of Odd Rode and into the parish of Betchton. Higher Roughwood Corn Mill is located about half a mile west of the Chellshill aqueduct on the canal. The earliest evidence found so far of this mill is the symbol on Burdett's map of Cheshire published in 1777, however the mill may well be much older as there exists a deed for a mill in Betchton dated 1623.[1]

The land tax returns for the end of the 18th century show that the Higher Roughwood Mill was owned by George Wilbraham who lived at Rode Hall and the occupier of the mill from 1783 to 1814 was John Foden.[2] In 1811 the lease of a mill in Betchton was offered for sale but it is not possible to determine if this was Higher or Lower Roughwood Mill (*q.v.*). The lease of the mill had 12 years to run and the annual rent was £60 for each of the first two years and £70 for each of the remaining ten years. The lease was being sold because the premises had been seized by the Sheriff, suggesting that the miller was probably bankrupt.[3]

It is not possible to say if this sale was successful as it was not until 1816 and 1817 that a Samuel Hassall was recorded as the new occupier.[4] He was not very fortunate in his occupation of the mill as he was declared bankrupt in June, 1817,[5] with the mill offered for sale two months later, as follows:-

"All that water corn mill or grist mill with newly erected messuage, barn, shippon, stable, and outbuildings, and the French stones, wheels, machinery and fixtures belonging to the mill situated at Betchton near Sandbach in the occupancy of Samuel Hassall.

Also all that other newly erected mill formerly used as a flint mill, with the French stones, wheels, machinery and fixtures therein, situated in Betchton also in the occupancy of Samuel Hassall."[6]

From this advertisement it would seem that Samuel Hassall operated both Higher and Lower Roughwood Mills, the only mills in the parish of Betchton, but it is not clear which was the corn mill and which was the flint mill. The only clue is that the flint mill was *"newly erected"* which tends to suggest that this mill was in fact Lower Roughwood Mill which was said to be *"new"* in 1783.[7]

In response to this advertisement a new miller, called Robert Joynson, came to Higher Roughwood Mill.[8] Unfortunately, he fared only slightly better than Samuel Hassall, remaining in business for a total of four years before he was also declared bankrupt in 1822.[9] The next miller, John Holland, was to stay rather longer than the two previous occupants. By 1841 John Holland and his wife lived in the mill house with their nine children, all under the age of fifteen years. He employed another miller, Charles Watson, to assist in working the mill and so the Watson family including five children also had to be accommodated at the mill.[10] Ten years later John Holland described himself as a *"master miller and farmer"* employing two jouneymen millers, James Proudlove and Thomas Stanyer. John Holland's eldest son, George, also worked as a miller in his father's mill.[11]

Around 1860 the Holland family took over Lower Roughwood Mill at which point John Holland retired to his farm at Hassall Green nearby, leaving the milling business at the two mills to his three sons George, John and William.[12]

George Holland lived and worked at Higher Roughwood Mill employing four men and a boy. Also living at the mill were James Proudlove and his family, who was now employed as a carter but his eighteen year old son, William, worked as a miller.[13] George Holland continued to operate the mill until the end of the century. At some stage during the last quarter of the 19th century the water supply at the mill was augmented by the addition of a steam engine.[14] By 1902 the Holland family no longer lived at the mill, which was operated by Frederick Stonier of Rode Mill (*q.v.*). He worked Higher Roughwood Mill in conjunction with Rode Mill until his death in 1921. He left a total

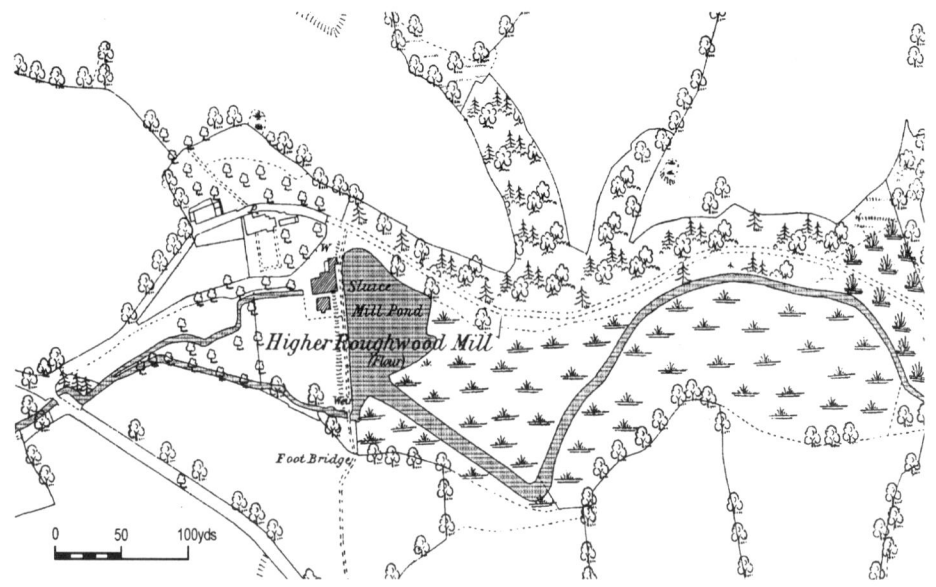

FIGURE 187. ORDNANCE SURVEY MAP, 1873, SHOWING HIGHER ROUGHWOOD MILL. AT THIS TIME THE LARGE MILLPOOL WAS MAINLY SILTED UP.

of £8751-17s-6d to his wife, which was a considerable amount in 1921 and a far cry from the bankruptcies at the mill of a century earlier.[15]

Higher Roughwood Mill building is still standing although none of the machinery has survived. It is a four storey brick building with a slate roof. The original waterwheel was mounted internally with the water flow taken through the mill lengthways. For some time a small iron waterwheel was used to generate electricity.[16]

References

1. Norris, J. H., "The Water-Powered Corn Mills of Cheshire", *Transactions of the Lancashire & Cheshire Antiquarian Society*, Vol. 75, 1965.
2. CRO, Land Tax Returns for Betchton.
3. *Macclesfield Courier*, 11th May 1811.
4. CRO, Land Tax Returns for Betchton.
5. *London Gazette*, 14th June, 1817.
6. *Macclesfield Courier*, 16th August 1817.
7. Bott, O., "Cornmill Sites in Cheshire 1066-1850, Part 4", *Cheshire History*, 14, 1984.
8. CRO, Land Tax Returns for Betchton.
9. *London Gazette*, 22nd June, 1822.
10. CRO, Census Returns, Betchton, 1841.
11. CRO, Census Returns, Betchton, 1851.
12. Morris's Directory of Cheshire, 1864.
13. CRO, Census Returns, Betchton, 1861.
14. Kelly's Directories of Cheshire, 1892, 1896.
15. Stonier Probate Records, Stonier family Internet site.
16. Norris, J. H., 1965, *op. cit.* 1.

---

67. ROUGHWOOD SALTWORKS   SJ 780579

Roughwood Saltworks were located on the southern bank of the Trent and Mersey Canal just below lock 58 at Hassall Green. The earliest confirmation of the existence of saltworks at Roughwood is found in the will of Thomas Barlow who was described as a salt proprietor of Roughwood in 1798.[1] A few years later the saltworks were described as having a waterpowered pump for raising brine.[2] This was confirmed when the works were offered for sale in 1821, by Messrs Broughton, Sutton & Co. who were also the owners of Lawton Salt Works, in the following terms:-

"*For Sale - All those desirable salt works called Roughwood Salt Works now carried on by Messrs. Broughton, Sutton & Co. situated in the township of Betchton near Sandbach in the County of Chester comprising of four pans containing about 2000 feet of superficial measure, together with four stoves with rooms over the same and two store houses, also a brine pit from whence brine of a superior quality is raised by a waterwheel at a trifling expense and three reservoirs, a dwelling house for an agent and all other conveniences suitable for carrying on the salt trade, with land if required.*

The above works are eligibly situated on the banks of the Trent and Mersey Canal within five miles from the Staffordshire collieries and well suited for both the export and Duty trade and from the superiority of the Salt Manufactory at this are peculiarly adapted for the London trade

For details contact Mr. Kent, solicitor, London and Mr. Hostage, solicitor, Northwich."[3]

FIGURE 188. BRYANT'S MAP OF CHESHIRE, 1831, SHOWING THE LOCATION OF "ROUGHWOOD NEW MILL". TO THE EAST OF THE NEW MILL A "BRINE ENGINE" IS SHOWN WHICH WAS PART OF ROUGHWOOD SALTWORKS. THE SALTWORKS THEMSELVES ARE INDICATED ALONDSIDE THE CANAL. HIGHER ROUGHWOOD MILL AND ITS LARGE POOL ARE ALSO INDICATED FURTHER EASTWARDS. (CCALS, M5.2)

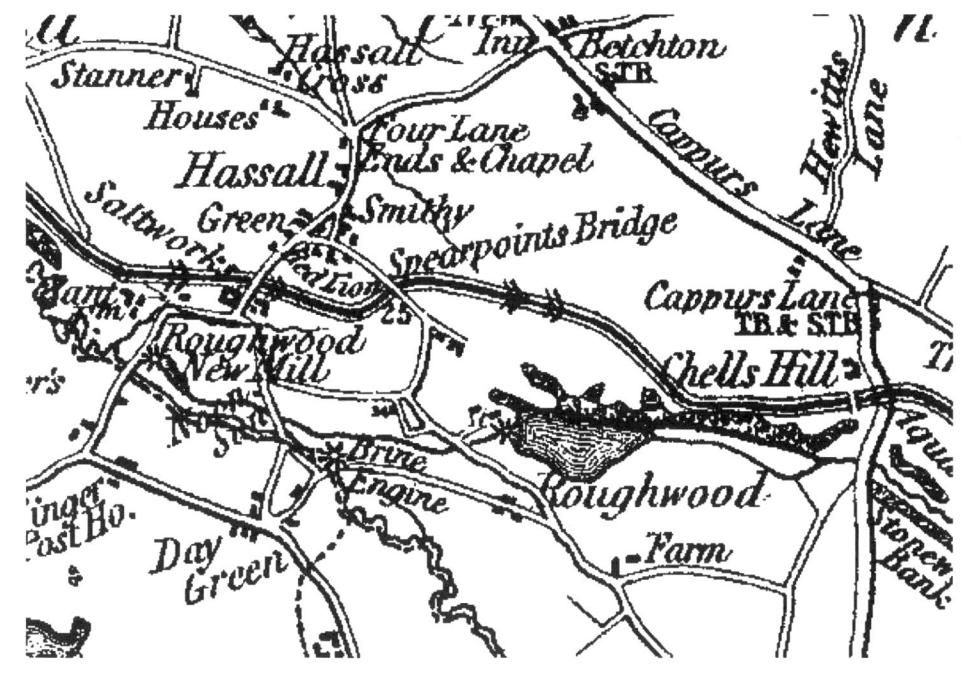

Although the advertisement specifies the use of waterpower to raise the brine, the site of Roughwood Saltworks alongside the canal is entirely devoid of any possible source of water. In fact, the brine pit was located about 500 yards to the south-east, down in the valley of the infant River Wheelock, close to its confluence with a small tributary stream that joins it from the south (see Figure 188).[4] The brine was pumped up to the Saltworks in a pipeline. The location of the saltworks alongside the canal was dictated by the need for coal from the Staffordshire collieries to evaporate the brine. The siting of the saltworks confirms that it was not established until after the building of the canal in the 1770s as prior to that time the supply of coal would not have been economically viable due to the high cost of transportation.

Today the site of the saltworks has been completely cleared and landscaped so no trace of it remains. At the site of the brine pit it is possible to discern the track of what could be the remains of leat, used to supply the waterwheel driven pump, in one of the gardens south of the lane passing the confluence of the two streams.

References

1. CRO, Will of Thomas Barlow, 1798.
2. Holland, H., *General View of the Agriculture of Cheshire*, 1808.
3. *Macclesfield Courier*, 12th May, 1821.
4. CRO, Bryant's map of Cheshire, 1831.

---

68. LOWER ROUGHWOOD MILL   SJ 775582

Lower Roughwood Mill stands beside the road from Hassall Green to Hassall, about a mile downstream from Higher Roughwood Mill (*q.v.*). The earliest evidence found so far of this mill is on Burdett's map of Cheshire, published in 1777, where the mill is labelled as "*Betchton Mill*". However, the mill site may have been in use for some time prior to this date as there is in existence a deed of 1623 referring to a mill at Betchton which may refer to either Lower or Higher Roughwood Mill.[1]

The mill has been noted as being completely rebuilt in 1783 which has some bearing on later information.[2] In 1811 the lease of a water corn mill in Betchton was offered for sale but again it is not clear if this referred to Lower or Higher Roughwood Mill.[3] However, in 1817, when Samuel Hassall was the miller at both mills he was declared bankrupt[4] and both mills were advertised for sale. In the sale notice one mill is referred to as a corn mill and the second mill was described as "*all that other newly erected mill formerley used as a flint mill*" but which was then also operating as a corn mill.[5] As Lower Roughwood Mill is said to have been rebuilt in 1783 then it is likely that this second mill referred to is at Lower Roughwood. It would seem that the mill had been used for flint grinding at some time but whether that took place before the rebuilding in 1783, or afterwards, is not clear.

The mill was part of the Wilbraham estate, as was Higher Roughwood Mill, and was occupied by Thomas Holbrook from 1820.[6] Interestingly the mill was called "*Roughwood New Mill*" on Bryant's map of 1831 and just "*New Mill*" on the first edition of the Ordnance

FIGURE 189. LOWER ROUGHWOOD MILL (*on the right*) AND MILLHOUSE IN THE EARLY 20TH CENTURY. (*S.B. Publications*)

Survey map of 1840. In 1841 Thomas Holbrooke employed two miller's labourers and a servant who lived with him and his wife at Hassall House.[7] Ten years later he was employing Joseph Stanyer, a journeyman miller, and Joseph's thirteen year old son James as a miller's apprentice.[8] Thomas Holbrook continued to occupy the mill until his death in 1853 when he left the mill to George Moore, son of his brother-in-law John Moore. His will was quite explicit that nothing was to go to his brother, George Holbrook, who owed him money, or to George's children.[9] George Moore was the miller at Lower Roughwood Mill until about 1860[10] when the mill came into the hands of the Holland family. John Holland, who had operated Higher Roughwood Mill since 1823, retired to his farm about this time leaving his three sons, George, John and William, to run both of the Roughwood Mills. John Holland operated Lower Roughwood Mill with George Holland running Higher Roughwood Mill.[11]

John Holland lived with his wife and his six children at Hassall Villa, moving later to Ivy Cottage, and continued running the mill until 1885 when he was declared bankrupt.[12] After 1885 the mill was still run by a John Holland who came from another branch of the family. This third John Holland ran the mill in conjunction with George Holland at Higher Roughwood Mill until 1902.[13] When George's interest in Higher Roughwood Mill came to an end in that year, John took over the mill at Unkinton Lodge near Tarporley which he operated together with Lower Roughwood Mill until 1939 in spite of the mill being offered for sale in 1931. At the time of this sale the mill was described as

"A well fitted Valuable CORN MILL built in brick and slate, driven by water power through an eleven foot wheel and containing three floors and a gear room; the fittings include four pairs of millstones, a wheat screen, an elevator and other fitments, the whole being in a very fair state of repair."

The rent payable for the mill at this time was quoted as £50 per year.[14]

The building of Lower Roughwood Mill still survives and has recently been converted into residential accommodation. It is a four storey high brick building with a slate roof. It appears to have been built in two phases, the first phase, built using English garden wall bond, was three storeys high and measured 30 feet by 21½ feet. The later second phase, which used Flemish bond, extended the building to 45½ feet long and raised the building by another storey with the height to the eaves being 27 feet and the roof ridge another ten feet higher. The window frames are made of cast iron. Although most of the machinery has been removed from the mill there were indications of the position of the upright shaft and the millstones. The remains of the waterwheel were still to be seen in the lean-to at the rear of the building. This was an overshot wheel with a diameter of 11 feet and a width of 6½ feet. The wheel was mounted on octagonal castings on an 8 inch diameter iron shaft. It had two sets of eight wooden arms bolted to iron hubs and shrouds. there were 34 buckets with wooden risers and iron bucket boards. The sole boards were also made of wood. There was also the remains of the sack hoist mechanism and one of the beams had the name "Thomas Holbrook" painted on it.[15]

References

1. Norris, J. H., "The Water-Powered Corn Mills of Cheshire", *Transactions of the Lancashire & Cheshire Antiquarian Society*, Vol. 75, 1965.

FIGURE 190. A RECENT VIEW OF LOWER ROUGHWOOD MILL (*on the right*) PRIOR TO CONVERSION INTO LIVING ACCOMMODATION.

2. Bott, O., "Cornmill Sites in Cheshire 1066-1850, Part 4", *Cheshire History*, Vol. 14, 1984.
3. *Macclesfield Courier*, 11th May, 1811.
4. *London Gazette*, 14th June, 1817.
5. *Macclesfield Courier*, 16th August, 1817.
6. CRO, Land Tax Returns, Betchton.
7. CRO, Census Returns, Betchton, 1841.
8. CRO, Census Returns, Betchton, 1851.
9. CRO, Will of Thomas Holbrook, 1853.
10. CRO, Betchton Parish Register.
11. Various Directories of Cheshire, Kelly 1857 & 1865, White 1960, Morris 1864.
12. *London Gazette*, 11th August, 1885.
13. Kelly's Directory of Cheshire, 1892, 1896.
14. CRO, Sale Catalogue, 1931, SC/BETC/1.
15. Congleton Borough Council, Survey Report attached to Planning Application 31434/3.

---

69. BANK MILL   SJ 843572

Two of the streams that converge at Wheelock to form the River Wheelock rise on the western slopes of Mow Cop. Bank Mill was situated about half way up the Cheshire side of Mow Cop where the southernmost of the two streams rises from a spring. This spring was immediately impounded in a pool to provide the necessary water supply for Bank Mill.

The position of Bank Mill is not ideal for a waterpowered mill being located almost at the source of the stream of water that it used. The flow would have been quite small and subject to seasonal fluctuation. This probably accounts for the fact that this is not a particularly old mill site, the first evidence of the existence of the mill being indicated on Bryant's map of Cheshire in 1831. Certainly the mill does not appear on the earlier maps of Cheshire published in the late 18th and early 19th century. Evidence from the 1840s shows that the mill was owned and worked as a corn mill by John Ford[1] who attempted to sell the mill in 1840 when it was described as follows:-

"A good corn mill worked by water and steam power with four pair of stones, two Frenches, one meal and one shelling mill situated in the township of Odd Rode near Lawton in the parish of Astbury in the county of Chester about five miles from Congleton, six miles from the Staffordshire Potteries and half a mile from the Macclesfield Canal.

Apply to the owner Mr. John Ford, Odd Rode."[2]

The details of the machinery of the mill show that it was a very conventional mill for its time and location, producing flour from the two pairs of French burr millstones and animal feed and oatmeal from the other two pairs of stones. Also, the steam engine may well have been installed when the mill was built considering the nature of the water supply. A year later John Ford was recorded as being a miller and farmer employing three men but whether they were employed in the mill or on the farm is not clear.[3]

It is most likely that the mill was not sold at this time because ten years later it was operated by William and Samuel Ford, presumably sons of the original John Ford.[4] In the intervening years from the first unsuccessful attempt to sell the mill the Ford family had made considerable improvements and additions to the mill to overcome its shortcomings and increase the mill's output. The nature of these improvements can be gauged from the description of the mill in the advertisement that was published when the mill was offered for sale in 1853, as follows:-

"All that excellent and truly valuable steam corn mill with three pair of French stones, meal and shelling mills, dressing & smut machines, bean splitter, malt mill and drying kiln, driven by a condensing steam engine of 24 horse power, the cylinder and nozzles of which are quite new, with 30 horse power boiler, together with about four acres of land attached to the mill situated at Bank aforesaid and in the occupation of Mr. John Ford.

The mill is in good repair and capable of getting up 200 sacks of flour each week exclusive of meal, etc. Situated within five miles of Congleton and the Staffordshire Potteries, one mile from Mow Cop and two miles from Harecastle in all of which there is great demand."[5]

The flour production capacity had been increased by the addition of a third pair of millstones for this process and the quality of the flour produced had been improved by initially passing the grain through a smut machine and finally using a dressing machine to separate the flour and bran produced by the millstones. Also other products had been introduced to supplement the output of flour, meal and oatmeal. This is highlighted by the installation of a malt mill (used in the preparation of beer) and a bean splitter. The figure given of 200 sacks per week for the flour output is somewhat optimistic. At the rate of one sack per hour for one pair of millstones, then three pairs of millstones would have to be grinding for nine hours per day, seven days per week to produce this total output of 200 sacks per week. Theoretically the mill was "capable" of this level of output but it would not be very likely in practice.[6]

Once again this proposed sale was unsuccessful, but

four years later, in 1857, the Fords were more determined than ever to sell the mill and went into even more detail when they described the contents of the mill as follows:-

"All the valuable machinery, meal and flour arks, new sacks, patent weighing machine, trucks, horses, wagons, carts, gears, harness, large weighing machine, iron rails, Delph wagons, corves, air pipes, anvils, bellows, tools, etc belonging to Mr. John Ford who is retiring from the business comprising cast iron drive shaft 7 feet by 8 inches with two bevil [sic] wheels, one of which is 4 feet 3 inch diameter, the other is 3 feet 3 inches and both 2¾ inch pitch, cast iron upright shaft with spur gear, three pairs of excellent French stones 4 feet 2 inch; meal mill, shelling mill, dressing machine, smut ditto, bean splitter, malt mill and drying kiln with brick quarries and cast iron bearers, cast iron pedestal, pair of bevil wheels to drive dressing machine etc, wrought iron upright shaft to drive shelling mill 10 feet 6 inch by 3½ inch diameter; cast drum to drive machine 4 feet 6 inch diameter; ditto to drive smaller 4 feet ditto; stone staffs, steel prover, elevated meal sifter with wood framework complete to all the machinery. A quantity of flour and meal arks, new sacks in lots, driving straps of various sizes, machine pulleys, patent weighing machine, beam with scales, weights, three flour trucks, lifting tackle, brass bushes & pedestals, about 80 yards of wrought iron rods in lengths of 15 feet with universal joints used in driving machinery at a distance; wood drum with cast iron arms 6 feet diameter; capital weighing machine as good as new to weigh 6 tons with cast iron frame and top; office desk with mahogany top, stool, large quantity of cast iron rails about 1/2 cwt each, wood and iron corves, 15 wood (air) pipes, delph wagons, cast iron moulding boxes, large water buckets, ropes, chain, two anvils, large pair blacksmith's bellows, small ditto, bench & large vice, various tools, sundry lots of iron, miscellaneous timber, various carts, grindstone, etc."[7]

It is quite clear from this advertisement that the mill gearing was organised around a conventional upright shaft and spur wheel configuration. The bevel gear pitch quoted in the advertisement as 2¾ inch is commensurate with gearing produced around 1830 which confirms to some extent the probable date when the mill was first constructed.[8] Apart from listing all the bits and pieces used in running a corn mill, the advertisement mentions a set of rods with universal joints used to convey the drive from the mill for some other machinery and application up to 80 yards from the mill. Unfortunately no details are given about what this remote drive facility might have been used for. Bryant's map of 1831 might give a clue in so far as there are coal pits marked quite close to Bank Mill that are linked to the mill by a short tramway. The power from the mill might have been used in connection with these coal pits or even for operating the tramway as the ground covered has quite a steep slope.

This time John Ford was successful in selling the mill to Jas. Goodwin & Co. who converted the corn mill into a flint grinding mill.[9] In 1861 Samuel Chappell was listed as a flint grinder at Bank, working the mill

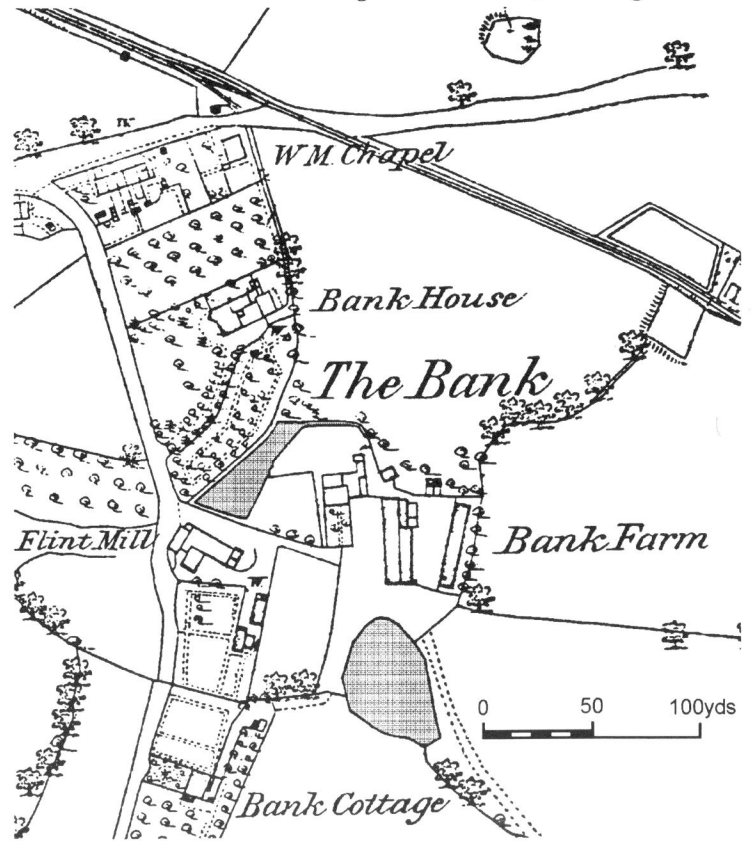

FIGURE 191. PART OF THE ORDNANCE SURVEY MAP, 1873, SHOWING BANK MILL WHEN OPERATING AS A FLINT MILL. NOTE THE SPUR TO THE MILL FROM THE TRAMWAY JUST TO THE NORTH. THIS TRAMWAY RAN FROM STONETROUGH COLLIERY ON THE FAR SIDE OF THE MOW COP ESCARPMENT TO A WHARF ON THE MACCLESFIELD CANAL AT KENT GREEN. THE TRAMWAY PASSED THROUGH A TUNNEL UNDERNEATH THE SUMMIT OF MOW COP.

with the help of his three sons John, Herbert and William who were 21, 16 and 14 years old respectively.[10] Samuel Chappell only remained at the mill until 1865 when he was replaced by George Stonier, the cornmiller at Rode Mill (*q.v.*) who tried his hand at operating the flint mill at Bank in conjunction with his cornmilling business.[11] It is possible that the flint mill was still in work in 1871 when John Chappell was recorded as a flint grinder living as a lodger at the Mill House at Bank.[12]

This is the last evidence of Bank Mill being used as some form of powered mill. In 1874 only a coal merchant was recorded living at Bank[13] but in 1881 George Baddeley, a fustian cutter noted as employing sixty hands, was living there, although it is not clear if the actual cutting of fustian was taking place at Bank Mill or elsewhere.[14] From 1892 until 1906 Edward and George Knapper were listed as fustian cutters at Bank Mill which tends to confirm that the mill was used for this activity.[15] At this time fustian cutting was purely a hand process and so the waterpower available at the mill would not have been used after the early 1870s. After 1906 no further industrial use was made of the site. At some stage the mill building was demolished and today the site is occupied by a house.

References

1. CRO, Odd Rode Tithe map and apportionments.
2. CRO, Census Returns, Odd Rode, 1841.
3. *Macclesfield Courier*, 7th November, 1840.
4. Bagshaw's Directory of Cheshire, 1850.
5. *Macclesfield Courier*, 24th September, 1853.
6. Davies, D. L., *Watermill, Life Story of a Welsh Cornmill*, Ceiriog Press, 1997.
7. *Macclesfield Courier*, 21st January, 1857.
8. Stoyel, A., *Perfect Pitch: The Millwright's Goal*, Society for the Protection of Ancient Buildings, 1997.
9. Kelly's Directory of Cheshire, 1857.
10. CRO, Census Returns, Odd Rode, 1861.
11. Morris's Directory of Cheshire, 1864; Kelly's Directory of Cheshire, 1865.
12. CRO, Census Returns, Odd Rode, 1871.
13. Morris's Directory of Cheshire, 1874.
14. CRO, Census Returns, Odd Rode, 1881.
15. Kelly's Directory of Cheshire, 1892, 1896, 1906.

---

70. RODE CORN MILL   SJ 822578

The stream that rises near Bank Mill (q.v.) on the western slope of Mow Cop quickly falls off the high ground, running through Kent Green and Scholar Green before flowing along the north side of the grounds of the Wilbraham family seat at Rode Hall. Rode Mill is situated quite close to the northern end of the ornamental lake which is in the grounds of Rode Hall.

The earliest indication of a mill at Rode is a reference in 1313 to a "*molendinum de Rode*" when Richard de Rode inherited his estate.[1] After this date there are a number of 14th century references starting with a lease dated 3rd May, 1328, for a watermill of Rode from Richard de Morton to Richard de Rode for 16 years at 20s per year.[2] Just over fifty years later the moiety of the mill of Rode was leased by Richard de Rode back to Richard de Morton at a rent of only 10s per year.[3] This drop in the value of the annual rent shows the effect of the Black Death which struck the country in the middle of the 14th century. The population was decimated and demand for milling capacity was greatly reduced thereby lowering the rents landlords could charge for their mills. Throughout the 16th and early 17th century the manor of Rode including its mill appears to have been leased out a number of times.[4]

In 1647 Randle Rode paid 29 shillings for the right to stop the water of the brook running between "*Little Wood and Selsacreto*[?]", to make a pool and maintain its height with a dam and sluices, and to let water out into the brook or watercourse at the mill.[5] Four years later, the miller at Odd Rode was Ralph Peever who entered into an agreement with Robert Wilkinson of Congleton, described as a "*paynter*".[6] When the Poll Tax assessment was made in 1660 two millers were recorded at Rode, namely Richard Peever with his wife Elinor and Thomas Hodgkinson with his wife Hannah. Both millers were assessed for a poll tax of 1s-0d each.[7] It is not clear whether both these millers worked together at Rode Mill or whether they operated separate mills. Nearly a decade later, Anthony Arrowsmith was recorded as the "*milner at Odde Rode*" together with his wife Sara when his sons William and John were baptised.[8] However, Richard Peever was still the miller at this time when he arranged to rent a tenement near to the mill. It seems likely that the Peever family were the resident millers at Odd Rode but they probably employed various journeymen millers to assist with the workload.[9] A few years later millers in Odd Rode had the surnames of Hodgkinson in 1673, and Burkitt in 1675, as recorded when their children were baptised.[10]

In 1669 Randyll Rode leased the manor of Rode including the water corn mill called Rode Mill to

Thomas Wichstead in trust for Roger Wilbraham of Nantwich, who was probably related to the Rode family.[11] During the second half of the 17th century Odd Rode mill was using Mow Cop millstones which were obtained on two months notice from the quarrymasters on Mow Cop at a special price of 20s each.[12] This was a considerable discount on the normal price for a millstone at this time of around £3-10s-0d.[13] The discount was due to the fact that lord of the manor was the landlord of both the mill and the millstone quarries. This custom would seem to signify that Mow Cop stones had been used at the mill since the time when 20s was the normal price for a millstone, probably in the 14th century.

During the 18th and into the 19th century the mill at Odd Rode featured in a number of transactions. In 1722, the pre-marriage settlement of Randle Wilbraham, the second son of Randle Wilbraham, and Dorothea, daughter of Andrew Kendrich, included the manor of Rode and watermill[14] and a century later Randle Wilbraham leased the manor and mill of Rode to Edward Bootle Wilbraham and Wilbraham Egerton.[15]

It was around this time that the Stonier family began to appear in connection with Rode Mill when John Stoneyer[sic] was noted as the miller at Rode Mill with his wife Mary when they baptised their son Thomas in 1825.[16] The Stonier (sometimes spelt Stonehewer or Stoneyer) family had a long connection with the area around Mow Cop, having been involved in the millstone trade since at least 1377.[17] Although the mill was considered to be occupied by Henry Yates in the 1839 as part of the Wilbraham estate,[18] in practice the Stonier family lived at the mill house and worked the mill. George Stonier was the miller at Rode Mill in 1841 when he was only 20 years of age.[19] His parents had, no doubt, passed on the miller's job to their son at this time as he had a wife ten years his senior and two small children to support. It is not known what happened to his wife and two children but by 1851 George Stonier had remarried and started another family. At this time he described himself as a miller and farmer, employing two men of which one was a miller's labourer.[20]

By the time of the next census in 1861 George Stonier's family had increased to four children. He employed a journeyman miller, John Jepson, who lived with his wife and family in the Mill Cottages.[21] In 1865 George Stonier tried his hand at running the flint mill at Bank Mill (*q.v.*) near Mow Cop in conjunction with his business at Rode Mill, but this venture did not last long.[22] The 1871 census shows that the Stonier family now consisted of six children and the eldest two sons, Francis and Frederick were assisting their father in the

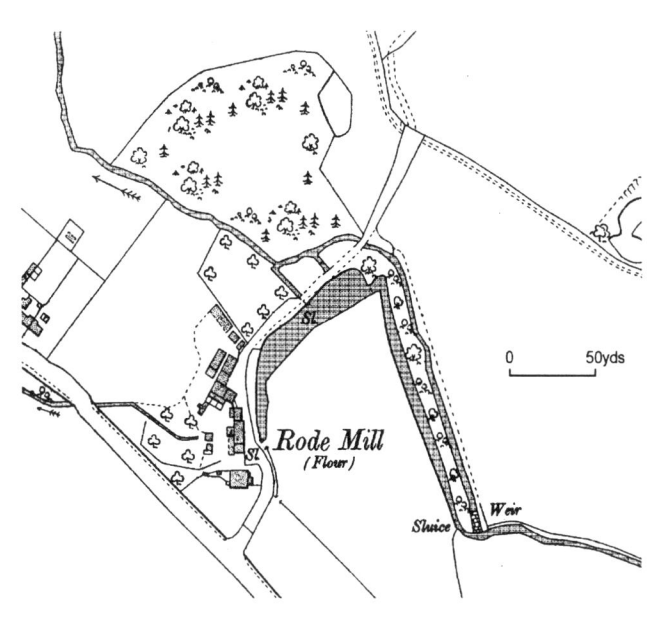

FIGURE 192. ORDNANCE SURVEY MAP OF RODE MILL, 1908, SHOWING THE SICKLE SHAPED MILLPOOL. NOTE THAT THE TAILRACE AND THE STREAM FLOW IN DIFFERENT CHANNELS FOR A CONSIDERABLE DISTANCE DOWNSTREAM OF THE MILL.

mill. The journeyman miller living with his family in the Mill Cottages was now James Robinson.[23]

Although George handed the milling business over to his sons Francis and Frederick Stonier in the late 1870s, George continued to live at the mill and probably took part in the daily work. In an attempt to widen the product range of the mill Frederick Stonier turned to the trade that his family had worked in for centuries, namely that of supplying millstones and in 1880 he advertised these wares in the trade papers.[24]

FIGURE 193. FREDERICK STONIER'S ADVERTISEMENT IN *The Miller* FOR THE SALE OF MOW COP MILLSTONES.

It is probable that he was not producing these millstones himself but was just marketing those produced in the quarries owned by Jamieson.

In 1881 Frederick Stonier was in sole charge of the milling business, living nearby with his wife Mary, his new born son and one servant, while at the mill lived his father George, his younger brother, also called George, both described as millers, his sister Emily and

FIGURE 194. A RECENT VIEW OF RODE MILL. THE INTERNAL WATERWHEEL WAS AT THE FAR END OF THE BUILDING WITH THE ENGINE ROOM AND CHIMNEY BEYOND. THE KILN IS IN THE BUILDING ON THE RIGHT SET BACK FROM THE FRONT OF THE MILL.

a lodger who worked as a carter. Frederick Stonier also had two other employees to work in the mill.[25] This was a considerable number of mouths to feed from the business of the mill and it is likely that the rather limited water supply was augmented by the addition of steam power sometime in the late 19th century.

Just after the turn of the century, after the death of his wife at the early age of 44,[26] Frederick Stonier took over running Higher Roughwood Mill (*q.v.*) as well as Rode Mill and was joined in the business by his son Frederick William Stonier.[27] It is not clear which Frederick Stonier operated which of these two mills but the Frederick Stonier of Rode Mill who died in 1914 leaving £4,784-9s-9d was probably the elder, with the Frederick Stonier of Roughwood Mill who died in 1921 leaving £8,751-17s-6d, probably being the younger.[28] Notwithstanding these two deaths an F. Stonier continued to be listed as the miller at Rode Mill until 1939.[29] After the Second World War the mill was used for other purposes and at one time the buildings were used to sterilise straw for use in the Potteries as packing. This activity caused the area to the rear of the mill and adjacent stables to be enclosed with roofing (which has since been removed).[30]

The small streams from the slopes of Mow Cop no longer feed into Rode Mill pool which is completly dry these days. The three storey brick mill building with its slate roof and central lucam is built into the dam of the mill pool such that it appears to be only two storeys high at the front. There used to be a footpath along the dam but this has been made into a road in the mid-20th century. The elongated pool lies alongside this road in a north-south orientation with the overflows located at the northern end flowing north-westwards to join the stream from Little Moreton Hall near what is known as the Pump House. The tailrace flows south-west along the western side of Rode Pool. There is no connection with Rode Pool as the tailrace veers to the west to rejoin the stream.

The mill building shows the signs of numerous modifications and extensions. The wall of the lowest floor which is built into the dam consists mainly of massive stone blocks, however, these provide only about two thirds of the wall, probably representing the foundations of an earlier structure. This stone wall has been extended in brick to the south to include the current wheelpit. The rear brick wall of the ground floor is built in English garden wall bond with alternating header and stretcher courses. This wall at first and second floor level is built in English garden wall bond but with three courses of stretchers to every course of headers, suggesting that the mill was also extended upwards when the mill was enlarged.

Most of the mill machinery was removed in the 1950s but the size of the wheelpit indicates that the waterwheel would have been about five feet wide and between 12 and 15 feet diameter. The wheel would have been breastshot or possibly overshot. In the mill itself there are indications that there was a standard spur wheel layout. The positions of two millstones alongside the wheelpit wall are indicated by the remains of their cast iron pans. Holes in the first floor suggest the position of a third pair of stones. The first floor is supported by iron beams at the end where the millstones were situated giving the impression that the hursting was renewed sometime in the latter half of the 19th century. The remains of what might have been the earlier wooden hursting can be seen as lintels and floor supports in the mill. The first floor is now bereft of any machinery but there are indications of the position of a lineshaft running the whole length of the mill. On the top floor all the storage bins have been removed except for one which still has the top part of a loading elevator in position. In the roof are the remains of the sack hoist. This consists of an oak shaft, about one foot diameter and about 16 feet long with a three feet diameter wooden pulley with four compass arms at the southern end which was originally driven by a slack rope from the floor below. About a third of the way along the shaft there are the signs of wear caused by the chain that passed over a pulley in the lucam to raise sacks from carts at the front of the mill. Further along the shaft there is another worn part caused by the internal sack hoist chain which ran upwards to a pulley then horizontally to another pulley before passing down through the floors of the mill.

At some time, possibly early in the 19th century, the mill was extended to the north by the addition of a kiln. The brickwork on this part of the mill is again English garden wall bond but with five rows of stretchers between the rows of headers. This construction has not been keyed into the earlier building. On the ground floor the furnace is still *in situ* with its cast iron grate. The first floor is the grain floor, which is made of ceramic kiln tiles laid on cast iron supports. This floor appears to be complete. Lying around in the kiln there are a number of broken kiln tiles showing three distinct types. The first type has round holes of about one inch diameter on the underside with each hole having ten small perforations in the upper surface. The second type is very similar in style but the round holes in their undersurface are about ¾ inch diameter with five perforations to each hole. The third type has square holes in the undersurface with five perforations per hole. All three types appear to have been commercially made. The space above the kiln is connected to the adjacent stable block by an enclosed walkway over the cart entrance to the rear of the mill, which links the second floor of the mill to the first floor of the stables (these floors are at the same level). In the late 19th century a further extension has been built onto the mill at its southern end to house a steam engine and boiler. This extension is now ruinous but the small chimney, about 30 feet tall, is still in good condition.

Some of the millstones used in the mill have also survived. Two French burr stones are built into the garden wall of the millhouse close to the mill. Elsewhere in the curtilage of the millhouse are the remains of a pair of Mow Cop millstones, each made up of a number of pieces of gritstone and with evidence of iron bands around their circumferences. The mill buildings are now used as a store for building materials.

FIGURE 195. THE REMAINS OF THE SACK HOIST INSIDE RODE MILL

References

1. CRO, Grant in Fee of 1313, DBW/A/A/B/16.
2. CRO, Lease of 1328, DBW/A/A/C/2.
3. CRO, Lease of 1389, DBW/A/A/B/46.
4. CRO, Leases of 1593, 1609 and 1625, DBW/A/B/1, DBW/M/A/7 and DBW/A/D/1-3.
5. CRO, Agreement of 1647, DBW/M/A/11.
6. CRO, Declaration of 1651, DBW/A/B/4.
7. Northwich Hundred Poll Tax Return, 1660.
8. Congleton Library, Astbury Parish Register.
9. CRO, Lease of 1670, DBW/M/B/1.
10. Congleton Library, Astbury Parish Register.
11. CRO, Lease of 1669, DBW/M/A/33 & 34.
12. CRO, Leases of 1647 - 1692, DBW/A/C/2-8.
13. Congleton Corporation Accounts, Congleton Town Council.
14. CRO, Pre-marriage settlement, 1722, DBW/A/C/27 & 28.
15. CRO, Lease of 1822, DBW/K/A2/34.
16. Congleton Library, Astbury Parish Register.
17. Bonson, T., "The Millstones of Mow Cop", *Proceedings of the 19th Mills Research Group Conference*, 2001.

18. CRO, Tithe Map & Apportionments, Odd Rode.
19. CRO, Census Returns, Odd Rode, 1841.
20. CRO, Census Returns, Odd Rode, 1851.
21. CRO, Census Returns, Odd Rode, 1861.
22. Kelly's Directory of Cheshire 1865.
23. CRO, Census Returns, Odd Rode, 1871.
24. *The Miller*, 4th October 1880.
25. CRO, Census Return, Odd Rode, 1881.
26. The Internet, Stonier family history web site.
27. Kelly's Directory of Cheshire, 1902.
28. The Internet, Stonier family history web site.
29. Kelly's Directory of Cheshire, 1939.
30. Information from Mr. Gardner whose family currently own Rode Mill.

---

71. LITTLE MORETON CORN MILL   SJ 833589?

Shortly after leaving Rode Mill the mill stream is joined by another stream which rises not far away to the north near Little Moreton Hall. Quite close to the source of this stream, in the fields alongside the south side of the hall, evidence has been found of the use of waterpower.

A possible indication of when the mill at Little Moreton Hall was built may be deduced from a quitclaim of 1320 from Richard son of Robert de Rode to his brother to make a water dyke over land at "*le Wodehouzes*" for a mill pond with a dam raised by his father.[1] Later maps of the area show Wood House was just south east of the hall. The stream and a pool are shown between "*Wood Ho.*" and another building called "*Pool Head*".[2] Mention of a watermill in Moreton was made at the beginning of the 15th century but in 1464 reference was made to two watermills and a smithy in the parish.[3] This situation had not altered by 1530 when two watermills and a bloomsmithy were at the centre of a dispute over water rights heard in the Star Chamber Court.[4] In 1655 the Moreton family of Little Moreton sold one watermill to John Bellot of Great Moreton.[5] During the latter half of the 17th century the owners of Moreton Mill were able to purchase millstones from the quarries on Mow Cop for the much reduced price of 20s each,[6] the normal price being around £3-10s-0d at that time.[7]

Burdett's map of Cheshire published in 1777 clearly shows a mill symbol close to Little Moreton Hall on a stream running to the south west. When Greenwood published his map of Cheshire in 1819 the mill was still indicated just south of Little Moreton Hall. In 1825 the Reverend Moreton, who was the absentee landlord of the estate, proposed to his agent that the mill pool should be drained and the mill demolished. The agent was not in agreement with this proposal as he claimed the machinery of the mill had been renewed within the last three years, that the building could be repaired for the expenditure of £10 and that estimates were being sought for dredging the pool, all of which

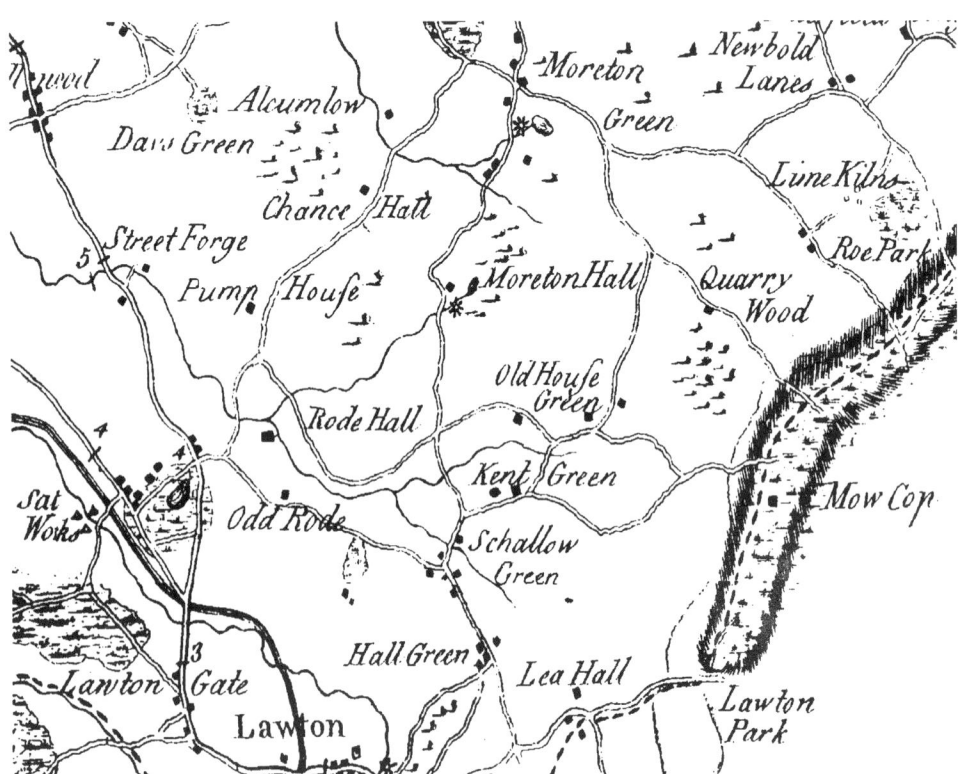

FIGURE 196. BURDETT'S MAP OF CHESHIRE, 1777. MORETON HALL IS SHOWN CENTRALLY WITH A MILL SYMBOL JUST TO ITS SOUTH. NOTE THE SYMBOL FOR THE SECOND WATERMILL IN MORETON, TO THE NORTH OF THE HALL. STREET FORGE IS ALSO SHOWN WEST OF THE HALL. (CCALS, PM12/16)

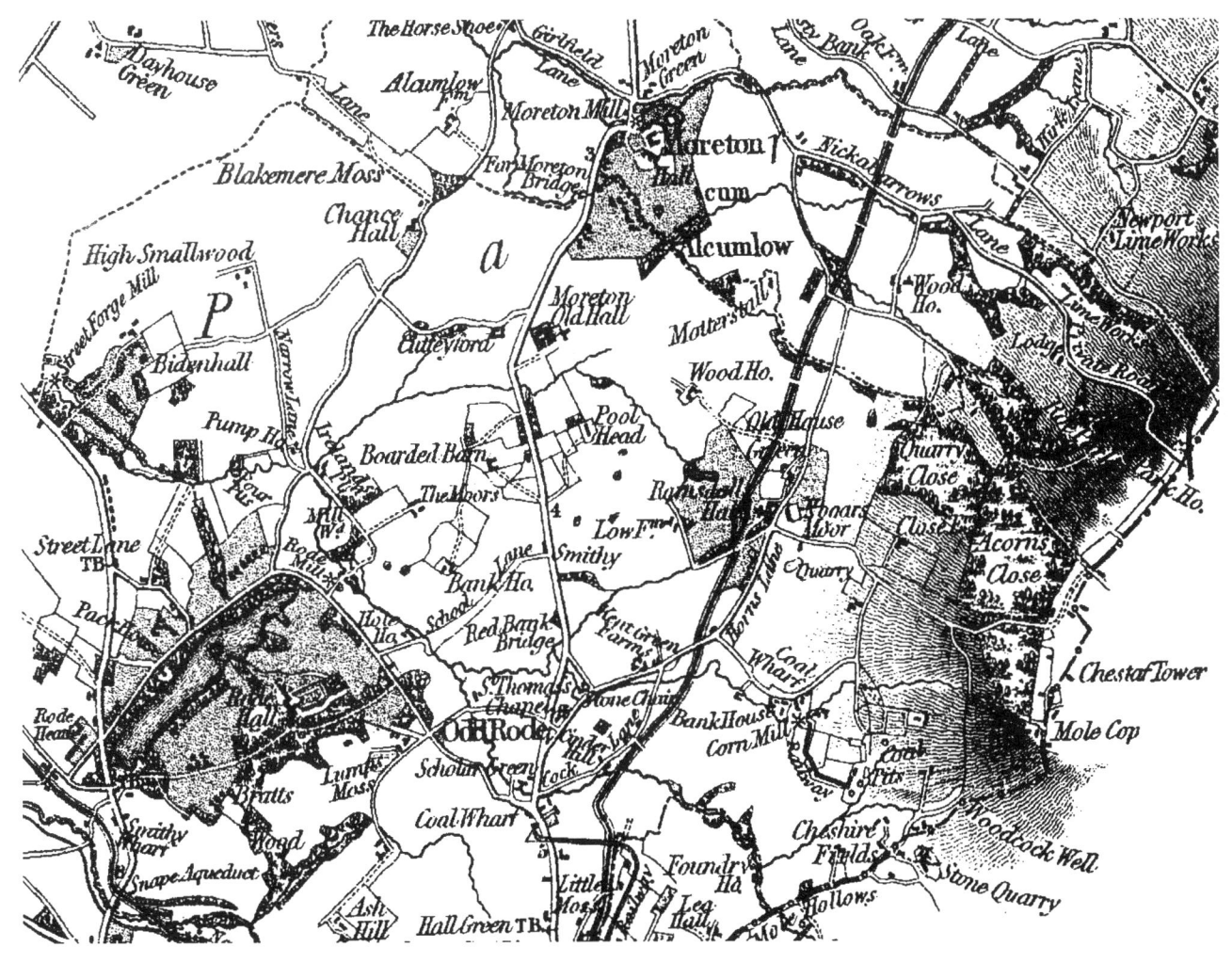

FIGURE 197. BRYANT'S MAP OF CHESHIRE, 1831. MORETON OLD HALL IS SHOWN CENTRALLY BUT THERE IS NO MILL SYMBOL SHOWN NEAR THE HALL, HOWEVER THE PROPERTIES CALLED WOOD HOUSE AND POOL HEAD ARE IDENTIFIED. OTHER MILLS TO BE SEEN IN THE AREA AROUND THE HALL ARE MORETON MILL TO THE NORTH, STREET FORGE MILL TO THE WEST, RODE MILL TO THE SOUTH-WEST AND THE CORN MILL AT BANK HOUSE TO THE SOUTH-EAST. (CCALS)

would enable the mill to be let to a tenant.[8] Whose proposals were followed is not certain but the mill was not indicated on either Swire & Hutchins map of 1830 or Bryant's of 1831 so it must be presumed that the landlord's view prevailed and the mill was in fact demolished around 1825.

Some fieldwork took place in 1987 at Little Moreton Hall to see if any trace of the mills remained. The main findings were two substantial banks (at SJ 833588 and SJ 835586) which ran across the line of two small streams. These were thought to have been the remains of millpool dams, a supposition supported by the field names on the tithe survey of the 1838 such as *"Mill Pool Meadow"* and *"Higher Pool"*. The field work also unearthed two large lumps of slag which were found near to the Higher Pool bank which would suggest that this was probably the site of the bloomsmithy. Close to the other bank, just south of the hall, some timbers were found in the stream which could have been the remains of a sluice gate. The site of this bank near to the hall is compatible with the position of the mill symbols on Burdett's and Greenwood's maps and probably signifies the most likely location of the corn mill.[9]

References

1. CRO, Quitclaim of 1320, DBW/A/A/B/26.
2. CRO, Bryant's Map of Cheshire, M5.2.
3. Dodgson, J. McN., *The Place Names of Cheshire, Part 2*, Cambridge University Press, 1970.
4. Lancashire & Cheshire Record Society, Vol. 71, quoted in Bradley, A. C., "Industry at Little Moreton", *Journal of the Congleton History Society*, Vol. 7, 1988.
5. Bradley, A. C., "Industry at Little Moreton", *Journal of the Congleton History Society*, Vol. 7, 1988.
6. CRO, Leases of 1647 - 1692, DBW/A/C/2-8.
7. Congleton Corporation Accounts, Congleton Town Council.
8. Bradley, A. C., 1988, *op. cit.* 5.
9. *Ibid.*

72. STREET FURNACE/FORGE/ MILL SJ 806586

The stream that powered Bank Mill (*q.v.*) and Rode Mill (*q.v.*) is joined just to the west of Rode Hall and its ornamental lake by the stream that powered the mill at Little Moreton Hall. Downstream of this confluence there is a low weir that provided the input to a half mile long leat which follows the stream in a north westerly direction to provide the necessary supply for the waterpowered site at Street.

The early history of this site is obscure but a furnace is likely to have been built around the time of the restoration of the monarchy in the 17th century. This was a time when the iron industry was being re-established after the depredations of the Civil War. In the late 1650s both Lawton Furnace (*q.v.*) and Cranage Forge (*q.v.*) had been established and by 1664 Cranage Forge was being operated by a partnership of William Fletcher and George Whyte. It was in this year that "*Mr. Fletcher & his partners*" bought hematite ore from Sir Thomas Preston's Stainton mine in Furness.[1] This was obviously destined to be used in a blast furnace, probably mixed with Staffordshire ore, which was a practice at Lawton Furnace around this time. As Fletcher and Whyte were not connected with Lawton Furnace, it is more likely that this ore was destined for use at the only other blast furnace in the area at this time, namely Street Furnace.

Information concerning Street Furnace in the 17th century is non-existent but it must have competed for charcoal and other supplies with Lawton Furnace as the two sites were within two miles of each other. This was probably the reason why, in 1700/1, Thomas Hall and the "Cheshire Works" partnership of ironmasters acquired Street Furnace. A comment in the Warmingham Forge (*q.v.*) accounts states that "*Street Furnace being only two miles from Lawton...and the woods in the country growing scarce, the intent of Thomas Hall taking this furnace was for the service of Lawton Furnace*".[2] In 1701 Street Furnace was in the process of being dismantled with stock and equipment being sent to the furnaces at Vale Royal and Lawton. Although the initial intention was to convert the furnace at Street into a wire works, the site was, in fact, converted into a plating forge specialising in producing salt pans, domestic pans and saws.[3] Both Edward Hall, son of Thomas Hall of the Cheshire Works, and Obadiah Lane of the Staffordshire Works, each advanced £400 for the running of Street Forge and Warmingham Forge, with Edward Hall as their manager. This was the situation until 1706 when there was a reorganisation of responsibilities with the Cheshire Works partnership. Edward Hall moved to the management of Vale Royal Furnace and Bodfari Forge (near Wrexham) and William Vernon took over at Street Forge and Warmingham Forge.[4]

Obviously the salt pan trade was confined to Cheshire but the trade in domestic pans was much more widespread with products travelling to places such as Manchester, Liverpool, Bristol, Uttoxeter, and London. Saw irons were delivered to a John Podmore of Kidderminster in large quantities for finishing as the final product. The accounts of this period attribute only about four tons per year of output directly to

FIGURE 198. A WATER POWERED HELVE HAMMER SIMILAR TO THAT USED AT STREET FORGE. (*Sheffield Chamber of Commerce*)

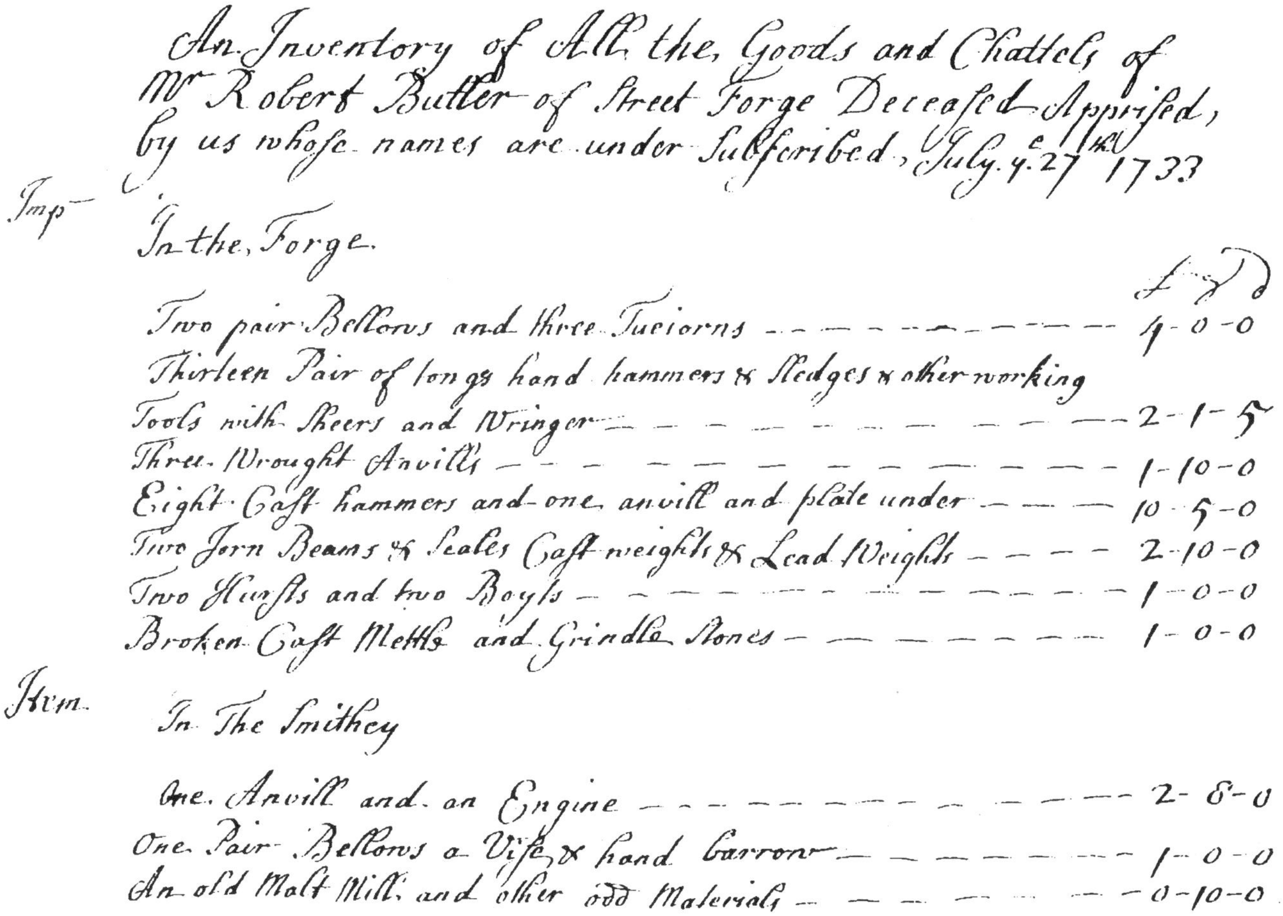

An Inventory of All the Goods and Chattels of Mr Robert Butler of Street Forge Deceased Apprised, by us whose names are under Subscribed, July 27th 1733

Imp~ **In the Forge**

	£ s d
Two pair Bellows and three Tueiorns	4-0-0
Thirteen Pair of tongs hand hammers & Sledges & other working Tools with Sheers and Wringer	2-1-5
Three Wrought Anvills	1-10-0
Eight Cast hammers and one anvill and plate under	10-5-0
Two Jorn Beams & Scales Cast weights & Lead Weights	2-10-0
Two Hursts and two Boyls	1-0-0
Broken Cast Mettle and Grindle Stones	1-0-0

Item **In The Smithey**

One Anvill and an Engine	2-8-0
One Pair Bellows a Vise & hand barrow	1-0-0
An old Malt Mill and other odd Materials	0-10-0

FIGURE 199. THE INVENTORY OF ROBERT BUTLER'S WILL LISTING THE ITEMS IN THE FORGE AND THE SMITHY AT THE TIME OF HIS DEATH IN 1733. THE CONTENTS INDICATE THAT STREET FORGE HAD REVERTED TO BEING JUST AN ORDINARY COUNTRY FORGE BY THIS TIME. (CCALS, WS 1733)

Street Forge with over 100 tons being produced at Warmingham.[5] This amount attributable to Street Forge seems to be so negligible that it would not justify the forge's operation, consequently it has been speculated that most of the output from Street Forge was included in the Warmingham figures as both forges were under the same manager during this period. When Stonier Parrot was installing an atmospheric steam engine at Park Colliery in Newcastle under Lyme in 1718 he proposed that the boiler should be made from salt pan plates. The possibility exists that these plates were to come from Street Forge as it was the nearest supplier of this type of iron product.

Sometime towards the end of the first quarter of the 18th century Street Forge was dispensed with by the Cheshire Works. In 1714 an *"iron plating mill"* was recorded as being part of the Moreton estate on the death of John Bellot in 1714 when it provided the estate with a yearly income of only £5.[6] The tenant during this period was Robert Butler who was producing much the same type of products as previously i.e. salt pans, pans and saws. When these changes took place is not known but the forge was definitely in Robert Butler's possession on his death in 1733 when there was a forge and a smithy at Street. The forge had *"two pair of bellows and three tueiorns"*, three wrought anvils with thirteen pair of tongs and various hand hammers. There was also eight cast hammers, an anvil with a plate under it, scales and weights, *"two hursts and two boyls, brokenncast mettle and grindle stones"*. In the smithy there was one anvil and an engine with one pair of bellows and a vice, and an old malt mill.[7] This inventory gives the impression of a small country forge rather than a place for large scale production. The old malt mill included in the inventory probably indicates that the forge was also making beer which, no doubt, was consumed in large quantities by the forgemen.

After Butler's death the forge was run by James Paddy who was a witness of Robert Butler's will and was probably married to Butler's daughter. The continuity of product line is attested to in 1750 when James Paddy was shipping consignments of saws down the River Weaver.[8] In the early months of 1779 James Watt was installing a steam engine at Lawton Saltworks (*q.v.*). This would have been an event of great interest to many people in the locality, none more so than a forge master such as James Paddy whose forge was less than a mile away. It is almost inconceivable that Paddy did not meet James Watt at Lawton and may even have done some work for him during the steam engine's erection. James Paddy certainly had an interest in the performance of the engine because, once the engine was working satisfactorily, he wrote to James Watt, as follows:-

> "I have a forge situated within about a mile of Lawton Salt Works where you erected an engine for Mr. Salmon...though we do a great deal of business are not withstanding in the summer time much distressed for water. I have some thoughts of erecting a fire engine upon your instructions to return at least a part of my water into the pool again."

Unfortunately, owing to the pressure of other work, Boulton and Watt had to decline the possible order from James Paddy.[9]

In 1781 the owners of the forge were registered as Thomas & John Powes Esq. who had taken over the Great Moreton Estate.[10] By this time, the registered occupier, James Paddy, must have been the son of the original James Paddy. This new generation had ideas of expanding their business interests. Martin and Charles Paddy who were probably James's brothers, became involved with Warmingham Forge and a partnership in cotton manufacture respectively. Details noted in 1791 show that James Paddy held the lease of Street Forge for one life aged 51 (probably himself), that the house, garden, plating & forge shop occupied just over three acres with the pool covering one acre. The rent on the forge was £30 per year and that on the pool was £6-17s-3d. The forge consisted of a chafery with one hammer and was short of water in summer.[11]

In 1793 James Paddey [sic] of Street Forge was declared bankrupt[12] and two years later the partnership between James Paddey of Street Forge and Martin Paddey of Warmingham Forge was also declared bankrupt.[13] Around this time the landowner of the forge changed when, in 1794, Holland Ackers purchased the Great Moreton Estate from Thomas and John Powes. The following year Holland Ackers sold the forge property to John Twiss who was probably a forgeman or ironmaster as he continued to operate the forge business himself.[14] In 1802 Charles Paddy and his partners in cotton manufacture were also declared bankrupt so ending all the Paddy business interests.[15] Meanwhile John Twiss continued to operate the forge until 1815 when Thomas Mellor and John Mabery were two of the men working at the forge.[16]

In 1815 John Twiss sold the forge to William Pointon who converted the forge into a corn mill.[17] William Pointon was a member of the family operating Colley Mill (q.v.) and his brothers also later operated Cranage Mill (q.v). William Pointon employed a miller called Thomas Pass to assist in working the mill.[18] When they moved to Street, William Pointon was 41 years of age and Thomas Pass was 23 years of age.[19] Over the following years the two families lived and worked together bringing up their various children at the mill until William Pointon died in 1842.[20] During this period the name of the mill changed, being recorded as "Street Forge Mill" in 1831 and eventually to just "Forge Mill" by 1841. After William Pointon's death the mill was run by his heirs William and George Pointon, who were presumably his sons, but Thomas Pass and his family left the mill in order to take the lease of Sandbach Mill (q.v.). In 1851 William Pointon's widow was still living at the mill house with two lodgers. One of the lodgers was John Holland, listed

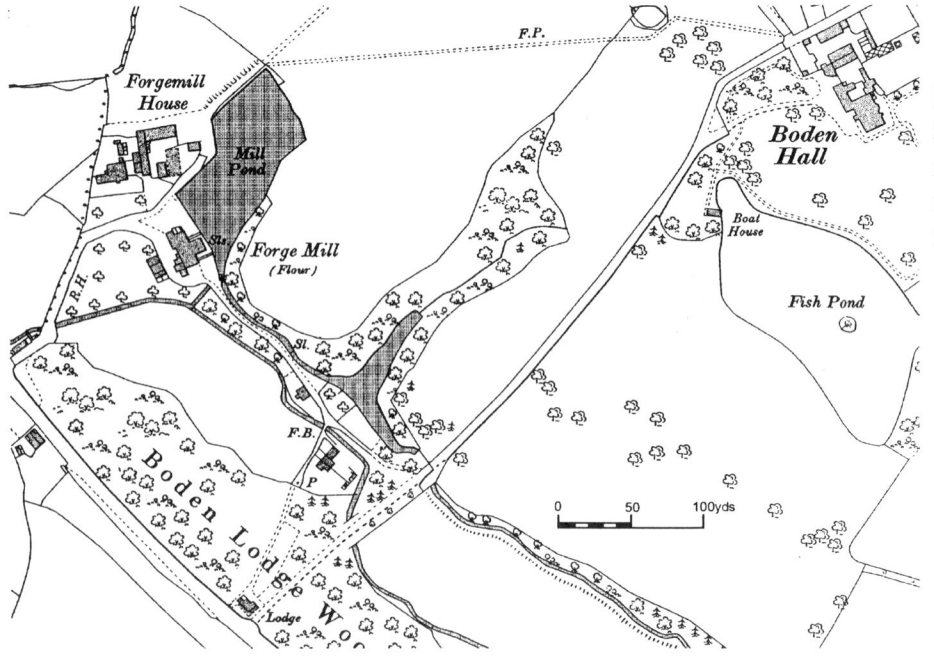

FIGURE 200. ORDNANCE SURVEY MAP OF 1908. THE MILL LEAT RUNS NORTH-WESTWARDS INTO THE MILL POND BEHIND FORGE MILL.

as a miller aged 22 years, who was in some way related to the Pointons, and the other lodger was Charles Condler, a journeyman miller.[21]

By 1860 George and William Pointon were operating both the Odd Rode Steam Mill as well as Forge Mill. The two brothers concentrated their efforts in the steam mill on the banks of the Trent & Mersey Canal with Forge Mill in the hands of their agent, John Holland who was also milling with his two brothers at Higher Roughwood Mill (*q.v.*).[22] However, by 1861 John Holland was no longer concerned at Forge Mill, which was being operated by William Pointon who lived with his wife Ann and eight children and a kitchen maid at Forge Villas. Also living at Forge Mill was John Dean, a carter, whose sons George and John Dean were both employed at Forge Mill as millers, as was John Goodwin.[23] Around 1874 both Forge Mill and the Odd Rode Steam Mill were acquired by Thomas Hall of Congleton Town Mill (*q.v.*) who then ran all three mills until about 1880 when he relinquished Congleton Mill.[24] At the height of his business empire, in 1881, Thomas Hall employed 16 men in his mills where his two sons Thomas and Cecil were both millers. At Forge Mill, John Jones was the miller living at Forge Mill Lodge with his wife Margaret and their four children. Thomas Hall lived at Boden Hall, quite near to Forge Mill, with his four sons and two servants.[25]

In the mid 1890s Thomas Bibby became the miller at both Forge Mill and Odd Rode Steam Mill but he only operated Forge Mill until around 1902. After this date no millers are listed at Forge Mill, only farmers, so it is to be assumed that Forge Mill ceased to operate as a mill around this date.[26]

Today the leat can be followed from its weir to the mill. The leat is about half a mile long and part way it has to cross a side valley on a small embankment. At the mill the water supply arrangements are quite unusual for the area as the leat has to pass the mill to enter the pool which was quite large but is now empty. The mill used to have two waterwheels and at one time was four storeys high but at some stage was reduced to only two storeys. There was also a drying kiln and at some time there was also a steam engine.[27] All the machinery was eventually removed around about 1940 but some of the brick built mill remains to be seen as a roofless, ruined shell.

References

1. PRO C 6. 182/18 quoted in Awtry, B. G., "Charcoal Ironmasters of Cheshire & Lancashire, 1660-1785", *Transactions of the Lancashire & Cheshire Antiquarian Society*, Vol. 109, 1957.
2. PRO C 9. 372/11 quoted in Awtry, B. G., 1957.
3. Johnson, B. L. C., "The iron industry of Cheshire & North Staffordshire: 1688-1712", *Transactions of the North Staffordshire Field Club*, Vol. 88, 1954.
4. Awtry, B. G., "Charcoal Ironmasters of Cheshire & Lancashire, 1660-1785", *Transactions of the Lancashire & Cheshire Antiquarian Society*, Vol. 109, 1957.
5. *Ibid.*
6. CRO, Inventory and Survey of John Bellot's estate, 1714, DSS 1/3/210.
7. CRO, Will of Robert Butler, WS 1733.
8. Awtry, B. G., 1957, *op. cit.* 4.
9. Birmingham City Library, Boulton & Watt Collection quoted in Chaloner; W. H., "Cheshire activities of Matthew Boulton & James Watt, of Soho, near Birmingham, 1776-1817", *Transactions of the Lancashire & Cheshire Antiquarian Society*, Vol. 61, 1949.
10. CRO, Land Tax Returns, Odd Rode.
11. Documents at Brereton Hall, Cheshire.

FIGURE 201. THE RUINS OF FORGE MILL, 1996, LOOKING TO THE SOUTH-EAST.

12. *London Gazette*, 13th July 1793.
13. *London Gazette*, 7th March 1795.
14. CRO, Land Tax Returns, Odd rode.
15. CRO, Document DCB/1595/5/13.
16. Congleton Library, Astbury parish Register.
17. CRO, Land Tax Returns, Odd Rode.
18. Congleton Libary, Astbury Parish Register.
19. CRO, Census Return, 1841.
20. CRO, Will of William Pointon, 1842.
21. CRO, Census Returns, Odd Rode, 1851.
22. White's Directory of Cheshire, 1860.
23. CRO, Census Returns, Odd Rode, 1861.
24. Morris's Directory of Cheshire, 1874.
25. CRO, Census Return, 1881.
26. Kelly's Directories of Cheshire, 1896, 1902, 1906, 1910, 1934.
27. Norris, J. H., "The Water-Powered Corn Mills of Cheshire", *Transactions of the Lancashire & Cheshire Antiquarian Society*, Vol. 75, 1965.

---

73. MORETON CORN MILL   SJ 836599

Another stream, called the Arclid Brook, rises on the western slope of Mow Cop running westwards through the parishes of Moreton and Smallwood and then the township of Sandbach before joining the River Wheelock. The first water mill on this stream is in Moreton where the stream was dammed at the point where it was crossed by the original turnpike road from Church Lawton to Congleton close to Great Moreton Hall.

The earliest reference to a mill on this stream in Moreton was in 1464 when Moreton was described as having two water mills and a smithy.[1] One of the mills and the smithy were at Little Moreton Hall, the other mill being at Great Moreton. Both these corn mills were owned by the Moreton family, who resided at Little Moreton Hall, until in 1655 one of the watermills was sold by the Moretons to John Bellot of Great Moreton.[2] In 1714 the inventory of the late John Bellot's estate showed that Moreton Mill was leased for £5 per year to Hugh Marton.[3] Between 1781 and 1786 the owner of Moreton Mill was recorded as J. Powys Esq. and the occupier as George Cartwright, but from 1787 to 1793 the occupier was recorded as Thomas Cooper.[4] Notwithstanding this record there is also evidence from 1791, that John Evans was actually at the mill which was described as having *"Three pairs of stones in tolerable repair"*.[5]

In 1794 the Great Moreton estate was sold to Holland Ackers after which Thomas Cooper continued to be recorded as the occupier of the mill until 1812.[6] The following year saw the start of eight years with Timothy Booth as the miller at Moreton Mill during which time he and his wife Ester raised three daughters.[7] Unfortunately, after eight years at the mill Timothy Booth was declared bankrupt in 1823, as follows:-

"Whereas Timothy Booth of Moreton in the County of Chester, miller and corn dealer, hath by indenture dated the 7th day of May, 1823, assigned over all his real and personal estate and effects to Jonathon Broadhurst of Congleton, corn merchant, and John Galley of Congleton, aforesaid draper, in trust for the equal benefit of such creditors as shall execute the same before 7th August 1823."[8]

After this episode the mill stood idle for a year being recorded as *"late Booth's mill"*, but in 1825 a new miller, called William Massey arrived.[9] William Massey, who was born in 1796, had married Sarah Sutton at Gawsworth in 1815 and was 29 years of age when he took over the mill.[10] William Massey was still at the mill in 1841 when it was having problems with its mill pool causing the following advertisement to be issued:-

"To contractors and others, to be let by ticket, the mudding or cleaning of the Moreton Pool or Mill Dam.

The work will be let at a Gross Sum and is to be completed on or before the 30th day of April, 1841.

The contractor at his expense to find all implements and materials necessary for the purpose and give satisfactory security for the due performance of the work.

Moreton Pool is about three miles from Congleton and near the turnpike from there to Newcastle.

Apply to John Yates, Smallwood, and Thomas Lawton, Bradley Green, Biddulph."[11]

In 1844, George Ackers, who was the landlord of Moreton Mill and owner of the Great Moreton estate, decided that he wished to construct parkland around Great Moreton Hall. To ensure his privacy he proposed moving the turnpike road to a new route to the west and removing the village of Moreton altogether.[12] Although the work on improving the mill pool had only taken place a few years earlier and that William Massey and his wife, both aged 49 years, and their five children depended for their livelihood on the mill, they were all evicted and the village with its mill demolished.[13] The turnpike road was diverted at George Ackers expense to the line now occupied by the A34. The site of the village and its mill can be found as the mill pool was kept as a landscape feature which can still be seen today in the grounds of Great Moreton Hall.

References

1. Dodgson, J. McN., *The Place Names of Cheshire, Part 2*, Cambridge University Press, 1970.
2. Bradley, A. C., "Industry at Little Moreton", *Journal of the Congleton History Society*, Vol. 7, 1988.
3. CRO, Inventory and Survey of John Bellot's estate, 1714, DSS 1/3/210.
4. CRO, Land Tax Returns, Moreton.
5. Valuation of Rode township, 1791, at Brereton Hall.
6. *Op. cit.* 4.
7. Congleton Library, Moreton Parish Records.
8. *Macclesfield Courier*, 17th May 1823.
9. *Op. cit.* 4.
10. Congleton Library, Gawsworth Parish Record.
11. *Macclesfield Courier*, 13th February, 1841.
12. Bradley, A. C., 1988, *op. cit.* 2.
13. CRO, Census Return, Moreton, 1841.

---

74. SMALLWOOD CORN MILL   SJ 800612

About 2½ miles downstream from the site of Moreton Mill, and about 2½ miles east of the town of Sandbach lies Smallwood Mill. This is another mill that originated in the medieval period as there is a reference to "*a water mill in Smalwode*" in 1294[1] with another recording of "*Smalwode milne*" in 1404.[2] No other evidence has been found of the mill until the mill was noted as being part of the Moreton estate in 1714 with an annual rental of £10.[3]

In the ten years between 1781 and 1791 the land owner of Smallwood Mill is recorded as Thomas Jeff. Powys of Great Moreton Hall and the occupier was Josiah Holland.[4] Further evidence from 1791 shows that Josiah Holland held the mill on a lease for two lives of 38 and 35 years of age. The rent paid for the mill and mill pond appears to have been £25 per year and the mill is described at this time as having two waterwheels and four pairs of stones.[5] For the two years 1792 and 1793 the occupier is recorded as being Daniel Dale but in 1794 the Great Moreton estate was sold to Holland Ackers and the mill reverted to Josiah Holland until 1798.[6]

In 1799 the occupier of the mill became John Holland who was, in all probability, Josiah Holland's son. John Holland continued to hold the mill until his death in 1824. From 1825 to 1830 the mill was held by John Holland's widow who employed a journeyman miller called Richard Ince.[7] This arrangement kept the mill in the Holland family until John Holland, probably the son of the original John Holland at Smallwood Mill, became the miller in 1831.[8] Why he did not take over the mill on his father's death is not clear as he was aged around 30 at the time and would already have been a trained miller. He continued to operate the mill, with the assistance of a miller called John Wilkinson whom he employed,[9] until about 1839 when the Manor and Mill of Smallwood was advertised for sale.[10] Around about this time John Wilkinson died and John Holland moved to Higher Roughwood Mill (*q.v.*).[11]

In 1841, at the age of 51, Jane Wilkinson, the widow of John Wilkinson was the miller in her own right at Smallwood Mill, where she lived with her two daughters.[12] It is possible that her son, George, was apprenticed to another mill to learn the trade as he eventually took over from his mother at Smallwood Mill by 1850 where he lived with his wife Mary, five young children and a servant.[13] George Wilkinson remained at the mill until 1857 when Ambrose Dean took the lease of the mill and operated it until about 1880.[14]

By 1881 John Condliffe was living at Yew Tree House in Smallwood with his wife Helen, his four children, one servant and one carter when he was described as a miller and farmer.[15] John Condliffe worked the mill until just before the First World War by which time he would have been about 70 years old. In the last years of the 19th century he also leased Somerford Booths Mill (*q.v.*) having gained a contract to supply the local poor house at Arclid. After John Condliffe the mill was operated by Frederick William Hall,[16] one of the Hall family of millers that had

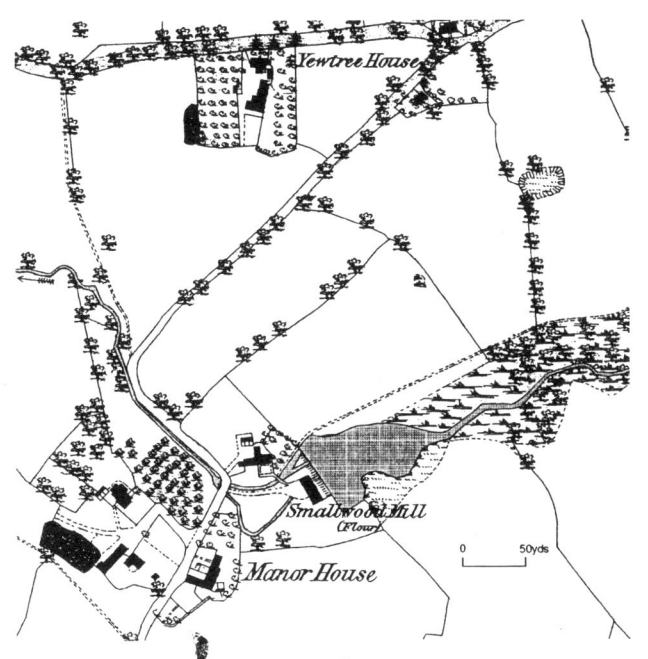

FIGURE 202. ORDNANCE SURVEY MAP, SHOWING SMALLWOOD MILL AND POOL.

previously been involved with Congleton Town Mill (*q.v.*), and both Forge Mill (*q.v.*) and the steam mill in Odd Rode. Although F. W. Hall was still occupying Yew Tree House in 1939, Harold Baker was listed as the miller at Smallwood Mill using steam and water power. Obviously at some stage in the early 20th century a steam engine had been acquired at the mill to supplement the water power.[17]

In 1953 the mill pool was drained and the water power was replaced by electricity.[18] In the 1960s a new building was erected on the site *"around the old mill"*.[19]

References

1. PRO, Close Rolls, quoted in Dodgson, J. McN., *The Place Names of Cheshire, Part 2*, Cambridge University Press, 1970.
2. PRO, Ministers Accounts, quoted in Dodgson, J. McN., *The Place Names of Cheshire, Part 2*, Cambridge University Press, 1970.
3. CRO, Inventory and Survey of John Bellot's estate, 1714, DSS 1/3/210.
4. CRO, Land Tax Returns, Smallwood.
5. Valuation of Rode and other townships, 1791, held at Brereton Hall.
6. *Op. cit.* 4.
7. Congleton Library, Smallwood Parish Register.
8. *Op. cit.* 4.
9. *Op. cit.* 7.
10. *Staffordshire Advertiser*, 3rd August 1839.
11. CRO, Will of John Wilkinson, 1839.
12. CRO, Census Returns, Smallwood, 1841.
13. Bagshaw's Directory of Cheshire, 1850; CRO, Census Returns, Smallwood, 1851.
14. Various Directories of Cheshire, 1857 - 1880.
15. CRO, Census Returns, Smallwood, 1881.
16. Kelly's Directory of Cheshire, 1914 & 1934.
17. Kelly's Directory of Cheshire, 1939.
18. Norris, J. H., "The Water-Powered Corn Mills of Cheshire", *Transactions of the Lancashire & Cheshire Antiquarian Society*, Vol. 75, 1965.
19. Meeke, M., *More Old Smallwood*, 2000.

---

75. BROOK SILK MILL   SJ 759605

About 300 yards south-east of the centre of Sandbach there is a slight valley in which the Arclid Brook flows on its way from Moreton and Smallwood to join the River Wheelock. Just south of where the stream is crossed by the road from Sandbach to the Potteries (now the A533) a silk throwing mill, called Brook Mill, was established in 1825 by Ralph Percival on land belonging to Lord Crewe.[1] To provide power to the mill the stream was dammed to form two mill pools, either side of the road. The mill was of considerable size being 480 feet long, probably four storeys high, with a five bay wide pedimented section towards its southern end, giving it an appearance typical of the Georgian period.

Although the mill was recorded as being occupied by Bull & Co between 1825 and 1831,[2] the firm was also known as Percival and Bull at this time.[3] By the mid-1830s Thomas Percival had taken over from his father Ralph in running the silk throwing business at the mill.[4] It was at this time that a Parliamentary Report of 1836 showed that there was only one waterpowered silk mill in Sandbach, namely Brook Mill, and that it had a single waterwheel capable of producing 6 h.p.[5] As Brook Mill was such a large mill it is possible that it was originally constructed to use steam power to augment the power available from its waterwheel.

Percival and Bull continued silk throwing at the mill until Ralph Percival's death in 1853,[6] after which the firm of Malkin, Walker & Hope took over the business. It was around this time that the mill pool north of Sandbach Bridge was partly filled in at the end of the pool nearest the bridge.[7] In 1869 Malkin, Walker & Hope were throwing silk at Town Mill in Sandbach as well as at Brook Mill.[8] By 1874, though the operation at Town Mill had ceased, they were empoying a manager, John Warrington, to supervise the work at Brook Mill.[9] In 1887 the firm ceased trading and the mill was offered for sale, as follows:-

"To be sold or let - The Brook Mills, Sandbach. The mill together with a compound steam engine of about 30 h.p., two steam boilers of 25 h.p. each, a waterwheel of about 5 h.p., gas fittings, etc, contains about 3600 swifts, 2600 drawing spindles, and 5256 throwing spindles. In good working order and of good construction, together with a full supply of all needful articles would be sold or let, or leased with option to purchase, and any offer would be considered; a third of the mill is now working and early possession could be had."[10]

The long term deleterious effect of the Free Trade Act of 1860 can be seen in that only a third of the mill was in use by this time. Although the waterwheel only provided a small part of the total power available it was possibly quite important at this time as its use would be beneficial in minimising running costs at a time when foreign competition was curtailing the activities of many of the country's silk mills.

This advertisement appears to have been unsuccessful, the mill being operated between 1887

FIGURE 203. ODNANCE SURVEY MAP, 1909, SHOWING THE CENTRE OF SANDBACH WITH BROOK MILL AND ITS POOL. THIS MAP SHOWS EVIDENCE OF SILTING TAKING PLACE IN THE MILL POOL.

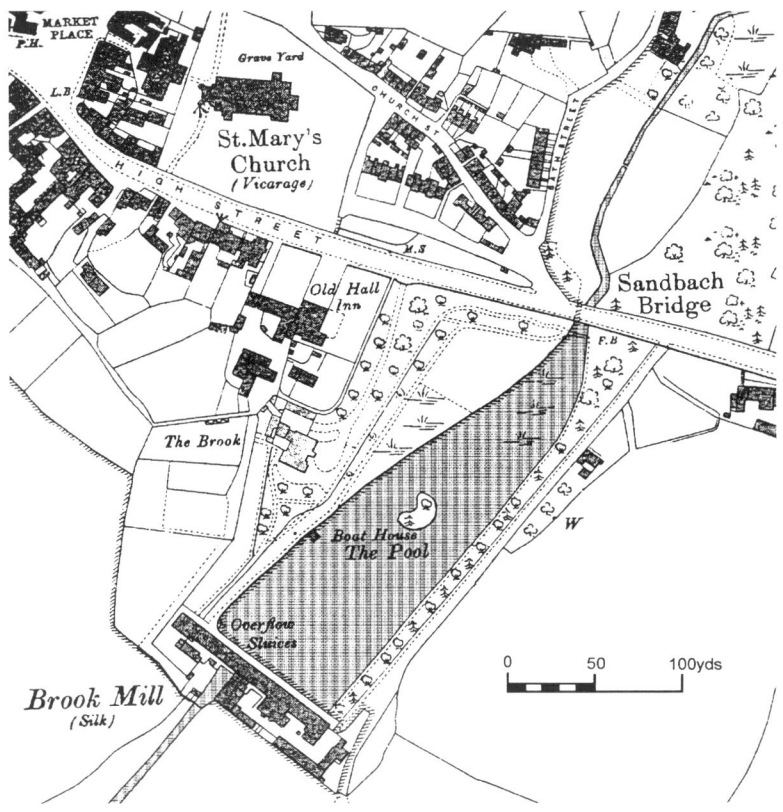

and 1890 by Edward Percival, a descendant of the mill's original builder.[11] In 1892 the mill was being used for silk throwing by Thomas Skerratt, but his business was taken over by a Macclesfield based company, J. H. Heath & Co. in 1908.[12] Although J. H. Heath & Co. persevered with silk throwing at the mill until at least 1939, after the Second World War they changed over to making man-made fibres.[13] This business lasted until 1973 when J. H. Heath & Co. Ltd. closed Brook Mill and moved production of man-made fibre to Kidsgrove.[14] The mill was eventually demolished in the 1980s.

Today the area once occupied by Brook Mill has been "redeveloped" with the building of the Sandbach inner relief road. The site of the mill itself is where the roundabout is situated outside the Safeway supermarket. The mill pool nearest to the mill has been filled in and is covered by the inner relief road. However, the other mill pool, located to the north of the A533 road, can still be seen.

References

1. Earwaker, J. P., *The History of the Ancient Parish of Sandbach*, 1890, (reprinted by E. J. Morten, 1972.)
2. CRO, Land Tax Return, Sandbach.
3. Pigot's Directory of Cheshire, 1828.
4. Pigot's Directory of Cheshire, 1834.
5. Parliamentary Papers, XLV, 1836.
6. CRO, Will of Ralph Percival, 1853.
7. Sandbach Library, Plan of Lord Crewe's Estate, 1849; OS 25 inch to 1 mile map, 1875.
8. Slater's Directory of Cheshire, 1869.
9. Morris's Directory of Cheshire, 1874.
10. *Macclesfield Courier*, 2nd June 1887.
11. Slater's Directories of Cheshire, 1888 & 1890.
12. Kelly's Directories of Cheshire, 1892, 1896, 1902, & 1906.
13. Kelly's Directories of Cheshire, 1910, 1914, 1934, & 1939.
14. *Congleton Chronicle*, 13th February 1998.

---

76. SANDBACH CORN MILL SJ 754598

Sandbach Mill is situated over half a mile south of the town, on the Arclid Brook, about a quarter of a mile from the stream's confluence with the River Wheelock. This is a relatively ancient site as it is said that the mill was founded in 1423.[1] In 1660 there were four millers listed as residing in Sandbach, namely, William Shaw, Charles Blore, and Thomas and Wharton Greene but the mill or mills where they worked are not known.[2] Another miller who probably worked at Sandbach Mill was Peter Chear who died in 1772.[3]

The landowner of the Sandbach Mill was Lord Crewe, and in 1781 and 1782 the miller was an Irishman called Edward Riley. Not only is Riley an Irish name but he presumably spoke with an Irish accent as he was recorded in 1782 as "Royley" [sic]. From 1784 until the early 1820s the miller was John Henshall.[4] He called himself a corn factor and miller when he insured a brick and slate water cornmill and its machinery in Sandbach for £200 in 1798. He also insured his utensils for £200 at the same time.[5] John Henshall had handed over the mill to Samuel Henshall by 1822 and Samuel continued to work the mill until he was succeeded by Thomas Henshall around 1840.[6] The Henshall family were followed at the mill by Thomas Lea in 1841 who then operated the mill for about ten years.[7]

By 1851 Thomas Pass had taken over as the miller at Sandbach Mill and was being assisted in the mill by his son Henry.[8] He and his family had been involved previously with Forge Mill (q.v.) in Odd Rode. Sometime in the next ten years Henry Pass had got married and was living nearby but still assisting his father at the mill.[9] Between 1861 and 1871 both Thomas and Henry became widowers so Henry and his five children moved in with his father and unmarried sister at the mill house.[10] It has been claimed that Sandbach Mill was rebuilt in 1872 but no confirmation of this rebuilding has been found although the appearance of the mill is consistent with this possible construction date.[11] Somewhere around this time the mill pond, which was shown as being heavily silted on a map published in 1875, was cleaned out.[12] This could have occurred when the mill was rebuilt.

Thomas Pass died between 1878 and 1881 at an age in excess of 86 years and so Henry Pass became the miller assisted by his two young sons, Henry aged 20, and Thomas aged 17.[13] Shortly after 1881 the Passes were replaced at the mill by the Lea family of well known local millers. Initially the mill was operated by John H. Lea, but by the mid 1890s it was occupied by his nephew, James Thomas Lea.[14] Prior to his arrival at Sandbach Mill he had been the miller at Buckley Mill and his parents operated Wistaston Mill in the Crewe area. James Thomas Lea's family was quite large and out of six daughters at least two married into other milling families (the Pass and Holland families). However, his four sons were not as lucky. Thomas and James both died in infancy, Thomas was drowned in the mill pool on 31st March 1895, and James drowned in the stream when it was in flood. John was killed in the First World War in 1916, leaving only Frederick to follow his father into the milling business.[15]

In 1917 the landowner, Lord Crewe, needed to raise money and so proceeded to sell a large part of his estate, including Sandbach Mill, which was described as having a 16 h.p. waterwheel with a grain cleaner, four pairs of French millstones and a flour dresser (see Figure 204).[16] The mill was bought by Brunner Mond & Co. for £1350. They had interests in Wheelock Salt Works just downstream of the mill and no doubt needed to safeguard this interest by being able to control the stream through their premises. It is interesting to notice that, even at this date, the mill was set up to produce flour, having only French burr stones, although, it is possible that animal feed was also being produced..

After his father's death in 1935, Frederick continued as miller at Sandbach Mill until 1965. Frederick bought the mill from ICI sometime after its formation as the successor company of Brunner, Mond & Co. In 1963

LOT 86
Coloured Green on Plan No. 4.

A DESIRABLE WATER CORN MILL
WITH HOUSE AND LAND

Known as SANDBACH MILL, occupied by Mr. J. T. Lea, and Field No. 766 occupied by Mr. J. E. Holland with The Fields Farm (lot 76) on Yearly Tenancies

The Rent and apportioned Rent being £87 0s. 0d.
The Tenant of the Mill paying Tithe in addition

The Premises are most substantially built and equipped, and are in good order and condition throughout and comprise:

**4-STOREY MILL WITH DRYING AND STORE ROOM
16-H.P. WOOD WATER WHEEL WITH IRON BUCKETS
4 PAIRS FRENCH STONES, DRESSER AND SCREEN on First Floor
also ALL THE SHAFTING AND PULLEYS.**

Adjoining is a Brick and Slate Engine House used as a Store, Cart House with Room over, and 2 good Piggeries.

FIGURE 204. THE DETAILS IN THE 1917 AUCTION CATALOGUE FOR THE SALE OF SANDBACH MILL. (CCALS)

FIGURE 205. A RECENT VIEW OF SANDBACH MILL AND MILL HOUSE. THE WATERWHEEL WAS POSITIONED AT THE FAR END OF THE BUILDING.

the mill stopped using water power because it could not meet the condition required by the Factories Act to be able to stop the machinery from all floors. Consequently, the mill was converted to electric power. After Frederick Lea's retirement in 1965 the mill was let to C. J. Brown & Sons, seed and corn merchants, whilst the mill cottage became the home of Frederick's nephew Henry William Lea. He continued to work in the mill for a while but after Frederick's death in 1976 the mill was sold for house conversion.[17]

Unfortunately, nothing is known about the mill itself prior to its rebuilding in 1872. The existing mill building is built of brick and is three storeys high. It has a tiled roof with a small lucam under which was a series of loading bays which are now converted into windows. The mill cottage adjoins the mill at its western end. Initially the mill had a steam powered beam engine with an eight feet diameter flywheel fitted alongside the main gearing, with the beam operating parallel to the front of the building. No doubt this rebuilding in 1872 represented a considerable capital investment in order to meet the increasing competition in the milling industry. However, the days of country corn mills were doomed and the steam engine was removed before the end of the 19th century. The waterwheel, which was used until 1963, was overshot, being 10 feet diameter and 6 feet wide, and was located at the eastern end of the building. This drove through conventional gearing of pitwheel, an eight feet diameter spur wheel and stone nuts to the millstones. Originally the mill had four pairs of stones as per the 1917 sale advertisement but later on this was reduced to three pairs. There used to be a drying kiln at one side of the mill and the old boiler house and chimney at the other end.[18] Unfortunately all the machinery has been removed from the mill which is now used mainly for storage.

References

1. Norris, J. H., "The Water-Powered Corn Mills of Cheshire", *Transactions of the Lancashire & Cheshire Antiquarian Society*, Vol. 75, 1965.
2. Dodgson, J. McN., *The Place Names of Cheshire, Part 2*, Cambridge University Press, 1970.
3. CRO, Will of Peter Chear, 1772.
4. CRO, Land Tax Returns, Sandbach.
5. Royal Exchange Fire Insurance Policy.
6. Pigot's Directories of Cheshire, 1822, 1828; CRO, Sandbach Tithe Map & Apportionments.
7. CRO, Census Returns, Sandbach, 1841.
8. CRO, Census Returns, Sandbach, 1851.
9. CRO, Census Returns, Sandbach, 1861.
10. CRO, Census Returns, Sandbach, 1871.
11. Norris, J. H., 1965, *op. cit.* 1.
12. CRO, OS 25 inch to 1 mile map, 1875.
13. CRO, Census Return, Sandbach, 1881.
14. Various Directories of Cheshire, 1880, 1888, 1890, 1896.
15. Lea, Herbert, *Swettenham Mill*, 1984.
16. Sandbach Library, Sale Catalogue, 1917.
17. Lea, H., 1984, *op. cit.* 15.
18. Norris, J. H., 1965, *op. cit.* 1.

Section 7. The River Wheelock and the River Dane to Northwich

The River Wheelock runs north-west from the township of Wheelock to meet the River Dane about one mile north-west of Middlewich. On the way it is joined by the Fowle Brook which rises just south of Wheelock and flows westwards for a few miles before turning northwards to meet the Wheelock just south of Warmingham. From Middlewich the River Dane continues north-westwards until it joins the River Weaver at Northwich. The rivers are accompanied in their north-western journey across the flat, fertile, agricultural Cheshire plain by the Trent and Mersey Canal.

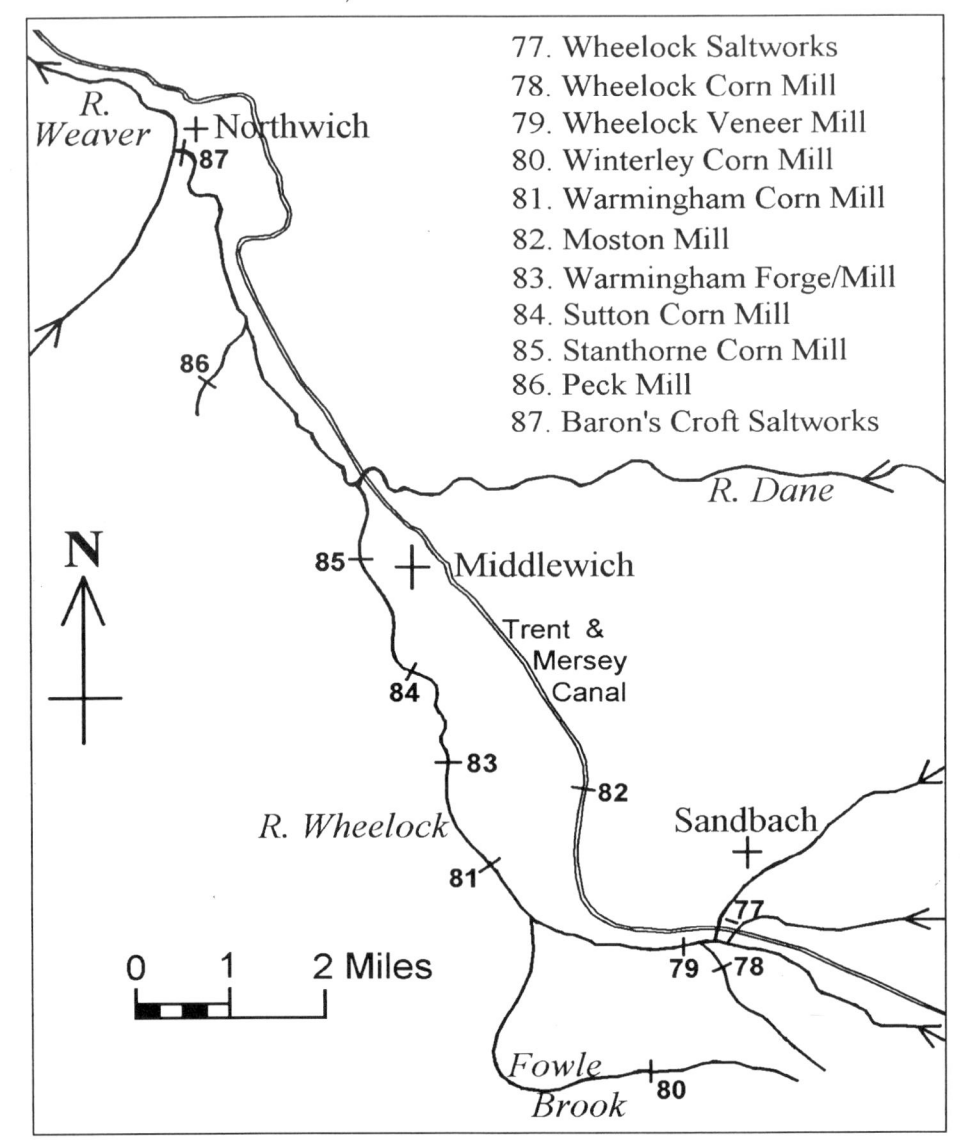

77. Wheelock Saltworks
78. Wheelock Corn Mill
79. Wheelock Veneer Mill
80. Winterley Corn Mill
81. Warmingham Corn Mill
82. Moston Mill
83. Warmingham Forge/Mill
84. Sutton Corn Mill
85. Stanthorne Corn Mill
86. Peck Mill
87. Baron's Croft Saltworks

FIGURE 206. MAP OF THE RIVER WHEELOCK AND THE RIVER DANE FROM WHEELOCK NEAR SANDBACH TO THE CONFLUENCE OF THE RIVER DANE WITH THE RIVER WEAVER AT NORTHWICH.

77. WHEELOCK SALTWORKS   SJ 753593

North of Wheelock there is a promontory of higher ground between the valleys of two of the streams which converge to form the River Wheelock. It was on this promontory that Wheelock Old Saltworks was situated, bounded on the south by the Trent & Mersey Canal and on the north by the railway.

These salt works may well have existed prior to the 18th century but how the brine was accessed is not known.[1] The only evidence of waterpower being used occurred in 1808 when Holland mentioned Wheelock Old Saltworks as one of the five saltworks "*...which admit of the assistance of a stream of water, the brine is raised by this means.*"[2] Although the saltworks were offered for sale in 1815, only seven years after Holland's statement, no mention was made of the means of pumping the brine, although the proximity of the two streams would have made the use of a waterwheel extremely feasible.[3] Later in the 19th century, information indicates that the pumping was carried out by two steam engines.[4]

Today the saltworks do not exist and the whole site has been cleared and returned to agricultural use.

References

1. Earwaker, J. P., *East Cheshire*, 1887/8.
2. Holland, H., *General View of the Agriculture of Cheshire*, 1808.
3. *Macclesfield Courier*, 29th April 1815.
4. *Macclesfield Courier*, 3rd March 1852.

---

78. WHEELOCK CORN MILL   SJ 755588

One of the four streams that converge to make the River Wheelock rises at a place known as "Bottomless Mere", about two miles south-east of Wheelock, near Hassall. On this stream, less than a quarter of a mile from the confluence with the other streams at Wheelock is the site of Wheelock Mill.

The earliest reference to a mill at Wheelock so far found occurred in 1316 when a "*Molendinum de Quelock*" is mentioned in a deed.[1] In the 15th century a "*a milne domme...in Whelock*" is referred to in a charter of 1440.[2] Very little is known about the mill in the 16th and 17th century but it is mentioned as being in operation in 1656.[3] Later in the 18th century there is an entry by the engineer and millwright, James Brindley, in his notebooks, that he did some work at Wheelock Mill in 1755.[4] The mill was part of the Great Moreton estate owned in the late 18th century by Thomas Powys Esq. and the miller recorded at the mill between 1781 and 1799 was called John Large.[5] However, John Large died in 1797 so either he was succeeded by another John Large (presumably his son) or the tax recorders mistakenly continued with their habit of listing John Large as the miller for the two extra years.[6]

Over the first seven years of the 19th century four occupiers of the mill were registered. They were John Horkeman in 1800, William Colclough between 1801 and 1803, William Lownds in 1804 and 1805, and Ralph Kirk in 1806.[7] It is unlikely that these four individuals were in fact millers, certainly William Colclough was a solicitor based in Sandbach, so they were probably leasing the mills to rent to a miller at a profit. This was confirmed in 1807 when the mill was offered for sale as follows:

> "*The corn mill is held by the bankrupt Mr. John Carter on a lease of nine years. It has four pair of grinding stones and is turned by a stream of water running through the premises.*
>
> *Apply to Mr. Colclough, Sandbach who is the mortgagee.*"[8]

Obviously John Carter was the miller at Wheelock and may have been there since the death of John Large, but his business had not prospered.

When the corn mill was for sale the same advertisement offered a cotton factory for sale. A lane ran across the mill dam with Wheelock corn mill located to the north of the lane at the western end of the dam. The cotton factory was situated across the lane from the corn mill but still below the dam (see Figure 206). The cotton mill is only likely to have been built in the 1780s or 1790s and, from its location, the mill would appear to have also been powered by water from the mill pool. The cotton mill was quite small, being only 71 feet long by 31 feet wide, but was four storeys high. It contained about 3000 throstle spindles, so it could easily have been driven by waterpower. However, around the end of the 18th and beginning of the 19th century there was a great increase in the number of steam engines brought into use to replace, or assist, waterpower especially in the textile industry. It is noticeable that the advertisement of 1807 makes no mention of water power, only stating that the cotton mill had a steam engine of 24 horse power. As the annual rent being paid for this small mill was a hefty £600 per year it is perhaps not surprising that only two years of its lease had elapsed prior to this sale.

After this sale in 1807 Wheelock Corn Mill was occupied by John Coomer but he seems to have been no more successful as the corn mill and the cotton mill were offered for sale only four years later in 1811 when both properties were held on a 99 year lease.[9] The millers that succeeded John Coomer are not known until both mills were put on sale again in 1828. By that time the miller was Thomas Arden who was paying an annual rent of £116 for the corn mill which was described as having four pairs of grinding stones of which one pair were French burrs. The cotton factory had been operated for many years by Jesse Drakeford who was paying a rent of £112 per year for the factory, a much more sensible figure than the £600 of 1807.[10]

Whether the cotton factory was sold a this time is not known but by 1831 it was described as a silk mill.[11] Wheelock Corn Mill does not appear to have been very successful during the first half of the 19th century for it was once more offered for sale in 1834. This time the sale advertisement, which named William Mosley as the miller, gives greater detail about the mill. It is

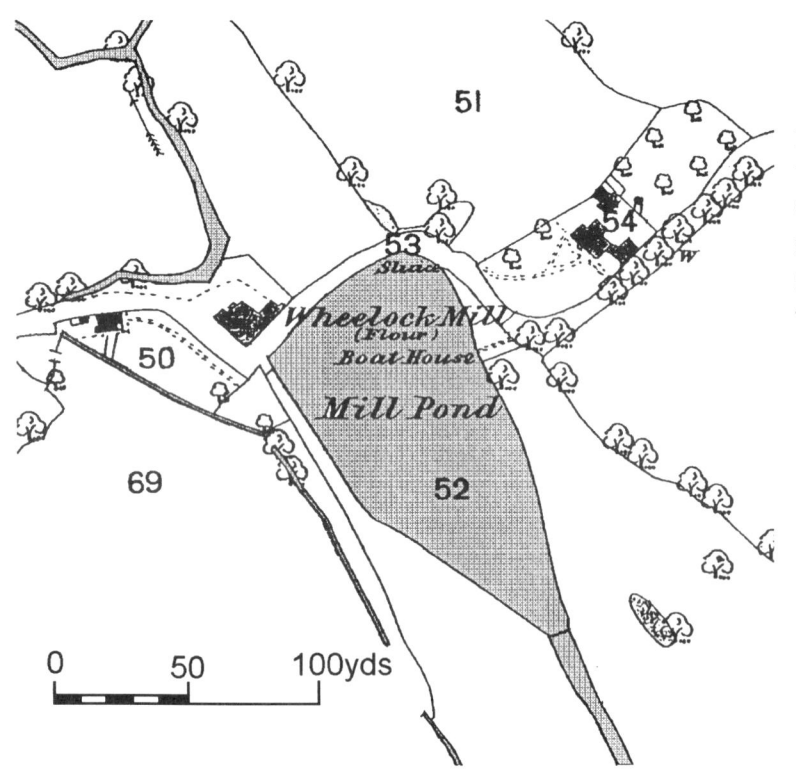

FIGURE 207. AN ORDNANCE SURVEY MAP OF 1875 SHOWING THE LOCATION OF WHEELOCK CORN MILL AT THE WESTERN END OF THE MILLPOOL DAM. THE COTTON/SILK MILL WAS LOCATED ON THE PLOT OF LAND INDICATED BY THE NUMBER 50.

described as being

> "...worked by one large breastshot wheel and contains two pair of French stones, two pair of Grey stones, a large dressing machine, drying kiln and all other necessary utensils and machinery for carrying on an extensive trade."[12]

By this time there were two pairs of French burr stones and there was now a dressing machine for grading the resultant product. It is noticeable that this advertisement does not mention the textile factory, which previously had always accompanied the corn mill in all the sale advertisements. This may signify that the factory had ceased to operate by 1834 and was no longer seen as a saleable asset.

William Mosley was replaced immediately by Samuel Henshall as the miller at Wheelock Mill in 1834 but a few years later the mill was held by Joseph Fox.[13] In 1841 Richard Pollitt and his wife arrived at the mill accompanied by his son, who was also a miller, and his wife.[14] Sometime after 1841 Richard Pollitt's son left Wheelock Mill leaving his parents to operate the mill with the help of an employee.[15] The Pollitt family stayed at Wheelock Mill until the mid 1850s when the mill was taken by the Massey family. First of all Samuel Massey was the miller in 1856 but he had handed over to James Massey by 1857.[16] Four years later James Massey was living at the Mill House with his wife, Amelia, six young children and a servant. He was assisted in the mill by Charles Wilkinson, a journeyman miller, who lived with his wife and family at the Mill Cottages.[17] In 1871 the Mill House was occupied by James Massey, who was only 24 years of age. He was the eldest son of James and Amelia Massey. He was described as miller and corn factor and was living at the Mill House with his younger sister Harriet, described as the housekeeper, his younger brother William, an engine fitter, his three other young siblings and a servant. At this time the young James Massey was employing another miller and a carter.[18] This state of affairs indicates that possibly some family tragedy overcame James and Amelia Massey sometime in the 1860s to leave their young family to carry on the milling business on their own.

Ten years later, although James Massey junior still lived at the Mill House, he was only accompanied by his youngest brother, Herbert, who was an apprenticed miller aged 15. He employed a young miller called Alexander McKenzie, who had only just completed his apprenticeship, and a house keeper.[19] James Massey only remained at the mill another couple of years after 1881. In the 1880s a steam mill began operating at Wheelock, alongside the Trent & Mersey Canal, which seems to have caused the demise of the water mill.[20]

Today the location of the mill pool can be traced as a depression in the fields devoid of any water and the only sign of the water mill is a few stone blocks lying in the undergrowth.

References

1. PRO, Catalogue of Ancient Deeds, as quoted in Dodgson, J. McN., *The Place Names of Cheshire, Part 2*, Cambridge University Press, 1970.
2. British Museum, Additional Charters, as quoted in

Dodgson, J. McN., *The Place Names of Cheshire, Part 2*, Cambridge University Press, 1970.
3. Ormerod, G., *The History of the County Palatine and the City of Chester*, Routledge & Sons, 1882.
4. Birmingham Library, James Brindley's Notebooks, as quoted in Boucher, C. T. G., *James Brindley - Engineer*, Goose, 1968.
5. CRO, Land Tax Returns, Wheelock.
6. CRO, Will of John Large, 1797.
7. *Op. cit.* 5.
8. *Staffordshire Advertiser*, 6th June, 1807.
9. *Macclesfield Courier*, 2nd October, 1811.
10. *Macclesfield Courier*, 4th October, 1828.
11. CRO, Bryant's Map of Cheshire, 1831.
12. *Macclesfield Courier*, 18th January, 1834.
13. CRO, Tithe map and Apportionments, Wheelock.
14. CRO, Census Returns, Wheelock, 1841.
15. CRO, Census Returns, Wheelock, 1851.
16. Bagshaw's Directory of Cheshire, 1850; Kelly's Directory of Cheshire, 1857.
17. CRO, Census Returns, Wheelock, 1861.
18. CRO, Census Returns, Wheelock, 1871.
19. CRO, Census Returns, Wheelock, 1881.
20. Kelly's Directory of Cheshire, 1892.

---

79. WHEELOCK VENEER MILL   SJ 751591

Just west of the bridge carrying the Sandbach to Crewe road in the village of Wheelock, Bryant's map of Cheshire, published in 1831, shows a watermill symbol. Alongside the symbol appears the caption "*Veneer Mill*". Possibly this waterpowered timber mill had only a short life as no other information has been found concerning this mill and it does not appear on any other map of the area.

FIGURE 208. BRYANT'S MAP OF 1831 SHOWING THE "VENEER MILL" JUST TO THE WEST OF THE ROAD THROUGH WHEELOCK. ALSO SHOWN ARE WHEELOCK CORN MILL & SILK FACTORY TO THE EAST OF THE ROAD AND SANDBACH CORN MILL. (CCALS, M5.2)

---

80. WINTERLEY CORN MILL   SJ 747571

Winterley Mill was located about a mile from the source of Fowle Brook and two miles south of Sandbach, in the township of Haslington. Because of the meagre water supply, being sited so close to the stream's source, the whole of the stream was dammed to form a very large mill pool.

Although there are records mentioning Winterley Mill dating from the last quarter of the 16th century they do not give any indication of the nature of the mill or the names of any of the millers.[1] The earliest known miller is John Smith who is recorded as occupying the mill from 1783 to 1814.[2] However, the probate of John Smith's will occurred in 1809 so presumably from that date to 1814 the mill was occupied by another John Smith, probably his son.[3] During the tenure of the John Smiths, the mill was owned by Samuel Scott. In 1815 the mill was occupied by Benjamin Scott & Company but the relationship between this company and Samuel Scott is not known, nor whether they were a milling company.

In 1816 the mill had new owners, namely Robert Timms & Company, who offered the mill to let as follows:-

"*A wind and water mill and a newly erected house at Winterley in the township of Haslington in the County of Chester with about three statute acres of land lying distant about one mile from Wheelock Wharf which is a capital road to Manchester and other market towns.*

The miller will show the premises, apply to Mr Lewis, New Wood House, Whitchurch, Mr. Robert Timms, Weston, Wynbury, or Mr. John Whitingham, Wynbury."[4]

Just when or why the windmill was built is not certain but the most likely explanation is that the meagre water supply was inadequate for the output required in spite of the existence of the very large mill pool. The windmill was just to the south of the watermill, close to where the road bends to the west towards Crewe.

It appears that the owners were not immediately successful in finding a tenant for their mills as no occupier is recorded until James Simpson in 1818. James Simpson only stayed at Winterley Mill for a couple of years until 1820 and set a pattern of occupation for the next decade. William Bate succeeded Simpson from 1821 to 1823, being followed by William Ollerenshaw in 1824. Slightly greater stability occurred from 1825 to 1830 when the occupier was Thomas Smith but he was replaced by Samuel Dale in 1831 for just the one year.[5] The windmill appears to have been in operation throughout the 1820s and it is possible that this fact caused the turnover in millers as it did at Kinderton Mill (*q.v.*) in Middlewich. Although indicated on a map of 1831 it had disappeared by 1840 with the area where the mill once stood being identified as "*Windmill field*".[6]

The year of 1832 saw the arrival of William Mosley as the miller at Winterley who was probably the same person as the William Mosley recorded at Wheelock Mill (*q.v.*) around this time. He was still at the mill in 1840, aged 70, when he was assisted by his 20 year old son Edward.[7] It was at this time that the owners of the mill were recorded as Messrs Cook, Lewis and Timms.[8] During the 1840s the Mosley family left the mill, presumably on the death of William, and a new miller called John Hallmark took over at the mill from at least 1847.[9] In 1851 John Hallmark, who was 39, describing himself as a master miller, lived at the mill with his wife, five children and a servant and employed an apprentice miller in the business.[10]

The Hallmark family stayed at the mill until sometime after 1857. By 1861 Joseph Astbury had succeeded John Hallmark at the mill, living there with his wife and five children.[11] The exact composition of the Astbury family is uncertain as in the mid 1860s it was a Thomas Astbury who was listed as the miller.[12] However, in 1871, Joseph Astbury was still recorded as a miller and farmer living at the mill. By this time he had six children, with his eldest son, John, assisting in the mill. Ten years later John Astbury had taken over from his father entirely as the miller and had installed his own young family at the mill.[13] What happened to John Astbury is unclear because Thomas Astbury was again listed as the miller in 1888. This situation only lasted a short time until Joseph Astbury, the youngest son of the original Joseph and brother of John Astbury took over by 1892. Again the effectiveness or otherwise of the mill's natural power source is highlighted by Joseph Astbury using steam power to augment the water supply. In 1906 when a John Astbury lived at the mill he was listed only as a farmer, most likely indicating that the mill went out of commercial production around this time.[14] However, it is possible that the mill continued to grind on an occasional basis, possibly for animal feed, as it was only reported to have ceased entirely in 1926.[15] After this date the mill was used as an agricultural and then an industrial store until its demolition in 1998. Since then the site has been used for a modern housing development.

The mill dam across the Fowle Brook carries what used to be the main Sandbach to Crewe road before the new Wheelock and Haslington by-passes were built.

FIGURE 209. WINTERLEY MILL SEEN FROM THE ROAD ON TOP OF THE DAM. NOTE THE LUCAM FOR LOADING DIRECTLY FROM ROAD LEVEL INTO THE TOP STOREY OF THE MILL.

The three storey brick building with its tiled roof and lucam was built against this dam with direct loading into the top floor from the road. The amount of fall provided by the dam is commensurate with supplying an overshot waterwheel. Although the mill itself no longer exists the extremely large mill pool is still to be seen and is the home of large numbers of water fowl of many varieties.

References

1. Dodgson, J. McN., *The Place Names of Cheshire, Part 2*, Cambridge University Press, 1970.
2. CRO, Land Tax Returns, Haslington.
3. CRO, Will of John Smith, miller, Winterley, 1809.
4. *Macclesfield Courier*, 20th July 1816.
5. *Op. cit.* 2.
6. CRO, Bryant's Map of Cheshire; CRO Tithe Map, Haslington.
7. CRO, Census Return, Haslington, 1841.
8. CRO, Tithe Map and Apportionment, Haslington.
9. Sandbach Library, Haslington Parish Register.
10. CRO, Census Return, Haslington, 1851.
11. CRO, Census Return, Haslington, 1861.
12. Morris's Directory of Cheshire, 1864.
13. CRO, Census Returns, Haslington, 1871, 1881.
14. Kelly's Directories of Cheshire, 1888, 1892, 1902.
15. Norris, J. H., "The Water-Powered Corn Mills of Cheshire", *Transactions of the Lancashire & Cheshire Antiquarian Society*, Vol. 75, 1965.

---

81. WARMINGHAM CORN MILL   SJ 709612

The earliest reference to a watermill in Warmingham is dated 1289, however there are no further details of this medieval mill other than its mere existence.[1] It is not until the late 18th century that more information becomes available.

The miller at Warmingham Mill from before 1781 until his death in 1793 was Thomas Penlington, and the mill was recorded as being owned by Lord Crewe.[2] After Thomas Penlington's death the mill was operated by his widow until she was joined in 1800 and 1801 by another Thomas Penlington who was presumably her son.[3] In late 1801, Lord Crewe leased the mill to Abraham Rosson for 21 years, but in 1807 he left Warmingham Mill in favour of Hough Mill, near Wynbunbury, just south of Crewe.[4] He sublet Warmingham Mill to William Rosson but the family relationship between William and Abraham Rosson is not known. Unfortunately, this state of affairs only lasted until Abraham Rosson was declared bankrupt in 1815 which allowed Lord Crewe to rescind the lease thereby evicting William Rosson from the mill. Lord Crewe then found another tenant for the mill called James Cookson.[5] Although James Cookson operated the mill in 1841 with the assistance of a 15 year old apprentice called John Robinson, he was recorded at that time as being only 40 years old.[7] It would appear that the James Cookson who came to the mill in 1815 was probably his father.

Although James Cookson was still at the mill in 1850 the next year a new miller moved into Warmingham Mill. This was John Arden, who at the age of 29, moved into the Church House with his elder sister as housekeeper together with his much younger sister who was only 16 years of age, and another servant.[8] John Arden was related to the family which had been involved in operating Moston Mill (*q.v.*) by the side of the Trent and Mersey Canal nearby for almost the previous half century. In the middle of the 1850s he got married and settled down at Warmingham Mill to raise his family. By 1871 he had five daughters and employed one house servant and an 18 year old apprentice miller called Samuel Basford.[9] Ten years later his two eldest daughters had left home but he carried on the tradition of employing an apprentice miller in the guise of 17 year old William Dobson.[10] The Arden family moved from Warmingham Mill in the early 1890s, possibly on the death of John Arden who was by then over 70 years of age.

By the 1890s the prospects for a small country watermill such as that at Warmingham were not very good. Certainly after the long tenure of John Arden the mill was occupied by a succession of millers with Thomas Hilditch in the last few years of the 19th century followed by Frank Wilding in the first years of the new century and then William Kinsey up until the start of the First World War.[11] The last miller to be recorded at Warmingham Mill was John Moreton in 1914, so presumably the mill ceased to grind not long after the start of the war.[12]

In the normal course of events, as far as water powered country corn mills were concerned, that would have been the end of its story, however, the mill was to have a reprieve and found a new application in the 1930s. It was occupied in 1934 by James Moores & Son, a firm of mechanical engineers, and the entry in a directory of that year states:-

"*Of interest are the model works of Messrs. J. Moores & Son, manufacturers of precision tools; the power is supplied by the original waterwheel reputed to be over 100 years old.*"[13]

This clearly shows that this firm was not just occupying the space provided by the mill buildings but was actively utilising the waterpower available at the site. They must have found this source of power quite attractive because by 1939 the directory had the following statement "..*the power was, until about 1937, supplied by the original waterwheel reputed to be about 100 years old, but was replaced by a modern turbine.*" thereby indicating that they had actually invested in improving the efficiency and power output obtainable from the fall of water at the Warmingham Mill site.[14] Unfortunately, no information has been found so far about this turbine that appears to have been installed in 1937. However, in May 1940 Gilbert, Gilkes & Gordon Ltd. of Kendal sent a quotation to Messrs. J. Moores & Son for two possible turbine installations. The first was for a 16½ inch Series "C" vertical shaft Francis turbine designed to provide 29.8 h.p. at 279 r.p.m. from a flow of 1610 cu. ft./min. on a fall of 12 feet together with an "A" size oil pressure governor, for a total price of £507-12s-0d. The second quotation was for a 12 inch Series 'C' vertical shaft Francis turbine designed to provide 14.4 h.p. at 382 r.p.m. from a flow of 798 cu. ft./min. on the same fall of 12 feet and with the same governor for a price of £365-10s-0d. These quotes contained the services of one of Gilkes skilled erectors but the purchaser had to supply labouring assistance and lifting tackle as well as preparing the turbine foundations. Due to "*abnormal conditions prevailing*", namely the effects of the Second World War, Gilkes could not accept any liability for not meeting the delivery quoted of five weeks nor could they guarantee not to change the price.[15]

James Moore & Son decided to go ahead with the smaller of the two turbines and a machine with serial number 4493 was duly delivered later that year.[16] As a governor was purchased it is likely that this installation was meant to generate electricity for the works. The turbine equipment was installed in a purpose built turbine house which carries the inscription

This stone was laid by
Gillian & Graham T. Moore Oct 1941
to the memory of RAF airmen
in the Battle for Britain
Aug - Oct 1940
Never was so much owed by so many to so few

The mill buildings at Warmingham still exist although now without any of the mill machinery. They are used by a number of small businesses. The turbine and its turbine house still exist, as does the inscription over its door, but is not now in use. All the water courses and the weir that provide the water supply are also still intact.

References

1. Dodgson, J. McN., *The Place Names of Cheshire, Part 2*, Cambridge University Press, 1970.
2. CRO, Will of Thomas Penlington, 1793.
3. CRO, Land Tax Returns, Warmingham.
4. *London Gazette,* 21st January, 1815.
5. *Op. cit.* 3.
6. Pigot's Directory of Cheshire, 1834; CRO, Tithe map & apportionments, Warmingham.
7. CRO, Census Returns, Warmingham, 1841.
8. CRO, Census Returns, Warmingham, 1851.
9. CRO, Census Returns, Warmingham, 1871.
10. CRO, Census Returns, Warmingham, 1881.
11. Kelly's Directories of Cheshire, 1896, 1902, 1906, 1910.

FIGURE 210. THE TURBINE HOUSE AT WARMINGHAM MILL.

12. Kelly's Directory of Cheshire, 1914.
13. Kelly's Directory of Cheshire, 1934.
14. Kelly's Directory of Cheshire, 1939.
15. Gilbert Gilkes & Gordon Ltd. quotation dated 10th May 1940 in the Turbine file at Bosley Wood Treatment Co. Ltd.
16. Information from Gilbert Gilkes & Gordon Ltd. 13th March 1998.

---

82. MOSTON MILL　　　　　　　　　SJ 731621

Moston Mill is situated about two miles north-west of Sandbach and three miles south-east of Middlewich. However, Moston Mill was built alongside lock 67 at Crowsnest Bridge on the Trent and Mersey Canal and derived its waterpower from the canal, utilising the waste water passing over the lock by-pass weir, rather than from a natural stream. This mill can be justified as being within the River Dane catchment area because much of the water in the Trent & Mersey Canal derives from the River Dane. Water is extracted from the river just downstream of the site of Wincle Paper Mill (*q.v.*) and flows via Rudyard Reservoir and the Caldon Canal into the Trent & Mersey Canal. Another source of the water in the canal comes from the Bosley Brook, a tributary of the River Dane, via Bosley Reservoir and the Macclesfield Canal and thereby to the Trent & Mersey Canal. It is most unusual for a watermill to be powered solely from a navigable canal as the canal companies would normally jealously guard their water supplies for use by the navigation.

Although the Trent and Mersey Canal was opened along all its route in 1777 it is not until 1805 that the earliest entry for Moston Mill in the returns for payment of the land tax occurred. As these tax records exist from 1781 and there is no entry for the mill prior to 1805, it is reasonable to assume that the building of the mill was complete in that year. Between 1805 and 1808 the land owner of the mill was Edward Vernon and the occupier was Henry Arden. In 1808 Laurence Armitstead acquired the estate of Edward Vernon so becoming the land owner at Moston Mill. Henry Arden continued to be recorded as the occupier of the mill until 1822 when he was replaced by Thomas Arden. This state of affairs only lasted for one year and then Henry Arden was recorded as the occupier from 1823 until these records ceased in 1831. The relationship between these various members the Arden family are not known but it is quite possible that there were two Henry Ardens involved, probably father and son.[1]

Baptismal entries in the parish register show that Ralph Merrill was also a miller living in Moston in 1828 so it is possible that he was employed by Henry Arden at the Mill.[2] In 1829 the Arden family attempted to sell Moston Mill, publishing the following advertisement:-

> "*To be sold by auction, Moston water corn and bone mill in the occupation of Thomas Arden. The corn mill consists of four pair of stones with an excellent granary above and every convenience for carrying on an extensive business, the whole is entirely new. The bone mill is of new construction and calculated to crush an almost unlimited quantity of bones.*"[3]

This shows that as well as producing flour the mill also made bonemeal which was popular at this time as a manure for improving the land. The bonemeal machinery would have been separate from the

FIGURE 211. AN OLD PHOTOGRAPH OF MOSTON MILL FROM THE END OF THE 19TH CENTURY SHOWING THE ENGINE HOUSE AND ASSOCIATED CHIMNEY. (CCALS)

millstones producing the flour. This could have consisted of a set of stamps, raised by cams and allowed to fall under gravity onto the bones being processed, or a set of edge runners, or even ordinary millstones similar to those used for producing flour. This auction in 1829 appears to have been unsuccessful as Thomas Arden continued milling at Moston Mill well into the 1830s.[4] By the late 1830s it is possible that the bonemeal business was not doing as well as previously because the following advertisement appeared in the local paper in 1839

"*The Original Cattle Bone Mill, Moston, near Sandbach and Middlewich.*

Samuel Arden in returning his sincere thanks to the landowners & farmers who have patronised his establishment for so many years and in soliciting for a continuance of support begs to remind them that he keeps on hand a large stock of bones purchased from the best market and which may be ground to any degree of fineness to suit the wishes of the purchaser. Price £7 per ton with liberal allowance for cash."[5]

Certainly this type of direct product advertisement was quite uncommon in the early 19th century.

Although the above advert indicates that the Ardens were working the bonemill themselves there is evidence that they were probably also employing a corn miller to operate that side of the business. In 1838 there is the first indication that William Dutton was working as a miller at Moston Mill presumably as an employee of the Ardens.[6] The entries in the 1841 census shows that, in fact, there were two millers living at the mill, Ralph Merrill who had been there since at least 1828, and William Dutton.[7] The mill was still leased to the Ardens who offered it for sale in 1845.[8] In response to this advertisement it appears that the mill was sold to William Dutton as his is the only name associated with the mill after this date. After this sale there is no further mention of the bonemill so presumably William Dutton just concentrated on the trade of corn milling.

By 1851 William Dutton had been joined in the mill by his two sons, Robert aged 18, and John aged 15, and it is likely that Ralph Merrill, who still lived nearby, also helped out as he was only aged 59 and was still recorded as being a miller.[9] William Dutton continued to work Moston Mill until about 1880. He branched out into other retail activities, being listed as a beer retailer and shop keeper as well as a miller in 1864.[10] His two eldest sons who had been working in the mill in 1851 left to start their own families and were replaced in the mill by 1871 by his two younger sons, William Henry and Walter.[11] William Dutton died around 1880 and William Henry Dutton took over the business but he also moved away leaving Walter Dutton to run Moston Mill from around 1888.[12]

The waterpower at Moston Mill had always been dependent on the canal company, a situation that was ameliorated by the addition of a steam engine around 1890. It is probable that Walter Dutton died around 1904 because for the next thirty years or so the mill was run by Mrs. Caroline Dutton. Howver, she had been succeeded by W. Robert Dutton before 1939.[13]

Moston Mill is a three storey brick building that housed a single waterwheel. There is a fairly large pool that was fed from the canal pound above Crowsnest lock. After powering the waterwheel the water was returned to the canal below the lock thereby giving the mill's water supply a head of about eight feet. There was a two storey brick built engine house on the eastern side of the mill with a tall square brick chimney (see Figure 211). The mill building was converted into a house in the later part of the 20th century but the engine house has been demolished.

FIGURE 212. MOSTON MILL IN THE 1980s SHOWING THE TAILRACE RETURNING TO THE CANAL WITH CROWSNEST LOCK ON THE RIGHT.

References

1. CRO, Land Tax Returns, Moston.
2. Sandbach Library, Moston Parish Records.
3. *Macclesfield Courier*, 7th November 1829.
4. Pigot's Directory of Cheshire, 1834.
5. *Macclesfield Courier*, 26th October 1839.
6. *Op. cit.* 2.
7. CRO, Census Returns, Moston, 1841.
8. *Midland Counties Herald*, 20th November 1845.
9. CRO, Census Returns, Moston, 1851.
10. Morris's Directory of Cheshire, 1864.
11. CRO, Census Returns, Moston, 1871.
12. Slater's Directories of Cheshire, 1883, 1888.
13. Kelly's Directories of Cheshire, 1902, 1906, 1934,

---

83. WARMINGHAM FORGE/MILL   SJ 705625

Warmingham Forge was located on the River Wheelock about ¾ mile north of Warmingham Mill (*q.v.*). The exact date when this forge was built is not known, however, in the 1670s the tenant of the forge was Richard Foley of Longton, son of Richard Foley of Stourbridge, who leased the forge from Lord Crewe. In 1677 Richard Foley (the younger) was in debt to Lord Crewe for £300 on account of rent and other charges, most probably for charcoal. Richard Foley had widespread interests in ironworking, being involved with Mearheath Furnace and probably with the forges at Consall and Oakamoor, all in Staffordshire. Richard Foley died in 1678 and his heir and eldest son, also called Richard, died in 1681. As this Richard Foley was the only son, the estate then passed to John Foley who was the youngest half-brother of Richard Foley of Longton. John Foley died in 1684 when the estate was inherited by Henry Glover, who had married a Priscilla Foley. During Henry Glover's tenure he appointed John Wheeler of Wollaston Hall as manager of the ironworks. At the time of Henry Glover's death in 1689 Warmingham Forge no longer appeared in the accounts of the Foley enterprises. Sometime in the late 1680s the forge was leased to Dennis Heyford and William Cotton, who had also leased Lawton Furnace (*q.v.*) and Cranage Forge (*q.v.*). These three sites at Lawton, Cranage and Warmingham then became the nucleus of what came to be known as the "Cheshire Works".[1]

Thomas Hall, who was the cousin of William Cotton and also the general manager of the Cheshire Works, took up a new lease for Warmingham Forge from Ann Crewe Offley in 1694. He was also involved in the establishment of a furnace at Vale Royal about this time. This furnace became the principal supplier of pig iron to Warmingham Forge whose output towards the end of the 17th century was around 100 tons per year. In 1696 there was a re-organisation of responsibilities in the Cheshire Works with Edward Hall, Thomas Hall's younger brother, taking over the day to day management at Warmingham Forge. Apart from the normal work of a forge in converting pig iron into bar iron and rod iron, a specialist trade developed at Warmingham Forge in the last years of the 17th century. This was the manufacture of plates for making salt pans which were in demand from the local Cheshire saltworks.[2] In 1701 the partners in the Cheshire Works acquired Street Furnace (*q.v.*) which they eventually converted into a forge. This purchase was financed by both Edward Hall and Obadiah Lane investing £400 each in the new forge at Street. Edward Hall then became the manager of Street Forge as well as the one at Warmingham. Obadiah Lane was one of the partners in the "Staffordshire Works" group of ironmakers that eventually merged with the Cheshire Works in 1707.[3] The salt pan plating that had been carried out at Warmingham Forge was transferred to

FIGURE 213. A WATERPOWERED FORGE HAMMER SIMILAR TO THE ONES USED AT WARMINGHAM FORGE.

Street Forge and saw making became the speciality at Warmingham. Most of the Warmingham output of saws went to John Podmore of Kidderminster. In the 1703/4 season Warmingham Forge shipped him 39 dozen saws out of the 50 dozen produced and the following season, 1704/5, they shipped 50 dozen out of the 53 dozen produced, to Kidderminster.[4] The normal forge output of bar and rod iron was sold to nearby towns such as Nantwich, Wrexham, Congleton, Northwich and also to Warrington for onward sale to Lancashire.

In 1706 the management of the Cheshire Works was re-organised again and Edward Hall was replaced at Warmingham and Street Forges by William Vernon who had been one of the partners in the Cheshire Works since 1700. Two years later William Vernon further cemented his position in the partnership by marrying William Cotton's eldest daughter. Eventually, in 1732, William Vernon was succeeded by his son Ralph at Warmingham Forge. Fortunately, the Cranage Forge accounts for 1732 have survived to show some indication of the production at Warmingham Forge. At this time the forge was sending about 10 tons of bar iron per month to Cranage for slitting. About 25% of this bar iron was coldshort made from Staffordshire ores but the rest of the shipments were "mixt" bar made using a mixture of Cumberland hematite with Staffordshire ore. The amount paid to the Warmingham Forge account for these shipments averaged just over £18 per ton.[5]

In 1739 Ralph Vernon took out a new lease on Warmingham Forge which was producing a total of around 300 tons of bar iron per year at this time. Ralph Vernon continued to play a central role at Warmingham Forge up to the late 1760s when he retired from the business and the lease of the forge was then taken by Jonathon Kendall. The output of the forge appears to have been maintained through the middle of the 18th century and was 283 tons in the year when Jonathon Kendall superseded Ralph Vernon. In 1759, when the long standing partners belonging to the Cotton and Hall families died, Jonathon Kendall became a leading member of the Cheshire ironworks partnership, along with Samuel Hopkins.[6] Warmingham Forge continued under this management until the partnership was dissolved in 1776 when the notice dissolving the partnership gave details of the extent of their holdings in the iron trade as follows:-

> "Partnership between Jonathon Kendall, Henry Kendall, John Hopkins and Thomas Hopkins in the iron trade at Doddington Furnace and at Cranage, Warmingham, and Lea Forges in the county of Chester, at Rugeley Slitting Mill, Rugeley, Cankwood and Winnington Forges and Mearheath Furnace in the county of Stafford and at Dovey Furnace in the county of Cardigan was this day (15th April 1776) dissolved by mutual consent and that the Trade will for the future be carried on by the said John Hopkins and Thomas Hopkins in partnership and except Winnington Forge which is to be worked for one year from Lady Day 1776 by the said John & Thomas Hopkins and afterwards by the said Jonathon & Henry Kendall. All debts to be paid to Jonathon Kendall at Drayton in Shropshire, Henry Kendall at Ulverston in Lancashire, John Hopkins at Rugeley in the county of Stafford or Thomas Hopkins at Cankwood Forge in the same county"[7]

The forge continued to be leased to Jonathon Kendall until 1784 but in that year Martin Paddy, a member of the family that had taken over Street Forge some years earlier, became the leaseholder of Warmingham Forge.[8] By the late 18th century output at the forge was in decline and it was no longer part of a large integrated ironmaking business including furnaces, forges and slitting mills, but was reverting to an ordinary country forge, albeit with some specialist lines such as pans and saws. This decline in the volume of business left the forge business of the Paddys vulnerable and in 1793 James Paddy, who was based at

FIGURE 214. A DIE PRESS FOR CUTTING SAW TEETH. MACHINERY SIMILAR TO THIS WOULD HAVE BEEN USED AT WARMINGHAM FORGE.

Street Forge, was made bankrupt.[9] Less than two years later the partnership between Martin Paddy, who was operating Warmingham Forge, and James Paddy was also declared bankrupt.[10] The lease of Warmingham Forge was then taken by Charles Whittington (or Whittaker) who continued to operate the forge until 1804 when he also succumbed to bankruptcy.[11]

In 1807 the site of the forge was occupied by Samuel Percival, who was a corn miller, so it is most likely that Warmingham Forge was converted to a corn mill between 1804 and 1807 when it became known as Forge Mill. This conversion was probably carried out at the behest of the land owner Lord Crewe. Between 1814 and 1825 the occupier of the mill is recorded as Joseph Percival who presumably took over from his father Samuel. There were a number of millers in the Percival family, one of whom was John Percival who died in Warmingham in 1815.[12] However, from 1825 the occupier was listed as William Gresty although Joseph Percival still appeared to be working there as a miller, so it is possible that Joseph Percival and William Gresty were related in some way, possibly as brothers-in-law.[13] In the early 1840s the mill was being worked by both William Gresty, Joseph Percival and another miller called William Edgerton, showing that the mill was able to support three families of millers at that time.[14] In 1843 William Gresty died and was replaced by his son Charles who was only 23 years old and unmarried.[15] It was probably Charles Gresty's young age that caused the leaseholder to be recorded again as Joseph Percival at this time, although he considered himself to be a corn factor rather than a miller.[16]

In 1851 Charles Gresty was the master miller at Forge Mill living at the mill house with his two sisters, two servants and two employed journeyman millers together with Joseph Percival, described as a retired miller at the early age of 63 years. William Edgerton, who lived with his family nearby, was also still working as a miller at Forge Mill.[17] Charles Gresty continued to operate the mill until around 1878. He never married, living at the mill house with his sister Elizabeth as housekeeper, and by 1871 was only employing one apprentice miller, although he continued to employ two house servants and a groom.[18]

By the time of the 1881 census the miller at Forge Mill was Arthur Plant who was living in the mill house with his young family, an apprentice miller, one servant and one nurse for his young child. Undoubtedly Samuel Barrow who lived in Forge Mill Lane and was described as a corn and flour miller also worked at Forge Mill.[19] Although Forge Mill appears to have been extremely prosperous since its conversion from a forge, by the end of the 19th century signs of the effect of competition were to be seen in a lack of maintenance. A committee of local councillors inspected the dam and weir at Forge Mill in 1900 and recommended that the land owner, Lord Crewe, be requested to build a fence between the road and the weir as a safety measure and to repair the fencing around the mill pool. They also proposed to draw his attention to "*the dilapidated state of the weir*" which was felt to be unable to withstand flood conditions. This report was accompanied by a drawing of the road along the mill dam which shows that Forge Mill was driven by two waterwheels.[20]

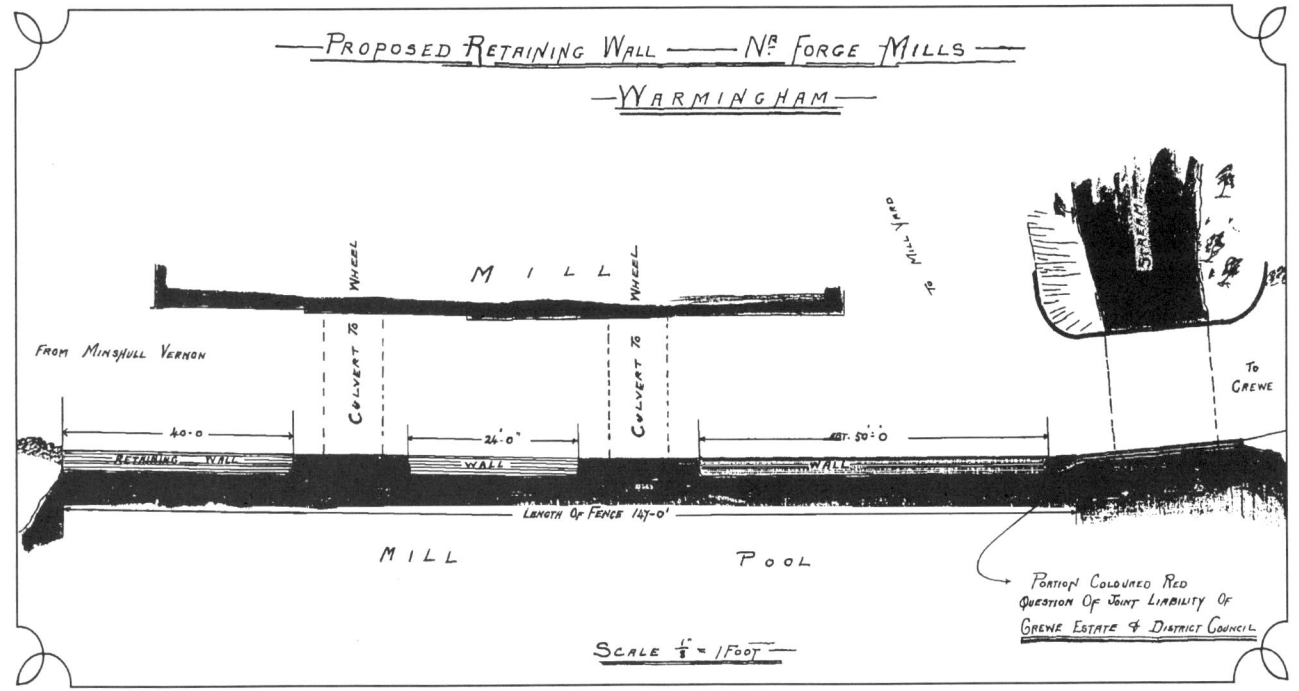

FIGURE 215. PLAN OF FORGE MILL, 1900, SHOWING THE CULVERTS LEADING TO TWO WATERWHEELS. (CCALS)

Arthur Plant continued to be listed as the miller at Forge Mill up to the First World War and was probably the last miller to operate the mill which ceased to grind around 1920. Certainly in the 1930s the occupiers of Forge Mill were only described as farmers, confirming that the mill had stopped working by then.[21] Today there is no sign of the mill building or its location and even the large pool that existed on the line of the river has disappeared.

References

1. Awtry, B. G., "Charcoal Ironmasters of Cheshire & Lancashire, 1660-1785", *Transactions of the Lancashire & Cheshire Antiquarian Society*, Vol.109, 1957.
2. Johnson, B. L. C., "The iron industry of Cheshire & North Staffordshire: 1688-1712", *Transactions of the North Staffordshire Field Club*, Vol. 88, 1954.
3. Awtry, B. G., 1957, *op. cit.* 1.
4. Johnson, B. L. C., 1954, *op. cit.* 2.
5. CRO, Cranage Forge Accounts, DTO 1/4.
6. Awtry, B. G., 1957, *op. cit.* 1.
7. *London Gazette*, 15th April 1776.
8. CRO, Land Tax Returns, Warmingham.
9. *London Gazette*, 13th July 1793.
10. *London Gazette*, 7th March 1795.
11. CRO, Land Tax Returns, Warmingham; *London Gazette*, 11th September 1804.
12. CRO, Will of John Percival, 1815.
13. CRO, Land Tax Returns, Warmingham.
14. CRO, Census Returns, Warmingham, 1841.
15. CRO, Will of William Gresty, 1843.
16. CRO, Tithe Apportionments, Warmingham.
17. CRO, Census Returns, Warmingham, 1851.
18. CRO, Census Returns, Warmingham, 1871.
19. CRO, Census Returns, Warmingham, 1881.
20. CRO, Report & Plan, DCR 59/37/24.
21. Kelly's Directories of Cheshire, 1910, 1914, 1934.

---

84. SUTTON CORN MILL SJ 700642

The site of Sutton Mill is about 1¾ miles south of Middlewich. A road running southwards from the A533 at Newton in Middlewich passes Sutton Hall and ends just across a bridge over the River Wheelock. The mill site is just upstream of this bridge. The earliest reference to a mill at Sutton found so far occurred in 1410 but no information on the mill or its miller was recorded.[1] The earliest miller at Sutton Mill whose name is known was William Cranage, who paid 1s-0d poll tax in 1660.[2]

For many years Sutton Mill was part of Lord Vernon's estate, appearing on an estate map in 1770.[3] Later, in 1792, the miller was Samuel Kinnerley who insured the *"utensils and trade in his water corn mill and drying kiln situated in Sutton"* for £300.[4] In the first quarter of the 19th century the millers at Sutton ran into financial problems. In 1826 John Dutton was declared bankrupt[5] and was replaced at the mill by Joseph Darlington, but he was also declared bankrupt in 1828.[6] It is possible that Joseph Darlington was followed at the mill by George Woolrich who was recorded as being the miller there in 1834.[7]

At some time in the 1840s Samuel Pickering became the miller, living at Sutton Mill with his wife and family, and by 1851 he was being assisted in the mill by his two sons, John and Thomas Pickering who were aged 17 and 15 years old respectively.[8] By 1857 Samuel Pickering, who had also become the miller at Kinderton Mill (*q.v.*), decided that he and his wife would live there, leaving his son John at Sutton Mill.[9] John Pickering then got married and by 1871 he was living at the mill with his wife and young family of four children, together with two young servants.[10] John Pickering and his family stayed at the mill until about 1882.[11] Between this date and the end of the century a number of millers worked Sutton Mill but did not stay more than a few years. James Marsh followed John Pickering at the mill in 1883, by 1888 the miller was Frederick Williamson and he had been replaced by John James Slater by 1892.[12] In 1902 John Sheffield was operating the mill but was probably not living there, as a farmer

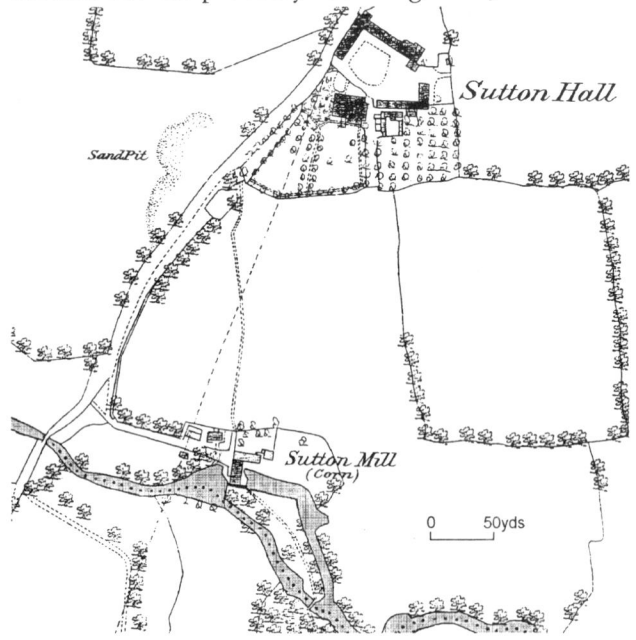

FIGURE 216. ORDNANCE SURVEY MAP, 1880, SHOWING SUTTON MILL ON THE RIVER WHEELOCK, SOUTH-WEST OF MIDDLEWICH.

was listed as residing at Sutton Mill. Thomas Edward Cave was listed as the miller at Sutton between 1906 and 1910 but after 1910 only farmers, rather than millers, are recorded as living at the mill.[13]

The mill building at Sutton survived for a considerable time as it formed part of the farm. It was a three storey brick building with a covered cartway for loading directly into the mill. The wooden gearing and upright shaft also survived until well into the late 20th century but the ironwork was removed during the Second World War. Unfortunately the waterwheel was an early casualty.[14] Today the mill building has been demolished and nothing remains on the site.

References

1. Lawrence, C. F., *Annals of Middlewich*, 1911.
2. Lawton, G. O., ed., *Northwich Hundred Poll Tax 1660 and Hearth Tax 1664*, Record Society of Lancashire & Cheshire, Vol. 119, 1979.
3. CRO, Lord Vernon's Estate map, 1770, DDX 471.
4. Royal Exchange Fire Insurance, 1792.
5. *London Gazette*, 11th March, 1826.
6. *London Gazette*, 3rd March, 1828.
7. Pigot's Directory of Cheshire, 1834.
8. CRO, Census Returns, Sutton, 1851.
9. Kelly's Directory of Cheshire, 1857.
10. CRO, Census Returns, Sutton, 1871.
11. CRO, Census Returns, Sutton, 1881.
12. Slater's Directories of Cheshire, 1883, 1888, 1890; Kelly's Directory of Cheshire, 1892.
13. Kelly's Directories of Cheshire, 1902, 1906, 1910, 1914, 1934.
14. Norris, J. H., "The Water-Powered Corn Mills of Cheshire", *Transactions of the Lancashire & Cheshire Antiquarian Society*, Vol. 75, 1965.

---

85. STANTHORNE CORN MILL   SJ 694662

Stanthorne Mill is located on the River Wheelock about ½ mile due west of Middlewich. Although very little is known concerning the early history of this mill there is a record of a mill at *"Little Stanthurl"* in 1350 which may well have been on the site of the current mill.[1]

The first indication of the mill in more "modern" times occurred when it was offered for sale in 1813 as follows:-

> *"To be sold at auction, A lease for 21 years of the newly erected Stanthorne Corn Mill consisting of five pairs of stones, drying kiln, and every other requisite with a house & stable, carthouse, cowhouse, pigsty, etc, situated with good roads in every direction and within a short distance of the Staffordshire Canal at Middlewich: in the occupation of Mr. Brereton, tenant, who will show the property."*[2]

The phrase *"newly erected"* shows that the mill could well have been rebuilt anytime in the previous fifty years! The status of Mr. Brereton is not clear but the fact that he was available to show the property suggests that he was the miller.

Twelve years later, in 1825, the mill was advertised to let. Unfortunately the advertisement does not add very much to the details given in 1813 except that a dressing machine had been added to the list of *"other requisites"* and the comment that the mill *"might be advantageously applied to the purpose of silk manufacture provided sufficient inducement should be offered to the proprietor to allow it, to be converted to that use."*[3] Although this comment indicates that the lease holder considered that he might get a better reward letting for silk throwing rather than corn grinding, it also shows that the mill must have been of a considerable size. The miller around this time was Charles Davenport who died in 1829 and was succeeded by his son Thomas.[4] In 1835 Thomas Davenport attempted to let the mill in much the same terms as in 1825, again including the possibility of converting the mill to some other use. However, the phrasing used, namely that *"the proprietor would not object to the mill being converted with respect to manufacturing if it was advantageous to him."* is perhaps not as optimistically stated as in 1825.[5] This advertisement must have been unsuccessful because Thomas Davenport was still the miller at Stanthorne in 1840 when the actual landowner was recorded as William Gilpin.[6]

At some time between 1844 and 1850 John Bostock took over the mill from Thomas Davenport and was to remain at the mill until 1882.[7] In 1851 he employed William Carter, a journeyman miller, plus a house servant and an errand boy.[8] Ten years later John Bostock was still employing one servant and a journeyman miller, James Stonier.[9] Even in 1871, when his eldest son was old enough to have joined John Bostock in the mill, he had to employ a journeyman miller because his son preferred the occupation of a bank clerk rather than that of a miller.[10] With the passing of another ten years all John Bostock's children had left home leaving him as the only miller at Stanthorne at the age of 68.[11]

It is perhaps not surprising that in 1882 there was a new miller at Stanthorne Mill. This was William B. Rigby who continued to work the mill into the 20th

FIGURE 217. A POSTCARD VIEW OF STANTHORNE MILL IN THE EARLY 20TH CENTURY. NOTE THE WEIR ON THE RIGHT AND THE HEADRACE RUNNING DOWN TO THE MILL. THE CHIMNEY INDICATES THAT STEAM POWER WAS ALSO USED BY THE MILL, PROBABLY IN THE LATE 19TH CENTURY. (*S. B. Publications*)

century, finally ceasing around the time of the First World War.[12] After the war F. Buckley was recorded as the miller at Stanthorne until the mill finally finished grinding in 1939.[13]

The mill building and most of its watercourses have survived until the present day. Some 100 yards upstream of the mill the low semi-circular weir is still in position on the River Wheelock although it was damaged in 1947 during a flood due to Sutton Mill, which lies upstream, not operating its flood gate correctly. The mill pond, upstream of the weir, is indistinguishable, but the dry mill leat can be traced running down to the mill where it used to supply two undershot or low breast shot waterwheels. The shafts and spoke stubs of these two wheels were still *in situ* in 1965. At one time there was an additional waterwheel which was used for electricity generation and pumping. The mill building, with its whitewashed brick construction and stone roof, appears to have been built in two stages. At the western end the road gave direct loading access onto the top two floors. The mill building was converted into living accommodation when the internal machinery was removed in 1955.[14]

References

1. Lawrence, C. F., *Annals of Middlewich*, 1911.
2. *Macclesfield Courier*, 1st May 1813.
3. *Macclesfield Courier*, 24th September 1825.
4. CRO, Will of Charles Davenport, miller, 1829.
5. *Macclesfield Courier*, 19th September 1835.
6. CRO, Tithe map and apportionments, Sutton, 1840.
7. Bagshaw's Directory of Cheshire, 1850.
8. CRO, Census Returns, Stanthorne, 1851.
9. CRO, Census Returns, Stanthorne, 1861.
10. CRO, Census Returns, Stanthorne, 1871.
11. CRO, Census Returns, Stanthorne, 1881.

FIGURE 218. A RECENT VIEW OF STANTHORNE MILL, NOW CONVERTED INTO A HOUSE.

12. Slater's Directories of Cheshire, 1883, 1888; Kelly's Directories of Cheshire, 1892, 1896, 1906, 1914
13. Kelly's Directories of Cheshire, 1934, 1939.
14. Norris, J. H., "The Water-Powered Corn Mills of Cheshire", *Transactions of the Lancashire & Cheshire Antiquarian Society*, Vol. 75, 1965.

---

86. PECK MILL, BOSTOCK   SJ 666698

About two miles south of Northwich a small stream flows in a north-easterly direction to join the River Dane. On this stream about 3½ miles north-west of Middlewich, in the parish of Bostock just south of Davenham, lies the site of Peck Mill. There is mention of a "*Molendinium de Bostok*" in a document of 1354 which could well have been on the site of Peck Mill.[1]

No further information concerning the existence of Peck Mill has been found until the mill was indicated on Burdett's map of 1777. In 1787 the owner was recorded as Jonathon Stubbs. It is possible that Jonathon Stubbs died in 1788 because by 1789 the owners were noted as being Mrs. Stubbs and Edward Tomkinson, and the occupiers were Edward Tomkinson and George Mainwaring. The records show that George Mainwaring was the miller at Peck Mill but the relationship between Mrs. Stubbs and Edward Tomkinson is not clear. Between 1790 and 1795 the only owner recorded was Edward Tomkinson while he was still listed as the occupier along with George Mainwaring. However, from 1795 to 1803 the owner was again listed as Mrs. Stubbs whereas the occupier during these years was still George Mainwaring but now in association with John Bennett Jun.[2]

Mrs Stubbs died in 1820. Her demise is noted by the fact that the ownership of Peck Mill between 1821 and 1826 is stated to be "*the late Mrs. Stubbs*", during which time the occupier of the mill was James Brooks. In 1827 the mill was purchased by James France France Esq., who lived at Bostock Hall, but James Brook stayed on as occupier for another year. In 1828 no tenant was recorded at the mill but from 1829 to 1831 the occupier was noted as Thomas Williamson.[3]

Peck Mill continued to appear on various maps up to 1840 but the tithe map and apportionment of 1841 shows that the mill pool had virtually ceased to exist and that there was again no tenant at the mill.[4] After 1840 only farmers are listed as living at Peck Mill rather than millers, consequently it seems reasonable to assume that the mill went out of use sometime in the ten years between 1831 and 1841.[5]

In spite of Peck Mill ceasing to function as a mill over 150 years ago, its continued use as a farm has meant that the mill house has survived to the present day. It is possible that this existing building once contained not only the mill house but also the mill which would not have taken much space in the building.

References

1. Dodgson, J. McN., *The Place Names of Cheshire, Part 2*, Cambridge University Press, 1970.
2. CRO, Land Tax Returns, Bostock.
3. *Ibid*.
4. CRO, Tithe map & apportionments, Bostock, 1841.
5. CRO, Census Returns, Bostock.

---

87. BARON'S CROFT SALTWORKS   SJ 659738

Although Northwich was a major centre of the salt industry in Cheshire most of the saltworks were located to the north of the town, some distance from the River Dane. However, just before the confluence of the River Dane with the River Weaver there was a salt works situated on the north bank of the River Dane known as Baron's Croft Saltworks.

When Holland was writing in 1808 he mentioned that one of the saltworks in Northwich used waterpower and this reference was probably to Baron's Croft.[1] Just two years later, in 1810, the salt works was offered for sale giving some indication of the scale of operations at the beginning of the 19th century

"*Lot 1. A desirable leasehold salt work called Baron's Croft, extending more than 700 feet on the bank of the river Dane, where vessels of 30 tons burthen may lie, and having an immediate communication with the river Weaver. The salt work consists of two stoves, three warehouses, and three salt pans, of 2500 superficial feet measurement or thereabouts, with suitable pan houses, two large cisterns for melting rock salt, suitable quays, and weighing machines for shipping salt and landing coals, with ground plot sufficient to increase these works to 20000 feet without incurring any increase in rent. An undivided moiety of half part of a leasehold piece of ground, situated at Northwich and called the Cotton Works extending more than 200 feet on the banks of the navigable river Weaver. In this ground is a capital brine pit, from which the brine is raised by a waterwheel of large power, turned by an abundant supply of water, brought by a canal from the River Dane, through a new-built stone race of excellent quality. the wheel has been lately made at a considerable expense, and beside raising the brine, has sufficient water and power to turn several pair of stones for grinding corn. All the beforementioned premises are held under lease for 99 years from March 25, 1801."*[2]

This advertisement confirms the use of waterpower at the works although the use of the surplus power for corn grinding did not materialise. The power canal mentioned in the advertisement continued to Baron's Quay Saltworks which was located on the bank of the River Weaver just downsream from the Dane confluence. Consequently it is possible that this saltworks also used waterpower at some time, but no evidence has been found except the location of this power canal.[3]

When the works was offered for sale again in 1828 by the De Tabley estate there was no mention of the waterwheel but the existence of the weir on the River Dane was confirmed.[4] Whether the waterwheel continued in use during the later part of the 19th century or was replaced by steam power is not known. Being so close to the town centre when the works closed the site was redeveloped into the fabric of the town so no trace remains of Baron's Croft saltworks today.

References

1. Holland, H., *General View of the Agriculture of Cheshire*, 1808.
2. *The Times*, 1st May 1810.
3. Calvert, A. F., *Salt in Cheshire*, SPON, 1915.
4. *Ibid.*

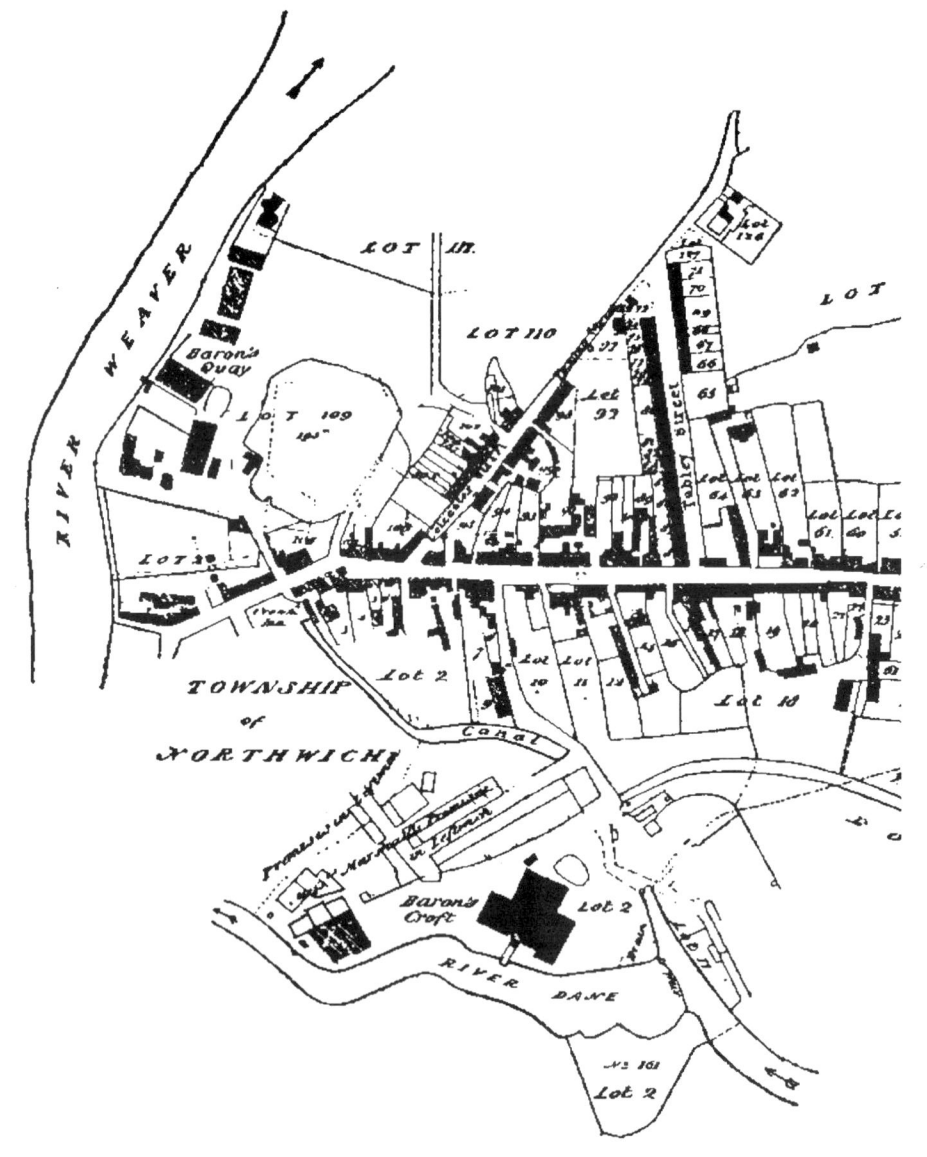

FIGURE 219. A PLAN OF BARON'S CROFT SALTWORKS, LOT 2 IN THE SALE OF 1828. NOTE THE WEIR ON THE RIVER DANE AND THE POWER CANAL RUNNING TO BARON'S QUAY SALTWORKS ALONGSIDE THE RIVER WEAVER. (*A. F. Calvert*)

TECHNOLOGY

Waterwheels

Although there is much discussion concerning the use of horizontal or vertical waterwheels in post-Roman Britain, no evidence of the type of construction of any waterwheel used on the River Dane and its tributaries is available until that dating from the 17th century. Due to the lack of any evidence to the contrary it will be assumed that all the waterwheels used in the area were of the vertical type. For the medieval times assumptions are used, based on the locations of the mills and on contemporary and later illustrations, to indicate the likely properties of these early waterwheels. Illustrations indicate that medieval waterwheels were either undershot, where the wheel was turned entirely by the impact of water striking the lower paddles of the waterwheel, or overshot where the wheel was driven by the weight of water which fell onto the waterwheel from above. When the sites of the thirteen watermills known to have existed on the tributaries of the River Dane before 1350 are examined, seven were located on what can be considered a typical overshot mill site with a substantial fall of about 10 to 15 feet, with the remaining six sites all having falls of less than seven feet. These latter sites with quite low falls, more typical of breastshot waterwheels, could have been used with undershot waterwheels where the fall was used to increase the impulse provided by the flow of the water (see Figure 219). This data presupposes that the medieval mills occupied the same sites as those used in later centuries.

Similarly, the mills first recorded between 1350 and 1700 show the same pattern with six mills on obvious overshot sites and six on falls of less than seven feet and hence possibly undershot sites. The medieval mill builders, who were usually the manorial lords, avoided the main River Dane itself concentrating their efforts on the smaller tributaries. This was due to the costs and engineering difficulties involved with building weirs on a major river and the cost of maintenance likely to be necessary due to repeated flood damage. Prior to 1350 the only mill known to be using the full flow of the River Dane itself was the King's Mill located in Congleton.

The waterwheels built prior to the 18th century would have been constructed almost entirely from wood. The wheelshaft and arms were probably made from oak, with possibly elm used for the paddles, or "ladles" as they were sometimes called. One method of waterwheel construction used compass arms which were mortised directly into the waterwheel shaft. Alternatively clasp arms, fastened across the sides of a squared-off wheelshaft, were used. The undershot wheels would have had simple radial paddles with either one or two sets of arms and rims. The overshot wheels would have had timber boards set at an angle to the radius to form buckets, housed between wooden shrouds, and with a wooden sole to keep the water in

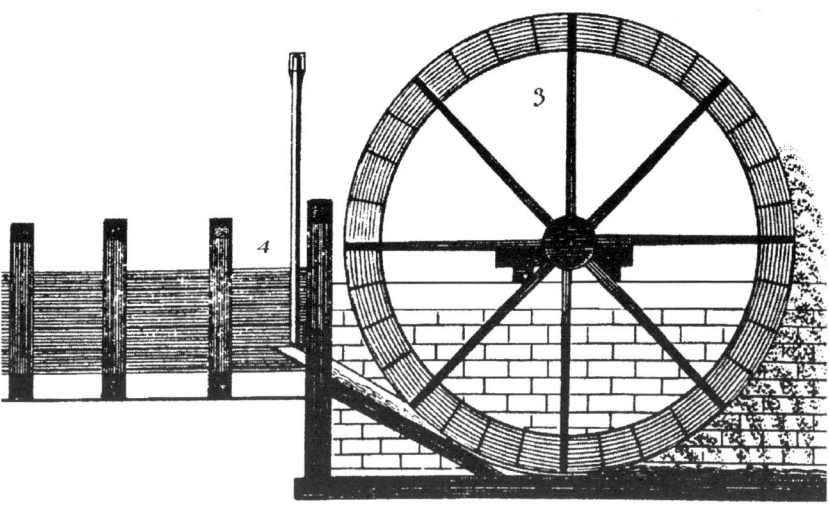

FIGURE 220. AN UNDERSHOT WATERWHEEL WHERE THE HEIGHT OF THE HEADRACE (*left*) IS USED TO ACCELERATE THE FLOW OF WATER AND SO INCREASE THE IMPULSE GIVEN TO THE BOTTOM OF THE WHEEL. (T. Ellicot, *The Practical Millwright*, 1795)

the buckets. By the 17th century, records show that iron was being used for the gudgeons on the waterwheel shafts and brass was being utilised for the bearing material.

In the second half of the 18th century, with the expanding industrial use of waterpower, new materials and a more scientific approach improved the design and performance of waterwheels. The research and testing undertaken by John Smeaton showed that waterwheels driven by the weight of water were much more efficient than those relying on the impact of the current. This led to the replacement of many undershot wheels by low breastshot wheels.[1] Also at this time, the rudimentary weirs that had consisted of a row of poles with pebbles piled up behind them and any gaps filled with moss, etc. were replaced by more permanent structures built of stone masonry so enabling greater heads to be achieved, especially on the larger rivers. Smeaton was also responsible for introducing iron for the construction, firstly, of waterwheel shafts, and then for other parts of the waterwheel. The use of iron enabled larger wheels to be constructed that produced more power as well as being a lot more durable.

One problem that occurred in the early textile mills, due to the great amount of machinery, gearing and shafting, was how to overcome the inertia of all the machinery connected to the waterwheel when starting the system. One solution was to use an impulse waterwheel where the power developed immediately on starting the wheel is equal to the power developed when fully running and, providing the flow is sufficiently strong, is therefore capable of overcoming the inertia of the load. When a breastshot waterwheel was used, the power developed on starting only increased slowly as one bucket after another became full of water. Unless the power developed by the buckets that could be filled without turning the wheel was sufficient to turn all the machinery then the waterwheel would not rotate and further buckets could not be filled. This condition was countered by disconnecting all the machinery until the waterwheel was fully running and only then gradually increasing the load on the wheel. This second option was implemented at Congleton Silk Mill in 1752, where the curved breast of the wheelpit shows that a breastshot waterwheel depending on the weight of water in the buckets was used. This appears to have been the first time this solution of disconnecting each machine had been implemented. The earlier Derby Silk Mill, built in 1721, had all the power of the River Derwent available for an impulse driven, undershot, waterwheel and so did not need to disconnect machinery. This problem of overcoming inertia may well be why some of the early textile mills had breastshot waterwheels that actually used the impulse generated by the flow of water, rather than its weight, to provide the necessary power. Providing the water flow was sufficient then the expense of having arrangements to disconnect/connect each machine could be avoided.

On the River Dane and its tributaries, of the 24 locations where the type of waterwheel has been recorded, 18 were breastshot, three overshot, one pitchback and only two undershot. Of course, in most cases this information relates to the last type of waterwheel in use at a particular site, but does show that the breastshot waterwheel became the de facto standard in this area. Although geography probably played a significant part in determining this statistic it is noticeable that breastshot waterwheels were just as

FIGURE 221. A WATERWHEEL WITH A RING GEAR ATTACHED TO ITS ARMS SO THAT POWER IS NOT TRANSMITTED BY THE WHEELSHAFT. (*Fitz Steel Overshot Water Wheel*, 1928)

popular in the Pennine foothills as on the Cheshire plain. The possibility of generating more power from a particular fall using a breastshot as opposed to an overshot waterwheel also played a part.[2] Gradbach Mill is one example of where this took place when a 24 feet diameter overshot waterwheel was replaced by a 38 feet diameter breastshot wheel.

All the surviving waterwheels on the River Dane and its tributaries, at Washford, Swettenham, Lower Roughwood and Park Mill in Brereton were entirely made of iron. The waterwheel at Bearda Mill, which is a modern construction, is probably more typical of the waterwheels that would have been in use in this area in the 19th century in that it is of composite construction with iron shaft, arms and shrouds, but having wooden buckets and sole.

With a traditional waterwheel, torque was transmitted from the buckets at its periphery, through the arms, to the waterwheel shaft where it was taken by gearing to the mill's machinery. In the first years of the 19th century, Thomas Hewes built some very large wooden waterwheels for the cotton mills at Belper in Derbyshire. These waterwheels were 40 feet wide by around 12 feet diameter and were constructed with a hollow wooden shaft made up from planks like a barrel. This type of hollow shaft, which was over five feet diameter, was incapable of transmitting the power generated by such a large waterwheel without twisting itself to destruction. Hewes used a gear ring mounted around the circumference of the waterwheel to transmit the power which relieved the arms and shaft of any torque (see Figure 221).[3]

Although the ring gear arrangement had been known for many years, Hewes realised the implications for the construction of iron waterwheels. Because of the inherent strength of iron, waterwheels could be constructed using very thin tension rods for the arms, rather like a bicycle wheel, as they merely had to support the weight of the waterwheel itself rather than transmit power as well. This method of construction allowed even larger iron waterwheels to be built, generating much greater power than had previously been possible. Because of their method of construction they were known as "suspension" waterwheels (see Figure 222). The use of a ring gear on a waterwheel provided a much higher speed on the first set of gearing of a mill compared to the rotational speed possible at the wheelshaft. This factor was beneficial in the textile industries where the desirable high speeds required were attainable with far less gearing, thereby reducing losses and machinery costs.

In the 1820s and 30s, the design of suspension waterwheels was brought to perfection by William Fairbairn, a Manchester based engineer, who designed

FIGURE 222. A TYPICAL SUSPENSION WATERWHEEL DESIGNED BY WILLIAM FAIRBAIRN. THIS IS SIMILAR TO, BUT MUCH LARGER THAN, THE WATERWHEEL AT PARK MILL, BRERETON.

and built some of the largest and most powerful waterwheels ever constructed.[4] On the River Dane and its tributaries ring gear was used at Bearda Saw Mill and a suspension waterwheel was installed at Park Mill, Brereton (designed and manufactured by William Fairbairn himself). The size and use of the waterwheels at Gradbach, with a 38 feet diameter breastshot wheel, and at Bosley, with a large wheel operating on a fall of 33 feet (probably installed by Fairbairn), indicate that these also might have been suspension type waterwheels.

References

1. Smeaton, J., "An experimental enquiry concerning the natural powers of water and wind to turn mills and other machines, depending on a circular motion." *Philosophical Transactions of the Royal Society*, Vol. 51, 1759.
2. Buchanan, R., *Practical Essays on Mill Work and other Machinery*, ed G Rennie, 3rd Edition 1841.
3. Gifford, A., "The Waterworks at Strutt's Mills at Belper and the first Suspension Waterwheel", *Wind and Water Mills*, 13, 1994.
4. Fairbairn, W., *Treatise on Mills and Millwork*, 1863.

Gearing & Millwork

In the medieval period, and later, the teeth of mill gearing were made using round staves of wood. These staves, or "cogs", could be arranged around the periphery of a gear wheel; either perpendicular to the wheel's axis, in which case it was known as a spur wheel because of its likeness to the rowel of a spur; or parallel to its axis, when it was known as a face wheel or a crown wheel depending on whether it was mounted on a horizontal or vertical shaft. These types of gear wheel meshed with a "lantern pinion" which consisted of two circular boards fixed some distance apart on a shaft, known as "rounds", which were joined together by a number of staves such that it resembled a cage or a lantern (see Figure 223). This was the basic set of gears used to transmit power from one shaft to another from Roman times through the medieval period and well into the post-medieval time

Although the design for helical gears had been published as early as 1666[1] and that for epicycloid gears in 1694[2] they had no impression on the practical craft of millwrighting. Some forty years later rules for making lantern pinions were still being promulgated.[3] It was not until the beginning of the second half of the 18th century when the need to transmit greater amounts of power provided a practical impetus to the replacement of the existing crude gearing. The first design of a pair of bevel gears was published in 1752[4] and it was shown in 1754 that the theoretical epicycloid gear shape could be substituted by the more easily manufactured shape based on the arc of a circle.[5] This period of a more mathematical approach to gearing coincided with the introduction of a plentiful supply of cheap iron which became extensively used in gear making. After about 1770 cast iron was especially favoured for the new bevel gears which became the standard means of transmitting power from one plane to another, so much so that in this country the lantern pinion disappeared completely, although its use continued in continental countries even in to the 20th century (see Figure 223). In the 19th century iron was used extensively for making gears. Iron gears could be made with their teeth cast integrally with the gear wheel, or else the casting could provide mortises around the gear wheel periphery so that wooden cogs could be inserted into them. The use of wooden cogs meshing with iron gear teeth provided a smooth and quiet motion and, of course, replacing damaged wooden cogs was easier and cheaper than repairing iron teeth. The wooden cogs were usually made from hornbeam, applewood or some similar timber. However, iron gears manufactured by an engineering company were exact enough not to need wooden teeth to "iron out" any inaccuracies. Park Mill at Brereton has a fine set of all iron gears, made by the firm of

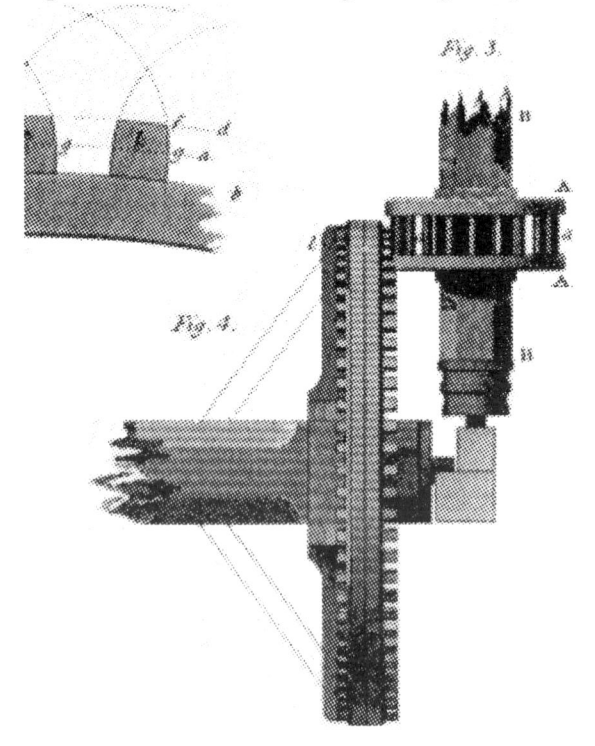

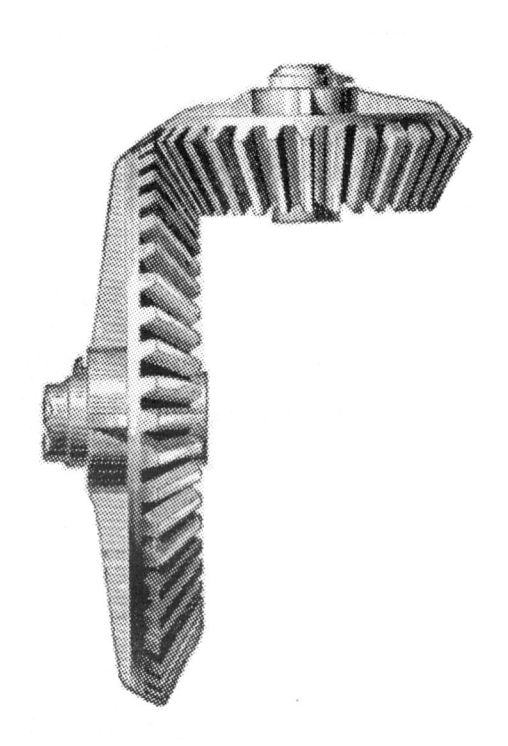

FIGURE 223. UNTIL THE LATE 18TH CENTURY THE UNIVERSAL METHOD OF TRANSMITTING POWER FROM ONE PLANE TO ANOTHER WAS BY A COG WHEEL AND LANTERN PINION (*left*). THESE WERE REPLACED BY BEVEL GEARS (*right*), FROM THE END OF THE 18TH CENTURY IN THIS COUNTRY. (*Rees's Cyclopedia*, 1819)

William Fairbairn, which were installed in 1835, whereas Swettenham Mill has a mixture of iron, iron with wooden cogs, and all wooden gearing dating from various times in the 19th century.

Prior to the 18th century, the only millwork required in corn mills was for a pair of gears, a face wheel and lantern pinion, to take the power from the horizontal wheelshaft to drive the iron stone spindle on which was mounted the runner millstone. Other types of mills, where a reciprocating motion was required, used cams mounted directly on the extended wheelshaft, or on another camshaft geared to the wheelshaft, to operate hammers or stamps. Throughout the medieval and early post medieval period only one machine or "mill" was driven from one waterwheel. With the advent of the early silk throwing mills in the first half of the 18th century, the power of the waterwheel had to be transmitted to a large number of machines throughout the whole mill. The first silk mill, which was built in Derby in 1721, used a vertical shaft, driven via a pair of gears from the wheelshaft, to distribute power to all five floors of the mill. This vertical shaft then drove horizontal shafts on each floor to take the power to the various machines. It is likely that Congleton Silk Mill, built in 1752, followed a similar arrangement of gearing and shafting to distribute the power over the 290 feet length of the mill on all five of its floors. As can be appreciated, the design and construction of the transmission mechanics for this mill represented a quantum leap in the complexity of millwork compared to that which had been seen previously on the River Dane and its tributaries.

One innovation, said to have been first installed at the Congleton Silk Mill, was a facility to engage or disengage any machine with the power being derived from the waterwheel. This facility was needed in order to overcome the inertia of the whole system when the waterwheel was started. It is possible that the driving gear for any particular machine would have been constructed so as to "free-wheel" upon its drive shaft. Fixed to the shaft, close to the drive gear, were a couple of paddles that were always rotating with the shaft. These paddles could be slid along the shaft until they made contact with a couple of pegs fixed to the rear of the free-wheeling gear. The drive was then transmitted from the shaft via the paddles and pegs to the drive gear for the machine. To disengage the machine the paddles were slid away from the gear so that they no longer made contact with the pegs on the rear of the drive gear (see Figure 224). In the 19th

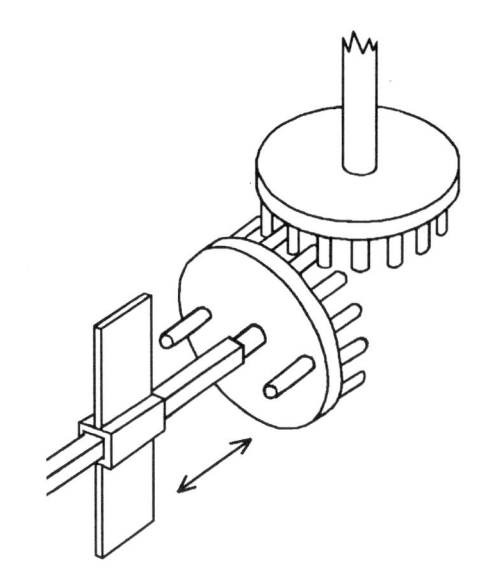

FIGURE 224. A SIMPLE METHOD OF ENGAGING/DISENGAGING THE POWER FROM A PAIR OF COG WHEELS.

century, when machines were mainly belt driven rather than having a direct connection via gearing, this function of disengaging a machine was provided by the ubiquitous "fast and loose" pulley, a system that was first patented in 1797.[6]

The shafting in the early textile mills would have been made of timber, possibly with iron hoops to reduce deformation due to the twisting forces. Gradually iron would have been introduced as a stronger material for transmitting power. In the first part of the 19th century, William Fairbairn made his fame and fortune by the introduction of wrought iron shafting. This enabled much thinner, lighter and longer shafts to be used, which could then be rotated at much higher speeds. This had a great effect on the efficiency and size of the textile mills that could then be built at that time. When Bosley Works ceased cotton spinning in 1853 the millwork was described as being "mill gearing, wrought and cast shafting, wheels, pulleys, pedestals, couplings, hangers, and brass steps." There is a possibility that the millwork at Bosley could have been installed by Fairbairn himself.

References

1. Hooke, R., *Collection of Lectures*, 1666.
2. de Lahire, P., *Mémoire sur les Epicycloides*, 1694.
3. Belidor, B. F., *Architecture Hydraulique*, 1737.
4. Camus, C. E., *On the Teeth of Wheels*, 1752.
5. Euler, L., *Comment*, Petropol, 1754.
6. Jenkins, R., "Fast and Loose Pulleys", *The Engineer*, 12th April 1918.

Corn Milling

Since mankind became farmers around 10,000 years ago the majority of people have depended mainly on plants for their basic food, which in the temperate part of Europe, including our own country, has been provided by the various types of seeds obtained from cereal crops. These seeds are protected by an outer casing which has to be broken through if the nutritious internal part of the seeds is to be utilised. The method of grinding grains using waterpower to drive a vertical waterwheel had been introduced into this country in Roman times and the same configuration of machinery was used throughout the medieval period and later until the late 17th century and up to the middle of the 18th century.

These simple machines had a vertical waterwheel mounted on a horizontal shaft on which was also mounted a face wheel gear called the pitwheel. The cogs of the pitwheel meshed with a lantern pinion mounted a vertical shaft, or spindle, that passed through the centre of the fixed bottom millstone before being connected to the top millstone. The millstones were surrounded by a wooden enclosure, or tun, above which a hopper was supported such that grain could be fed from it into the eye, or central hole of the top millstone. This arrangement of waterwheel, shafts, gearing and millstones was known as the "mill", rather than the building that housed this machinery (see Figure 225). If two pair of millstones were required then the whole arrangement of waterwheel, gearing and millstones was duplicated. When more than one set of milling equipment was housed in the same building it was sometimes described as *"two mills under one roof"*. This was the situation at Congleton mill in the 16th and 17th centuries.

The miller was faced with the task of not only grinding a range of different grains but even the same type of grain could vary in characteristics from one batch to another. In order to control the milling process the miller had to be able to vary the gap between the two millstones and vary the rate of feed of the grain into the millstones. The gap between the millstones was varied by raising or lowering the footstep bearing of the upright spindle attached to the top millstone. However, the variations required in this gap were quite small, consequently a system of levers was used that allowed the upper millstone to be raised or lowered by a fraction of the adjustment made by the miller. This adjustment was known as "tentering" and was usually implemented using levers called the "bridgetree" and the "brayer arm". Initially these were operated by a third lever known as a "lighter staff" but this was later replaced by various methods of operating a screw arm (see Figures 226/7). The rate of feed of grain into the millstones was also controlled by setting the angle that the shoe makes with the grain hopper which could be varied by means of a "crook string".

It was essential when operating the mill that it was not run without any grain between the millstones as sparks could be generated that would set fire to the mill. To prevent this an additional spindle was seated on top of the drive spindle and projected above the top stone. It was made square or triangular where it passed through the mouth of the shoe, such that it knocked the shoe as the spindle rotated, the shoe being returned to its original position by a length of willow acting as a spring. The vibration caused by this knocking ensured that there was a free flow of grain into the eye of the millstone. This part of the machinery is called the "damsel" either because of its continuous clacking noise or because it replaced the job done by young girls while their mothers used a quern. The same problem could occur if the hopper ran out of grain, so some millers would arrange a leather strap attached to a bell to be held down by the grain, which, when released by a lack of grain would sound an alarm by ringing the bell.

In the late 17th century social stability and its associated growth in population after the English civil war produced pressure for greater output from the

FIGURE 225. A SIMPLE MEDIEVAL CORN MILL WITH A VERTICAL WATERWHEEL AND ONE SET OF GEARS. (*G.Agricola, De Re Metallica*, 1556)

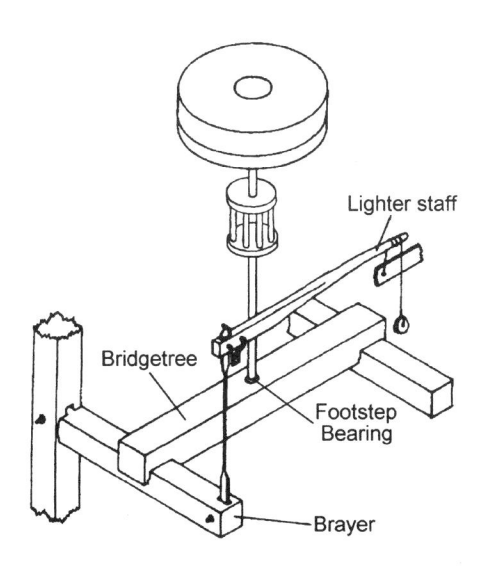

FIGURE 226. THE EARLY ARRANGEMENT OF LEVERS FOR ADJUSTING THE POSITION OF THE RUNNER MILLSTONE USING A LIGHTER STAFF.

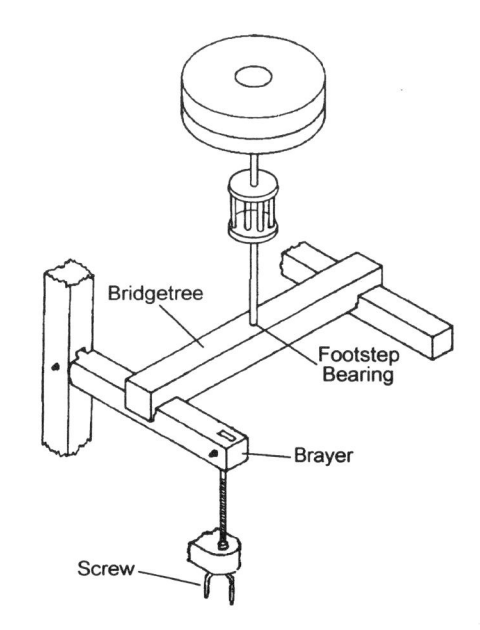

FIGURE 227. THE LATER ARRANGEMENT OF LEVERS FOR ADJUSTING THE POSITION OF THE RUNNER MILLSTONE USING A SCREW ARM ON THE BRAYER.

FIGURE 228. A DRAWING OF THE ARRANGEMENT OF THE GEARING OF A TREBLE MILL AT NUNEATON IN 1723 WHICH ALLOWED A SECOND PAIR OF MILLSTONES TO BE DRIVEN FROM THE ONE WATERWHEEL.

existing corn mills. This led to attempts to add another pair of stones to be driven from an existing waterwheel, sometimes called a treble mill. The earliest depiction of this configuration, showing a mill in 1723, has another lantern pinion gear mounted on a horizontal shaft meshing with the pit wheel. This layshaft then had another set of gears which drove a second pair of millstones (see Figure 228). There is evidence in the Congleton Mill accounts that in 1680 a "new mill", including a pair of millstones, was added but there is no indication of a new waterwheel. It is possible that this refers to a configuration similar to that depicted in 1723. Some years later, in 1750, a millwright called William Forster when estimating work to done at West Heath Mill claimed that he "never had less than six pounds for putting up a treble mill". Later this method

245

was used to drive millstones on either side of the pitwheel via layshafts running at right angles to the wheelshaft.

By the middle of the 18th century the introduction of the spurwheel system of gearing allowed a number of pairs of millstones to be driven from a single waterwheel. In this system the pitwheel turned the wallower which was mounted on an upright shaft that formed the centre of the system. On this upright shaft was fixed a large diameter spur gear around which were ranged the millstones. The millstones were driven by stone nuts or pinions mounted on the stone spindles which meshed with the spurwheel (see Figure 229). The spurwheel could be mounted on the upright shaft below the millstones, which was known as an "underdriven" mill, or alternatively if the spurwheel was above the millstones it was called "overdriven". With a multi-millstone configuration it was essential that a pair of millstones could be disconnected when not required so that they did not rotate without any grain between the stones. Various levers, screws, etc were devised to raise or lower the stonenuts so disengaging them from the great spurwheel. The upright shaft with spur gear became the most popular type of corn mill configuration from the 1750s onward. Many of the cornmills on the River Dane and its tributaries would have used this configuration but evidence only exists for its use at Somerford Booths Mill, with an overdriven system, and at Bank Mill on the slopes of Mow Cop. An example of an underdriven version this type of mill configuration has survived at Swettenham Mill which is possibly a replacement for a similar system initially installed in 1765.

Along the River Dane and its tributaries the millstones in use in the medieval period and later consisted of millstone grit that was quarried locally in places such as the Mow Cop ridge. The simple one-step gearing in use in these early mills meant that the millstones would rotate at a speed of 50 - 60 r.p.m. To produce reasonable flour the millstones in this type of configuration had to be five feet diameter or larger. A typical millstone of this type, which probably originated from a nearby outcrop of millstone grit called the Roaches, is still to be seen in Bearda Mill. There is evidence of millstones from Mow Cop being used in the medieval and post-medieval times at Rode Mill, Moreton Mill, Congleton Mill and Cotton Mill but no doubt they were used in most corn mills in the area. Eventually French burr stones were introduced for grinding flour which relegated the local millstones to producing animal feed. The French burr stones were made up from a number of pieces of stone cemented together. These stones originated in the Marne region of France and were imported by millstone makers who assembled the complete stones in this country. When the millstones were being driven through two sets of gearing they could be rotated at speeds of 100-120 r.p.m. which allowed millstones of only about four feet diameter to be used. The French burr stone became ubiquitous for flour production but local millstones from Mow Cop continued to be used for animal feed and oat processing. All the surviving millstones from Mow Cop date from the late 19th century when they were made of separate pieces of stone rather like a French burr. Some examples of this later type of Mow Cop Millstone can be seen at Somerford Booths Mill, Swettenham Mill and Rode Mill.

Increasing the number of millstones obviously increased the throughput of a mill which in turn

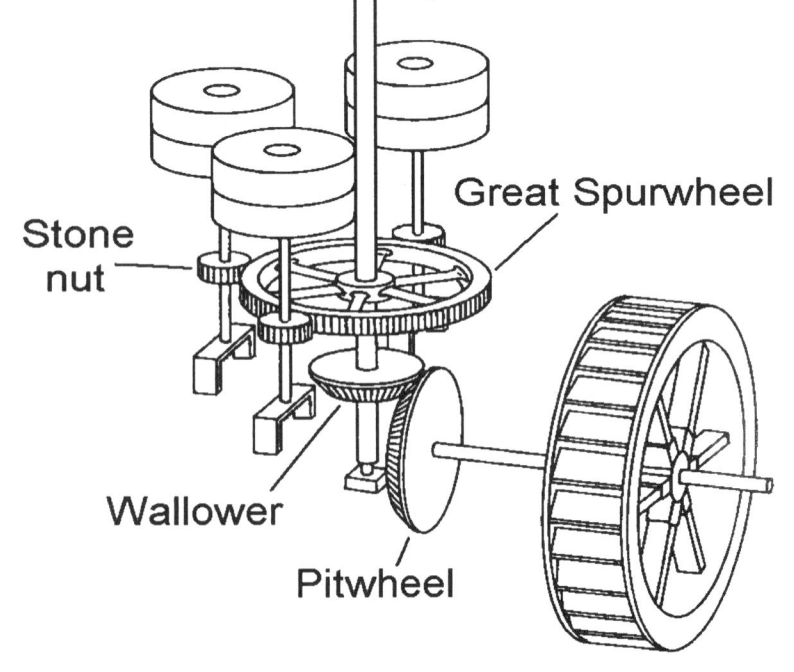

FIGURE 229. A TYPICAL CORN MILL CONFIGURATION USING AN UPRIGHT SHAFT AND GREAT SPURWHEEL.

required more storage area for grain. This type of corn mill usually had three floors with the top floor used for storage. This in turn meant that a sack hoist was needed to lift the sacks of grain up through the mill to this storage area. The sack hoist was usually driven from the waterwheel via a crown wheel situated near the top of the main upright shaft. There were many varied designs of sack hoists but the only types used in the region employed some method of friction drive that could be controlled by the miller.

Another configuration that came into use to drive multiple pairs of millstones placed the pairs of millstones in a straight line driven by bevel gearing from a horizontal lineshaft parallel to the wheelshaft (see Figure 230). This arrangement achieved great popularity in the middle of the first half of the 19th century when it was much favoured by William Fairbairn. A classic example of this layout is to be found at Park Mill, Brereton which has the only known set of corn grinding mill machinery supplied by William Fairbairn to have survived to this day. Another example of this configuration was installed at the Havannah in 1840 to drive five pairs of millstones. Although this configuration is often considered to have originated in the 19th century, the earliest drawing of this type of corn mill was made by John Smeaton in 1781.

Being in a fairly wet part of the country it was often necessary to dry the grain before milling so that it would not clog up the millstone surfaces. This was achieved using kilns where the grain could be spread on a floor made up of perforated tiles with a furnace supplying heat from below. The most complete surviving kiln was at Swettenham where the perforated kiln floor was made of cast iron tiles. Another example exists at Rode Mill where ceramic tiles were used. Also the remains of ceramic kiln tiles have been found at Bearda Mill and Park Mill. Kilns were probably widespread in the area as their use was also essential for the processing of oats for human consumption. The oats were first dried in the kiln then passed through a pair of millstones set quite wide apart. This enabled the oats to be cracked i.e the outer casing was split. The output from the stones was sieved and blown with a fan to separate the oats from their casing. The resultant groats were then passed through the process again with the millstones set lower. The meal was then sorted with the sieve and fan into different grades. The remains of oatmeal machinery were present at Swettenham Mill and in Park Mill at Brereton. A reminder of this practice is the on-going appetite for oatcakes by the local population in South Cheshire and North Staffordshire.

During the 19th century there was a demand for whiter bread which caused millers to improve the quality of their output. The initial stages of the milling process was improved by the introduction of grain cleaners to remove foreign bodies from the incoming grain. There are many varieties of grain cleaners, some types consist of horizontal oscillating sieves to separate the grain by size, coupled with a fan which would blow off any lighter material. Another type of grain cleaner contained some form of tubular screen or sieve, usually at a slight angle to the horizontal, inside which rotating brushes or beaters, turning at about 300 rpm, forced the grain against the mesh to separate by size. These machines also incorporated a fan to remove light particles such as dust, straw and chaff. A fine example of a grain cleaner, patented by Jonathon Booth, a local millwright, is still to be found in Park Mill at Brereton (see Figure 231).

After grinding, the meal from the millstones was then separated into flour and bran using one of the many types of flour separators such as a bolter or wire

Figure 230. A typical in-line corn mill configuration using a horizontal layshaft and bevel gearing.

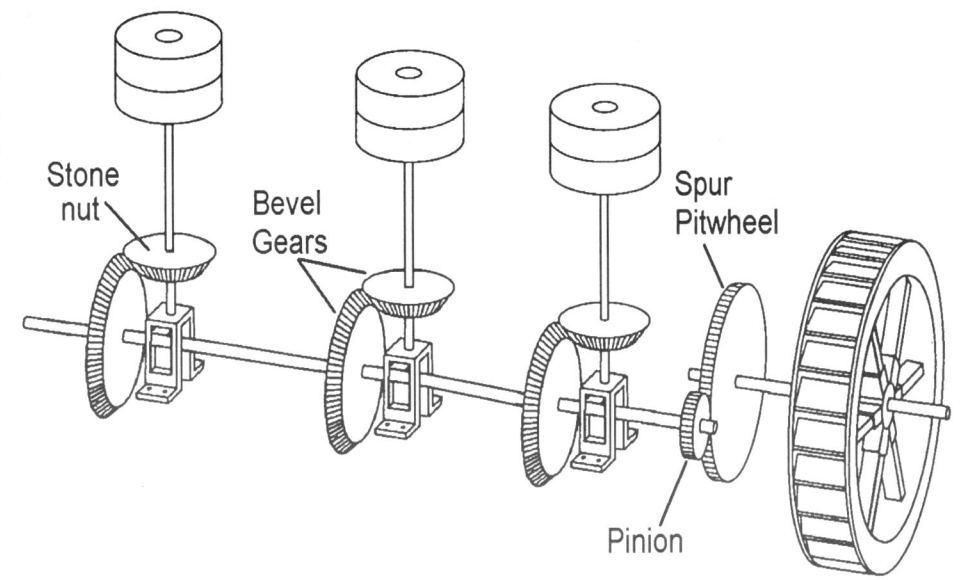

machine. The bolter consisted of a rotating cloth sleeve that knocked against fixed bars as the meal was fed into it. Centrifugal force accompanied by the knocking forced the finer flour through the cloth leaving the bran inside. The wire machine consisted of a cylinder of wire mesh angled to the horizontal, with the finest mesh at the upper end and the mesh increasing in size towards the bottom. Although this cylinder was stationary a number of brushes rotated at high speed inside the cylinder. Sacks to collect the various grades of flour were located under the cylinder. These machines eventually rotated at up to 500 rpm. Plans of Havannah Mill in 1840 show a grain cleaner and a flour sifter in an arrangement that would have been common to many mills in the area. With the introduction of these additional stages in the milling process the grain and meal had to be moved around the mill between the various machines involved. This was automated by the use of worms, like long screws, to move the material horizontally and elevators consisting of small buckets attached to a continuous loop of belting to raise the material up the mill (see Figure 232). When the corn mills were relegated to producing just animal feed the millers had no further use for their cleaners, separators and ancillary equipment and so most of them were removed from their mills. In spite of this trend elevators and worms can still be seen at Swettenham Mill and Park Mill.

The only water powered mill in the area that continued flour production on a large scale after the introduction of roller milling to this country was that of F. R. Thompstone at Bosley Works. Here machines housing pairs of vertical millstones were used in a high

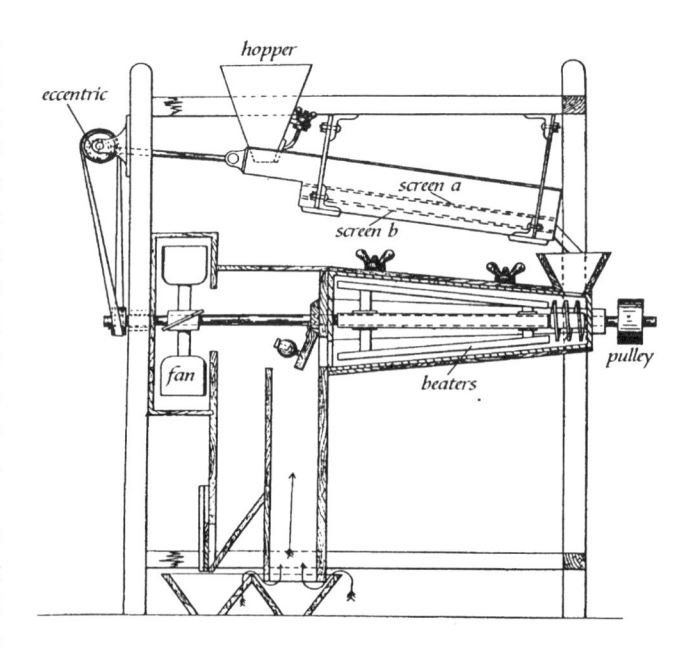

FIGURE 231. A PATENT GRAIN CLEANER AND OAT CLIPPER MADE BY JONATHON BOOTH OF CONGLETON. (Patent No. 10,057, 1902)

grinding system that passed the partial product of the grinding through the millstones a number of times sifting and separating the meal at various stages using machines called plan sifters. These plan sifters consisted of a number of interconnected horizontal flat sieves encased in a housing suspended on bamboo canes. The plan sifters were vibrated with a horizontal reciprocating motion which graded the meal and passed each separation on to further treatment. Two other sites have continued in the milling business upto the present day, albeit as provender millers producing animal feeds, at Rushton and Cranage Mills.

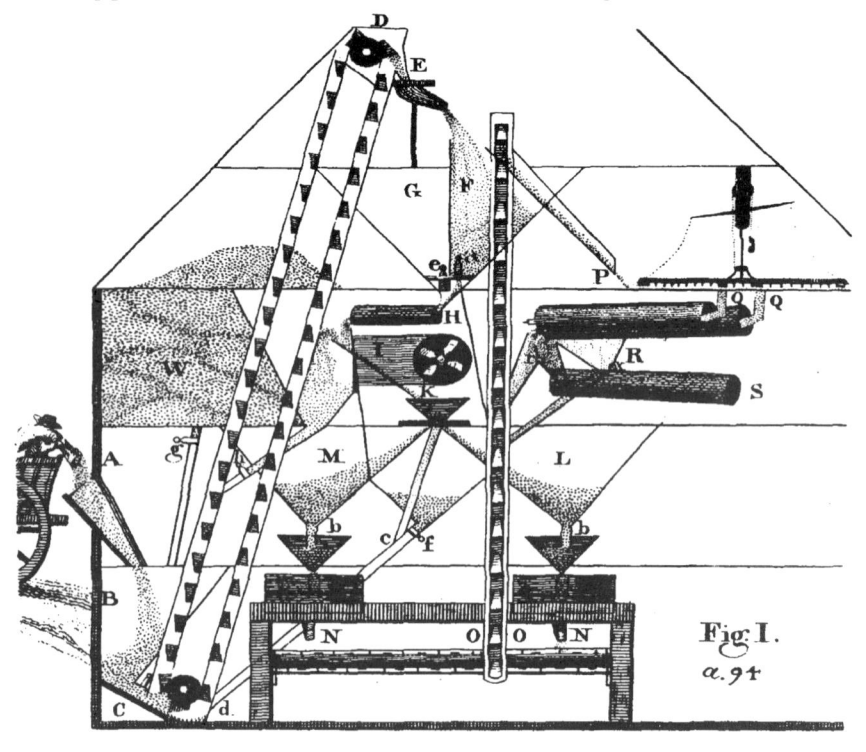

FIGURE 232. A CORN MILL USING ELEVATORS AND WORMS TO MOVE MATERIAL AROUND THE MILL. (O. Evans, A Young Millwright & Miller's Guide. 1795)

Paper Making

During the medieval period most of the paper used in this country, especially white writing paper, was imported from the continent. The first paper mill in this country which was built in 1490 was not successful, a pattern that seemed to set a precedent for the next century, due no doubt to the entrenched position in the market of the imports from the continent where labour costs were lower. However, by the end of the 17th century about a hundred paper mills had been established in various parts of the country, one of which was at Danebridge.

The manufacture of paper at this time was essentially a hand process. The raw materials were rags; fine linen rags were used to make white writing paper, with other rags, canvas and even old ropes being used in the manufacture of coarser papers that were used for wrapping. These rags were cleaned and then shredded to form a pulp or fibrous suspension in water contained in a vat. The "tool" which is used to make each paper sheet consists of a fine wire screen mounted in a timber frame. This is known as a "deckle" or mould. The mould was dipped into the vat of pulp so that an even coating of fibres was formed on the wire mesh. When the surplus water was drained off, the fibrous web of pulp lying on the mesh was turned out onto a woollen felt. Further webs and felts are stacked on top of one another and then transferred to a screw press where any excess water was removed by the action of the press. The webs were then removed from the felts before being hung in a loft or shed for drying. Better quality papers were dipped in size to give them a sheen and some stiffness before the final drying process.

The whole process is quite simple in principle but requires considerable skill on the part of the operative to obtain a consistent product. In the 17th century and earlier the only part of the process that used waterpower was the pounding and pulverising of the rags. This was performed by stamps or hammers driven by the waterwheel whereby cams fixed to the extended wheelshaft lifted the heads of a bank of hammers which were then released as the cams rotated, allowing the hammers to fall under gravity onto the material to be pulped (see Figure 233). The main requirements for a paper mill at this time was a good supply of rags together with a supply of clean water. Danebridge was situated some distance from its nearest supply of rags, the towns of Macclesfield and Leek, but did have an excellent supply of clean water for the vats and for power in the River Dane. The number of vats at Danebridge Paper Mill is not known but each vat would probably have been worked by five men and produced around six reams per day. Female labour would also have been used in sorting and cleaning the rags.

In the early 18th century a new waterpowered device was invented to speed up the processing of the rags. This was known as the "hollander", a name derived from the probable location of its invention in the Netherlands. The hollander consisted of an oval tub containing a cylinder with iron blades around its circumference set into one side of the tub. The oval tub was filled with rags and water and the cylinder, which was rotated by the waterwheel, macerated the rags to produce the paper-making pulp (see Figure 234). This process was much more effective than the stamps and hammers used previously. This was the new type of machinery that was installed by Abraham Bennett and his apprentice, James Brindley, at Wincle Paper Mill in 1738 and was most likely in use at Primose Vale Mill later in the century. There is no indication that any of the paper mills on the River Dane and its tributaries used water power for any of the finishing processes as they were mainly involved in producing brown and blue paper for wrapping and examples of locally made writing paper from this period have a coarse, grey, unglazed appearance.

FIGURE 233. AN ILLUSTRATION OF THE EARLY PAPERMAKING PROCESS. THE WATERPOWERED HAMMERS IN THE FOREGROUND BEAT RAGS TO PULP. THE WORKERS ARE USING A DECKEL AND A PRESS. THE WET SHEETS ARE HUNG IN THE ROOF TO DRY AFTER WHICH THEY ARE PARCELLED INTO REAMS THEN STACKED READY FOR SHIPMENT. (G. A. Böckler, *Theartrum Machinum Novum*, 1661.)

FIGURE 234. A 19TH CENTURY HOLLANDER. THIS EXAMPLE WAS BELT DRIVEN FROM THE POWER SOURCE. (*Centrum Industriekultur Nürnberg*)

Early in the 19th century the whole paper making process was mechanised with the introduction of a machine designed by the Fourdrinier brothers and developed at Stoke-on-Trent. This machine had pulp continuously fed on to a moving belt of gauze which was squeezed and drained then passed on to a series of steam-heated rolls for further drying. The introduction of this continuous process into the paper making industry sounded the death knell for the paper mills on the River Dane which had all ceased to operate by 1850. However, the continuous paper making process was introduced to the area in the 1890s by Carson & Bradbury at Eaton Mill who manufactured punched cards for Jaquard weaving looms amongst other products. By this time wood pulp had been substituted for rags as the industry's raw material.

Iron Making

Iron is the fourth most abundant element in the world but usually occurs as various types of ore that are a mixture of iron oxides and small quantities of other elements. To abstract the iron from the ores the oxides have to be reduced and the other impurities removed, a process that was achieved by the application of fire. In the early method of iron making, the iron ores were first roasted in layers with wood or pit coal which drove off any arsenic and sulphur from the ore. The roasted ore was then placed in a hearth with charcoal which could be raised to about 800 degrees centigrade by using hand or foot powered bellows. At this temperature the oxygen in the iron oxides combined with the charcoal to form carbon monoxide and carbon dioxide which escaped from the hearth leaving a spongy ball, or "bloom", consisting of iron and slag. This bloom was then hammered to consolidate its mass and remove excess slag. The resulting metal, a type of wrought iron, was malleable and could then be reheated to be worked into whatever shape was required. These bloomeries or bloomsmithies, as they were called, which could produce less than one ton of iron per week, were the only way of producing iron until the blast furnace was introduced into this country around 1500. Documentary and physical evidence indicates the presence of bloomeries in the area of the Biddulph Brook near to Biddulph Hall.

When the blast furnace was introduced into this country from the continent it revolutionised the manufacture of iron. The main factor in the blast furnace's development was the use of waterpower to provide the air blast. The amount of air and the pressure of the blast that could be provided by water-driven bellows increased the temperature achievable in the furnace and also increased the amount of material that could be processed. With this increased air blast, temperatures in excess of 1200 degrees centigrade could be attained and consequently a greater percentage of carbon was absorbed in the metal. This higher temperature and increased carbon content resulted in the iron produced being molten. This liquid iron could then be run off from the bottom, or hearth, of the furnace and cast directly into objects such as anvils and cannon. Although cast iron was itself a useful product there was a much greater demand for the malleable wrought iron and so the major part of the production from the blast furnaces was run into channels which had side channels arranged like a series of combs resembling a litter of piglets suckling at a sow. This "pig-iron" was then used for the manufacture of wrought iron.

FIGURE 235. A SECTION VIEW OF A CHARCOAL BASED BLAST FURNACE. WATERPOWER OPERATED A PAIR OF BELLOWS PROVIDING AN AIR BLAST INTO THE FURNACE CHAMBER ON THE FAR RIGHT. (*D. Diderot, Encylopédie, 1751*)

A typical blast furnace was about 25 feet high, square in external form but tapering towards the top. The internal shape of the blast furnace was circular with its greatest girth at about the middle of its height, tapering towards both the top and towards the bottom. The iron ore, charcoal and limestone (used as a flux) charge was introduced at the top of the furnace. The air blast entered via a "tuyere", or nozzle, near the bottom of the furnace, with the air provided by a pair of alternating bellows. The bellows were operated by cams on a shaft driven from a waterwheel with counterweights to provide the opposite motion to that provided by the cam (see Figure 235). Later the bellows were driven by crank arms and "rocking beams".

Whereas the old bloomery method of iron making was a batch process the use of the blast furnace turned iron making into a continuous process. A blast furnace in the late 17th and early 18th century would have operated continuously for a "campaign" which would have lasted for the winter months of a year. Raw materials of charcoal, limestone and roasted iron ore would be stockpiled and once the furnace was blown in it would operate 24 hours per day, seven days per week. The amount of iron produced was greatly in excess of the older bloomery method, with around ten tons per week or approximately 300 tons per campaign. Because water power was only used at a furnace for driving the bellows furnaces could be sited on fairly small streams such as at Lawton and Street furnaces. This technology had a profound effect on working practices. The nature of the process required constant supervision which dictated that the workers operated a two, 12 hour, shift system, seven days per week, throughout the campaign. This type of discipline, forced onto the workers by the technology, was one of the defining parameters of the factory system. The

FIGURE 236. PIG IRON BEING FORMED BY RUNNING OFF MOLTEN IRON INTO A ROUGH MOULD IN THE FOUNDRY FLOOR. (*Penny Magazine*, 1844)

quantities of iron produced not only cut the price of iron but forced the owners to consider markets beyond those just available locally.

For the next stage in the manufacture of wrought iron, the pigs produced by the furnace had to be taken to a forge. Although Street was short lived as a furnace, Lawton Furnace supplied pig iron not only to Cranage and Warmingham Forges (and later to Street Forge after it was converted from a furnace) but also supplied pig iron to the North Staffordshire forges as well. Because the forges were also operated by water power and needed charcoal for their operation they were normally situated some way from the furnace so as not to clash in their need for power and raw materials, as can be seen by the locations of Lawton furnace and its associated forges. At the forge the cast iron pigs, which

FIGURES 237-240. STAGES IN THE PRODUCTION OF BAR IRON. 1. HEATING THE METAL IN THE FINERY WITH WATER POWERED BELLOWS (*top left*). 2. WORKING THE HOT IRON UNDER THE WATERPOWERED HAMMER (*top right*). 3. REPETITIVE HAMMERING TO REDUCE THE CROSS SECTION AND SO LENGTHEN THE BAR (*bottom left*). 4. FINAL HAMMERING WITH COLD WATER THROWN ONTO THE HOT IRON TO SCALE OFF IMPURITIES (*bottom right*). (D. Diderot, *Encyclopédie*, 1751)

contained around 3 to 4 % of carbon, had to be refined to a state where they had less than 1% of carbon. To achieve this the pig iron was re-melted and stirred in a hearth called a "finery", again using an air blast produced by water powered bellows. As the metal was heated, the oxygen in the air blast converted the carbon into gases which were driven off, leaving a spongy mass of iron with the required carbon content. Hammering this lump of metal then drove off further impurities and consolidated the metal into a block. This block of wrought iron was then taken to another hearth known as the "chafery" where it was heated without any air blast until it was soft enough to work with a hammer. The output of this final hammering was usually a bar about one inch square which could then be sold to other forges and smiths for use in their local communities. These large forge hammers had to be water powered because the size of the blooms produced by this method were so much larger than those produced previously in the bloomeries.

Two types of water powered hammers were used. The first type was pivoted around half way along its shaft. Cams were arranged around the extended length of the wheelshaft such that they forced the non-striking end of the hammer stock downwards thereby raising the actual hammer head. When the cam's action released the hammer stock its head would then fall under gravity on to its anvil. Alternatively, the hammer could be pivoted at the non-striking end of the stock in which case the waterwheel-driven cams directly lifted up the hammer head prior to its release. The rate of striking of these hammers could usually be adjusted by the operator from his working position by altering the amount of water flowing on to the appropriate waterwheel.

One of the large markets for iron was the nail makers but they needed much thinner bar iron than could be produced in the chafery. The use of water powered hammers could not forge a very thin bar because the bar became too long and flexible to handle, which also caused problems in maintaining the temperature of the bar. Consequently the slitting mill was introduced to produce rod iron from the chafery bar iron. The design of a slitting mill was first patented in 1588 by Bevis

Bulmer who may have been the inventor, or alternatively the design was brought to this country through industrial espionage in Sweden or the Low Countries. The production of rod iron required two distinct processes. Firstly, the bar iron was rolled into a flat plate in the rolling mill which consisted of two smooth rollers, one above the other which were driven in opposite directions by a waterwheel. The hot bar iron was passed between these rollers repeatedly until it had achieved the thickness desired. Secondly, the resultant plate was then passed between two waterwheel-driven slitting rolls similar to those in the rolling mill but with cutters of various size along their length. When passed through these slitting rolls the iron was reduced to the size required (see Figure 241). Another water powered tool used in the forge/slitting mill was a pair of shears for cutting the metal bar and rod to length. This was operated by the action of a cam on the arm of the shears.

FIGURE 242. METAL SAW TEETH MADE WITH A PUNCH AND DIE. (*The Working Man*, 1866)

FIGURE 241. A ROLLING AND SLITTING MILL USED TO PRODUCE ROD IRON FOR SUPPLY TO NAIL MAKERS, ETC. (*D. Diderot, Encyclopédie*, 1751)

The operation of these forges, especially when they included a slitting mill, required large amounts of power. Typically a forge such as Cranage would have one waterwheel for the bellows, at least another wheel for the hammer(s), and one, or even two, waterwheels for the slitting mill. When the forges were built, in the last half of the 17th century, their power requirement was considerably larger than that normally provided by existing waterpowered sites, such as the local corn mills. The corn mills were largely confined to the tributary streams because of the difficulties experienced in harnessing the River Dane so the establishment of the forges, especially at Cranage, led to major improvements in weir design and construction, capable of withstanding the force of the whole river.

Because of the geographical distance between the furnaces and the forges, coupled with the complex partnership arrangements, the work of each forge became specialised in nature. Warmingham Forge's speciality was making saws which required the bar iron to be hammered into shape of the saw and then the teeth were punched into the metal, finally the teeth were offset. Bar iron made at Cranage and Warmingham Forges also went into the slitting mill at Cranage and was then supplied the nail making trade, mainly in South Lancashire. Street Forge was a plating forge concentrating on producing iron plates for making large evaporation pans for the salt trade. The introduction of fixed working hours at the furnaces a direct result of the technology in use. The specialisation of the workers in the forges was due to the efficient use of capital equipment and centrslised management. These characteristics of specialisation and fixed working time became some of the defining elements of a "factory system".

FIGURE 243. WATERPOWERED SHEARS FOR CUTTING BAR IRON TO LENGTH. (*R. R. Angerstein*, 1754)

Copper Making

When Charles Roe established his copper works at the Havannah and at Bosley in the 1760s the technology used was very similar to that used for the production of iron in the previous hundred years. Ingots of copper from the main smelting site at Macclesfield were sent to the two sites where they were then heated and hammered into shape. The water powered hammers used at both sites were operated in a similar manner to those used in the iron industry. Both sites also had a rolling mill for producing sheets of metal from the basic ingots which were again similar to the rolling mills used in the iron forges. Copper sheet was supplied to the Royal Navy for sheathing ship's hulls to prevent damage by marine life in tropical waters.

However, one new technique was introduced at the Havannah which had not been used in the area by the iron industry, namely wire making. Firstly, the brass ingots were heated in the annealing oven and then hammered out to as thin a bar as possible. This bar was then pulled through a drawing plate with a hole diameter such that it caused the metal to become thinner and hence longer. The resultant wire was then passed sequentially through ever decreasing diameter holes in the drawing plate until the required diameter of wire was achieved. Waterpower was used in the wire mill to pull the brass wire through the dies and to reel up the resultant wire. Normally there would be two horizontal reels between which was a fixed holder for the drawing plate with its various hole sizes. Whichever

FIGURE 244. WIRE DRAWING USING WATERPOWER. THE DRAWING PLATE IS JUST UNDERNEATH THE OPERATIVE'S RIGHT ARM. (*C. Tomlinson, The Usefull Arts and Manufactures of Great Britain*, 1848)

reel was empty was then powered from the waterwheel to pull the wire through the drawing plate. On changing the hole for a smaller size the wire would then be reeled back to the first reel, again using water power (see Figure 244). Brass wire was an important product of the Havannah copper works while it was active in the second half of the 18th century as it was used extensively in making the carding machines for the expanding textile industries.

Fulling

When woollen cloth had been woven it was necessary for it to undergo a process known as fulling whereby the cloth was cleansed of any dirt, grease and oil and was also pounded so that the cloth fibres would mat together providing a degree of felting and shrinkage. Fulling cloth can be traced back to antiquity when cloth was trampled, or walked on, in a bowl or trough of liquid, usually a mixture of fuller's earth and water but other liquors could be used that had a similar effect, such as stale urine. This practice gave rise to the term "walking" being used for fulling cloth and hence many later fulling mills were known as walk mills.

Although the vertical waterwheel and its associated gearing had been used since antiquity for corn grinding and raising water it was not until a means could be found to convert the rotary motion of the waterwheel into reciprocal motion that waterpower could be applied to a wider range of processes. In medieval times this problem was solved by the introduction of the cam. The cam is a very simple device being a projection fixed onto a rotating shaft which could then activate machinery as the shaft rotated. The simplest application of the cam is in providing vertical stamps where a number of cams are fixed around the circumference of an extended wheelshaft. The rotation of the shaft brings the cams into contact with a similar projection on the vertical stamp. As the shaft rotates the cam lifts the stamp until the rotational movement takes the cam out of contact with the projection on the stamp at which point the stamp falls under its own weight onto the material being processed.

The earliest method of mechanical fulling used vertical stamps. The ends of the stamps were notched so that when they fell onto the roll of cloth in the fulling liquor they caused the roll of cloth to turn so that a new surface would be exposed to the next beat

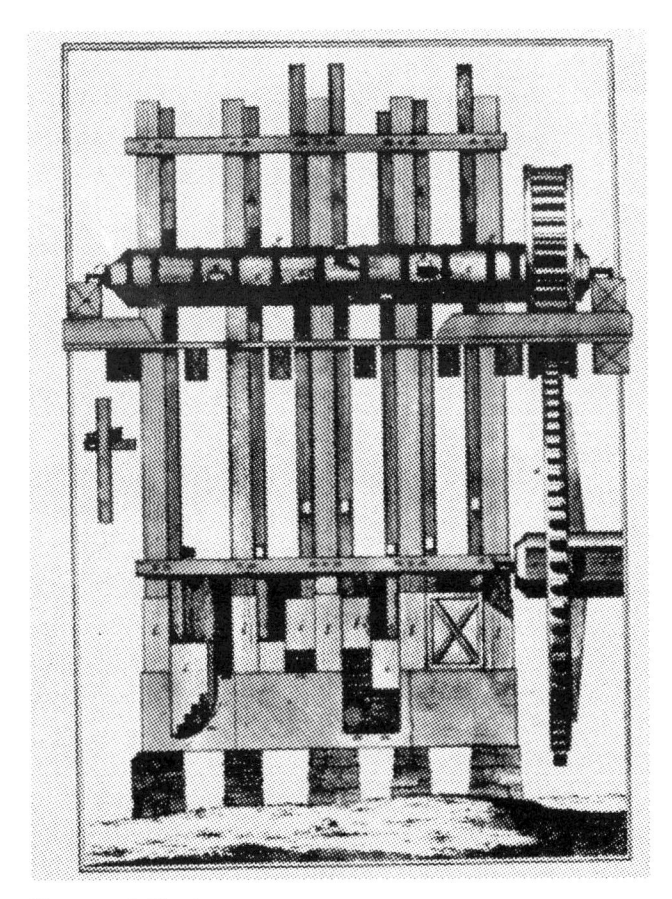

FIGURE 245. A SERIES OF WATERPOWERED STAMPS FOR FULLING CLOTH. NOTE THE NOTCHED ENDS OF THE STAMP SHOWN ON THE LEFT.

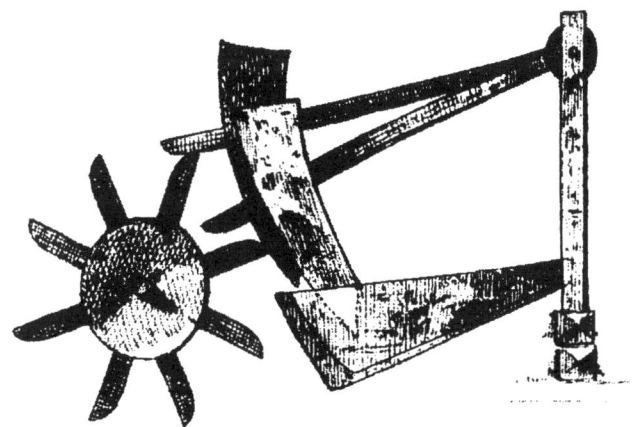

FIGURE 246. A PAIR OF FULLING STOCKS, THE CAMS ON THE SHAFT DRIVEN FROM A WATERWHEEL RAISES THE HEADS THEN ALLOWS THEM TO FALL ONTO THE CLOTH.

of the stamp (see Figure 245). It is quite possible that this was the method employed at Congleton Fulling Mill in the 14th century. This machinery was gradually replaced by the fulling stock. This usually consisted of a pair of hammers which hung from a pivot at the other end of the hammer shafts. The heads of the stocks were lifted by the cams on the main drive shaft and then fell when released into shaped recess or "box". The notches on the stock head and the curved shape of the cloth-containing box ensured that the cloth circulated freely and was turned over when it was struck (see Figure 246). This later arrangement would most probably have been the type of machinery used at Danebridge Fulling Mill when it operated in the 17th century.

The last fulling mill in the area, at Danebridge, ceased work in the early 18th century consequently later developments in the fulling industry were never introduced into any mills on the River Dane and its tributaries.

Silk Throwing

The raw material for the silk industry is the filament that the silk worm extrudes to make the case for its cocoon. To make a usable thread this filament has to be twisted and then a number of these twisted filaments have to be further twisted together. In spite of the apparent simplicity of this process a number of operations, using various machines, were required to produce a thread suitable for weaving into cloth.

The first stage of silk production is to kill the silk worm pupae and unravel the length of filament from the cocoon. This is accomplished by inserting the cocoons into near boiling water which not only kills the pupae but softens the gum holding the filament in shape. The filament, which is some 500 to 1000 yards long, is then wound onto large square reels such that the filament is not flattened or becomes stuck to itself.

Unfortunately, the nature of the silk worm is such that it does not thrive in this country so the factories, or "filatures", where this first process of reeling took place were located in the region where the silk worm was cultivated i.e. Italy, France, the middle and far east.

Because silk filaments were extraordinarily long, compared to other textile fibres where the staple, or length of fibres, is only a few inches, it was only necessary to twist the single filaments rather than join numerous fibres together as occurs in the spinning process. Consequently the process that the silk filaments underwent was known as "throwing" rather than spinning. A number of processes in the throwing mills were mechanised in the 16th and 17th century in Italy, but the design of these machines was kept as a close secret, despite some of the details being published as early as 1607 (see Figure 247). The acquisition of the design of these machines is a classic

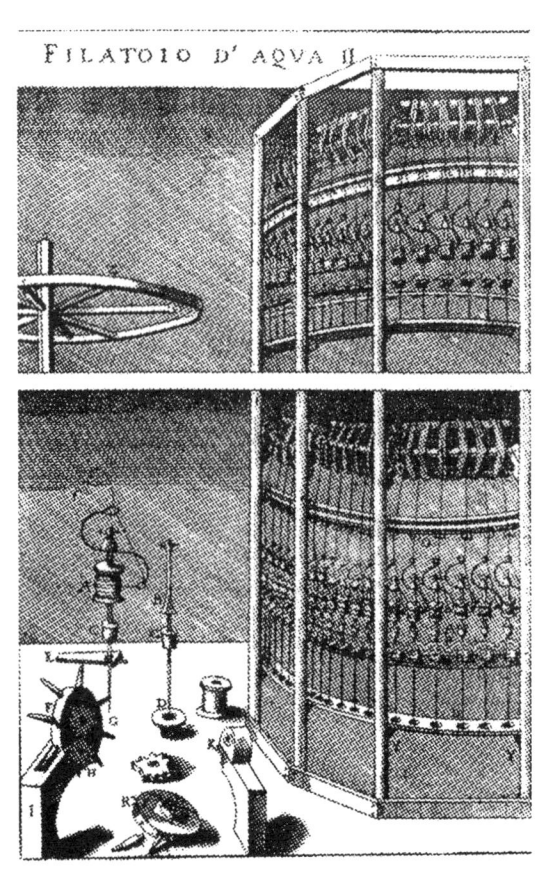

FIGURE 247. AN ITALIAN THROWING MACHINE, THE CENTRE PART ROTATED BY THE WATERWHEEL CONSISTED OF ACTUATORS TO OPERATE THE PARTS OF THE MACHINE (*left*) WITH ALL THE SPINDLES, BOBBINS AND REELS HOUSED ON AN OUTER CAGE (*right*). (*V. Zonca, Novo Teatro di Machine*, 1607)

FIGURE 248. A WINDING MACHINE. (*Macclesfield Museums Trust*)

FIGURE 249. A ONE THIRD SCALE MODEL OF AN ITALIAN SILK THROWING MACHINE. (*Macclesfield Museums Trust*)

story of industrial espionage by John Lombe, a native of Derby. The secrets of these silk machines having been stolen from the Italians, he and his brother, Thomas Lombe, built the first water powered silk throwing factory in this country in 1721. The large Congleton Silk Mill, the fourth to be built in this country, utilised machines built to the designs introduced from Italy by the Lombe brothers.

The first process, after grading and washing, was that of winding. The skein of silk filament was placed on a swift, a four (or six) armed, star-shaped, reel, and the filament was then wound from the swift, through an eye, onto a bobbin. The winding machine was driven by water power which turned not only both the swifts and the bobbins but also provided the oscillating horizontal movement of the eye which ensured that the filament was evenly wound over the full length of the bobbin (see Figure 248). If the filament broke it was rejoined using the natural gum on the filament. Any knots had to be removed from the filaments using a cleaning machine prior to the actual throwing process. The cleaners were almost identical to the winding machines but the filament passed through the gap between two metal blades arranged rather like of a pair of shears. Any knot would snag on these blades causing the filament to break, the knot was then removed and the filament rejoined together. The gap between the two blades could be adjusted to suit the fineness of filament being processed.

The silk throwing process itself involved three separate operations. Firstly the filament was given a twist throughout its length of up to 80 turns per inch; secondly, two (or more) of the twisted filaments were loosely wound together, or "doubled"; and finally, a number of doubled threads (four, six or eight) were twisted together in the opposite direction to which the individual filaments were twisted. The early filament twisting machines, known as "filatoes" (from the Italian filatoio), were circular in plan with a diameter of 12 to 15 feet, and were around 19 feet high. These machines were made entirely of wood and would usually occupy the two lowest floors in the throwing mill. These machines consisted of a fixed outer cage onto which were mounted all the bobbins and reels with their associated gearing, with an inner rotating framework of actuators mounted on a vertical shaft that was driven from the waterwheel. Near the base of the outer cage there was a circular row of spindles arranged in twelve groups, with six spindles per group. Each reel received six windings of twisted filaments from a group of six spindles. This arrangement of rows of spindles and associated reels occupied a height of between 4 and 5 feet, consequently four rows of spindles and reels could be accommodated altogether within the height of the machine i.e. 6 spindles per group x 12 groups x 4 rows giving a total of 288 spindles per machine (see Figure 249).

The rows of spindles and reels were rotated by the action of the waterwheel. For each row of spindles, the rotating upright shaft in the centre of the machine had four long arms attached to it. At the extremity of the arms there were curved wooden arcs making a horizontal "T" shape (see Figure 250). These wooden arcs were pivoted where they joined the arms and one end had a string attached to it which was led over a pulley and attached to a weight. This weight pulled one end of the wooden arc towards the centre of the machine which caused the other end of the arc to press against the inside of the spindles as it passed them and thereby made the spindles rotate due to the friction. To ensure a good frictional contact with the spindle the outside of the wooden arc was covered with leather or felt. Later versions of the machine eliminated the weights by using a spring to keep the arm pressed against the spindles.

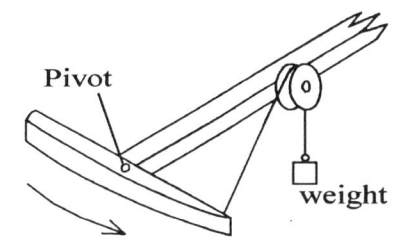

FIGURE 250. ONE OF THE ROTATING "T" SHAPED ARMS THAT RUBBED THE SPINDLES CAUSING THEM TO ROTATE.

The reels were organised in pairs on a common shaft which was driven via a couple of gears from a spur wheel mounted on the outer frame between the two reels. This spur wheel was constructed with a number of staves, or pegs, projecting from its rim (see Figure 251). For each row of reels the central rotating part of the machine had eight curved blades of wood, known as "serpents", attached to it. As these blades slid under the pegs in the spur wheel they caused the spur wheel to rotate. The upward curve of these wooden blades were designed so that each successive blade slid against successive pegs thereby turning each spur wheel continuously and hence also its associated reels. In between the bobbins and the reel each of the six filaments passes through an eye mounted on a horizontal piece of wood called a "layer". These layers were given a horizontal reciprocating motion, derived from the spur wheel driving each reel, through gearing and the use of an eccentric cam fixed to the layer. This motion distributed the filament evenly, in layers, on the reel (see Figure 252). The amount of twist given to the filaments depended on the differential speed between the reels and the flyers attached to the

FIGURE 251. HOW THE REVOLVING CURVED SERPENTS OPERATE THE SPUR GEARS ATTACHED TO THE REELS.

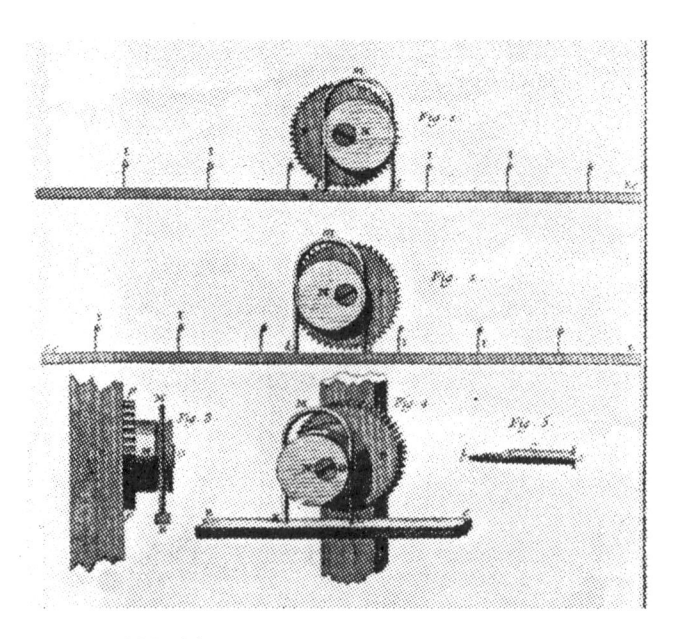

FIGURE 252. THE ECCENTRIC IS ROTATED VIA GEARING WHICH THEN MAKES THE LAYER OSCILLATE HORIZONTALLY

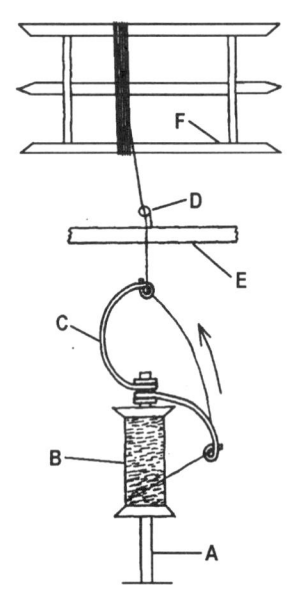

FIGURE 253. THE BASIC OPERATION INVOLED IN THE UPTWISTING OF SILK FILAMENTS.

spindles. The spur wheels could be exchanged for ones with different numbers of pegs thereby altering the speed of rotation of the reels which would alter the degree of twist in the filament.

The basic operation of the machine is shown in Figure 253. Each spindle, A, which was mounted in a glass footstep bearing attached to the outer frame, held a bobbin of silk filament, B, which was led through a flyer, C, mounted on top of the spindle. The flyer, a device which had been used on spinning wheels for many years, rotated with the spindle and imparted a twist to the filament. The filament then passed through an eye, D, before being wound onto a reel, F, to form a skein. During this process the layer, E, is slowly moved horizontally backward and forward to ensure that the filament is wound onto the reel in even layers. This method, which twisted the filament as it left the bobbin, was known as "uptwisting" due to the direction in which the filament moved and was a peculiarity of silk throwing as all other versions of textile spinning impart the twist at the receiving spindle or reel.

The next process was that of doubling. Here two, three or even four of the twisted filaments were folded together to receive just a slight amount of twist. This was a hand process, usually performed by women, who wound the thread onto a wheel (see Figure 254). Then a number of doubled threads (either four, six or eight) were twisted together in the opposite direction to that which the individual filaments were twisted on another type of machine called a "tortoe", (from the Italian torcitoio). This machine was extremely similar in size and operation to the filatoe but the spindles had to rotate in the opposite direction. In the case of the tortoes the "T" shaped arms that rubbed across the spindles stretched underneath and beyond the row of spindles. These arms then had a wooden arc mounted on their end such that the inside of the arc rubbed against the outside (as opposed to the inside in the case of the filatoes) of the spindles, thereby causing them to rotate in the opposite direction to those on the filatoes.

The resultant thread, called organzine, had a high degree of twist and could be used as the warp threads for weaving silk cloth. Until the building of these early silk mills all organzine used in this country had to be imported from places such as Italy. The nature of this technology pirated from Italy determined the shape of the mills built to house them. The two storey throwing machines occupied the two ground floors

FIGURE 254. DOUBLING FOUR FILAMENTS OF SILK TOGETHER WITH A SLIGHT TWIST. (*A. Ure, A Philosophy of Manufactures*, 1861)

with the lighter winding machines and other processes occupying the floors above. The number of winding machines with respect to throwing machines determined the height of the mill, usually five storeys. At the Derby silk mill, and quite possibly also at Congleton Silk Mill, all the machines were driven by lineshafts and gearing running above the throwing machines, a configuration that seems to have been quite common in the much smaller continental silk throwing establishments.

Although the first mill at Derby was wide enough to accommodate two rows of throwing mills, problems arose in dealing with the very fine silk filaments in the dark area between the two rows of machines. All the later mills were built to house the throwing machines in a single row giving maximum light for the operatives. This pattern of mill building, which was perfected in the Congleton Silk Mill, became the standard pattern for all future water powered textile mills.

On the River Dane and its tributaries the silk mill in Congleton was the only one to use this Italian technology. By the time further silk mills were built in the area at the beginning of the 19th century throwing machine design had changed. The machines were reduced in height to only one row of spindles etc. and instead of being circular the machines were built in a straight line. The frame work was made of cast iron and metal was used for all the gearing and shafting. These machines, which used ideas from the design of cotton spinning machinery, were easier to accommodate and as they were made with greater precision but increased strength they could operate at far greater speeds than the original Italian machines (see Figure 255).

FIGURE 255. A MID-19TH CENTURY, METAL FRAMED, RECTANGULAR THROWING MACHINE. (*A. Ure, A Philosophy of Manufactures*, 1861)

Silk Spinning

The whole process of producing silk thread from the cocoons of the silk worm created a large amount of waste silk filament. When the cocoons were unreeled the first and last part of a filament was discarded because it was not consistent in thickness with the majority of the filament. The various processes at the throwing mills also created waste silk. In 1765 it was noted that 50% of all raw silk became waste, with also a similar percentage being generated from the throwing process. As silk was expensive it is not surprising that attempts were soon made to turn this waste into a saleable product.

Initial attempts towards the end of the 18th century tried to spin this waste silk on cotton spinning machines. The waste silk was first scutched to remove impurities then combed and carded on the same machinery as used for cotton. The result of the carding process was then chopped into short staples of about 1½ inches by a machine resembling a chaff cutter. These were then boiled and rinsed before being scutched and carded again. The waste silk was then spun on throstles or mules in the same way as cotton. Unfortunately, although technically viable, this method, using short staples, produced a cloth that had lost the very qualities that allowed silk to command a premium in the marketplace.

In the early 19th century, when machines were being developed to spin long fibres such as flax, it was discovered that by using a long staple of around ten inches, similar in length to flax, these problems with silk spinning could be overcome. One important improvement was the use of a gill, which is a set of combs that move along between the two pairs of drawing rollers so carrying forward any short fibres. The thread produced by the spinning process was passed through a gas flame to singe away the feathering on the thread, which also helped to improve the lustre of the cloth produced. Silk spinning became a speciality in Congleton where a new six storey steam powered mill was built to extend Brook Mills when it changed from cotton spinning to waste silk spinning in 1835. Later in the 19th century both Dane Mill and Forge Mill concentrating on waste silk spinning. It is possible that flax spinning machinery was used to spin waste silk when the Dakeyne establishment at Gradbach changed from flax spinning to waste silk spinning around 1850. The "Equilinum" machine patented by the Dakeynes for flax spinning as early as 1794 was similar to the later silk spinning machines. This branch of the silk industry survived long after the throwing branch had ceased to exist in the area. Ring spinning of waste silk was introduced in the 1880s and both Forge Mill and Dane Mill continuing silk spinning until about the beginning of the Second World War.

Cotton Spinning

Raw cotton consists of short staples of fibre about 1½ inches long. To spin these short fibres mechanically into a continuous thread required a great deal of effort in preparing the fibres prior to the actual spinning process. Consequently, the whole process of spinning cotton was a sequence of operations requiring a number of machines, all driven by waterpower, arranged so that the cotton could proceed from one machine to another until it became the final weavable yarn.

Once the bales of raw cotton were opened, any foreign bodies had to be removed and the tight wads of fibre loosened and opened. During the late 18th and early 19th centuries this process, known as "batting" was carried out by hand. Having loosened the fibres the first preparation process was that of carding. Although a carding machine had been patented by Lewis Paul & Daniel Bourne in 1748 the machine needed many modifications to be able to operate continuously. The basic mechanism of the carding machine consisted of a large horizontal cylinder, known as the "swift", which is rotated inside a concave frame. The outside of the swift was covered with brass wire pins which meshed with similar brass wire pins mounted on the inside of the concave frame. The cotton was fed onto the outside of the revolving cylinder and the action of the brass pins combed the cotton fibres so that they were all lying in the same direction. The waterpowered carding machine patented by Arkwright in 1775 included an automatic feeder. Also another cylinder, called a "doffer", rotating in the opposite direction to the swift, removed the carded cotton from the carding cylinder. A comb working up and down, activated by a crank, then took the cotton fleece off the doffer (see Figure 256). These modifications had been developed by a number of people with only the comb and crank possibly being devised by Arkwright, although even this was debated at the time. On the finishing carder the fleece of cotton was contracted by passing through a funnel with rollers to form a continuous roll of cotton fibres, or "sliver", which was about one inch in diameter.

The sliver of cotton from the carding machine was too thick and irregular to be spun mechanically. To

FIGURE 256. LATE 18TH CENTURY CARDING MACHINES. THE MACHINE IN THE FOREGROUND IS A FINISHING CARDER PRODUCING SLIVER. (*Helmshore Textile Museum*)

remove the irregularities, lengths of sliver were fed into a machine called a drawing frame where up to six slivers were united to produce a single sliver. This process was repeated a number of times until a uniform sliver of consistent quality had been produced. The drawing frame had two pairs of horizontal rollers, with the second pair of rollers rotating faster than the first pair. The slivers were fed into the first pair of rollers where they were amalgamated and then the second pair, because of their faster rotation, drew out the resulting cotton sliver to regain the thickness of the original slivers (see Figure 257). This process of drawing was probably the greatest contribution of Arkwright to cotton spinning technology.

The sliver from the drawing frames was still too thick and fragile to be spun directly so it was converted into a thinner sliver which was given a small amount of twist to provide some strength. The machine used for this process was known as a roving frame. This machine again used the same principle of roller drawing as in the drawing frame, combining two slivers together and drawing the combined sliver much thinner, about a quarter of an inch thick, which was then known as roving. This roving was fed into cans called "lantern frames", which were circular containers that slowly rotated as the roving was fed into them. The rotation of the lantern frames produced a slight twist in the roving and centrifugal force caused the roving to coil up in the container (see Figure 258). These rovings then had sufficient strength to be wound by hand onto bobbins and taken for spinning. In later years the lantern frames were discarded and rovings were wound directly onto bobbins automatically using a flyer to provide the small amount of twist required.

FIGURE 257. A LATE 18TH CENTURY DRAWING FRAME AMALGAMATING TWO LENGTHS OF SLIVER TO PRODUCE A MORE UNIFORM SINGLE SLIVER. (*Helmshore Textile Museum*)

FIGURE 258. A LATE 18TH CENTURY ROVING FRAME FEEDING INTO TWO SLOWLY ROTATING LANTERN FRAMES TO IMPART A SMALL AMOUNT OF TWIST TO THE ROVING. (*Helmshore Textile Museum*)

FIGURE 259. A REPRESENTATION OF A WORKER AT ONE OF THE LATE 18TH CENTURY WATER FRAMES USED FOR THE SPINNING OF COTTON INTO WARP THREADS. (*Helmshore Textile Museum*)

The spinning machine, or waterframe, patented by Arkwright was a rectangular machine, built of wood, with the bobbins of roving arranged vertically in a horizontal row at the top of the machine. Underneath each of the roving bobbins were two pairs of horizontal rollers (see Figure 260). The top roller of each pair was weighted and covered in leather and the bottom roller fluted, all of which helped to grip the cotton staples. The main secret to successfully using rollers to spin cotton was the setting of the gap between pairs of rollers so that it was greater than the staple length but short enough to ensure that any staple was always being gripped by one of the pairs of rollers. The cotton roving was passed between the first pair of rollers which, by their rotation, drew the roving through and compressed it; whilst still passing through these rollers, it was caught by the second pair of rollers which revolved with three, four or five times the velocity of the first pair, and which therefore drew out the roving to three, four or five times its former length and degree of fineness. After passing through the second pair of rollers the reduced roving was passed through flyers attached to the row of vertical

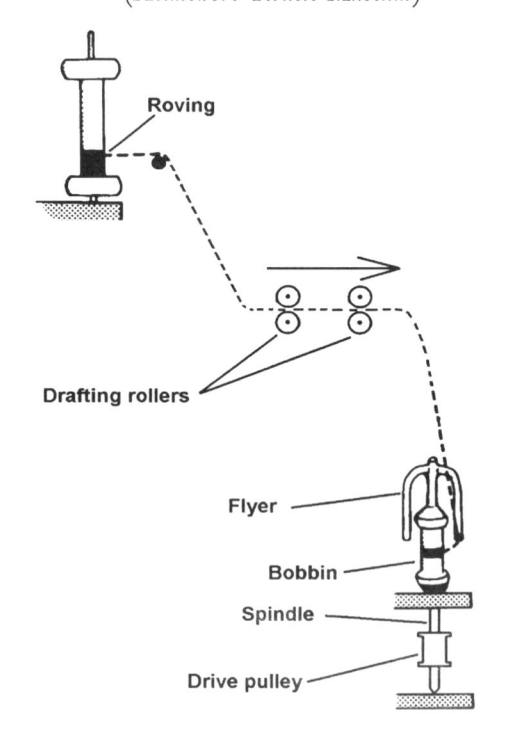

FIGURE 260. THE BASIC OPERATION INVOLVED IN SPINNING COTTON USING DRAFTING ROLLERS.

spindles at the bottom of the machine. These spindles were rotated by leather belts consequently the flyers twisted the roving into a thread and at the same time wound it upon bobbins mounted on the spindles (see Figure 261). The bobbins were seated on forks attached to a horizontal beam which was raised and lowered by the action of a crank arm. This caused the thread to be wound evenly over the length of the bobbins.

Although much has been made of Arkwright's "invention" of machine cotton spinning, this technology was also employed by many other spinners who were contemporary with Arkwright. One of these other spinners was John Barnes who operated at Lower Daneinshaw Mill during the 1770s and was one of the firms Arkwright sued unsuccessfully for breach of patent. Once Arkwright's patents were annulled this technology was quickly installed during the early 1780s in mills at Danebridge and at Stonehouse Green in Congleton. Over a period of time improvements to these machines were made. The waterframe soon became constructed of metal using cast iron for the frame, gears and shafting which allowed a much greater number of spindles to be driven at higher speeds. The noise made by these machines was responsible for their name of "throstles" (see Figure 262).

Although the introduction of the waterframe revolutionised cotton spinning it was limited to producing fairly coarse thread which was only suitable for use as warp threads for weaving. Threads suitable for weft threads still had to be produced on spinning jennies which were powered by hand. The production of weft threads was only achieved using powered machinery with the introduction of the spinning "mule" designed by Samuel Crompton in 1779. This machine combined the principles of the waterframe

FIGURE 261. A SET OF FOUR SPINDLES ON AN EARLY WATER FRAME. THE ROVINGS AT THE TOP ARE FED THROUGH THE DRAFTING ROLLERS IN THE MIDDLE (*note the large weights hung on the rollers to maximise the grip on the cotton*) AND THE THREAD IS WOUND ONTO THE BOBBINS AT THE BOTTOM. AS CAN BE SEEN EACH DRIVE BELT WAS IN CONTACT WITH FOUR SPINDLES. (*Helmshore Textile Museum*)

FIGURE 262. A METAL FRAMED SPINNING THROSTLE OF THE EARLY 19TH CENTURY WHICH REPLACED THE WATER FRAME. (*A. Ure, A Philosophy of Manufactures*, 1861)

FIGURE 263. A COTTON SPINNING MULE WHICH WAS SUPERIOR TO THE WATER FRAME FOR SPINNING THE FINER COUNTS OF THREAD. NOTE THAT OPERATING A MULE WAS A HIGHLY SKILLED JOB AND ALMOST ALWAYS MEN WERE EMPLOYED TO WORK THESE MACHINES. (*Penny Magazine*)

and the spinning jenny. It used the pairs of rollers as in the water frame to draw out the cotton roving but the spindles were mounted on a carriage that moved away from the rollers as in the jenny. As they moved away, the spindles rotated to give the yarn a twist and the movement away from the rollers stretched the yarn in a similar manner to the jenny. Once this spinning sequence was completed the rollers stopped rotating and the recently spun yarn was disengaged and wound on to the spindles as the carriage returned to the rollers. The early spinning mules were only semi-powered with much operator skill needed to ensure that the winding-on process was even (see Figure 263). A fully self-acting mule was only designed in 1830.

Although Arkwright stayed loyal to the water frame, the cotton mills on the River Dane and its tributaries updated mainly to mule spinning. Certainly by 1811, when a census of spinning machines was taken so that a suitable reward for Samuel Crompton could be determined, the records show that Bosley Works had 4404 mule spindles but only 720 throstle spindles, Lower Daneinshaw had as many as 13680 mule spindles compared to 1400 on throstles, and at Danebridge all 4296 spindles were on mules.

The cotton mills on the River Dane and its tributaries had all ceased production by 1865 and so the later improvements in the industry, such as ring spinning, were never introduced into this area.

Flax Spinning

The spinning of flax to produce a linen thread was a long established trade in England mainly because the raw material, the linseed or flax plant, was grown here. The material that was used to make the flax yarn was the fibres contained in the stalks of the plant which had to be extracted and prepared for spinning. The first process, known as "retting", was to soak the flax stems in water for about two weeks to soften the woody material in the stems. The stems were then beaten in a scutching machine to remove this woody material. As far as is known these processes did not take place at Gradbach Mill, where the Dakeyne family operated the only mill on the River Dane involved in spinning flax. Presumably these processes took place at the Dakeyne's other mill in Darley Dale, not far away in Derbyshire.

After the fibres had been extracted from the stems, the flax was heckled by passing the bundles of fibre through a series of combs to straighten all the long fibres, called "line", and to remove the short fibres, called "tow". The process of spinning was very similar to that used for cotton, with line used to produce the finest yarn and tow making a rather coarser thread. Firstly the fibres were formed into rovings on a form of roving machine and then spun using pairs of drawing rollers as in the waterframe. The main difference from cotton spinning was caused by the nature of the flax fibres. The staples, which had a length of about ten inches, were much longer than cotton and were more brittle. The machine used for these processes at Gradbach was developed by the Dakeynes themselves who took out a patent on the design in September 1793. In this patented machine the pairs of drafting rollers were set about a foot apart to suit the staple length and in-between the pairs of rollers was an endless belt containing a continuous row of blocks with protruding wire pins. As the belt

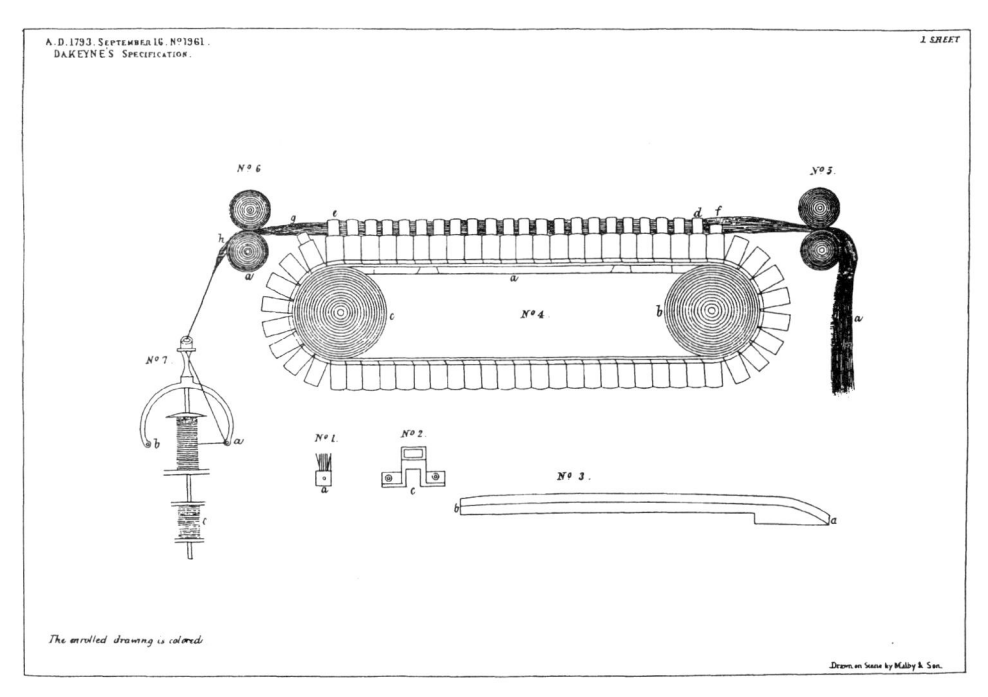

FIGURE 264. THE HEART OF THE "EQUILINUM" FLAX SPINNING MACHINE INVENTED BY THE DAKEYNE BROTHERS IN 1793 WAS THE MECHANISM WHEREBY METAL PINS WERE USED AS A GILL OR COMB BETWEEN THE TWO DRAFTING ROLLERS. (Patent No. 1961, 1793)

revolved these pins were forced up into the fibre emanating from the first set of rollers, the pins were then pulled through the fibre thereby giving a combing action similar to carding as well as supporting the fibres and assisting to move them into the second pair of rollers. This part of the machine was christened the "equilinum" by its inventors and was later known as a "gill" (see Figure 264). The equilinum straightened and levelled the fibres which produced a more even sliver. The machine was used in two modes. Initially it was used as a heckling machine with the rollers and the equilinum producing a roving of flax fibres which were then passed through the machine a second time when the output from the rollers was taken through a flyer attached to a spindle. This flyer imparted twist to the yarn which was then wound on to a bobbin mounted on the spindle.

Flax fibres are brittle which resulted in breakages during the production process. This problem was not solved by the Dakeynes' invention and was not overcome until the 1820s when a machine was invented by James Kay of Preston which passed the fibres through hot water before they were spun. This softened the gum in the flax so easing the drafting by the rollers (see Figure 265). Dry spun yarns proved to be more elastic but wet spun yarns were stronger and silkier in appearance. One machine using the wet spun process was being used at Gradbach when flax spinning came to an end there in 1837.

It is perhaps ironic that the principle of using a gill between the drawing rollers, patented in the Equilinum in 1793, was to become the solution to the problem

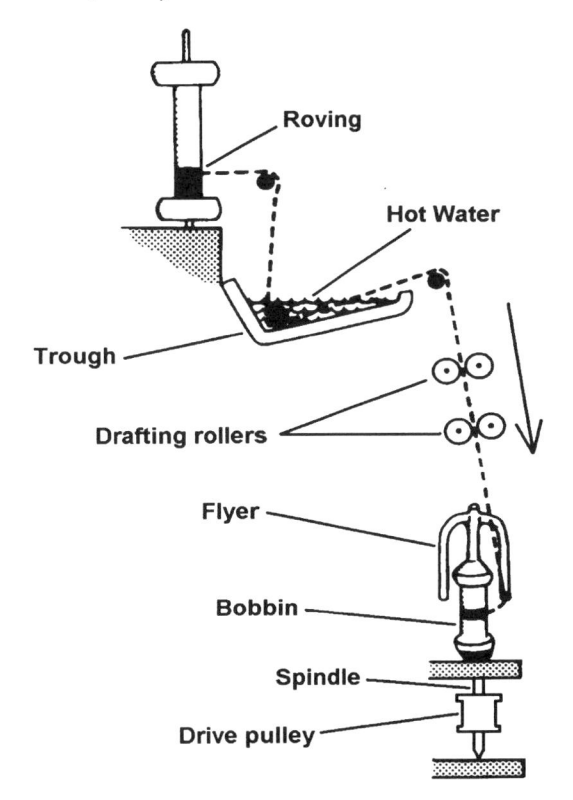

FIGURE 265. THE BASIC OPERATION OF THE WET PROCESS FOR SPINNING FLAX.

of waste silk spinning in the 1830s. The Dakeynes were 30 - 40 years ahead of their time but unfortunately they applied their idea to the wrong branch of the textile industry! It is possible that waste silk spinning replaced flax spinning at Gradbach Mill in the 1840s once it was realised that the Equilinum machines were more suited to the new process.

Cloth Printing

Printing a pattern on cloth used to take place using carved wooden blocks about a foot square, with a separate block being used for each colour. The design on the blocks was covered in dye and the block was pressed down onto the cloth. This operation had to be continually repeated along the length and width of the cloth. Although this hand process was extremely slow it did survive virtually to the present day for some specialised fabrics and patterns. However, the vast majority of cloth printing became mechanised by using rotary cylinders instead of blocks. These cylinders were covered by rolled sheets of copper engraved with the required pattern. The copper covered cylinders rotated at speed whilst the cloth was fed through the machine. This machinery could be driven by belts, or via gears and shafts, from a waterwheel as at Crag Works in the early years of the 19th century (see Figure 266).

In the 1840s Crag Works was taken over by Joseph Burch initially to manufacture his patent machine for printing up to six colours on heavy materials such as woollens and towelling but by 1850 he was using his patent machine for printing carpets. His machine had six printing blocks mounted three abreast face downwards in a reciprocating frame. As one set of blocks reached the central printing section of the machine it was pressed down onto the fabric by a cam mechanism. Simultaneously the second set of blocks was furnished with print paste by passing them over furnishing rollers. Initially, Burch used blocks moulded

FIGURE 266. A CYLINDER PRINTING MACHINE FOR CALICOES. (*C. Tomlinson, The Usefull Arts and Manufactures of Great Britain*, 1848)

in type metal which were attached to the main wooden print plates. Where large areas of colour were required the spaces between the raised metal edges of the pattern were filled with wool felt. The method of making the printing blocks was rather long winded and costly so Burch patented a much simpler method in 1843. In this system, profiled steel pins which could be moved by hand in both the vertical and horizontal planes were heated by a gas flame. The surface of a wooden block was then burned out by the hot pins to form the master mould. The block could then be used to make mouldings using low melting point type metal. Between 1843 and 1851 Joseph Burch took out five patents for improvements to this machine.

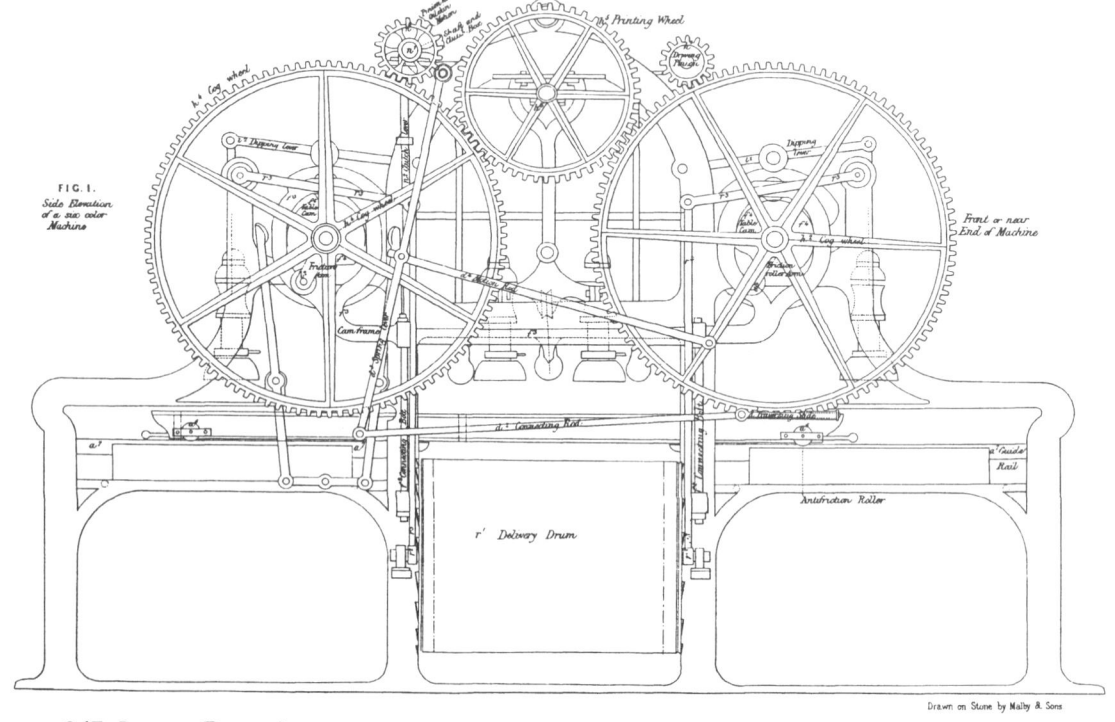

FIGURE 267. JOSEPH BURCH'S MACHINE FOR PRINTING HEAVY CLOTHS WITH UP TO SIX DIFFERENT COLOURS. (Patent No. 7937, 1839)

Fustian Cutting

The cloth used for fustian was usually woven in the Manchester region and then brought to the Congleon area for cutting to make fustian, a type of velvet. This cloth was woven so that the weft threads were raised in loops. Until the 1920s these weft loops were then cut by hand. The cloth was limed to stiffen it and then it was stretched out on a frame about 11 yards long. The fustian cutter then inserted the guide on his/her cutting tool into a "race", or line of weft loops, raising the loops to allow them to be cut by the blade of the cutting tool, so creating the velvet pile. This process was then repeated on all the lines of weft sequentially, with the fustian cutter walking along the length of cloth.

To automate this process a special machine was invented and eventually patented by George Rogers who, in 1924, entered into a partnership with G. J. Antrobus to manufacture these machines and use them to cut fustian at the Havannah Mills in Buglawton.

To prepare the cloth for machine cutting it was passed through a trough of lime, then dried by being carried round a five feet diameter steam-heated copper drum. This process was then repeated using stiffener as opposed to lime. The cloth was made into a piece 150 feet long, which was joined at the ends to form a complete loop, before being put on the cutting machines. The machines had three tools mounted across the width of the cloth. The guides still had to be set in a race by hand but, on turning a handle, the tools cut the line of weft and then returned to their start position ready to cut the next race. The machine then fed the cloth round thereby removing the necessity of the worker having to walk along the length of cloth (see Figure 268). Although these machines were designed to be driven by electricity, mains electricity was not available in the area until later in the 1930s. When these machines were installed at the Havannah, a water turbine coupled with an electric generator was also installed to provide the necessary electrical power, so in a sense the fustian cutting machines can be considered to have been water powered.

FIGURE 268. THE FUSTIAN CUTTING MACHINE PATENTED IN 1924 BY THE HAVANNAH MILLS COMPANY. LATER IT WAS PATENTED IN FRANCE, GERMANY AND THE U.S.A. (Patent No. 213,723, 1924)

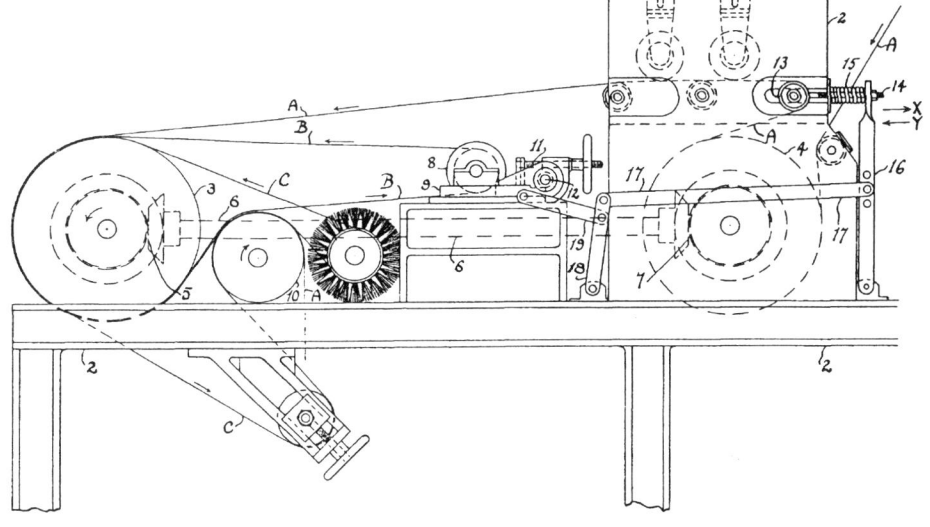

Pumping

Salt has been made in Cheshire from the naturally occurring brine springs since Roman times. This practice continued after the Romans retreated from England as the salt was a vital ingredient in preserving meat throughout the winter period. With the passage of time the level of the brine became lower due to extraction and so shallow wells or brine pits were dug to provide access to the brine. As the level of the brine continued to drop pumping became necessary to maintain the supply of this basic commodity.

The art of raising water by using a waterwheel for irrigation is probably older than the use of waterwheel to grind corn. In the Middle Ages the science and art of pumping received considerable attention due to the need for mines to be de-watered as the miners followed the mineral veins below the water table. A number of publications in the 16th century by authors such as Agricola and Ramelli describe and show a considerable array of different types of pumps.

The first reference to the use of a pump in the saltworks of Cheshire was recorded in 1636 but no details of its type or how it was operated were noted.

FIGURE 269. A RAG AND CHAIN PUMP
(*G. Agricola, De Re Metallica, 1556*)

FIGURE 270. FORCE PUMPS DRIVEN FROM A CRANK
ATTACHED TO A WATERWHEEL. (*Stadtarchiv Nürnberg*)

At what stage water power was introduced to drive brine pumps has not been recorded. However, in 1808, waterwheel driven brine pumps were noted by Holland as being present "*at Lawton, at Roughwood, at the Old Works at Wheelock and at one of the pits in Northwich and Middlewich*". Unfortunately, once again the type of pumps used were not mentioned.

Some idea of the technology that might have been used can be gained by noting the type of pumps used for raising drinking water. From the late 17th century onwards many water powered pumping systems were introduced for the provision of the public supply of drinking water.

Initially, the most likely type of pump used would have been the paternoster or "rag and chain" pump. The design of this very simple device probably dates from antiquity but was in wide use in the 17th and 18th centuries and was quite capable of raising the brine the required height. The main part of this pump consisted of an endless chain passing over a suitable pulley that was driven from the waterwheel. Balls consisting of some stuffing material enclosed in rags were fastened at regular intervals on the chain (see Figure 269). On its upstroke the chain and rag balls passed through a pipe such that the balls made a reasonable fit in the inside of the pipe. The bottom of the pipe would be positioned underneath the surface of the brine so that a certain amount of liquid would be raised to the surface by the passage of each rag ball up the pipe. The pipes themselves would have been bored using the power generated by a waterwheel but although this would probably have taken place locally no evidence has been found of pipe boring using the power of the River Dane or its tributaries.

As the techniques of boring metal cylinders improved in the later part of the 18th century piston pumps would have become suitable for use in brine pumping. Piston pumps consist of a piston enclosed in a metal cylinder with a pair of valves that allowed brine to be sucked into the cylinder and then forced out through a pipe to the surface. The valves would be operated by the movement of the piston which was driven by a crank from the waterwheel (see Figure 270). This type of force pump became universal in its application to moving liquids and was to be found not only in use in the various saltworks but also in tanneries such as the one in Park Street Congleton. Until quite recently waterwheel driven force pumps could be seen *in situ* at Washford Mill where they were used for moving the slop from one process to another.

Tanning

The art of tanning leather has a long history as it provided the basic materials for the provision of footwear and clothing as well as the essential ingredient for the making of saddlery and other tackle that enabled man to harness the power of the horse. In the Congleton area the leather industry in the Middle Ages not only catered for local needs but provided a product called "Congleton points", laces tipped with tin or silver that were used as fastenings for footwear and clothes, which were supplied far and wide.

The preparation of the hides involved soaking in milk of lime for a few days which softened the skin and allowed the hair and any fat to be scraped off the hide. After a couple of days in dilute sulphuric acid which opened the pores in the hide they were then left to soak in vats, or pits, of bark liquor for up to a year. During this period the hides would be regularly turned and gradually moved through a succession of pits with increasing strength of bark or other tanning liquor. When the tanning process was complete the hides were removed from the liquor and hung on poles to dry.

Due to the length of time taken by the process, a tan yard would have contained several sets of pits or vats containing hides at various stages of the process so that a reasonably constant output could be maintained. The pits would be interconnected by watercourses or pipes, with pumps or gravity used to distribute the water to where it was required. The tanhouse would have been a two storey building with the top floor used as a drying shed. This would probably have been constructed of wood with slats or louvres in the walls to allow for the circulation of air.

Most of the work in the tannery was carried out using manual labour but the preparation of the bark used as a source of tannin was sometimes mechanised in the larger tanneries. The bark was chopped and crushed to a powder for use in the vats and the intermittent nature of this preparation was very suitable for the application of animal power. Traditionally the bark was crushed by an edge runner stone rotating on a circular stone pavement. A horizontal beam, forming the axle of the vertical runner stone, was pivoted on a post in the centre of the pavement circle and the draft animal was harnessed to the other end of the beam (see Figure 271). As the animal walked in a circle around the pavement, bark was shovelled in the path of the runner stone to be crushed.

In the larger tanneries animal power was sometimes replaced by water power as was the case at the tannery in Park Street, Congleton, and at the tannery on the Midge Brook Farm, just downstream from Somerford Booths, in the first half of the 19th century.

FIGURE 271. A HORSE POWERED BARK MILL.
(*W. H. Pyne, Microcosm*, 1808)

Unfortunately, the exact nature of the bark mills used at these two sites was not specified. It is possible that the waterwheel at Midge Brook Farm drove a pair of edge runner stones as it was also used for grinding bones (see Figure 276). In the 19th century a number of patent designs for bark mills were introduced which used steel cutters and rollers in an apparatus similar to a chaff cutter combined with an oat roller (see Figure 272). It is possible that this type of technology was used at the Congleton Tannery, but the traditional edge runner cannot be ruled out.

FIGURE 272. A BELT-DRIVEN BARK MILL. THE SECTION DRAWING SHOWS THREE ROTARY CUTTERS WHICH GRADUALLY REDUCED THE SIZE OF THE BARK.
(*H. Dussauce, Treatise on the Art of Tanning*, 1865)

Flint Grinding

The pottery industry became established in North Staffordshire during the 17th century because the raw materials of clay and coal were to be found in the vicinity. Coarse red clay ware was produced from the local marls but the demand for finer and whiter wares caused a search for new materials. The addition of flint was discovered to give strength and whiteness to the product as well as preventing shrinkage during firing. Initially, after calcining the flints to make them more friable, attempts were made to crush the flints with conventional millstones and with stamps. These methods created considerable dust which was inhaled by the workers reducing their life expectancy to only two or three years. This was unacceptable even in the 18th century.

In an effort to solve this problem Thomas Benson took out a patent in 1726 for grinding flints under water to reduce the dust produced. Unfortunately, this patent used iron vessels and balls which introduced iron particles into the mixture which caused brown flecks when the pottery was fired. Thomas Benson improved his method with another patent in 1732, similar to that of 1726, except stone balls were used in a wooden pan with a stone floor.

The calcined flints were placed in circular pans about 12 feet in diameter and covered with water. The floor of the pans were built up of chert blocks about one foot thick called "pavers" which were sealed with clay. There was a vertical shaft driven from a waterwheel in the centre of the pan which had four arms or "sweeps" carrying vertical hanging pieces of timber which pushed loose blocks of chert, called "runners" around the pan (see Figure 273). The action of the runners tumbled the flints into contact with the pavers thereby reducing the flint, after about 24 hours, to very small particles suspended in the water. Any material in the suspension from the chert runners or pavers was a similar type to that provided by the flint and so contributing to the final mixture rather than contaminating it.

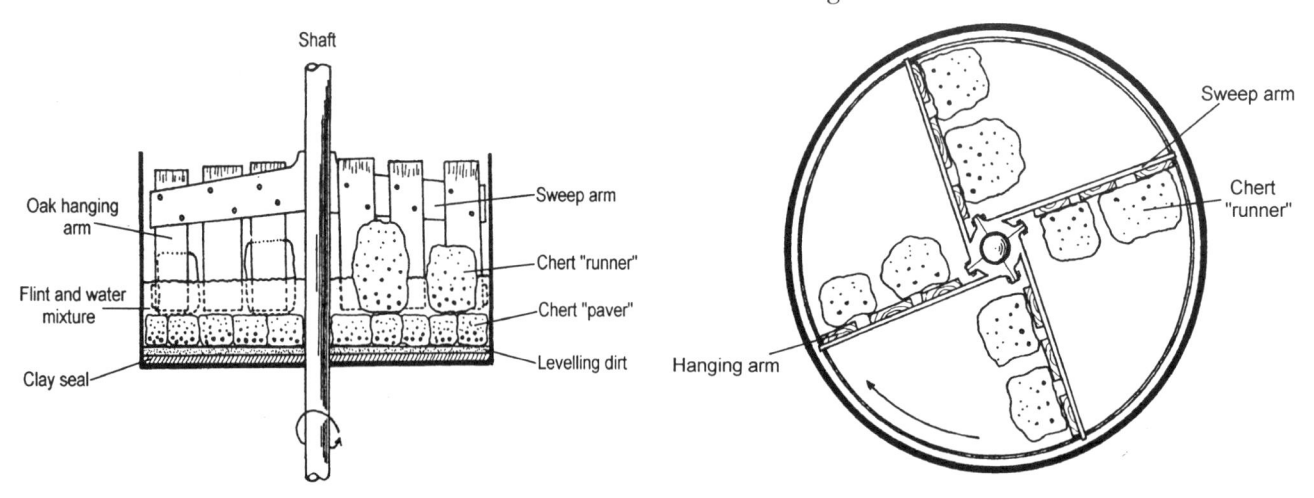

FIGURE 273. A SECTIONAL AND PLAN VIEW OF A FLINT GRINDING PAN.
(B. Job, *Watermills of the Moddershall Valley*, 1995)

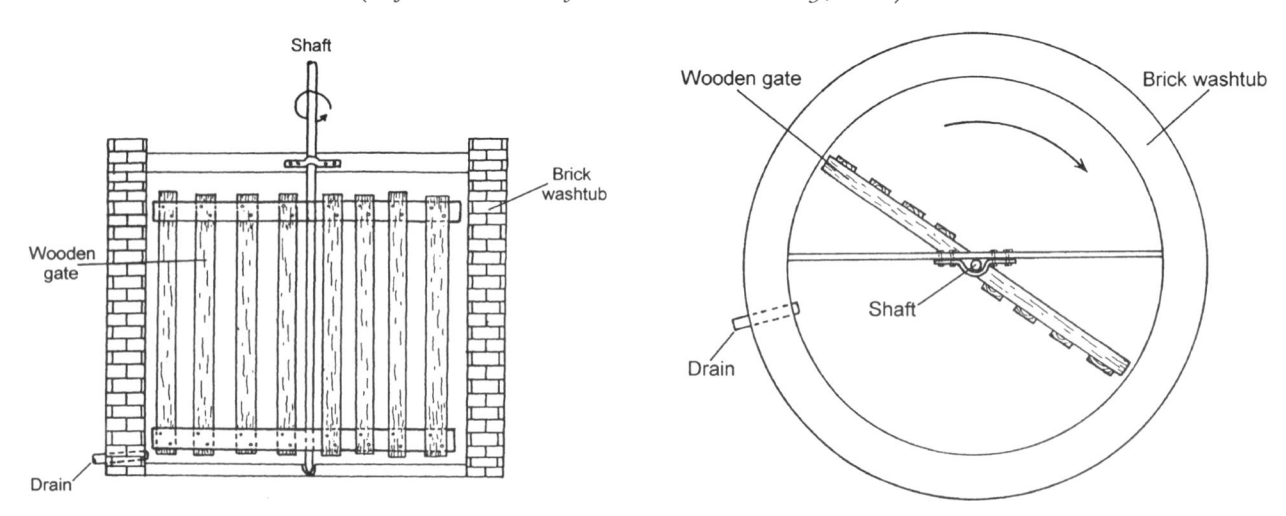

FIGURE 274. A SECTIONAL AND PLAN VIEW OF A WASHTUB USED TO AGITATE THE SLOP FROM THE GRINDING PAN.
(B. Job, *Watermills of the Moddershall Valley*, 1995)

Once the flint had been ground the mixture was transferred to a washtub where it was agitated. The washtub was another circular pan with a vertical shaft on which was mounted a vertical wooden gate. The shaft and hence the gate were rotated to disperse the flint particles throughout the mixture (see Figure 274). The coarser particles tended to sink to the bottom with the finer particles rising towards the surface. The top part of the contents of the tub, with the finer particles, was then run off into an ark where it was left to settle. The heavier particles left in the washtub were then returned to the grinding pan to be further reduced in size.

In one side of the settling ark there was a plate, known as a "plug plank" which had a vertical row of holes containing wooden bungs. As the materials settled, the bungs were removed, one at a time starting at the topmost, to run off clear water from above the concentrated suspension. After a day of settling and running off, the residue was a thick slop which was then transferred to a drying bed to remove the rest of the moisture. The drying beds had low side walls surrounding a brick pavement beneath which were flues running from a coal fire to a chimney. After about 24 to 36 hours the material had dried such that it could be bagged and sold.

The flint mills that were established in the 18th century at Buglawton and at Biddulph used water power to turn their grinding pans and washtubs as well as operating the various pumps to move the liquid slop from place to place. This same technology was used at Fleet Mill, which was sited alongside the Trent & Mersey Canal north of Middlewich giving it waterborne transport to the potteries, when it became operational in the early 19th century. When increasing demand in the 1850s resulted in Washford Mill, in Buglawton, and Bank Mill, on the slopes of Mow Cop, being converted from corn mills to flint mills, the technology involved was still basically as patented in the early 18th century.

In 1870, J. R. Alsing invented the cylinder or ball mill for grinding flint. This consisted of a horizontal cylinder lined inside with chert blocks. The cylinder had a door in its side through which it was loaded with water and the raw material to be ground. The grinding medium in the cylinder was either steel balls or flint pebbles. The cylinder was then rotated about its longitudinal axis so that the contents were tumbled together causing impacts and hence crushing (see Figure 275). This method was not immediately taken up by the north Staffordshire potters and it was not until 1910 that the first cylinder mill was installed by George Goodwin & Sons in Hanley. This type of grinding method was introduced at Washford Mill in the 20th century where the cylinder grinder was driven by a water turbine. However, a grinding pan was still retained, driven from the mill's waterwheel. When Lower Washford mill ceased being used for fustian cutting it was converted to cylinder grinding of potter's material using seven cylinders, all driven by electricity. Amongst these cylinders was the first one that had originally been installed in Hanley in 1910. When Mr. Goodwin, the owner of Lower Washford Mill, retired in 2000 this flint grinding cylinder was removed to the Stoke Industrial Museum at Jesse Shirley's Etruscan Bone and Flint Mill to become an exhibit.

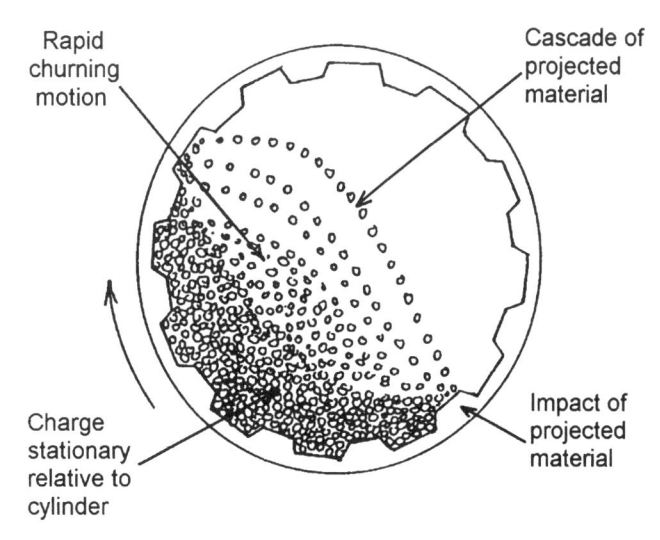

FIGURE 275. THE ACTION IN A GRINDING CYLINDER WHILE ROTATING.
(B. Job, *Watermills of the Moddershall Valley*, 1995)

Coal Grinding

In 1870 Danebridge Mill was modified to grind coal to produce a black dye material for the silk industry. The raw material was coal slack which was ground between millstones as in corn grinding. The town of Leek was famous in the late 19th century for its black silk, a colour known as "Leek raven", made popular by Queen Victoria's long mourning period for her husband, Prince Albert. Towards the end of the 19th century the ground coal was also used to give its black colour to polishes for boots, shoes and stoves, as well as providing one of the ingredients for patented black "lead" pencils.

Bone Crushing

The use of crushed animal bone had two applications in the area of South Cheshire and North Staffordshire. The potteries in the Stoke-on-Trent area used ground bone in the manufacture of bone china but the most common use was in agriculture where bonemeal was used as a fertiliser to improve the productivity of poor soils. Of the various manufacturers of potters' materials there is only Washford Mill in Buglawton where the grinding of bone as well as flint is documented. In all probability this mill ground bone using the same technology as that used for grinding flint.

Another way of grinding bone was to use a pair of edge runner stones to crush the bone into powder. This type of mill usually consisted of a pair of broad stone cylinders, similar to millstones, but positioned vertically instead of horizontally, although sometimes a single runner stone was used. They were connected by a short shaft which was joined in its centre to a rotating vertical shaft driven from the waterwheel. The rotation of the shaft caused the stones to rotate and travel in a circular path. These edge runners were usually positioned on a raised circular platform onto which the material to be crushed could be placed. The material was crushed by the weight of the edge runners as they rolled over the material and also by the rotating of the runner stone as it progressed around its very tight circular path. There is no direct evidence to determine how bones were ground at Moston Mill, Park Mill at Brereton or at Midge Brook Tannery, all of which were crushing bone for agricultural use.

FIGURE 276. A SET OF EDGE RUNNERS COULD BE USED FOR CRUSHING BONE FOR AGRICULTURAL USE OR SOMETIMES FOR CRUSHING BARK IN A TANNERY.

These bone mills may well have used edge runners although it is also possible that the bones were broken by hand and then reduced to powder using normal horizontal millstones with larger eyes than those stones used for corn grinding.

Wood Flour Milling

During the late 19th and early 20th century Bosley Works had concentrated on commercial corn grinding. Due to unfavourable trading conditions in the flour industry the owners decided in 1933 to enter the trade of inert filler materials, starting with wood flour. The raw material was sawdust, which was obtained from the many saw mills in operation. This was then ground between vertical millstones left over from the corn grinding days at Bosley Works. The grading of the product took place using plansifters which were likewise remaining from the flour milling days. The wood flour was used as a filler in the manufacture of Bakerlite, a new plastic material becoming popular in the 1930s. Since then many other materials have been ground to powder at Bosley Works such as peach stones, coconut shells, and even mica.

FIGURE 277. PLAN SIFTERS FOR GRADING INERT POWDERS AT BOSLEY WORKS.

Water Turbines

The water turbine is a species of water motor where the action, either by impulse or by reaction of the water on a series of curved blades mounted on the rotating part, or runner, causes the motor to rotate. Water turbines are normally classified by their type, either impulse or reaction, then by the direction of flow of the water through the turbine, radially outward, radially inward, axially or mixed flow, and then by the orientation of their drive shaft, vertical or horizontal.

The first successful turbine was designed by Benoit Fourneyron, a French engineer, and was installed in 1827. This invention was not immediately taken up in this country as waterpower was seen to be fully exploited anyway, and the steam engine was seen as the future universal prime mover. Consequently, much of the early development of the water turbine in the 19th century occurred in the United States or on the continent in France, Germany and Switzerland. This was a short-sighted attitude because the water turbine had many advantages for factories that wished to continue using water power.

Whereas the vertical waterwheel was limited to heads of about 70 feet, impulse turbines could work on a wide range of heads, up to many hundreds of feet. When exploiting lower heads, reaction turbines were able to utilise the total head available and were also capable of running submerged. Both types of turbine ran faster than waterwheels thereby reducing the gearing needed, were more efficient and hence could generate much more power from a given head, were much smaller than a waterwheel using the same head, but most of all they were considerably cheaper. However, turbines were designed to give maximum efficiency for a particular set of circumstances of head and flow. If the flow conditions varied then the efficiency of a turbine could fall dramatically. This factor made the use of a second hand turbine on a new site rarely practical.

The first water turbine to be installed on the River Dane was at the Higher Works at Bosley Works in 1888 when the owners, F. R. Thompstone, purchased a vertical shaft, Girard turbine from W. Günther & Sons of Oldham. This firm had been established in 1881 selling water turbines based on continental designs. The Girard turbine was an impulse turbine with an outward flow configuration. In an impulse turbine, such as the Girard, the wheel passages were never completely filled with water, which is at atmospheric pressure. The pressure caused by the head of water was converted into kinetic energy by passing it through a number of small orifices. The resultant jets of water were fired at the buckets of the runner causing it to rotate. During its flow through the turbine, the absolute velocity of the water was reduced as the water gave up its kinetic energy to the runner. In effect the Girard turbine was akin to a horizontal waterwheel with the water fed to the inside circumference of the runner and discharged from its outer circumference (see Figures 45 & 278). When working, a Girard turbine is quite spectacular as water is sprayed outwards from all around its circumference. The impulse type Girard turbine might be considered more suitable for high heads but proved to be most effective using the medium height head of 31 feet available at Bosley Works. This turbine produced 130 h.p. using a runner of only 48 inches diameter, a vastly superior performance than achievable by any design of ordinary waterwheel using the same head.

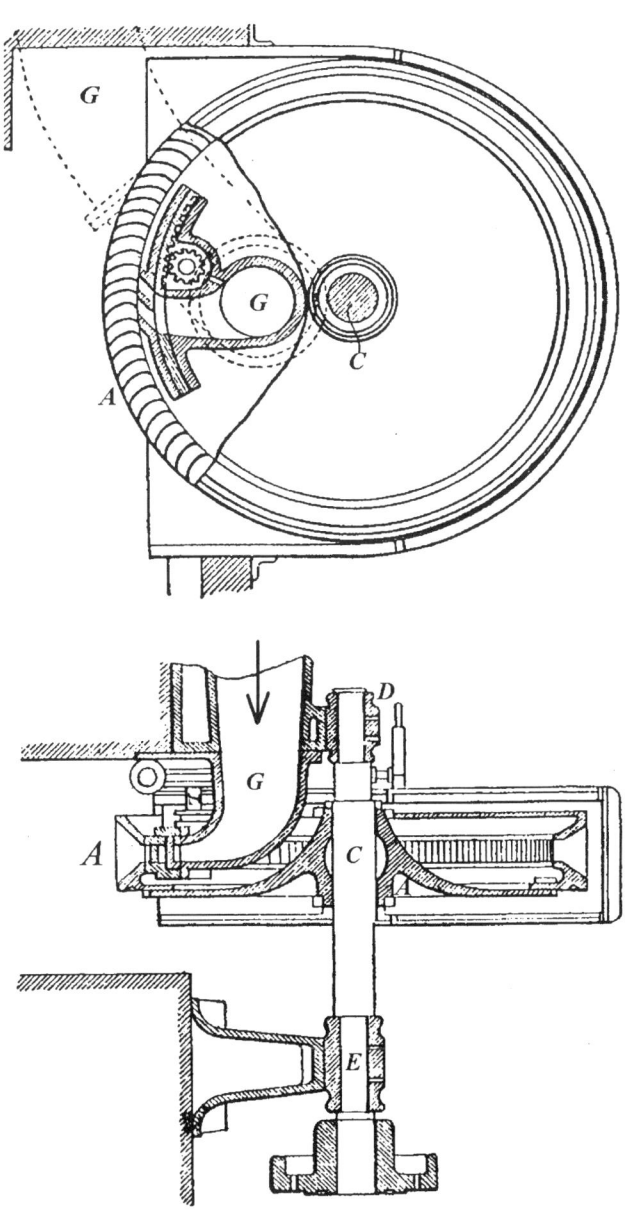

FIGURE 278. PLAN (*top*) AND SECTION THROUGH A GIRARD TURBINE. WATER ENTERS DOWN TUBE G AND EXITS AT THE OUTER EDGE OF THE RUNNER AT A.
(*P. R. Björling, Hydraulic Motors, 1894*)

Six years later, in 1894, Günthers supplied Bosley Works with another vertical shaft Girard turbine of 50 h.p. together with a horizontal shaft Jonval turbine of 20 h.p. The Jonval turbine, which was a reaction turbine with an axial flow, was used to drive a pump for the new sprinkler system. In a reaction turbine the wheel passages were completely filled with water under pressure, which passed through the machine transmitting all its energy by the pressure on the blades forcing the runner to rotate. To ensure that the water entered the runner with no shock or disturbance a reaction turbine had a series of vanes which guided the water into the buckets of the runner at the correct angle. The flow of water through this type of axial flow turbine passed through a set of guide vanes and then through the runner such that the water remained at a constant distance from the drive shaft (see Figures 46 & 279). The Jonval turbine was a surprising choice at this time, being a design that had largely been superseded, but then it was only needed to work intermittently to pump water to the sprinkler system when a fire occurred, so it was presumably the cheapest solution available.

In 1901 F. R. Thompstone decided to modernise the Lower Works at Bosley so he purchased a 25 h.p. horizontal shaft Vortex turbine from Gilbert, Gilkes & Co. of Kendal for this mill. This company had been founded in 1853, as Williamson Brothers, to exploit the design of turbine patented by James Thomson of Belfast in 1850. The Williamson company had been taken over by Gilbert, Gilkes & Co in 1881. The Vortex type of reaction turbine had inward flow with double discharge and adjustable guide vanes. The movable guide vanes gave good efficiency at low flow with its spiral case maintaining the uniform velocity of the water through the turbine. There were two discharge elbows, one each side of casing, and the suction tubes carried the discharge water from each bend to below tail water level. The Vortex design was very flexible, giving good efficiency over a wide range of heads and flows. The example installed at Bosley Works operated with a 16 feet head and had a runner of 30 inch diameter. Another example at Biddulph Grange Saw Mill operated with a 89 feet head whereas the one installed at the Havannah used only a 13 feet head.

As well as the Vortex turbine, Gilbert, Gilkes & Co.

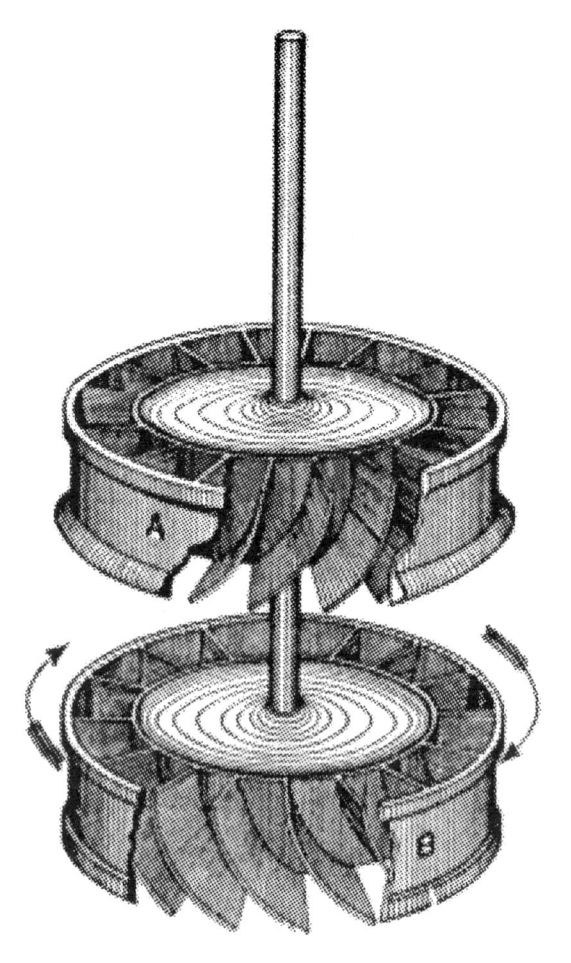

FIGURE 279. THE WATER ENTERS THE JONVAL TUBINE GUIDE VANES FROM ABOVE AND PASSES DOWNWARDS INTO THE RUNNER. (*D. W. Mead, Water Power Engineering*, 1908)

FIGURE 280. THE DOUBLE-SIDED RUNNER OF A THOMPSON VORTEX TURBINE (*Gilbert, Gilkes & Co. Ltd., Catalogue*, 1901)

also introduced a range of reaction type turbines based on the American, mixed flow, Francis design. In this type of turbine the water entered the wheel radially around its outer circumference, flowing inward, and then, during its flow through the runner, turning to be discharged axially. Losses incurred in changing the direction of flow abruptly had to be minimised. As a smooth flow was essential in all the passages so the curvature, angle, and cross-section of the blades of the guide vanes and runner buckets were most important. This type of turbine design had been developed in the United States by the trial and error of a number of designers and manufacturers. Although the various designs all differed, mainly in the shape of the blades and buckets, all of these mixed inward/axial flow turbines were named after the first engineer to experiment with the type, namely James Bichenor Francis of Lowell, Massachusetts.

In the late 1920s and 1930s the firm of Gilbert, Gilkes and Gordon Ltd., as it had become in 1921, manufactured a complete range of Francis type turbines with both horizontal or vertical shafts. They were originally known as the Series Y, 0, I, III, IV, and V, but in later years the Series 0 and I were replaced by the Series C and R. The different series merely represented the six different speed ranges available. In total seven of Gilbert, Gilkes & Gordon's Francis type turbines were installed at various sites on the River Dane and its tributaries. In 1927 a 21 h.p. Series I turbine, with a 12 inch diameter runner and a horizontal shaft, working on a head of 20 feet, was installed at Daneinshaw Silk Mill. In 1928 a Series IV was installed in Bosley Corn Mill to drive the compressor of an ice making machine. It had a 13½ inch diameter runner working on a head of 25 feet delivering nearly 15 h.p. When the firm of F. R. Thompstone changed from grinding corn to grinding sawdust they installed a Series R turbine of 33 inch diameter running at 230 r.p.m. on a head of 30 feet which generated 268 h.p. to power the whole factory. This horizontally shafted turbine was the most powerful to be installed on the River Dane. This was followed by two more Series R turbines at Washford Mill and the Lower Works at Bosley in 1935 and 1936 respectively. The turbine installed at Washford Mill was used to power a flint grinding cylinder. It had a horizontal shaft and generated about 16 h.p. with a 27 inch diameter runner using a head of only six feet. At the Lower Works at Bosley a return to the vertical shaft configuration for the Series R Francis turbine produced 110 h.p. for the generation of electricity. To maintain a constant speed this turbine was fitted with an oil pressure governor. At Warmingham Mill on the River Wheelock the water wheel was replaced with a Series C vertical shaft turbine in 1940 to drive their model shop. This was a relatively small machine having only about a 12 inch diameter runner generating nearly 15 h.p. from a 12 feet head. This turbine was also supplied with an oil pressure governor. The final turbine of this type to be installed was again at the Higher Works, Bosley, when a second-hand machine was purchased from Scotland in 1966. This had a 15 inch diameter runner and was to be used on the 30 feet head but the installation was never completed due to the river authorities introducing extraction charges for installations generating power from a fall of water, which made the practice uneconomical and the turbines at Bosley Works came to a standstill.

FIGURE 281. A FRANCIS TURBINE RUNNER. THIS ONE IS SOMEWHAT LARGER THAN THOSE INSTALLED ON THE RIVER DANE BUT THE DESIGN OF THE BLADES, ETC. WAS THE SAME.

Turbines at Bosley Works, Warmingham Corn Mill and eventually Daneinshaw Silk Mill had oil pressure governors fitted, all of which were supplied by Gilbert, Gilkes & Gordon Ltd. These governors have a very sensitive centrifugal pendulum driven from the turbine via a belt. This pendulum operates valves that allow an oil pump to distribute the oil under pressure to one side or the other of a piston. The movement of this piston then operated the turbine's guide vanes. As very little force is needed, this type of governor was extremely sensitive and therefore very suitable for controlling the electrical output generated from the turbine. The complexity of design and sensitivity of operation of these governors means that all their working parts are contained within a sealed casing.

In the last quarter of the 19th century, a range of impulse turbines were developed in the mining fields of the western United States that had many similarities to the traditional waterwheel. They consisted of a vertical wheel on a horizontal axle with buckets distributed around the periphery of the wheel. Water under the pressure of hundreds of feet of head was projected through a nozzle at the buckets to power the turbine. The efficiency of this type of turbine depended very much on the shape of the buckets. These were cup shaped and various manufacturers tried different profiles. The most successful was designed by the Pelton Waterwheel Manufacturing Company around 1880 and thereafter "Pelton Wheel" became the name for all of this type of turbine regardless of manufacturer. The buckets' shape consisted of a double cup joined in the middle at a knife-sharp centre ridge. The water jet was fired directly at this centre ridge and was split into two streams that circulated around the two cups delivering their power with minimum shock to the rotating wheel (see Figure 282).

Although impulse type Girard turbines had been used at Bosley Works, the heads generally available on the River Dane and its tributaries would not normally have been considered high enough to drive Pelton Wheels. However, one of this type of turbine was installed at Biddulph Grange in 2000 to generate electricity. This installation, supplied by Derwent Hydropower Ltd. is the only site on the whole river system of the Dane that is still operational.

Although the design and manufacture of turbines did not take place in the region of South Cheshire and North Staffordshire their utility and effectiveness were recognised by the users of water power in the area. In

FIGURE 282. THE PELTON WHEEL TURBINE LOOKS VERY SIMILAR TO A WATERWHEEL.
(*Gilbert, Gilkes & Co Ltd., Catalogue*, 1901)

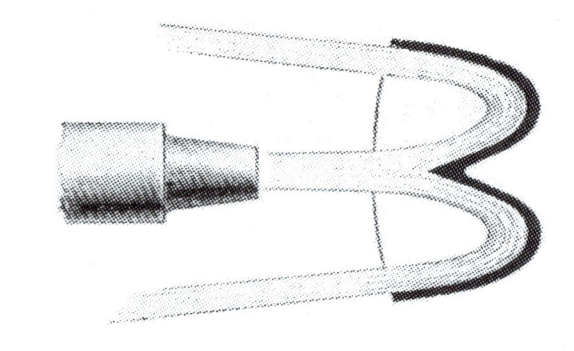

FIGURE 283. THE NOZZLE (*left*) AND DOUBLE BUCKETS OF A PELTON WHEEL.
(*Journal of the Franklin Institute, Vol.* 140, 1895)

total fourteen turbines were installed in the area from 1888 onwards. This acceptance of turbine technology enabled waterpower to be commercially used until 1966 when extraction charges made waterpower uneconomic. Even so it shows that waterpower is still a viable source of power that can be used efficiently in the modern world if freed from artificial financial handicaps. The site at Biddulph Grange, which is once again generating electricity, shows that waterpower can be a environmentally friendly source of power today and in the future.

CHRONOLOGY

Domesday - 1700

Where and when the waters of the River Dane or its tributaries were first used to drive machinery is impossible to determine. The earliest record of such a use occurs in the Domesday Book of 1086. On the whole of the River Dane and its tributaries there was only one mill recorded in this survey, namely at Brereton, where the corn mill was valued at 12d (or one shilling). As the Domesday Book was produced mainly for tax raising purposes no further information about the mill was noted.

Although the Domesday Book provides a "snapshot" of England in 1086, the view of Cheshire that is provided is somewhat abnormal. After the Norman Conquest of 1066 there was a large uprising of the northern shires in 1069/70 which was suppressed by the new Norman king, William, with great ferocity. The destruction caused in pacifying this insurrection resulted in many of the manors and land holdings in Cheshire being recorded as "waste" sixteen years later in 1086, especially in the eastern and central parts of the county. Certainly in the area under consideration, Bosley, Marton, Somerford, Kermincham, Cranage, and Kinderton were all "waste". In total there were only 18 corn mills recorded in the whole of Cheshire compared to around 70 in the neighbouring county of Derbyshire, with most of these Cheshire mills being located to the west of the River Weaver. Although the eastern moorlands of the county and their wood clad western slopes were virtually uninhabited even prior to the rebellion it is possible that other corn mills had existed on the Dane and its tributaries during Saxon times but had been destroyed during the conflict and hence were not recorded in the Domesday Book.

There has been considerable discussion as to what type of mills are mentioned in Domesday. Were they, in fact, all water powered, and if so did they utilise horizontal or vertical waterwheels to turn their millstones? The only information recorded is the values for the mills. In other counties these values seem to fall into three categories. Firstly, there are mills valued in terms of pence, such as Brereton at 12d, which are not uncommon but usually make up less than 10% of the total. Secondly, there are many mills valued at a few shillings, perhaps averaging around 5s per mill. These form the bulk of the mills recorded. Then, lastly, there are a few mills valued in terms of a number of pounds, usually found in the larger centres of population. It has been suggested that the low value mills might have been the less sophisticated, horizontal waterwheel powered, variety of mill, possibly only operating seasonally, whilst the more expensively valued mills had vertical waterwheels. Alternatively they may all have had vertical wheels with the difference in value being determined mainly by the number of people that they served. Archaeological evidence dating from Saxon times shows that both horizontal and vertical waterwheels existed before the Norman Conquest. This begs the question of when did the horizontal wheel cease to be used as it was the vertical waterwheel that was to become ubiquitous throughout the country. This controversy can only be solved by the archaeological excavation of a number of mill sites that can be positively identified as Domesday mills. Unfortunately, good mill sites were used over and over again throughout the centuries and consequently no

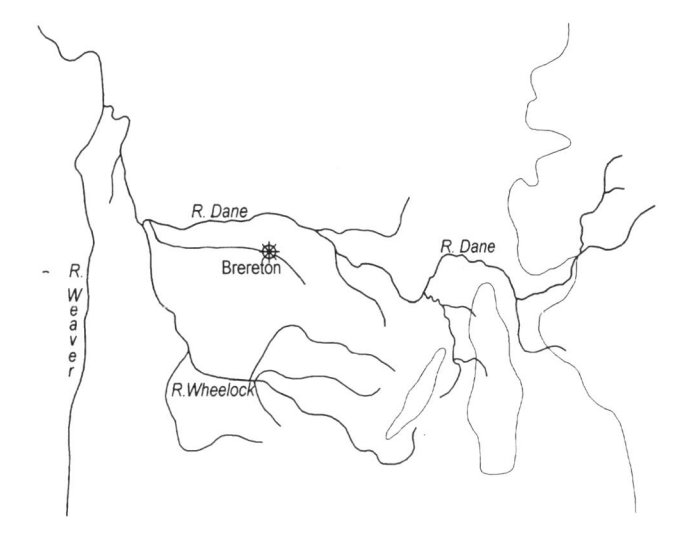

FIGURE 284. CORN MILLS ON THE RIVER DANE AND ITS TRIBUTARIES AT THE TIME OF THE DOMESDAY BOOK, 1086.

277

site attributable directly to the Domesday survey has yet been excavated.

In the three hundred years after the Norman Conquest there was a large expansion in the population of the country from about 1.4 million in 1086 to over 3.25 million in 1300. During this period the manorial system became consolidated, including the practice of "soke" whereby the peasantry had to take their grain to the lord of the manor's mill for grinding, paying a percentage of their grain in toll or "multure" for the privilege. As the ability to grind grains was essential to life itself, most manors were willing to make the capital investment in a waterpowered corn mill given this guaranteed, captive, customer base. In the whole of the country the number of mills increased from about 6,000 at the time of Domesday to over 10,000 by 1350. In eastern Cheshire the same pattern was experienced, of the 27 corn mills in existence in 1750 almost half of them were definitely recorded as being in existence by 1350. Whether the remaining fourteen corn mills were also built before this time cannot be established (nor can it be ruled out) as their history has only been traced back to the 15th century (4), the 16th century (2), the 17th century (5) or no further back than 1700 (3). So, from the devastation of 1086 with only one mill in the whole of the River Dane region, it is probable that by 1350 the pattern of corn milling and the attendant communities had already been established.

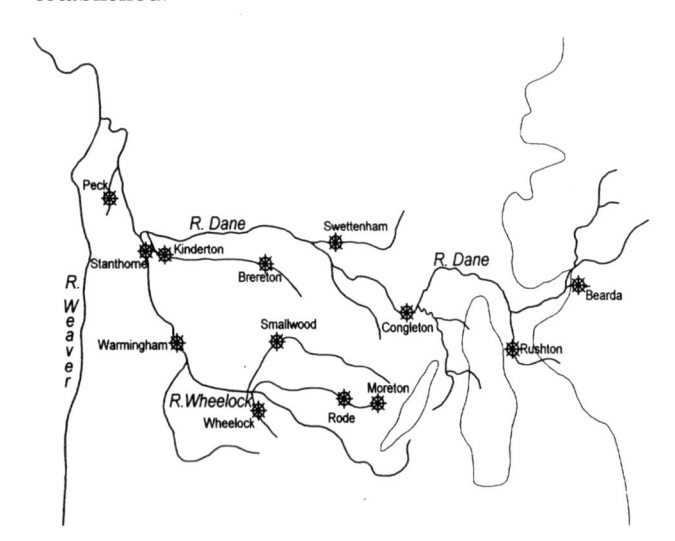

FIGURE 285. CORN MILLS KNOWN TO EXIST ON THE RIVER DANE AND ITS TRIBUTARIES BY 1350.

In 1349/50 the country was in the grip of the black death, or bubonic plague, which decimated the population, killing off around a third of all inhabitants. The consequence of this reduction in population after 1350 was to decrease the demand for flour so curtailing the profits to be made from operating the corn mills. This then adversely affected the rents that the mills could command. In the case of the River Dane and its tributaries the two mills where rental values are recorded show that in the latter half of the 14th century Rode Mill's annual rent was reduced from 20 shillings per year to 10 shillings and Congleton Mill's from £10 to £6 per year. Also any incentive to build more corn mills was greatly reduced. The only mill that is known to have been built in the medieval period after 1350 was at Congleton in 1451. However, this was only the replacement of an existing mill which had been severely damaged by floods. As the town had a population of about 400 at this time, the building of a replacement mill would have been essential to maintain the income of the lord of the manor.

Although it has been postulated that there was some kind of industrial revolution in the Middle Ages with waterpower being applied to a whole range of applications, there is not much evidence to support this theory on the River Dane and its tributaries. The only mill known to exist in this period, apart from the corn mills, was a fulling mill in Congleton, which existed prior to 1351 but had disappeared from the records by 1514. It is also conceivable that the fulling mill and paper mill recorded at Danebridge in the mid 17th century may have been built somewhat earlier, in the Middle Ages. Certainly the financial returns made from this type of venture does not seem to justify any widespread application of waterpower at this time. The rent paid for the fulling mill in Congleton was only £2-6s-8d per year in 1362, but that paid for the corn mill was just over £10.

This pattern of corn milling in the area continued through the 15th, 16th and 17th centuries with very little change in the technology used until towards the end of the 17th century. Although the practice of the soke was unpopular it was rigorously enforced during the 15th century. In 1451 the king, who was lord of the manor of Congleton, instructed his steward not to allow the townspeople to grind their grain anywhere but at the town mill. However, nearly two hundred years later, although the soke was still valid it was not so inflexible; the evidence from Lawton Mill shows that a number of people were regularly paying a fine for non-compliance with this custom which, in effect, allowed them to grind their corn at the mill of their choice. By the end of the 17th century, 25 water powered corn mills on the River Dane and its tributaries had been mentioned in various records, although many of these may well have been established much earlier, possibly even before 1350. There was a corresponding support industry for these corn mills, consisting of millwrights, blacksmiths, whitesmiths, carpenters, etc., but most importantly, the manufacture of millstones quarried from the rocky ridge just to the south-east of Congleton, especially at Mow Cop. Another significant group of workers needed by the mills of this period, who are often overlooked, were the labourers who repaired the weirs and watercourses for the mills and provided their muscle when required, such as at the installation of new millstones.

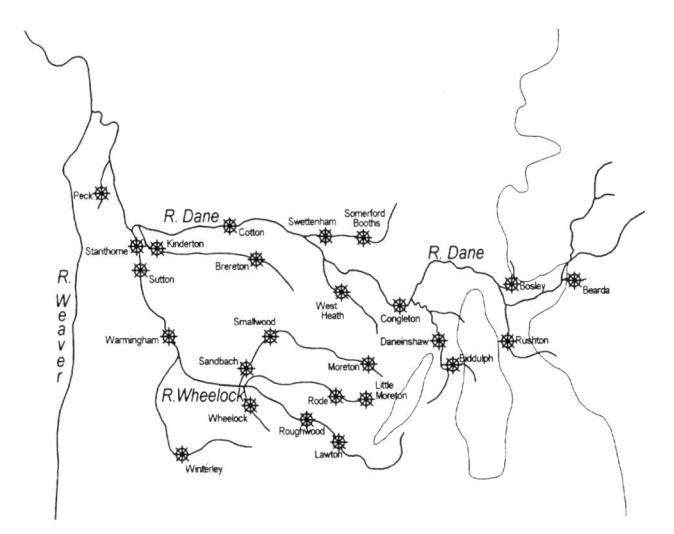

FIGURE 286. CORN MILLS KNOWN TO EXIST ON THE RIVER DANE AND ITS TRIBUTARIES BY 1700.

In the mid 17th century much damage was done to the iron industry in southern England during the English Civil War. When more stable conditions prevailed, after the Restoration in 1660, ironmasters began to look to re-establish the industry in new areas that could supply the necessary raw materials. At this time, apart from the iron ore itself, charcoal was the most important ingredient used to make iron. To exploit the local iron ores in North Staffordshire, and the wood available for charcoal making from the Cheshire forests, furnaces and forges were built utilising the waterpower available. In the area under consideration, furnaces were built at Lawton and Street in the parish of Odd Rode (also at Vale Royal just outside the Dane catchment area) together with forges at Cranage and Warmingham. Waterpower was used to operate the bellows at the furnaces and forges, the hammers in the forges, and the rollers and cutters in the slitting mill, which was situated alongside the forge at Cranage.

Initially, the iron industry had an inauspicious start in Cheshire characterised by arguments among the various partners and the production of very poor quality iron. By the 1690s these problems had been overcome and the iron industry in Cheshire was showing some of the signs of a modern business. At this time all the furnaces and forges in south-east Cheshire came under the control of one partnership to become known as the "Cheshire Works" and some

rationalisation took place. The iron was being produced in large quantities for the regional and national markets rather than just satisfying local demand and the various sites operated as an integrated business with each site concentrating on a particular product or operation. Similarly, a partnership in North Staffordshire, called

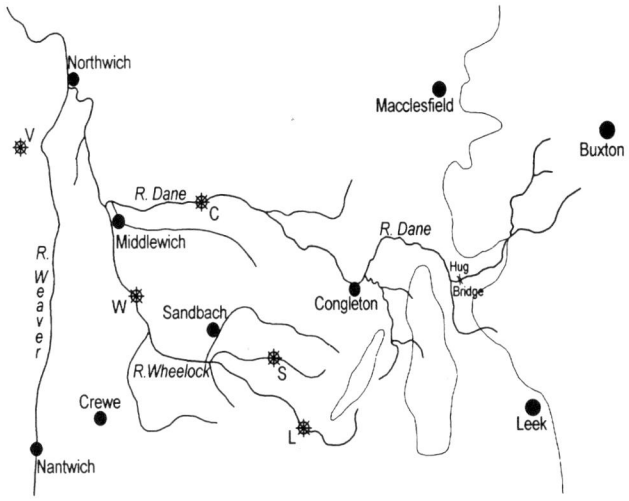

FIGURE 287. THE "CHESHIRE WORKS" IN THE LATE 17TH/EARLY 18TH CENTURIES.
C: CRANAGE FORGE & SLITTING MILL, L: LAWTON FURNACE, S: STREET FURNACE/FORGE, V: VALE ROYAL FURNACE, W: WARMINGHAM FORGE.

the "Moorland Works" was also in operation at this time. At the end of the 17th century, the Moorland Works amalgamated with two more forges and a slitting mill and became known as the "Staffordshire Works". Then, in 1701, the Cheshire Works and the Staffordshire Works joined forces in what looked like a take-over by the Staffordshire concern, although eventually it was the Cheshire partners that became the controlling influence. The mergers or take-overs that were a facet of the iron trade at this time tied the industry in Cheshire into links with the trade in North Staffordshire in particular, but also with South Staffordshire, North Wales, the Forest of Dean, Lancashire, the Lake District, Hampshire and elsewhere. These take-overs and mergers would have been instantly recognisable in today's commercial activities. It was the financial side of these businesses that were somewhat "old fashioned" being based on partnerships, which varied for each establishment. In many cases these partnerships were based on marital allegiances with the same group of owners having differing numbers of shares in the various sites. By 1706, the various partners in the Cheshire and North Staffordshire iron works were operating three furnaces, nine forges and three slitting mills, all operating as one single business.

The 18th Century

The charcoal based iron industry in South Cheshire and North Staffordshire had its heyday in the first half of the 18th century, declining during the latter half of the century and was totally defunct by 1800. The main reasons for this decline and disappearance of the iron industry was the growing shortage of wood from the diminishing forest for charcoal production and the introduction of coal as the fuel used for ironfounding. These factors caused the industry to move to areas where this fuel was readily available, such as South Staffordshire and South Wales. The great achievement of the charcoal based, waterpowered, iron industry in South Cheshire and other areas was to provide a ready and cheap supply of iron for use in industry from the beginning of the 18th century. This, in turn, made it possible to use iron in the design and manufacture of waterwheels, gears, and shafts used in waterpowered mills, thereby allowing the generation and transmission of more power than was previously possible using only wooden construction. During the 18th century this ability to generate and transmit greater amounts of power enabled many other processes and industries to utilise waterpower. In many ways the water powered, charcoal based, iron industry, of which the furnaces and forges in the Dane area were a prime example, was an enabling technology, one of the fundamental elements which made the "industrial revolution" of the 18th century possible.

One of the industries to expand into the area early in the 18th century was that of paper making. The first recorded paper mill was at Danebridge in the 17th century which would have used waterpower to operate stamps or hammers to pulp the rags that were the raw material of paper making. A new technology in the shape of the hollander, which was used for preparing rags instead of stamps or hammers, was introduced into this country early in the 18th century. A new mill was built at on this principle at Wincle, just downstream of Danebridge, by the young James Brindley in 1738 when he was serving his apprenticeship as a millwright. This mill made the older Danebridge paper mill redundant and within a few years it was derelict. Other paper mills were built later in the century at Primrose Vale near Congleton and at Folly Mill on the Clough Brook, which runs into the Dane from Wildboarclough.

The industry which was to benefit most from the harnessing of waterpower during the 18th century was that of textile manufacture. Prior to the 18th century the production of all type of threads and cloth was basically a hand powered domestic operation. This was certainly true of the production of silk cloth which was produced from imported thread by hand loom weavers in centres such as Spitalfields in London. It was impossible to cultivate silk moths in this country due to the climate, consequently the thread, which was made from the filaments from the silk moth cocoons, was imported from places like France and Italy where they had devised machinery to "throw" silk thread. The design of this machinery, which was a closely guarded secret, was stolen from the Italians by John Lombe in a classic case of industrial espionage. Having brought the secrets of the machinery back to this country their designs were patented and he and his brother Thomas built a factory in Derby in 1721 to capitalise on this patent machinery. Although Thomas Lombe pretended otherwise, this mill was eventually a great success. After the patent ran out, one of Lombe's workers moved to Stockport in 1733 where he was able to negotiate the lease of the town mill from the Corporation in order to establish the second silk mill in the country. In 1744 Charles Roe and Samuel Lankford of Macclesfield also built a silk mill in their town using the Italian machinery of the Lombes.

This then was the extent to which silk "manufactories", or factories for short, had reached when two partners arrived in Congleton with an offer they hoped the town could not refuse. One of the partners was Nathaniel Pattison who was familiar with the technical know-how of silk throwing, having lived with Thomas Lombe in Derby and worked in the silk mill there. Samuel Clayton, the other partner, had been mayor of Stockport when the silk mill had been established at that town's mill. He was able to persuade the Congleton Corporation to do likewise and lease the town corn mill and its water rights to the partners. One of the investors in the partnership was the father-in-law of Samuel Lankford, a partner of Charles Roe in the Macclesfield silk mill. When the partners started building their new silk mill alongside the town corn

FIGURE 288. TABLET COMMEMORATING JAMES BRINDLEY'S APPRENTICESHIP AT SUTTON, NEAR MACCLESFIELD.

mill in Congleton in 1752, they were combining the experiences derived from all the preceding silk mills that had been built at that time.

The mill that was built at Congleton was of a considerable size being 240 feet long by 24 feet wide with five storeys. The waterwheel that provided its power was situated in the middle of the mill fed by a leat taken from the town corn mill pool. The mill was designed to accommodate eleven of the Italian silk throwing machines, each of which was about 19 feet high by 14 feet diameter. These throwing machines or "engines" occupied the bottom two stories of the mill, with the other three storeys holding winding machines and other ancillary processes. The engineer responsible for building this machinery was James Brindley who introduced some new ideas to the design of the machinery that were to become standard in the textile industry. Firstly, he was able to arrange to connect or disconnect the power to any one machine without disturbing the supply of power to all the other machines, and secondly he arranged some form of governor so that the waterwheel would operate at a constant speed. In constructing the mill's machinery he also utilised a method of constructing gears that gave a fourteenfold saving in labour.

Although the processes involved in throwing silk are intrinsically not particularly complex, there is no doubt that the work in this mill had to be carried out in an organised and systematic manner with the members of the workforce, numbering around 400, specialising in each particular activity and the material being progressed from one process to another. Although this organisation and specialisation can be seen as the beginning of the factory system it was, in fact, merely extending the type of organisation that had previously existed for some time in the iron industry. The major development that was achieved by the design of these early silk mills was to house all the processes of manufacture in one building. In building the machinery for these early silk mills, the millwrights had to be able to implement systems of gearing that enabled more than one machine to be driven from just a single waterwheel, as well as being able to transmit that power over considerable distances and over many floors in height. These advances in millwrighting skills were then available to be applied to other industries.

Since the civil war period in England the population had been growing once more. This was especially true once the drive to industrialisation got under way, causing migration to take place from the countryside into the towns to work in the new factories. All these people needed to be fed so there was a demand for greater flour production at the same time that all the remaining waterpowered sites were being occupied as more and more industries needed to generate power. In the corn grinding mills traditionally one waterwheel had driven one pair of millstones so that where greater output was required, such as in Congleton, the town mill was built originally with two waterwheels to drive a total of two pairs of millstones. Just building extra corn mills was not an option so the design of the traditional corn mill had to alter to provide a greater output from the same mill sites. If a mill site had the potential to generate more power then the advances made in mill gearing and the judicious use of cheap, good quality, iron were used to redesign and rebuild the corn mill so that one waterwheel could drive many pairs of millstones. Examples of this occurred at Congleton in 1758 and Swettenham in 1765. Greater power could be achieved from the use of overshot and breastshot waterwheels but corn mills located where the terrain dictated the use of undershot waterwheels had little potential for improved power generation at this time and could not compete with the new mills that had a number of pairs of millstones. So although many corn mills were rebuilt and improved in the mid 18th century, some of the less commercially viable mills went out of use, such as Cotton Mill, an undershot corn mill on the River Dane itself near what is now called Holmes Chapel.

FIGURE 289. A PORTRAIT OF JAMES BRINDLEY, THE ENGINEER RESPONSIBLE FOR THE BUILDING THE MACHINERY FOR THE SILK MILL IN CONGLETON IN 1752. (S. Smiles, *Lives of the Engineers*, 1862)

As the pace of industrialisation speeded up in the second half of the 18th century waterpower was applied to a variety of uses. Just to the south of the area under consideration the infant pottery industry was beginning to become established around the Stoke-on-Trent area. One of the operations involved in this industry was the grinding of flint, a process that had been introduced in 1732. By the 1760s the demand for ground flint was increasing to such an extent that new water powered mills were being established in many areas surrounding the main centres of pottery production. In an attempt to cash in on this expanding market, two flint mills were built in 1751 and 1761 at Buglawton just outside Congleton. Unfortunately, the high transport costs to the Potteries meant that this application of water power only played a minor role on the River Dane and its tributaries. However, the Nearer Daneinshaw flint mill was operated by a member of the Wedgwood family at one time and the tradition of grinding potter's materials continued at the Washford Mills in Buglawton until the 1990s!

With the success of the silk industry after 1750, many silk mills were built in various parts of the country and supply began to outstrip demand. In the Congleton area the production capacity of Clayton and Pattison's silk mill was so large that even 20 years after its opening there was only one other silk mill in the area, and that was a small, animal driven, mill. It is possible that Charles Roe, who was responsible for the construction of the silk mill in Macclesfield, was able to foresee the impending collapse in silk prices because he sold out his share in his silk mill in Macclesfield to his partners in 1762, getting an excellent return for his original investment. Charles Roe had turned his attention to the possibility of exploiting the copper ores that were to be found at Alderley Edge not far from Macclesfield. The capital that he had acquired in the silk industry helped to finance the Macclesfield Copper Company which constructed three sites for the manufacture of brass and copper bolts, sheet, wire, etc. The crushing and smelting of the copper ore took place in Macclesfield where he used a windmill to power the crushing of the ore. To process the raw copper and brass into its saleable products Charles Roe acquired two "greenfield sites" to utilise the waterpower available in the River Dane itself. The first site was near Eaton, not far from Congleton, which was commenced in 1762 and so was given the name of "Havannah" to commemorate the Royal Naval victory at Havannah, Cuba, in the same year. Charles Roe installed five waterwheels at Bosley and five at the Havannah to power the bellows, the banks of hammers, the drawing of wire and the cutting of screws and bolts. The scale of these installations indicate that

FIGURE 290. CHARLES ROE, FOUNDER OF THE MACCLESFIELD COPPER COMPANY WHICH ESTABLISHED BRASS AND COPPER FACTORIES AT THE HAVANNAH AND AT BOSLEY IN 1762 AND 1766 RESPECTIVELY.
(*Macclesfield Team Parish*)

Roe was intent on becoming a manufacturer capable of supplying on a nation-wide basis.

As the British Empire began to expand in the second half of the 18th century, especially into tropical regions, the Royal Navy had an increasing need for copper sheeting to sheath the hulls of their ships as protection against penetrating marine organisms that lived in the tropical seas. With this large volume market to supply Roe's company expanded into South Wales and Liverpool, bringing copper ore to Cheshire for processing from Anglesey and other places when the local mines were worked out. The success of the brass and copper industry also played a part in the mechanisation of other industries as brass was an ideal material for making the complex small parts needed by some of the machines being invented at this time. In a similar way to the iron industry, the large-scale production of brass was an enabling technology for further advances in industrialisation.

One industry to make use of the plentiful supply of brass was that of cotton spinning. In 1769 Richard

Arkwright patented the invention, later called the waterframe, for spinning cotton using pairs of rollers to simulate one of the operations previously performed by hand. These rollers and their associated gearing and transmission utilised brass in their manufacture and his later patent for the carding machine also used brass pins in large numbers. The idea of using rollers to mechanise the cotton spinning process had been originally patented by John Wyatt and his partner Lewis Paul in 1738. Their machine had a circular form that had possibly been derived from the idea of the large Italian silk throwing machines introduced earlier in the century. Wyatt and Paul set up a factory in Birmingham and then later in Northampton to exploit this new invention. Although there are approximately the same number of processes involved in throwing silk and spinning cotton, the cotton spinning processes are more complex requiring more intricate machinery. To answer the requirement to lay all the cotton fibres in line prior to spinning Wyatt and Lewis patented a carding machine in 1748. Unfortunately, neither of the partners was an astute businessman which, coupled to the many technical problems that had to be overcome, led to them becoming bankrupt in 1760. The prototype carding machine used at the Northampton factory eventually ended up with a Mr. Morris in Wigan.

A few years later in 1767, Thomas Highs, a resident of Leigh, also designed a machine to spin cotton using rollers but his design abandoned the circular form of the previous patent. Thomas Highs employed a clockmaker in Warrington, called John Kay, to make the gearing and rollers for his machine. It was at this time that Richard Arkwright heard about this new machine, possibly from his wife's family who also came from Leigh. He persuaded John Kay to make a couple of machines for him, copying the techniques introduced by Highs, and eventually persuaded Kay to become his employee, working for him in Nottingham and then in Cromford where Arkwright built his first water powered cotton mills. In 1774, Arkwright took out a patent for the carding, roving and drawing machines. This carding machine was an improvement of the design patented by Wyatt and Paul including the refinements provided by a number of people. The roving and drawing machines were really just variations of Arkwright's original patent using the drafting rollers of the water frame. Although Richard Arkwright became extremely successful, building many cotton mills and also licensing other manufacturers to use his patents, there were many people who believed that he had no entitlement to his patents as they were the result of the work of other inventors. These people also built cotton mills using the carding engine and water frame but refused to pay royalties to Arkwright. One such

FIGURE 291. JOSEPH WRIGHT'S PORTRAIT OF RICHARD ARKWRIGHT SITTING ALONGSIDE A SET OF DRAFTING ROLLERS FROM HIS WATER FRAME PATENTED IN 1769. *(Internet, Joseph Wright homepage)*

cotton mill was established in Buglawton, near Congleton on the Daneinshaw brook, in a mill originally built as a flint mill, called Nearer Daneinshaw Mill (later called Davenshaw Mill). This waterpowered cotton spinning mill was operated by a consortium of four partners. Three of the partners, James Robinson, Henry Mather and John Barnes all came from Warrington and the fourth, Thomas Morris, was from Congleton. It must be almost certain that the three Warrington partners were familiar with Thomas High's work and it is tempting to think that the fourth partner, Thomas Morris, could in some way be related to the Mr. Morris who acquired the first carding engine from Wyatt and Lewis in 1760.

This state of affairs did not suit Richard Arkwright so he decided to sue the various "pirate" manufacturers, one of whom was the partnership at Buglawton, for breach of his patent for the carding machine. These pirate manufacturers were not just a few "cowboy" operators but represented 60% of the industry at that time. In 1781, in a victory famous in its day, the defendants won by demonstrating that the patent was deliberately obscure and misleading such that it was not possible to construct a machine solely using the

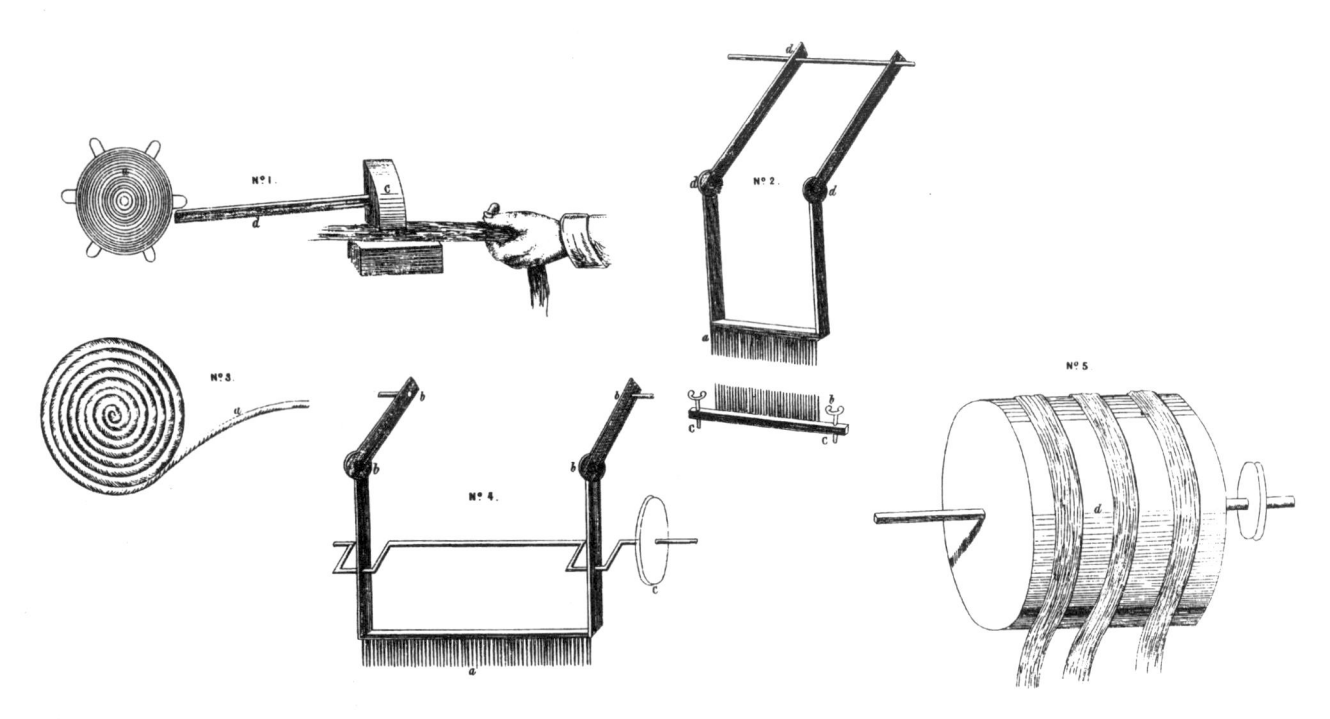

FIGURE 292. THE FIRST FIVE FIGURES IN ARKWRIGHT'S PATENT ARE SUPPOSED TO DESCRIBE THE CARDING MACHINE. NO. 1 IS A FLAX BREAKER; NO. 2 IS AN IRON FRAME WITH TEETH WORKING AGAINST A FIXED SET OF TEETH; NO. 3 IS A CLOTH WITH WOOL, FLAX, ETC. SPREAD THEREON; NO. 4 IS A CRANK AND FRAME WITH TEETH WORKING BACKWARD AND FORWARD AGAINST NO. 5 A CYLINDER WITH FILET CARDS, WHICH DISCHARGES THE COTTON, WOOL ETC. THE DRAWINGS AND DESCRIPTION OF THE ELEMENTS PURPORTING TO BE THE DRAWING AND ROVING FRAMES ARE EVEN MORE RUDIMENTARY. (Patent No. 1111, 1775)

patent information. As a consequence of this court case Arkwright's 1774 patent for the carding machine was annulled. However, as is often the case, it was the victors who failed to survive. They had invested heavily in plant and machinery in an embryo industry and could ill afford the cost and delays caused by the legal process. Arkwright, on the other hand, was "cash rich" having amassed over £60,000 from royalty fees so was in a much better position financially after the case. Today, of course, it is Arkwright who is remembered, not the many other men of principle who had the courage to stand up against him and win the day. The partnership in Buglawton had to sell their cotton mill in 1783, presumably to meet their legal fees. Arkwright has also been credited with developing the factory system but he was only applying the principles established earlier in the iron and silk industries. He has also received credit for introducing a typical style of textile mill design, today known as "Arkwright Mills", but this design of textile mill had been fully established and perfected by the construction of the silk mill in Congleton all of 25 years before Arkwright's first water powered mill at Cromford was built.

Once again it can be seen that this area of southeast Cheshire was in the vanguard of establishing the waterpowered textile industries that were to form the bedrock of the industrial revolution. As well as one of the very first waterpowered silk mills which was built in Congleton in 1752, one of the first waterpowered cotton spinning mills was established in the 1770s at Buglawton in opposition to and contemporary with Richard Arkwright and his Cromford Mills.

In 1785 Arkwright managed to get his 1774 patent reinstated by the courts but this led immediately to another trial. In the final patent trial, evidence from Thomas Highs and John Kay showed that Arkwright appeared to have stolen the design for his original patent and since the roving and drawing machines in the 1774 patent were but derivatives of this original design, their patent was annulled. Much evidence was also presented to show that Arkwright's claim to the carding machine design was also bogus and so by the end of the trial Arkwright lost not only his patents but his reputation. Once Arkwright's patents had been cancelled, there were many people willing to invest in the cotton industry, no doubt inspired by the large amounts of money mentioned in the patent case as being realised from this fledgling industry. Immediately after 1785, cotton spinning mills were being built in profusion in a "dash for cotton" as people scrambled to invest in this new technology. On the River Dane and its tributaries there were five cotton mills built in the late 1780s, at Danebridge, Further Daneinshaw, Pool Bank at Timbersbrook, Stonehouse Green in Congleton and Thomas Slate's Dane Mill also in

Congleton, followed by another five in the 1790s. This demand for waterpowered sites led to existing mills being taken over and converted to cotton spinning, such as at Danebridge and Pool Bank where existing corn mills were converted. Alternatively, cotton mills were built in remote areas, such as in the headwaters of the River Dane itself, where feasible waterpowered sites had not been occupied because of the lack of demand due to the sparceness of the population in the neighbourhood.

By the 1790s the growth of the cotton industry was so great that the market was saturated, consequently profits were far harder to achieve and some of the new mills failed to survive. This was especially the case when cotton mills were built far from good transport facilities such as at Gradbach and Crag, high in the foothills of the Pennines. Both Gradbach Mill and Crag Works were built specifically as cotton mills but Gradbach Mill operated for only two years spinning cotton before its builders were bankrupt in 1794, and at Crag Works the entrepreneurs went bankrupt before the mill could even be stocked with the necessary cotton spinning machinery. This is perhaps not surprising as Crag Works was a considerable undertaking being far larger than most of the cotton mills built at this time. It was built as a fire-proof construction using cast iron columns and beams for all its five floors. As this building began construction in 1793 it can be seen as one of the first iron framed buildings to be built, vying with the flax mill at Shrewsbury as the earliest building of this type.

For many years the salt industry in Cheshire had relied on using naturally occurring springs as a source of brine. However, as these springs decreased in flow or stopped altogether, the brine had to be pumped by mechanical means. The engineering involved in pumping had been known since antiquity and had been highlighted in learned publications in the 16th century by such as Agricola and Ramelli. In England, these ideas were put into use by millwrights such as George Sorocold of Derby who were using waterwheels to operate pumps for public water supplies at the beginning of the 18th century. At some stage during the century (or earlier), waterpower had been introduced into the salt industry for pumping brine which was then evaporated in iron pans to produce salt. Although the salt trade was the first industry to introduce steam power in Cheshire, at Lawton by Boulton and Watt at the early date of 1779, it did not replace waterpower in the industry for at least another fifty years.

At the end of the 18th century, someone looking back over the achievements of south-east Cheshire and north Staffordshire in that century could be justifiably pleased with the area's contribution to progress. A new method of working, the factory system, had been pioneered in the iron industry at the beginning of the century. This methodology had then been transferred to the textile industry and concentrated under a single roof with the establishment of the massive silk mill at Congleton in the middle of the century. This was the mill that set the standard for the design of textile mills for the next hundred years or so, not just in the silk industry, but in all branches of textile manufacture. The strategic copper and brass industry had followed the earlier lead set by the iron industry, its integrated manufacturing taking place over a number of sites, in this case wide-spread about the whole country. Towards the end of the century, the region had produced the pioneers in the cotton industry contemporary with Richard Arkwright and had successfully led the way in the fight against Arkwright's patent restrictions. The River Dane and its tributaries were well endowed with many of these new textile manufactories, mainly producing cotton thread, but with the Congleton silk mill also dominating the output of silk thread in the area.

No doubt the business community would have looked forward to the new century with a great deal of confidence and optimism in spite of a couple worrying signs. Although entrepreneurs were very willing to invest in new enterprises the mill owners maximised their profits without a thought for reinvestment in more modern equipment as it was developed and the boom and bust cycle apparent in the textile industry must have caused some anxiety for long-term prosperity.

The 19th Century

All this industrial activity in the area caused the population to rise considerably due to the increased prosperity and the inward migration of workers for the factories from other regions. This increase in population caused a corresponding increase in the demand for bread which the existing corn mills were hard pressed to satisfy. The early years of the 19th century saw a number of new corn mills established especially by using waterpowered sites no longer required by other industries such as the various metal trades. Three of the early iron forges at Cranage, Warmingham and Street, as well as parts of the two copper/brass factories at Bosley and the Havannah, became corn mills in the first twenty years of the new century. Also the flint mill at Washford in Buglawton was converted to a corn mill at this time. Entirely new waterpowered sites were very difficult to find but a new corn mill was built at Moston, near Sandbach, using the spare water in the Trent & Mersey Canal for its power, a source only available after the canal had been completed in 1777. Also, a new corn mill was built at the Bank, on the Cheshire slopes of Mow Cop to take advantage of a spring emanating from the hillside. The ever-growing demand for more power, mainly by the textile industry, caused engineers to improve the design of waterwheels, shafting and gearing. These improvements were also applied to some of the corn mills to increase their output, as was the case at Park Mill at Brereton in 1835, which was completely redesigned by William Fairbairn, the great Manchester based engineer. This period also saw a great reduction in toll milling where the miller was paid with a percentage of the customer's grain, and the greater introduction of merchant milling where the miller bought grain and sold the products of his mill to his customers.

Although the 19th century started with cotton spinning in the ascendancy in the textile mills on the River Dane and its tributaries, only two more waterpowered cotton mills were to be built in the 19th century, at both Timbersbrook and Biddulph in 1814.

Other regions in Britain also constructed cotton mills, which were further developed by the application of the steam engine. Cotton mills no longer were restricted in their choice of site by the availability of waterpower nor were they constrained in size except by the limits of steam engine design. As large steam powered cotton mills became the norm, the smaller waterpowered cotton mills on the Dane and its tributaries could not hope to compete. Eventually economic realities started to force them to close with many of them attempting to find some other trade that was not as desperately competitive. The earliest failures, at Gradbach and Crag at the end of the 18th century, turned their attentions to flax spinning and calico printing respectively. Of the other eleven cotton spinning mills in the area, six had converted to silk throwing or silk spinning by 1840, and two others, at Rushton Spencer and Slate's Mill in Congleton, became a dye house and a brewery respectively.

FIGURE 293. A PORTRAIT OF SIR WILLIAM FAIRBAIRN, MILLWRIGHT, ENGINEER, FOUNDER MEMBER OF THE INSTITUTION OF MECHANICAL ENGINEERS AND DESIGNER OF THE MACHINERY AT PARK MILL, BRERETON.

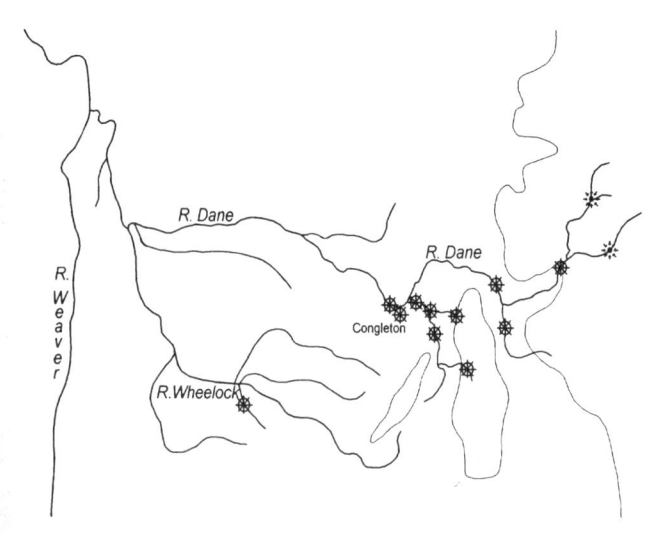

FIGURE 294. THE LOCATION OF WATERPOWERED COTTON MILLS IN THE EARLY YEARS OF THE 19TH CENTURY. NOTE THAT CRAG AND GRADBACH (*far right*) CEASED COTTON PRODUCTION PRIOR TO 1800.

Although silk throwing had been the earliest of the textile processes to become established in the area with the Congleton silk mill in 1752, its sheer size and capacity made it very difficult for any rival mills to be established in the area. However, by the beginning of the 19th century, the demand for silk thread had grown considerably and the small amount of power needed for silk throwing made the trade attractive. In the early years of the century, Nearer Daneinshaw Mill, Further Daneinshaw Mill and Pool Bank Mill all changed from cotton spinning to silk throwing and the corn mill at Washford in Buglawton had a wing built on to it for silk throwing, powered from the corn mill machinery. Demand for textile products had been high during the

THE OLD MILL.

FIGURE 295. CONGLETON SILK MILL AS IT APPEARED AFTER 1830. AT THIS TIME STEAM POWER WAS ADDED TO THE MILL AND PROBABLY THE WATERWHEEL WAS INCREASED IN WIDTH AT THE SAME TIME.
(R. Head, *Congleton Past and Present*, 1887)

Napoleonic Wars as the hostilities had curtailed foreign imports but once peace was established the industry went into recession. When trading conditions improved by the start of the 1820s, there was a "dash for silk" in the same way that there had been a "dash for cotton" in the 1780s. This expansion was fuelled by the release of capital caused by the reduction of duty on imported silk fibres with twelve water powered silk mills being established on the River Dane and its tributaries and many steam powered mills were built elsewhere in the area. Although four of these water powered silk mills were converted from other uses, the other eight were newly erected. Even the enormous Silk Mill in Congleton was extended by another 142 feet by Samuel Pearson when he purchased the mill in 1830, making it the biggest silk mill in the country. As might be expected, this massive growth in the silk throwing trade led very quickly to over-capacity, lower prices and a trade slump that lasted more than ten years. Even after this experience a further seven waterpowered silk throwing mills were established by 1840, four of them on completely new sites.

It would be fair to say that 1840 marked the "high water mark" for the use of waterpower on the River Dane and its tributaries. All available water powered sites had been pressed into service, many of the textile mills were also using steam power to supplement their waterwheels, as were some of the corn mills. Among the other industries of the period, tanning, which had a very long history, was using water, not only in the vats used to soak the hides, but to power bark mills for chopping the oak bark which was used as a source of tannin. Salt production still relied on water power, if only as back up for their steam powered pumps, and paper making was still present at two mills in spite of the continuous process of paper making invented by the Fourdrinier brothers in the 1830s at Stoke-on-Trent. At Crag, originally built to be a cotton mill, calico printing gave way to carpet printing in the 1840s. This

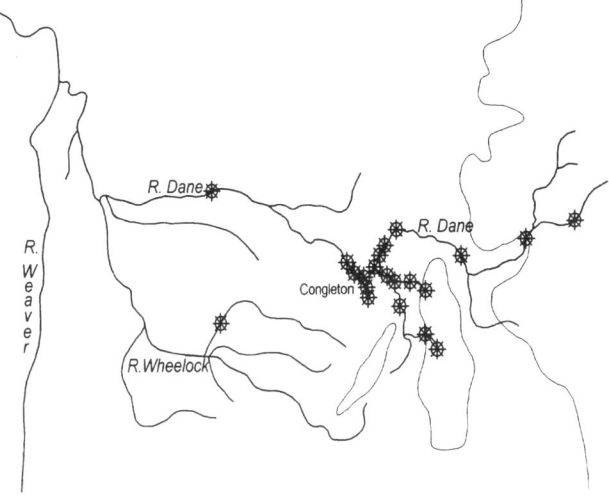

FIGURE 296. THE "HIGH WATER MARK" OF WATERPOWERED SILK MILLS AROUND 1840.

very successful business had its greatest success when one of its carpets was exhibited at the Great Exhibition of 1851 at the Crystal Palace.

Over the twenty years from 1840 to 1860 some of the more marginal sites ceased to function, such as a couple of silk mills on the moorland above Biddulph. The two paper mills ceased to function due to their remoteness and competition from the Fourdrinier process. Then in 1860 calamity befell the British silk trade with the passing of the Free Trade Act, which allowed foreign silk into the country without paying any duty. The accepted view is that this foreign competition decimated the British silk industry and virtually killed it off. In the River Dane area this is not exactly the picture that emerges. It is true that many of the steam powered silk mills did suffer greatly after the passing of this Act of Parliament but the water powered silk mills seemed to have fared much better, possibly because their power production was viewed as being "free". Although trade was never the same after 1860, the water powered silk mills continued in production, losing a few mills every decade. In the 1860s, only three water powered silk mills ceased work. One was a small mill located on Biddulph Moor, another converted to full time corn grinding by waterpower, and the third closed when its owner died and the mill could not be sold in the prevailing climate in the silk industry.

Times continued to be hard for silk manufacturing. In the 1870s two silk mills powered by the Howty in Congleton closed, together with Primrose Vale Mill on the Daneinshaw Brook. These mills were then used for other purposes such as fustian cutting and towel making but the use of waterpower at these sites ceased. The 1880s were a gloomy time with another five mills ceasing silk production with perhaps the biggest blow occurring with the death of Samuel Pearson, the owner of the Old Silk Mill in Congleton, which led inevitably to the cessation of silk throwing at this historic mill that had been in production since 1752.

If the demise of the silk industry in this area occurred quite slowly for the waterpowered silk mills then the passing of the country flour mills towards the end of the 19th century came much more quickly and unexpectedly. Until the 1880s flour production for bread making was still the province of the many waterpowered corn mills using the waters of the River Dane and its tributaries. However, in the 1880s a number of factors caused the corn trade to switch from British wheat to North American wheat. This coincided with the introduction of roller milling which involved a revolution in the grinding process. Instead of passing the grain once through a pair of millstones the roller process passed the grain through pairs of steel rollers many times, separating the resultant meal at each stage, giving a much greater yield of flour from the grain. The combination of these factors made it economically sensible to build large, efficient steam powered roller mills at the ports against which the small country mill could not compete. The inevitable outcome was not immediately obvious to the millers and initially their response to declining sales was to re-equip with improved machinery as at Somerford Booths Mill. However, this was not sufficient to stop the trend and by the end of the century most millers were facing the need to find alternative uses for their waterpowered mills to replace the loss of the flour trade or go out of business entirely.

FIGURE 297. HENRY SIMON OF MANCHESTER, ONE OF THE PROMOTERS AND SUPPLIERS OF ROLLER MILLING MACHINERY. (*G. Jones, The Millers, pub. Carnegie*, 2001)

So at the end of the 19th century any review of the situation of the waterpowered mills on the River Dane and its tributaries would have given the completely opposite view to that prevalent at the end of the 18th century. At the start of the 19th century the copper and brass industry had migrated to other parts of the country. By the middle of the 19th century the cotton spinning industry had likewise moved elsewhere when the last cotton mill, at Bath Vale, ceased spinning in 1865. By the end of the century the silk throwing business was in terminal decline due to effect of foreign competition, only three of the waterpowered silk mills continued production into the 20th century and they were probably using steam instead of waterpower. Also the traditional users of waterpower,

FIGURE 298. THE NEW BIRKENHEAD MILL OF MESSRS W. VERNON & SONS WAS ONE OF THE NEW ROLLER MILLING INSTALLATIONS AT THE ENTRY PORTS FOR NORTH AMERICAN GRAIN THAT STARTED TO COMPETE WITH THE SMALL WATERPOWERED MILLS IN THE LATE 19TH CENTURY. (*G. Jones, The Millers, pub. Carnegie*, 2001)

the corn millers, were facing annihilation due to the introduction of a radically new technology, the high grinding method using rollers, and the changing source of their raw material to North America.

By this time the salt industry was almost entirely steam powered and even the mineral grinding business had seen three of its four water powered mills close, Bank Mill on Mow Cop in 1870, Fleet Mill near Middlewich in 1892, and Danebridge Mill in 1898. Although waterpowered equipment had been scientifically improved over the 19th century, the only radical change had been the invention of the water turbine as early as 1838. Unfortunately this device, which was inherently more efficient than the traditional waterwheel, did not find much favour in this country compared to North America and continental Europe. On the River Dane and its tributaries the first water turbine was not installed until 1889 at Bosley Works. The use of water turbines provided the only small glimmer of hope for the future of waterpower on the River Dane and its tributaries as the 19th gave way to the 20th century.

The 20th Century

With the loss of the flour trade around the turn of the century the water powered corn mills had to find other uses. Many of them turned to provender milling, producing animal feed, but this market was not sufficient to employ all the mills. After 1880 corn mills closed or changed their role with regularity. Six mills ceased milling before the First World War with only one of them finding an alternative use for its waterpower, at the Havannah where the waterpower was used for generating electricity. The majority of the remaining corn mills were engaged in provender milling only. The advent of the First World War provided a stay of execution for a while as the need to reduce dependency on imported grain was a priority. However, in the two decades after this conflict a further ten mills ceased production or found alternative uses for their waterpower. For the remainder, again a stay of execution appeared in the guise of the Second World War, but once that conflict was over virtually all the remaining corn mills ceased to function leaving only two mills, at Rushton Spencer and Cranage, still manufacturing animal feed today, albeit without using waterpower.

In the textile industry, silk spinning continued at Dane Mill and Forge Mill, both in Congleton, until 1939 and Brook Mill in Sandbach introduced artificial fibre manufacturing and so survived until 1973, although in their later years none of these mills was still using waterpower. The waterwheel at the Old Silk Mill in Congleton was still in use in the 1920s, even though silk throwing has ceased and the mill was being used by Robert H. Lowe & Co. for hosiery manufacture. Also at Eaton Bank Mill, which was making punched cards for Jacquard looms, the waterwheel was still being used in the 1930s.

FIGURE 299. A PORTRAIT OF WILLIAM CHARLES SIGISMUND ALBANUS HIGGINSON LOWE WHO STARTRED HOSIERY MANUFACTURE AT CONGLETON SILK MILL WHEN SILK THROWING CEASED. (*Congleton Town Museum*)

In the 20th century, a certain amount of success was achieved with the introduction of turbines and sometimes imaginative changes to a particular mill's role. The corn grinding business at Bosley Works was able to compete using water turbines up to 1933. Then the business changed to producing wood flour as a filler material for the new plastics industry, necessitating even more turbines to be installed as the business progressed. The last turbine was installed at Bosley Works in 1967 but shortly after that the river authorities decided to charge waterpower generators on the same basis as abstractors and the use of waterpower finished at Bosley Works. Early in the century new water turbine installations provided power for ringing bells at Swythamley Church and a saw mill on the Biddulph Grange estate. In the 1920s turbines were installed at the Havannah, Daneinshaw Mill and Lea Forge in Biddulph, to generate electricity for a variety of uses. The one surviving flint mill on the River Dane at Washford installed a turbine to supplement its waterwheel in 1936 and continued working until the 1960s. Both the animal feed businesses that survive today installed turbines in their mills in the 20th century.

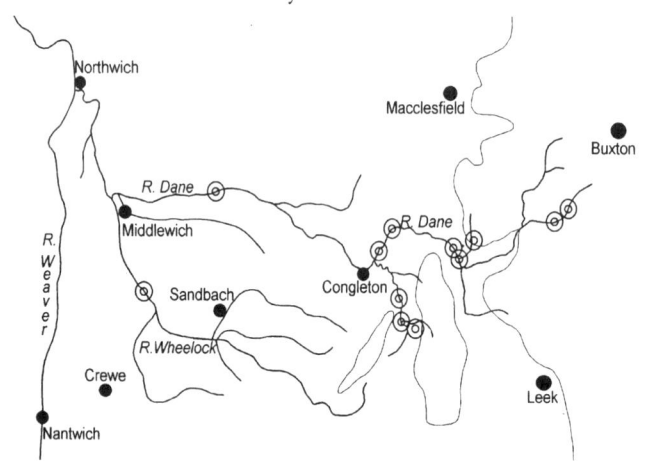

FIGURE 300. THE LOCATION OF MILLS USING TURBINES

In order to keep some mills in business a certain amount of ingenuity was needed. At Bosley corn mill a turbine was installed in the 1920s to power an ice-making machine and when the mill later became a garage the turbine was used to power an air compressor. Warmingham Mill became a light engineering and model workshop in the 1930s, initially using the waterwheel, but then installing a turbine in 1940. The use of waterpower on the River Dane or its tributaries came to an end at Swettenham Mill where the miller, having switched from flour production after the First World War, introduced electricity generators driven from the waterwheel to charge batteries, a service that was in demand for listening to the new-fangled wireless. At the same time he also installed a waterpowered ice-making plant for supplying restaurants and hotels as a mains electricity supply did not come to the area until the 1930s. He also added sawing and other woodworking machines driven by the waterwheel, initially making farm vehicles, but latterly garden furniture, a trade that continued into the 1980s. The miller's death in the mid-1980s brought to a close 900 years of continuously using the waterpower of the River Dane and its tributaries.

A postscript to this history occurred after a short gap of about fifteen years when a minor resurrection took place in 2000 with the installation of a small water turbine in the grounds of Biddulph Grange to generate electricity for the visitor centre. The turbine uses a small stream which is part of the River Dane catchment area and so the use of waterpower on the River Dane and its tributaries, albeit minor in nature, now continues into the 21st century.

BIBLIOGRAPHY

AIKIN, J., *A Description of the Country for Thirty to Forty Miles around Manchester*, 1795.
ALCOCK, J. P., The Corn Mill, *The Industrial Scene*, 1972.
ANGERSTEIN, R. R., (ed. Berg, T. & P.), *Illustrated Travel Diary, 1753 - 5, Industry in England and Wales from a Swedish Perspective*, Science Museum, 2001.
ASHMORE, O., *The Industrial Archaeology of North-west England*, Manchester University Press, 1982.
ASPIN, C., *The Cotton Industry*, Shire, 2000.
AWTRY, B. G., Charcoal Ironmasters of Cheshire & Lancashire, 1660-1785, *Transactions of the Lancashire & Cheshire Antiquarian Society*, 109, 1957.
BAINES, E., *The History of the Cotton Manufacture in Great Britain*, 1835, (reprinted by Cass, 1966.)
BAINES, P., *Flax and Linen*, Shire, 1998.
BONSON, T., The Millstones of Mow Cop, *Mills Research Group*, 2003.
BONSON, T., The History of Bosley Works, Cheshire, *Wind and Water Mills*, 14, 1995.
BONSON, T., The History of Washford Mill, Buglawton, Cheshire, *Wind and Water Mills*, 15, 1996.
BONSON, T., The Turbines Used at Bosley Works, Cheshire, *Wind and Water Mills*, 17, 1998.
BONSON, T., The First Iron Framed Building?, *Industrial Archaeology Review*, XXII, 2000.
BONSON, T., BOOTH, T., & JOB, B., Swettenham Mill, A History and Survey, *Wind and Water Mills*, 18, 1999.
BOTT, O., Cornmill Sites in Cheshire 1066-1850, Part 4, *Cheshire History*, 14, 1984.
BOUCHER, C. T. G., *James Brindley - Engineer*, Goose, 1968.
BRADLEY, C., Industry at Little Moreton, *Journal of the Congleton History Society*, 7, 1988.
BRADLEY, C., Stone Grinding at Washford Mill, Buglawton, *Wind and Water Mills*, 14, 1995.
BRADLEY, C., Potter's Milling - an early ball mill rescued, *Industrial Archaeology News*, 117, Summer 2001.
BROCKLEHURST, P., *Swythamley and its Neighbourhood*, 1874.
BUCHANAN, R., *Practical Essays on Mill Work and other Machinery*, ed G. Rennie, 3rd Edition 1841.
BUSH, S., *The Silk Industry*, Shire, 1991.
CALLADINE, A., & FRICKER, J., *East Cheshire Textile Mills*, Royal Commission on the Historical Monuments of England, 1993.
CALVERT, A. F., *Salt in Cheshire*, SPON, 1915.
CAPEWELL, A., et al., *A Journey through Time*, Intec Publishing Ltd., 1996.
CHALONER, W. H., Cheshire Activities of Matthew Boulton & James Watt, of Soho, near Birmingham, 1776-1817, *Transactions of the Lancashire & Cheshire Antiquarian Society*, 61, 1949.
CHALONER, W.H., Charles Roe of Macclesfield, *Transactions of the Lancashire & Cheshire Antiquarian Society*, 62, 1951/2.
CHALONER, W. H., Sir Thomas Lombe (1685-1739) and the British Silk Industry, *History Today*, 3, 1953.
CHALONER, W. H., Salt in Cheshire, 1600-1870, *Transactions of the Lancashire & Cheshire Antiquarian Society*, 71, 1961.
CHESHIRE WOMEN'S INSTITUTE, *A Village History: Swettenham*, 1952.
COPELAND, R., *A Short History of Pottery Raw Materials and the Cheddleton Flint Mill*, Cheddleton Flint Mill Industrial Heritage Trust, 1972.
CORRY, W., *History of Macclesfield*, 1817.
COSSONS, N., (ed.), *Rees's Manufacturing Industry*, David & Charles, 1972.
COSSONS, N., *The BP Book of Industrial Archaeology*, David & Charles, 1975.
DARBY, H. C., & MAXWELL, L. S., *The Domesday Geography of Northern England*, Cambridge University Press, 1962.
DAVIES, D. L., *Watermill, Life Story of a Welsh Cornmill*, Ceiriog Press, 1997.

DAVISON, C. ST. C. B., Geared Power Transmission, *Engineering Heritage*, 1963.
DAWSON, T. L., 150 Years of Carpet Printing, A Historical Retrospect, *Journal of the Society of Dyers & Colourists*, 115, 1999.
DAY, A. R., Congleton's lost Tobacco Industry, *Journal of the Congleton History Society*, 4, 1980.
DESGAULIERS, J. T., *A Course in Experimental Philosophy*, 1744.
DODGSON, J. McN., *The Place Names of Cheshire, Part 2*, Cambridge University Press, 1970.
EARWAKER, J. P., *East Cheshire*, 1887/8.
EARWAKER, J. P., *The History of the Ancient Parish of Sandbach*, 1890, (reprinted by E. J. Morten, 1972).
EARL, A. L., *Middlewich 900 - 1900*, Ravenscroft Publications, 1990.
EARLES, J., *Old Macclesfield*, 1915.
ENGLISH, W., A Study of the Driving Mechanisms in the Early Circular Throwing Machines, *Textile History*, 2, 1971.
EVANS, K. M., *James Brindley, Canal Engineer, A New Perspective*, Churnet Valley Books, 1997.
EVANS, O., *The Young Millwright and Miller's Guide*, 1795, reprinted 1990.
FAIRBAIRN, W., *Treatise on Mills and Millwork*, 1863.
FITTON, R. S., *The Arkwrights, Spinners of Fortune*, Manchester University Press, 1989.
GIFFORD, A., The Waterworks at Strutt's Mills at Belper and the first Suspension Waterwheel, *Wind and Water Mills*, 13, 1994.
GILES, K., *The Bromley Davenport Papers*, 1999.
GRAHAM, J., *The History of Printworks in the Manchester District from 1760 to 1846*, 1847.
GREENSLADE, M. W., ed., *Victoria County History of Staffordshire*, Vol 7, Oxford University Press, 1995.
HARDMAN, B. M., The Iron Industry of North Staffordshire and South Cheshire in the pre-coke Smelting Era, *North Staffordshire Journal of Field Studies*, 15, 1975.
HARRIS, H., *The Industrial Archaeology of the Peak District*, David & Charles, 1971.
HEAD, R., *Congleton Past and Present*, 1887, (reprinted Old Vicarage Publications, 1987).
HERTZ, G. B., The English Silk Industry in the 18th Century, *English Historical Review*, 24, 1909.
HEWITT, H. J., *Cheshire under the Three Edwards*, 1967.
HILLS, R. L., *Richard Arkwright and Cotton Spinning*, Wayland Publishing, 1973.
HOLT, R., *Medieval Mills*, Blackwell, 1988.
HOLLAND, H., *General View of the Agriculture of Cheshire*, 1808.
JENKINS, R., Fast and Loose Pulleys, *The Engineer*, 12th April 1918.
JENNINGS, T. S., *A History of Staffordshire Bells*, Privately published, 1968.
JOB, B., Drying Kiln Clay Tiles, *Wind and Water Mills*, 13, 1994.
JOB, B., *Watermills of the Moddershall Valley*, published by B. Job, 1995.
JOHNSON, B. L. C., The Foley partnerships: The Iron Industry at the end of the Charcoal Era, *The Economic History Review*, second series, 4, 1952.
JOHNSON, B. L. C., The Iron Industry of Cheshire & North Staffordshire: 1688-1712, *Transactions of the North Staffordshire Field Club*, 88, 1954.
JONES, D. H., The Waterpowered Cornmills of England, Wales and the Isle of Man. *Transactions of the International Molinological Society*, 2, 1969.
JONES, G., *The Millers, A Story of Technological Endeavour and Industrial Success*, 1870-2001, Carnegie Publishing Ltd., 2001.
KENNEDY, J., *Biddulph, By the Diggings*, University of Keele, 1980.
KING, P. W., The Vale Royal Company and its Rivals, *Transactions of the Lancashire & Cheshire Historical Society*, 142, 1993.
LAWRENCE, C. F., *Annals of Middlewich*, 1911.
LAWTON, G. O., ed., Northwich Hundred Poll Tax 1660 and Hearth Tax 1664, *Record Society of Lancashire & Cheshire*, 119, 1979.
LEA, Herbert, *Swettenham Mill*, 1984.
LEAD, P., The North Staffordshire Iron Industry 1600-1800, *Journal of the Historical Metallurgy Society*, 11, 1977.
LEAD, P., *Agents of Revolution*, Keele University, 1990.
LINDSAY, J., *Trent & Mersey Canal*, David & Charles, 1979.
LONGDEN, G., *The Industrial Revolution in East Cheshire*, Macclesfield & Vale Royal Groundwork Trust, 1988.

MASSEY, J. H., *The Silk Mills of Macclesfield*, B.A. Thesis for RIBA, 1959.
MORGAN, P., ed., *Domesday Book, Vol. 26*, Cheshire, Phillimore, 1978.
NEVILLE BARTLETT, J., *Carpeting the Millions*, pub. John MacDonald, 1977.
NORRIS, J. H., The Water-Powered Corn Mills of Cheshire, *Transactions of the Lancashire & Cheshire Antiquarian Society*, 75, 1965.
ORMEROD, G., *The History of the County Palatine and the City of Chester*, Routledge & Sons, 1882.
PELHAM, R. A., *Fulling Mills*, SPAB, 1954.
PHILLIPS, J., *Inland Navigation*, 1805, (reprinted by David & Charles, 1970).
PITT, W., *A Topographical History of Staffordshire*, 1817.
RATHBONE, C., *The Dane Valley Story*, 1954.
RENAUD, F., Church Lawton Manor Records, *Transactions of the Lancashire & Cheshire Antiquarian Society*, 5, 1887.
RENAUD, F., The Mineral Wealth of Lawton, *Lawton Chronicles*, Lawton Heritage Society, 5, 2000.
RICHARDS, R., *The Manor of Gawsworth*, Scholar Press, 1957.
RUSSELL, J., Millstones in Wind & Water Mills, *The Engineer*, 31st March 1944.
SCORE, S. D., Folly Mill, *Cheshire Life*, 1954.
SHERLOCK, R., *The Industrial Archaeology of Staffordshire*, David & Charles, 1976.
SHORTER, A. H., Paper Mills and Paper Makers in England, 1495-1800, *Paper Publishing Society*, Hilversum, 1957.
SMEATON, J., An Experimental Enquiry concerning the Natural Powers of Water and Wind to turn Mills and other Machines, depending on a Circular Motion, *Philosophical Transactions of the Royal Society*, 51, 1759.
SMILES, S., *Lives of the Engineers, Vol. I*, 1862.
STEVENS, W. B., *History of Congleton*, University of Manchester, 1970.
STOYEL, A., *Perfect Pitch: The Millwright's Goal*, Society for the Protection of Ancient Buildings, 1997.
STUBBS, M. J., ed., *A Short History of Odd Rode Parish*, Scholar Green History Society, 1989.
SUTCLIFFE, Rev. W., *Records of Bosley*, 1865.
THOMPSON, P., *The Archaeological Potential of a Town*, Cheshire County Council, 1981.
THOMPSON, W. J., *Industrial Archaeology of North Staffordshire*, Moorland, 1974.
TREVELYAN, G. M., *Life of John Bright*, Constable, 1913.
TUNNICLIFFE, W., *A Topographical Survey of the Counties of Staffordshire, Cheshire, and Lancashire*, 1787.
TURNBULL, G., *A History of Calico Printing in Great Britain*, 1851.
VARLEY, J., ed. A Middlewich Cartulary, *Chetham Society*, Vol. 1, 1941; Vol. 2, 1944.
WAILES, R., Water Driven Mills for Grinding Stone, *Transactions of the Newcomen Society*, 39, 1966-7.
WARNER, F., *The Silk Industry*, Dranes, 1921.
WATTS, M., *Corn Milling*, Shire, 1983.
WATTS, M., *Water and Wind Power*, Shire, 2000.
WATTS, M., *The Archaeology of Mills & Milling*, Tempus, 2002.
WHEELHOUSE, D. J., *Biddulph*, Chalford, 1997.
WILLIAMSON, F., George Sorocold of Derby, *Journal of the Derbyshire Archaeological and Natural History Society*, 57, 1936.
WIGFULL, P., The Dakeyne Mill and its Romping Lion, *Wind and Water Mills*, 16, 1997.

INDEX

NOTE. Those index entries in **bold** type refer to the principal location of a particular mill or technology

A

accounts
    Congleton Mill 112, 119
    Cotton Mill 174
    Cranage Forge 170
Ackers, Holland 214, 216, 217
air compressor 42
Andiamo Tobacco Company 60
Antrobus, Crawford John 60, 61
Antrobus, G. J. 267
Antrobus, Gibbs Craufurd 56
Antrobus, Philip 65, 104, 138
applications 1
Arden, Henry 229
Arden, John 227
Arden, Thomas 229
Arkwright, Richard
    104, 105, 132, 260, 263, 283, 284
Astbury, Joseph 226

B

Bage's flax mill 11, 285
Bank Mill (Buglawton) **102**
Bank Mill (Mow Cop)
    204, 207, 246, 271, 286, 289
bark mills 269
Barlow, Charles 101
Barnes, John 104, 263, 283
Baron's Croft Saltworks **237**
Bath Vale Mill **98**, 289
Bearda Mill 19, **32**, 246, 247
Bearda Saw Mill **30**
bell ringing 27
Belper 241
Benson, Thomas 270
Berresford, John & James 22
Berresford, Joseph 22, 44, 99
Betchton 200, 202
bevel gears 242
Bibby, Thomas 215
Biddulph Grange **83**, 274, 276, 290
Biddulph Mill **84**
Biddulph Moor Mill **78**, 288
bloomeries 250
bone crushing **272**

bone mills
    Midge Brook tannery 157
    Moston 229
    Park Mill, Brereton 177
Booth, Jonathon 120, 155, 160, 247
Booth, Timothy 216
Bosley Mill 32, **40**, 51, 159, 290
Bosley Reservoir 41
Bosley Wood Treatment Ltd. 47
Bosley Works
    42, 52, 89, 243, 248, 254, 272,
    273, 282, 289, 290
Bourne, Daniel 260
Bradford & Gent 137
Brereton Hall **175**
Bridge Street Mill 101, **136**
Bright, J. & Co. 13
Bright, John 13
Brindley, Henry 21, 88
Brindley, James
    1, 10, 20, 33, 43, 115, 124, 128,
    191, 223, 249, 280, 281
Brindley, Robert 46
Brindley, Samuel 88, 175
Brindley Snr, James 20
brine pumping 268, 285
British Waterways 36
Broadhurst & Cookson 116
Broadhurst, Hall & Co. 88
Broadhurst, Jonathon 65, 88, 95, 216
Brocklehurst, Lady 30
Brocklehurst, Sir Philip 23, 26, 35
Brook Mill **138**, 260
Brook Mill (Sandbach) **218**, 290
Brookes Mill 97
brooks
    Arclid 219
    Biddulph 77
    Bosley 40
    Chapel 150
    Clough 3, 10, 17
    County 87
    Daneinshaw 77
    Howty 109
    Loach 150
    Midge 150

    Swettenham 150
Brunel, Isambard Kingdom 14
Burberry Bros. 102
Burch, Joseph 11, 23, 266
Butler, Robert 213
Byfleet 112, 115, 123

C

calico printing
    Crag Works 11
carding machine 260
carpet printing 266
    Crag Works 13
Carson & Bradbury 64
cast iron
    beams 11, 15
    columns 7, 11, 15
    windows 179
chafery 252
chartists 129, 141
Cheshire Works
    167, 194, 212, 231, 279
Clayton, John 115, 123, 127, 132
Clayton, Samuel 280
Clayton's Mill 80
cloth printing **266**
clutch 125, 243
coal grinding 25, **271**
Colley Mill 41, **51**, 171
colour mills
    Danebridge Mill 24
    Lee Forge 87
Condliffe, John 156, 217
Congleton Engineering Company 107
Congleton Fulling Mill **110**, 255, 278
Congleton Mill
    59, **110**, 112, 156, 174, 239, 245,
    246, 278, 281
Congleton Silk Mill
    123, 145, 240, 243, 257, 259, 281,
    288, 290
Congleton Town Mill (see Congleton
    Mill)
Cook, James 38, 39
Cook, John 40
Cook, Jonathon 38

294

copper making **254**, 282
copper works
　Bosley Works 42
　Havannah 52
corn milling **244**, 278, 281
corn mills
　Bank Mill (Mow Cop) 204
　Bearda Mill 32
　Biddulph Mill 84
　Bosley Mill 40
　Bosley Works 46
　Colley Mill 51
　Congleton Mill 110
　Cotton Mill 174
　Cranage Mill 171
　Danebridge Mill 19
　Daneinshaw Mill 87
　Forge Mill (Street) 214
　Forge Mill (Warmingham) 233
　Hardingswood Mill 190
　Havannah 54
　Higher Roughwood Mill 199
　Kinderton Mill 184
　Lawton Mill 196
　Little Moreton Mill 210
　Lower Roughwood Mill 202
　Marton Mill 152
　Moreton Mill 216
　Moston Mill 229
　Park Mill, Brereton 175
　Peck Mill 237
　Pool Bank Mill 97
　Rode Mill 206
　Rushton (or Nether Lee) Mill 37
　Sandbach Mill 219
　Smallwood Mill 217
　Somerford Booths Mill 153
　Stanthorne Mill 235
　Sutton Mill 234
　Swettenham Mill 158
　Warmingham Mill 227
　Washford Mill 66
　West Heath Mill 150
　Wheelock Mill 223
　Winterley Mill 225
Cotton, Daniel 168
Cotton Mill 166,. **174**, 246, 281
cotton mills
　Bath Vale 98
　Bosley Works 42
　Crag Works 10
　Danebridge Mill 21
　Daneinshaw 91
　Davenshaw Mill 104
　Gradbach Mill 4
　Rushton 39
　Stonehouse Green Mill 138
　Stonier's Mill 78
　Wheelock Mill 223

cotton spinning 260, 283
Cotton, William 166, 194, 231
Crag Works **10**, 266, 285
　Lower Mill 15
　Upper Mill 15
Cranage Forge
　166, 174, 175, 194, 251, 253, 279, 285
Cranage Mill 166, 248
Crompton, Samuel 263
Crossledge Forge 148
Crossley, J. 13

D

Dakeyne, Bowden Bower 6, 23
Dakeyne, Daniel 5
Dakeyne, John & Co. 5
Dakeyne, Thomas 4
Dane Feeder 36
Dane Mill 129, **146**, 149, 260, 290
Dane Mill (Slate's) **144**, 284
Danebridge Mill
　6, **19**, 88, 249, 263, 264, 278, 280, 284, 289
Daneinshaw Mill **87**, 290
Daneinshaw Silk Mill **90**, 275
Davenshaw Mill
　55, 68, 91, **104**, 138, 283
Day, Abraham 17
de Lacy, Henry 111
De Luxe Manufacturing Co. 143
Dean, Ambrose 217
Derby silk mill 240, 259
Derwent Hydropower Ltd. 83, 276
Domesday Book 175, 277
Drakeford, Jesse 223
Drakeford, William 100
drawing frame 261
drying kiln
　116, 163, 176, 204, 209, 221, 247
Dutton, John 234
Dutton, William 230

E

Eaton Bank Mill 59, **63**, 250, 290
edge runners 269, 272
electricity generation
　Bank Mill (Buglawton) 103
　Biddulph Grange 83
　Bosley Works 49
　Brereton Hall 175
　Congleton Mill 120
　Daneinshaw 94
　Havannah 60
　Higher Roughwood Mill 201
　Lee Forge 87
　Stanthorne Mill 236
　Swettenham Mill 160
English Velvet Company 62

Equilinum 5, 260, 265

F

factory system 251
Fairbairn & Lillee 44
Fairbairn, William
　1, 176, 241, 243, 247, 286
Fielder, Charles 91
Fielder, John 140
filatoes 257
filatures 255
finery 252
fire-proof mill 11, 285
Fitton & Sons 59
flax mills
　Gradbach Mill 5
flax spinning 5, **264**
Fleet Mill, Croxton **188**, 271, 289
flint grinding **270**
flint mills 271
　Bank Mill (Mow Cop) 205
　Davenshaw Mill 104
　Fleet Mill, Croxton 188
　Lawton 195
　Lee Forge 85
　Washford Mill 65
flour dresser 57, 154, 205, 220, 248
Fodens Ltd. 178
Foley, Richard 231
Folly (Grove) Mill 10, **17**, 34, 280
Ford, James 148
Ford, John 204
Ford, Samuel 204
Forge Mill (Crossledge) **148**, 260, 290
Forge Mill (Street) 117, **212**
Forge Mill (Warmingham) **231**, 233
forges
　Cranage 166
　Crossledge 148
　Lee Forge 85
　Quarnford **9**
　Street 212
　Warmingham 231
Fourdrinier brothers 18, 250
Francis Brindley & Co. 46
Francis, James Bichenor 275
fulling **254**, 278
fulling mills
　Congleton 110
　Danebridge 19, 255
　Wincle 33
furnace, blast 250
furnaces
　Lawton 192
　Street 212
Further Daneinshaw Mill 91, 105, 284
fustian cutting **267**
　Bank Mill (Mow Cop) 206
　Havannah 60

G

Gaunt, Richard 39
gearing & millwork 126, **242**
Gent & Norbury 133
Gent, Charles 133, 137
Gilbert, Gilkes & Co.
    42, 47, 60, 83, 274
Gilbert, Gilkes & Gordon Ltd.
    71, 94, 228, 275, 276
Gilbert, John 191
Ginder, Richard Low 93, 98
Goodwin, J. M. 68, 271
Gosling, Francis 85
Gosling, Samuel Franceys 87
governor 15, 128
Gradbach Mill **4**, 23, 264, 265, 285
grain cleaner 57, 182, 220, 247
Great Eastern 14
Great Exhibition 13, 288
Gresty, William 233
grinding cylinder 71, 74, 271
grinding pan 69, 270
Günther, W. & Sons 46, 273

H

Hall & Co. 88
Hall & Johnson 88, 92
Hall, Edward
    167, 170, 194, 212, 231
Hall, Thomas (corn miller)
    117, 177, 215
Hall, Thomas (iron master)
    166, 194, 212, 231
Hall, Thomas (silk throwster) 92
Hall, William 194
hammers 249, 252, 254, 255, 280
Hardingswood Mill **190**
Hargreaves, John 37, 39
Hargreaves, Joseph 41
Hargreaves, Samuel 37
Hargreaves, William 38
Harpur, Sir Henry 5
Harpur-Crewe, Sir Vauncey 6
Harrison, Samuel 89
Harthern, Judith 81, 96
Harthern, Samuel 81, 96, 97
Harthern, William 82
Hassall, Samuel 200, 202
Havannah Mills
    43, **52**, 64, 119, 156, 247, 248,
    254, 267, 274, 282, 290
Havannah Mills Company Limited 60
Hawkins, Edward 52
Heapy, Arthur H. 42
Heapy, Edward 32
Heapy, Joshua 32, 41, 51
Heapy, William 38
Heath, J. H. & Co. Ltd. 219

Heath, Robert 83
Heaton 38
Henshall, John 220
Henshall, Samuel 220, 224
Henshall, Thomas 220
Hewes, Thomas 241
Heyford, Dennis 166, 194, 231
Hickson, Samuel 59
Higher Roughwood Mill
    199, 203, 208, 215, 217
Highs, Thomas 105, 283, 284
Hobson & Ford 54
Hogg, Capel Wilson 107
Hogg, Henry 63, 107
Hogg, James 55
Holbrook, Thomas 202
Holland, George 200, 203
Holland, John 51, 200, 203, 214, 217
Holland, Josiah 217
hollander 17, 20, 33, 249, 280
Hoole, Thomas 188
Hope, Peter 18
Hope, Richard 17
Hope, Thomas 17, 34
Hopkins, Jonathan 68, 107
Howard, John 175
Hunt, Peter 135
Hurst Mill 78, **81**, 96

I

ice making
    Bosley Mill 42
    Swettenham 160
iron making **250**, 279
Italian throwing machines
    126, 131, 257, 281

J

Jackson, Fred 61
Jeffries & Cockson 53
Johnson, John 66, 73
Johnson, Thomas 102

K

Kay, James 265
Kay, John 283, 284
Kendall, Edward 168
Kendall, Jonathon 232
kiln tiles 32, 162, 209
Kinderton Mill **184**
Kinderton Saltworks **188**
King Henry VI 112
King's Mills (see Congleton Mill) 111
Kinnerley, Samuel 234
Krinks, Simon 73, 100

L

Lancaster, Alf 162

Lancaster, Wilf 160
lantern pinion 242
Lawton Furnace
    113, 166, 169, **192**, 212, 251, 279
Lawton, John 152
Lawton Mill 192, **196**, 278
Lawton Saltworks **197**, 213
Lawton, William 152
layer 257
Lea, Frederick 221
Lea, Henry William 221
Lea, Herbert 178
Lea, Herbert James 160
Lea, James 159
Lea, John 40, 159, 178
Lea, John Herbert 220
Lea, Thomas 220
Lea, William 158, 159, 175
Lee Forge **85**, 290
Leek Raven 24
Little Moreton Mill **210**
Lombe, John 257, 280
Lombe, Thomas 123, 257, 280
Lowe, Robert H. 130
Lower Daneinshaw Mill 263, 264
Lower Roughwood Mill 200, **202**
Lower Washford Mill 68, 72, **73**, 271
Lucas, Peter 152
Lud's Church 9

M

Macclesfield Canal Company 41
Macclesfield Canal wharf 46, 99
Macclesfield Copper Company
    43, 52, 282
Machin, Joseph 86
machine tools 125
Malkin, Arthur 159
Malkin, Samuel 40, 46
Malkin, Walker & Hope 218
Marsuma cigarettes 60
Martin, Richard 88, 90, 91, 105, 138
Marton Mill **152**
Massey, Edward 172
Massey, James 224
Massey, Samuel 224
Massey, Stephen 41
Massey, William 216
medieval mills 244, 278
Midge Brook Tanyard **157**, 269, 272
mill accounts 112, 119, 170, 174
mill profits 115, 119
mill rental values 186
mill toll 110, 114
millers' wages 114
millstones
    32, 56, 67, 89, 112, 113, 116,
    121, 152, 154, 164, 177, 182, 204,
    220, 244, 246, 281

INDEX

millwrights
   John Edwards 57
   Jonathon Booth
      59, 69, 120, 155, 160
   William Forster 151, 245
   William Kennerley 171
Moores, James & Son 227
Moorland Works 279
Moreton Mill **216**, 246
Moston Mill **229**, 272, 285
Mow Cop 113, 204
Mow Cop millstones
   114, 154, 164, 174, 207, 209, 210, 246
multure 112, 278
Myott, Arthur 93

N

narrow gauge railway 46
Nearer Daneinshaw Mill
   91, 104, 105, 282, 283
Nether Lee Mill 37
Newton Saltworks **183**

O

Oakes, Daniel 89
oatmeal
   57, 116, 154, 164, 177, 204, 247
oil pressure governor
   49, 94, 228, 276
Old Mill (see Congleton Silk Mill)
organzine 258
Orme, John 59, 117, 155
overdriven 246

P

Paddy, James 213, 233
Paddy, Martin 214, 232
Palfreyman, Charles 11
Palfreyman, George 11
paper making 249, 278
paper mills
   Danebridge Mill 19
   Eaton Bank Mill 64
   Folly (Grove) Mill 17
   Primrose Vale Mill 100
   Wincle (or Whitelee) Mill 33
Park Mill, Brereton
   159, **175**, 242, 247, 248, 272, 277, 286
Park Street Tannery **138**, 268, 269
Pass, Thomas 214, 220
patent trials 13, 283
patents
   5, 11, 137, 248, 260, 263, 265, 266, 284
Pattison, James 128
Pattison, Nathaniel
   115, 123, 127, 132, 280

Pattison, Nathaniel Maxey
   116, 123, 128
Pattison, Samuel 115
Paul, Lewis 260, 283
Pearson, James 129, 146
Pearson, Samuel
   117, 128, 130, 146, 287, 288
Peck Mill **237**
Pelton Wheel 276
Penlington, Thomas 227
Penlington, William 197
Percival, Joseph 233
Percival, Nathan 39
Percival, Ralph 218
Percival, Thomas 38, 218
Pickering, John 234
Pickering, Samuel 186, 234
Pickering, Thomas 234
Pointon, George 214
Pointon, Ralph 51
Pointon, Samuel 51, 171
Pointon, Thomas 51, 171
Pointon, William 214
Polarcold 99
Pool Bank Mill **97**, 149, 284
Primrose Vale Mill
   74, **100**, 136, 280, 288
provender milling 119, 290
pulley, fast and loose 243
pumping **267**
   Bank Mill (Buglawton) 103
   Baron's Croft Saltworks 238
   Havannah Mills 60
   Kinderton Saltworks 188
   Lawton Saltworks 198
   Newton Saltworks 183
   Roughwood Saltworks 201
   Stanthorne Mill 236
   Washford Mill 71
   Wheelock Saltworks 222

Q

Quarnford 4

R

RAF memorial 228
Reade, George 91, 138, 140
Reade, John Fielder 140
Rennie, John 35
retting 264
Rigby, William B. 235
ring gear 15, 31, 176, 241
rivers
   Dane 1, 3, 37, 77, 150, 222, 240
   Weaver 1, 222
   Wheelock 1, 150, 190, 222
Rode Mill 196, 200, **206**, 246, 247
Roe, Charles
   42, 52, 123, 144, 254, 280, 282

Rogers, George 60, 267
Roldane Mill 130
roller milling 289
Roughwood Saltworks **201**
roving frame 261
Rudyard Lake 34
   feeder agreement 35
Rushton Dyeworks 38
Rushton Mill 32, **37**, 39, 248
rye mill 112

S

sack hoist
   121, 154, 164, 182, 186, 209, 247
Salmon, Edward 197
saltworks
   Baron's Croft 237
   Kinderton 188
   Lawton 197
   Newton 183
   Roughwood 201
   Wheelock 222
Sandbach Mill 214, **219**
saw making 232, 253
saw mills
   Bearda 30
   Biddulph Grange 83
   Gradbach 6
   Swettenham 160
   Wheelock (veneers) 225
serpents 257
shafting, wrought iron 243
Shakerley, Peter 151
shears 253
shulling mill 116, 154, 204
silk doubling 258
silk mills
   Bank Mill (Buglawton) 102
   Biddulph Moor Mill 78
   Bosley Works 45
   Bridge Street Mill 136
   Brook Mill 140
   Brook Mill (Sandbach) 218
   Cranage Mill 171
   Dane Mill 146
   Dane Mill (Slate's) 144
   Danebridge Mill 23
   Daneinshaw 93
   Davenshaw Mill 107
   Eaton Bank Mill 63
   Forge Mill (Crossledge) 148
   Gradbach Mill 6
   Havannah 54
   Hurst Mill 81
   Lower Washford Mill 73
   Pool Bank Mill 97
   Promrose Vale Mill 101
   Rushton 39
   Stonehouse Green Mill 140

Timbersbrook Mill 95
Vale Mill 133
Victoria Street Mill 137
Washford Mill 65
silk spinning 6, 23, 140, **260**
silk throwing **255**, 280, 287
Silver Springs Bleaching & Dyeing Co. Ltd. 96, 98
Simister, James 22
sites 1
Slate, Thomas 144
Slate, William 144
slitting mill 169, 252, 279
Smallwood Mill **217**
Smeaton, John 240, 247
Smith, Walter 8
smut machine 154, 204
Snelson, Henry 178
Snelson, James 117, 159
soke 110, 196, 278
Solly, Arthur Isaac 142
Solly, Edward Harrison 142
Somerford Booths Mill
    151, **153**, 217, 246
spinning mule 263
Staffordshire Works
    168, 194, 212, 231, 279
Stanley Pool 16
Stanthorne Mill **235**
steam engine 182
Stonehouse Green Mill **138**, 263, 284
Stonier, Frederick 200, 207, 208
Stonier, George 206, 207
Stonier, James 235
Stonier, William 78, 80, 81
Stonier's Mill **78**
Street Forge **212**
Street Furnace 167, **212**, 279
Sutton Mill 186, **234**
Swetenham, Clement 154
Swettenham Mill
    40, 117, **158**, 243, 246, 247, 248, 281, 290
Swettenham, Thomas 158
Swythamley Church **26**, 290
Swythamley Hall **26**

T

tanneries
    Midge Brook 157
    Park Street 138

tanning **269**
Templeton, Thomas 44, 97, 148
tentering 244
Thompstone, Francis Rathbone
    46, 273, 274
Thompstone, Frank 47
Thompstone, Mark 89
Thornton, John & Charles 59, 64
throstle 263
Timbersbrook Mill 81, **95**, 98
tortoe 258
Town Mill (see Congleton Mill) 112
treble mill 113, 151, 245
Trent & Mersey Canal 191, 229, 286
Trent & Mersey Canal Company 34
turbine
    Girard 46, 47, 273, 274
    Jonval 46, 274
    Pelton wheel 27, 83
    Series C (Francis) 228, 275
    Series I (Francis) 94, 275
    Series IV (Francis) 42
    Series R (Francis) 47, 71, 275
    Vortex 47, 60, 83, 274
turbines **273**, 290
    Biddulph Grange 83
    Bosley Mill 42
    Bosley Works 46
    Cranage Mill 172
    Daneinshaw 94
    Havannah 60
    Lee Forge 87
    Swythamley Church 27
    Swythamley Hall 26
    Warmingham Mill 228
    Washford Mill 70
Turner, John 193, 196

U

underdriven 246

V

Vale Mill **133**, 137
Vaudrey, John 98, 100
Vaudrey, William & Charles 99
vertical stamps 249, 254
Victoria Street Mill **137**

W

Walley, Thomas 84
Wallworth, Matthew & Charles 63, 65

Wallworth, Randle & Matthew 66
Ward, Arthur 154
Warmingham Forge
    167, 212, 214, **231**, 251, 253, 279, 285
Warmingham Mill **227**, 275, 290
Washford Mill
    63, **65**, 268, 271, 272, 275, 282
Washington, John 95
water turbines **273**
waterframe 262
waterwheel inertia 127, 240
waterwheels **239**
    breastshot
        7, 15, 25, 31, 32, 38, 42, 56, 69, 84, 103, 106, 121, 126, 134, 140, 154, 156, 169, 179, 186, 208, 224, 236, 240
    overshot
        6, 15, 19, 38, 47, 203, 221, 227, 239
    pitchback 163
    suspension 176, 241
    undershot 73, 145, 174, 239
Watt, James 52, 197, 213
West Heath Mill **150**, 245
wheat mill 112, 154
Wheelock Mill 178, **223**
Wheelock Saltworks **222**
Wheelock Veneer Mill **225**
Whitfield & Co. 88, 91
Whitfield, John 88, 91, 105
Whitfield, Thomas 88, 91, 105
Wild, Peter & Co. Ltd. 146, 149
Wildboarclough 10
Wincle Mill
    10, 17, **33**, 229, 249, 280
winding machine 257
windmills 185, 225, 282
Windsor Mill 56
Winterley Mill **225**
wire mill 254
Wood & Westhead 56
wood flour milling 47, **272**
Woodhouse Mill 97
Wyatt, John 283